Palaeozoic Vertebrate Biostratigraphy and Biogeography

PALAEOZOIC VERTEBRATE BIOSTRATIGRAPHY *and* BIOGEOGRAPHY

Edited by

John A. Long

The Johns Hopkins University Press
Baltimore

Printed in the United States of America on acid-free paper

First published in 1993 by Belhaven Press
Johns Hopkins Paperbacks edition, 1994

The Johns Hopkins University Press
2715 North Charles Street
Baltimore, Maryland 21218-4319

Library of Congress Cataloging-in-Publication Data will be found at the end of this book.
A catalog record for this book is available from the British Library.

CONTENTS

List of figures and tables vii
List of contributors xi
Preface *J. A. Long* xii

PART ONE
Morphology, Tectonic Setting and Extinctions of Palaeozoic Vertebrates

Chapter 1 – Morphological characteristics of Palaeozoic vertebrates used in biostratigraphy *J.A. Long* 3
Chapter 2 – Palaeozoic global reconstructions *Li, Z.X., C. McA. Powell and A. Trench* 25
Chapter 3 – Early–Middle Palaeozoic vertebrate extinction events *J.A. Long* 54

PART TWO
Palaeozoic Vertebrate Biostratigraphy

Chapter 4 – The Silurian–Devonian agnathan biostratigraphy of the Old Red Continent *P. Janvier and A. Blieck* 67
Chapter 5 – Silurian–Devonian vertebrate biostratigraphy of Siberia and neighbouring terranes *A. Blieck and P. Janvier* 87
Chapter 6 – Biostratigraphy of the Silurian and Devonian gnathostomes of the Euramerica Province *D. Dineley and E. Loeffler* 104
Chapter 7 – Silurian–Devonian vertebrate biostratigraphy of Western Gondwana and related terranes (South America, Africa, Armorica-Bohemia, Middle East) *H. Lelièvre, P. Janvier and A. Blieck* 139
Chapter 8 – Palaeozoic microvertebrate biostratigraphy of Eastern Gondwana *S. Turner* 174
Chapter 9 – Middle Palaeozoic macrovertebrate biostratigraphy of Eastern Gondwana *G.C. Young* 208
Chapter 10 – Vertebrate biostratigraphy of the Middle Palaeozoic of China *Wang, S.-T.* 252

Chapter 11 – Palaeozoic vertebrate biostratigraphy of South-East Asia and Japan *J.A. Long* 277

PART THREE
Palaeozoic Vertebrate Biogeography

Chapter 12 – Vertebrate faunal provinces in the Middle Palaeozoic *G.C. Young* 293

Chapter 13 – Biogeography of Palaeozoic tetrapods *A.R. Milner* 324

Systematic index 354

Stratigraphic and locality index 364

LIST OF FIGURES AND TABLES

FIGURES

1.1 Age ranges of major groups of Palaeozoic fishes 4
1.2 Osteostraci – basic structure and morphology 6
1.3 Heterostraci – basic structure and morphology 7
1.4 Thelodonti – basic structure and morphology 9
1.5 Galeaspida – basic structure and morphology 9
1.6 Chondrichthyes – basic structure and morphology 11
1.7 Acanthodii – basic structure and morphology 12
1.8 Placodermi – basic structure and morphology 13
1.9 Major groups of Palaeozoic osteichthyans, including tetrapods 15
1.10 Actinopterygian scale morphology and structure 16
1.11 Sarcopterygian scale types 19
2.1 Subdivision of continental blocks used in this chapter 26
2.2 Distribution of Palaeozoic palaeopoles for Gondwana 28
2.3 Cambro–Ordovician palaeogeographic reconstruction 34
2.4 Cambro–Ordovician reconstruction from a different view-point 34
2.5 Mid-Silurian palaeogeographic reconstruction 36
2.6 Early Devonian palaeogeographic reconstruction 41
2.7 Late Devonian palaeogeographic reconstruction 42
2.8 Late Carboniferous palaeogeographic reconstruction 43
2.9 Early to mid-Permian palaeogeographic reconstruction 45
3.1 Diversity of fish families in the Early and Middle Palaeozoic 55
3.2 Origination and extinction graphs for Early–Middle Palaeozoic vertebrate families 56
3.3 Major extinction events of higher taxonomic groups of Early–Middle Palaeozoic vertebrates 59
4.1 Silurian–Devonian agnathan-bearing localities of the Old Red Continent 69
4.2 Silurian–Devonian jawless vertebrates from the Old Red Continent 70
4.3 Stratigraphical distribution of the Silurian–Devonian agnathans of the Old Red Continent 73
4.4 The Silurian–Devonian sequence of the East Baltic area 77

4.5	Simplified stratigraphical distributions of vertebrate index-species, and genera, in the Silurian and/or Devonian of Euramerica	80
5.1	Post-Silurian continental blocks east of the Urals	88
5.2	Silurian and Devonian jawless vertebrates from Siberia and adjacent terranes	90
5.3	Silurian and Devonian gnathostomes from Siberia and adjacent terranes	91
5.4	Stratigraphical distribution of Lower Devonian Siberian vertebrates	94
5.5	Stratigraphical distribution of Emsian vertebrates of NE Salair	97
6.1	The Euramerican continent and associated mobile crustal belts	106
6.2	Stratigraphical ranges of Devonian gnathostomes in Euramerica	108
6.3	Map of Devonian gnathostome localities in Euramerica	111
6.4	Correlation of vertebrate and invertebrate zonations of the Silurian and Devonian of the Baltic area	114
6.5	Vertebrate succession in the Middle Old Red Sandstone of the Orcadian Basin	115
6.6	Faunal succession of gnathostomes in the Upper Old Red Sandstone of the Orcadian Basin	116
6.7	Faunal succession in the Old Red Sandstone formations of East Greenland	117
6.8	Faunal succession in the Silurian–Devonian of Spitsbergen	119
6.9	Ranges of gnathostomes in the vertebrate-yielding formations of the Devonian of the Baltic region	128
6.10	Ranges of vertebrate index genera in the Euramerican Old Red Sandstone	129
6.11	Late Devonian faunal comparison: Timan, Baltic, Poland, Rhineland, Québec	130
7.1	Middle Devonian reconstruction showing West Gondwana and related terranes	140
7.2	Distribution of the Silurian–Devonian vertebrate localities of West Gondwana	142
7.3	Silurian and Devonian vertebrates of South America	144
7.4	Devonian vertebrates of North Africa	149
7.5	Devonian vertebrates of South Africa	152
7.6	Silurian and Devonian vertebrates of Armorica and Bohemia	154
7.7	Devonian vertebrates from Turkey	158
7.8	Devonian vertebrates from Iran and the Arabian Peninsula	160
8.1	Ordovician microvertebrates from East Gondwana	178
8.2	Types of osteichthyan scales in East Gondwana	180
8.3	Thelodont scale ranges in East Gondwana	182
8.4	Shark scale types from East Gondwana	185
8.5	Placoderm body scale types from East Gondwana	188
8.6	Useful age indicators in the Palaeozoic of Eastern Gondwana	196
8.7	Correlation table for Silurian to Lower Devonian of New South Wales	199

9.1 Biostratigraphically significant Devonian vertebrate localities of Australia 209
9.2 Integration of East Gondwana Devonian conodont and palynomorph zonations 212
9.3 Stratigraphic distribution of key East Gondwana placoderm taxa 217
9.4 Endemic East Gondwana antiarchs 218
9.5 Stratigraphic distribution of various East Gondwana arthrodires 220
9.6 Some endemic East Gondwana arthrodires 221
9.7 Stratigraphic distribution of various East Gondwana agnathan, chondrichthyan, acanthodian and actinopterygian taxa 225
9.8 Some endemic East Gondwana acanthodians and osteichthyans 227
9.9 Stratigraphic distribution of various East Gondwana osteichthyans 231
10.1 Palaeogeography and distribution of Silurian and Devonian vertebrates in China 253
10.2 Stratigraphic sections and faunas for the Silurian and Early Devonian of China 256
10.3 Endemic galeaspid agnathans from the Siluro–Devonian South China terrane 258
10.4 Stratigraphic sections and faunas for the Middle Devonian–Lower Carboniferous of China 260
10.5 Late Silurian–Early Devonian microvertebrate scales useful in biostratigraphic correlations with China 261
10.6 Age ranges of sinolepid placoderms in China 263
10.7 Antiarch headshields from the Devonian of China 265
10.8 Age ranges of important biostratigraphic genera for the Middle Palaeozoic of China 268
11.1 Map showing major fossil localities and terrane boundaries for SE Asia and Japan 278
11.2 A, B, Late Silurian–?Early Devonian microvertebrates from north of Mae Sariang, Thailand 279
11.3 Palaeozoic chondrichthyans from Thailand 281
11.4 Vertebrate microfossils from the mid Famennian Xom Nha Formation, Binh Tri Thien Province, Vietnam 282
11.5 Early Devonian vertebrates from northern Vietnam (South China terrane) 284
11.6 The ?Middle–Late Devonian antiarch *Vietnamaspis trii* (northern Vietnam) 285
11.7 Palaeozoic vertebrates from Japan 286
12.1 Locality map for Devonian vertebrate occurrences 294
12.2 Major vertebrate faunal provinces with superimposed cladogram for various placoderms 296
12.3 Euramerica, Siberia, and Tuva (cephalaspid, amphiaspid and tannuaspid provinces) 298
12.4 Various endemic taxa of Euramerica and Siberia 300

12.5 Placoderms suggesting affinity between central and east Asia 301
12.6 East Gondwana, North and South China on the reconstruction of Young 302
12.7 Vertebrate localities analysed in relation to Armorica hypothesis 307
12.8 Alternative hypotheses of Palaeozoic continental collision sequences 310
12.9 Reconstructions based on palaeomagnetic data showing the relationship of Euramerica to West Gondwana 312
12.10 Alternative palaeogeographic reconstructions for the Late Devonian 313
13.1 Tetrapod distributions, Late Devonian–Late Carboniferous 326
13.2 Interrelationships of osteichthyan groups on Devonian reconstruction 329
13.3 Tetrapod distribution in the Permian 334
13.4 Diorama depicting lowland and freshwater elements of Edaphosaur–Nectridean Province of Texas 338
13.5 Diorama depicting intermontane and freshwater elements of Edaphosaur–Nectridean Province of Germany 339

TABLES

2.1 Selected palaeopoles used for palaeolatitude control in the reconstructions 30
3.1 Background extinctions and extinction ratios, per stage 58
3.2 Frasnian placoderm extinctions, ranges and habitats 60
6.1 Tectonic provinces of Devonian sedimentation on and around Euramerica 112
13.1 Pennsylvanian and Lower Permian terrestrial tetrapod families from the Edaphosaurid–Nectridean Province, west and east of the Hercynian–Appalachian 'barrier' 335
13.2 Pennsylvanian and Lower Permian aquatic families from the Edaphosaurid–Nectridean Province, west and east of the Hercynian–Appalachian 'barrier' 336

LIST OF CONTRIBUTORS

Alain Blieck, Sciences de la Terre, Université des Sciences et Technologies de Lille, URA 1365 du CNRS, F-59655 Villeneuve d'Ascq Cédex, France.

David Dineley, Department of Geology, University of Bristol, Bristol BS8 1TR, UK.

Philippe Janvier, URA 12 du CNRS, Institut de Paléontologie, 8 rue Buffon, 75005 Paris, France.

Hervé Lelièvre, URA 12 du CNRS, Institut de Paléontologie, 8 rue Buffon, 75005 Paris, France.

Li, Z.-X., Geology Department, University of Western Australia, Nedlands, Western Australia, 6009, Australia.

Elizabeth Loeffler, Department of Geology, University of Bristol, Bristol BS8 1TR, UK.

John A. Long, Department of Earth and Planetary Sciences, Western Australian Museum, Francis Street, Perth, Western Australia, 6000, Australia.

Andrew R. Milner, Department of Biology, Birkbeck College, Malet Street, London WC1E 7HX, UK.

Christopher McA. Powell, Geology Department, University of Western Australia, Nedlands, Western Australia, 6009, Australia.

Allan Trench, Geology Department, University of Western Australia, Nedlands, Western Australia, 6009, Australia.

Sue Turner, The Queensland Museum, PO Box 300, Brisbane, South Queensland, 4101, Australia.

Gavin C. Young, Australian Geological Survey Organisation, PO Box 378, Canberra, A.C.T., 2601, Australia.

Wang, S.-T., Institute of Geology, Chinese Academy of Geological Sciences, Baiwanzhuang Road, Beijing, China.

PREFACE

John A. Long

The study of Palaeozoic vertebrates has long been an integral part of understanding the evolution and anatomy of the lower back-boned animals. However the use of Palaeozoic fishes in geological correlations has also been long known, at least since the time of Hugh Miller, in the early nineteenth century, when Old Red Sandstone fish fossils were used to identify the same series of rocks, of approximately the same age, around different parts of Scotland. In the last twenty or so years there has been an upsurge in the study of Palaeozoic fishes for solving geological problems, both in areas of biostratigraphy and biogeography. This has resulted in an explosion of data, much of it so new that it will take years for all the recent discoveries to be published. This book has resulted to fill the need to provide up-to-date summaries of global work in progress showing the applications of both macroscopic and microscopic remains of Palaeozoic vertebrates to geological correlations, and to refinement of global palaeogeographic reconstructions. It is, of course, a bit ambitious to try and tackle the entire world perspective in this manner, so selections have been made to cover the major Palaeozoic terranes in which vertebrate fossils are proving to be very useful. Many of the chapters in this book are contributions to Unesco International Geological Correlation Project 328 (Project leaders: Sue Turner and Gavin Young), a testimony to the recognition of the international importance of developing studies of Palaeozoic vertebrate biostratigraphy.

The main problems with this book are its limitations in trying to provide both accurate summaries of the subject areas and ample illustrations showing the morphology of the most useful species. In many cases, work coming from China, Australia, South America, Siberia, and the Middle East is so new that schemes presented here are only progress reports of better things to come. However, the value of this data—to have in the hand now—will be invaluable to the many new workers around the world who are starting an involvement in IGCP 328, or to geologists who have turned to Palaeozoic vertebrates for solving problems of local correlation, age assessment or palaeoenvironmental interpretation.

The focus of the chapters in this volume is largely on the new discoveries of Palaeozoic vertebrates in successions that are well-constrained by other dating methods, and thus may be correlatable with either microfossil zonations or marine invertebrates and microfossil zonation schemes. Palaeogeographic uses of Palaeozoic vertebrates have been refined in recent years in accord with refinements in palaeogeographic reconstructions. By using distribution hypotheses, data from fossil fishes can be tested against other geological hypotheses to refine biogeographic and tectonic models.

Part One gives the background information necessary to utilise the summaries of biostratigraphic and biogeographic data provided in the rest of the book. Chapter 1 gives an introduction to the basic morphology of Palaeozoic vertebrates, providing a simple guide to the nature of their fossilised hard-parts (skeletal remains). In many cases the correlative value of a Palaeozoic vertebrate may lie in finding one scale or tooth from a drill-core sample. This chapter outlines the overall features of each major vertebrate group, and defines their biostratigraphically useful parts. Chapter 2 gives the latest Palaeozoic global reconstructions using new palaeomagnetic data, and using a 'filter system' on old palaeomagnetic data. This has a screening effect on preventing old, potentially unreliable palaeomagnetic data from distorting the global reconstruction. Chapter 3 is a short summary of major extinction events for Early–Middle Palaeozoic vertebrates. The heart of useful biostratigraphic methodology is identifying the constraints on age range of a taxonomically distinctive group or species. Extinction events, even minor ones, provide useful datum points to frame groupings of minor biostratigraphic events.

Part Two gives summaries of biostratigraphic schemes for the major Palaeozoic vertebrate provinces. I asked each author to provide the latest charts of age ranges, give a discussion of palaeoenvironmental settings and tie in vertebrate occurrences with other reliable dating criteria, such as marine invertebrate or microfossil occurrences. This has largely been achieved, and useful summary charts have been provided throughout Part Two. The classic localities of the Old Red Continent (Euramerica) have been divided into agnathan faunas (Chapter 4) and gnathostomes (Chapter 6). The faunas of Siberia and neighbouring terranes are summarised in Chapter 5, and those of Western Gondwana, covering South America, Africa, Spain and parts of the Middle East, are dealt with in Chapter 7. The wealth of new discoveries coming from Australia, Antarctica and Asia are covered by Chapters 8–11.

Part Three of this book gives two overviews of Palaeozoic vertebrate biogeography. Chapter 12 covers the definition of, and associated problems with, Middle Palaeozoic faunal provinces—why some groups of Siluro–Devonian fishes show high endemism while others are cosmopolitan. The age ranges of many biostratigraphically useful forms are shown to be intimately linked to their phylogenetic history and possible 'centres of diversification'. Chapter 13 is the only chapter of the book treating tetrapods, and discusses how land-dwelling amphibians and reptiles achieved global dispersal, and outlines the value of tetrapods to defining faunal provinces in the Late Palaeozoic.

I apologise to readers for the lack of coverage of Late Palaeozoic fish assemblages. Much of this work is in progress, and it was unfortunate that the several workers whom I approached to write on this subject were unable to contribute.

Biostratigraphy is an integral aspect of palaeontology that requires continual refinement and modification of existing schemes as new data comes to hand. Unlike many of the currently used invertebrate zone fossils, many Palaeozoic fish species can occur in transitional environments—the one species inhabiting marine, near shore and/or terrestrial facies. This means that the opportunity to link established marine invertebrate or microfossil zones with terrestrial sequences is largely the domain of Palaeozoic vertebrates and palynology (the study of spores and pollen). For this reason it is important to continually take stock of new work, and make an effort to summarise progress and outline problematic areas.

I hope that both researchers and students find this book useful in more ways than as a summary of latest work, and also trust that it will provide stimulus for ideas of how improvements can be made to future biostratigraphic schemes. It cannot be emphasised too much that good biostratigraphy relies on three factors: (1) a solid foundation of good taxonomy—correct identification of zone species and an understanding of the anatomical basis for biostratigraphic zone fossils; (2) careful collecting and preparation procedures—precise laboratory practises; and (3) an integrated geological/stratigraphical approach (understanding the geological and tectonic settings of the sample locations). The chapters in this book give numerous examples of the dating and correlation of geological strata throughout the world using Palaeozoic vertebrate remains. However, there is much work to be done before the numerous discoveries of new Palaeozoic vertebrates are described and then integrated into biostratigraphic schemes. I sincerely hope this book provides the necessary guidelines and inspiration for more students to become involved in studies on lower vertebrate fossils, and for petroleum and gas exploration companies to realise the huge potential for improved correlations and palaeoenvironmental interpretations that Palaeozoic fishes can contribute.

This book arose from a discussion with Ken McNamara who had been talking with Iain Stevenson of Belhaven Press. I would like to thank Ken for his encouragement and continual advice on this project as it unfolded, and Iain Stevenson for his receptive support in getting the book off the ground. The Trustees of the Western Australian Museum provided facilities for carrying out research for my chapters, and to edit this book. Special thanks to Kris Brimmell, W.A. Museum for her technical support. Finally, I give my heartiest thanks to each author for their excellent contributions and lack of tardiness in providing the chapters. If not for my three month trip to Antarctica over the summer season of 1991/92 this book would have been ready some months earlier, and I apologise to the authors for my own delay in this matter.

PART ONE

MORPHOLOGY, TECTONIC SETTING AND EXTINCTIONS OF PALAEOZOIC VERTEBRATES

Chapter 1

MORPHOLOGICAL CHARACTERISTICS OF PALAEOZOIC VERTEBRATES USED IN BIOSTRATIGRAPHY*

John A. Long

INTRODUCTION

The Palaeozoic Era saw the rise and diversification of the major vertebrate groups interspersed with several episodes of extinction (see Chapter 3). Figure 1.1 shows the age ranges for the major groups of Palaeozoic fishes. This chapter aims to provide a reference guide to the general morphology of these groups, with especial reference to those groups useful in biostratigraphic correlation or that are zoogeographically important in defining endemic faunal provinces. This encompasses mainly the fishes, with tetrapods not being further subdivided than the Amphibia. The zoogeographic provinces of Young (Chapter 12) are referred to in this chapter. However before outlining these groups some important points on fish anatomy, with respect to their fossilised remains, are explained. More detailed information can be found in Moy-Thomas and Miles (1971), Jarvik (1980) and in the recent references pertaining to each group.

FOSSIL BONES, SCALES AND BONY TISSUES

The skeleton of a fish consists of two major systems: the dermal skeleton, or bones and scales laid down within the dermis or skin; and endoskeletal bones, or those formed internally, often by replacing a cartilaginous framework through the secretion of bony tissues by osteocytes (bone cells). Bone layed down around cavities in cartilage to protect soft tissues, like nerves or arteries within the braincase, may form very thin layers or tubes which become preserved as fossils. This is called perichondral bone.

* A contribution to IGCP Project 328.

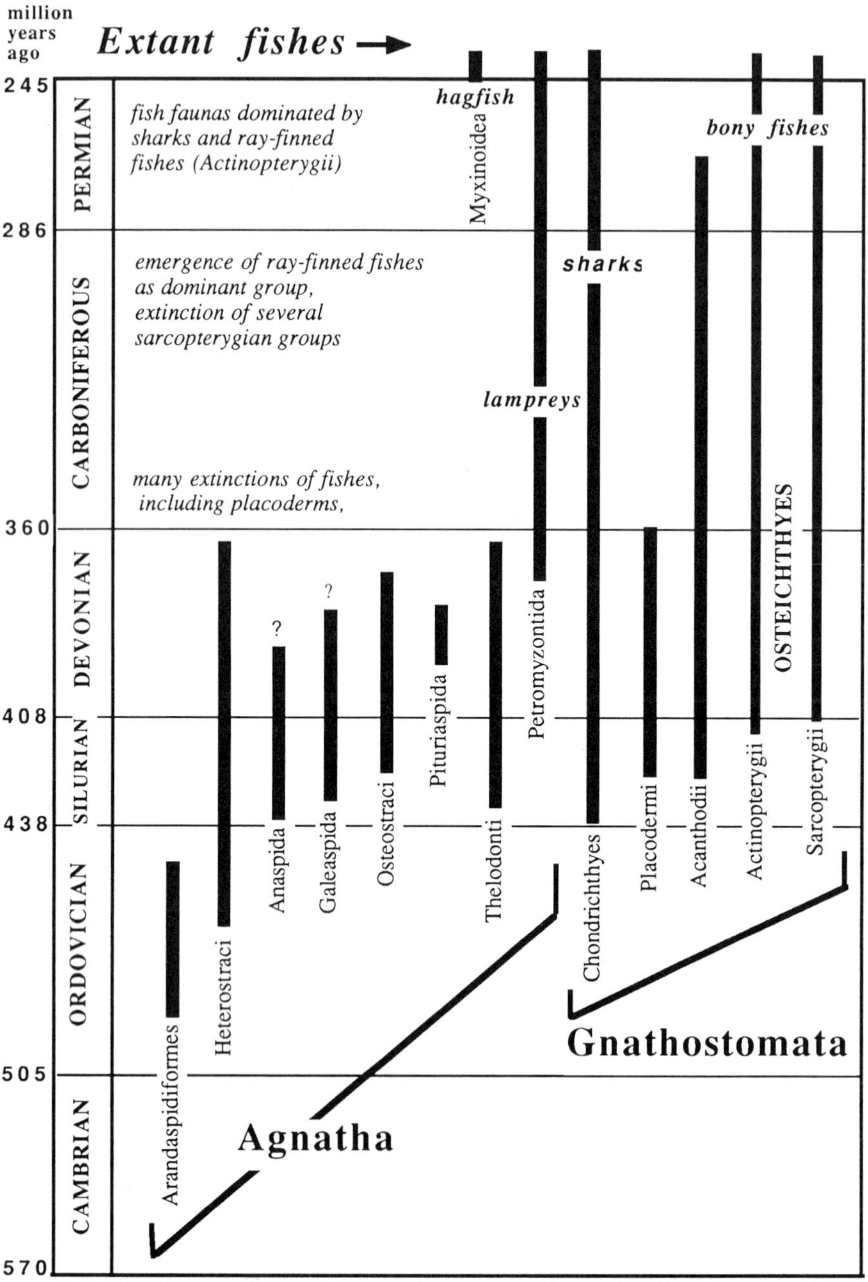

Figure 1.1. Age ranges of major groups of Palaeozoic fishes.

Within the skeleton of primitive fishes are found many different types of bony tissues. Early armoured forms may have armoured plates formed of basal lamellar bone with spongey trabecular bone inside and ornamented bone on the external surfaces. Such ornamentation, like tubercles, or ridges, may be capped with dentine, a more dense tissue with a shiny surface similar in appearance to tooth enamel but formed in a different manner. The external surface of many primitive vertebrates may show many small pores—these are interconnected by a pore-canal system within the surficial bony layers. In some forms a layer of dentine may enclose the pore-canal system, giving the surface a shiny look with regular small holes, and this tissue is called cosmine.

The skeleton of a fish consists of several systems, each with a discrete function to perform. The head includes a braincase (formed of endoskeletal and/or perichondral bones) which protects the brain and acts as a locus for cranial muscle and trunk muscle attachment. Covering the head in bony fishes and some acanthodians are sets of dermal bones: the skull roof and snout, the cheek unit, and the opercular series. The fins are supported by pectoral and pelvic girdles which may have ossified internal endogirdles like the scapulocoracoid. The backbone of the body is the axial skeleton—this consists of vertebrae, ribs and fin basal supports. The tail may have a well-developed caudal fin skeleton but in most primitive (Palaeozoic) forms there is no special development of axial skeleton tail bones. Fins of fishes are supported by ossified rods of bone or cartilage called lepidotrichia.

The most useful parts of the early fishes used for biostratigraphic correlation are their scales and teeth. In most fish these vary greatly in shape, size and complexity for each individual fish. Special scales may cover the inside of the mouth or gill arches, but most of the body is covered with regular trunk scales. The fins and ridges of the back and ventrolateral axis may have special scales aligned along them, such as fulcral scales or dorsal ridge scales. Palaeozoic sharks tend to show much less variation in tooth morphology within a single individual than for many modern shark groups. In all other Palaeozoic fish groups, the teeth tend not to vary much within individuals, although some acanthodians and higher osteichthyans (onychodontids and porolepiforms) may exhibit a different marginal biting dentition from the symphysial grasping teeth, developed as toothwhorls.

CLASSIFYING FISHES

All of the fishes can be divided into two superclasses: the Agnatha (having no jaws) and the Gnathostomata (or jawed vertebrates), the latter group also embracing the tetrapods as a clade within the Osteichthyes. The Agnatha were a diverse group back in the Palaeozoic, including many now extinct groups of armoured fishes covered in bony external plates. There are two living groups, represented by lampreys and hagfishes, both being eel-like parasitic forms.

For the purposes of using Palaeozoic fish remains as biostratigraphic tools, the nature of their remains as fossils is here outlined—general

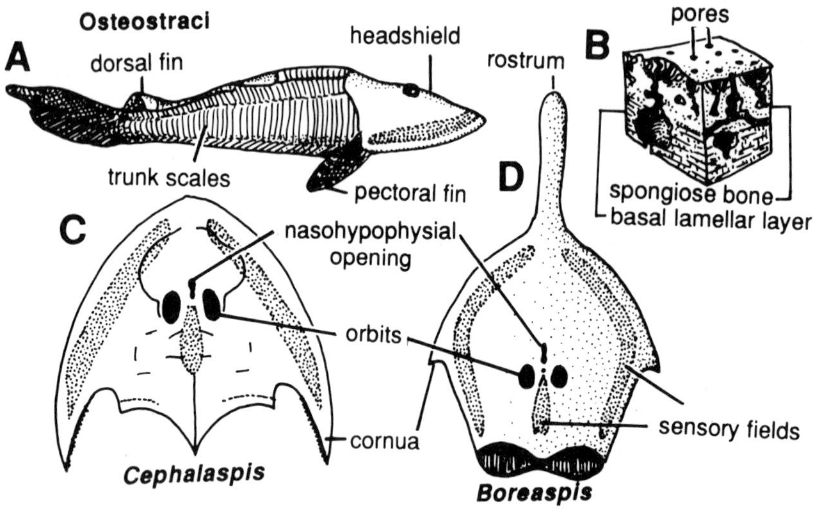

Figure 1.2. Osteostraci. A, body of *Hemicyclaspis*; B, osteostracan bone, C, D, cephalic shields showing major features; not to scale (after Moy-Thomas and Miles, 1971).

characteristics of each group, what tissue types they are composed of, and types of representation most commonly encountered as fossils.

AGNATHA

Osteostraci (Upper Silurian–Late Devonian)

These are armoured agnathans with a large bony cephalic shield covering the head region, the body being covered with large scales (Figure 1.2A). Osteostracans have bone consisting of a pore-canal system on the outer surface, a spongiose middle layer and a basal layer which may contain bone-cells (Figure 1.2B). The headshield or cephalic shield (Figure 1.2C, D) has two separate eye holes, a pineal opening within a small pineal plate between the eyes and a key-hole shaped nasohypophysial opening in front of the eye holes. There are sensory fields in the shield along its anterior edges (lateral field) and behind the eyes (dorsal field). The group has many variations on the shape of the cephalic shield, from simple, semi-circular forms to elongate oval shaped forms, some with long backwards protruding spines (cornua, e.g. *Cephalaspis*, Figure 1.2C). Others, like *Boreaspis* (Figure 1.2D), may have an anterior rostrum developed. The Osteostraci are restricted biogeographically to the Euramerican Province (Laurentia and Baltica), and are most useful as biostratigraphic indicators, as macrofossil remains of cephalic shields, during the Late Silurian–Early Devonian.

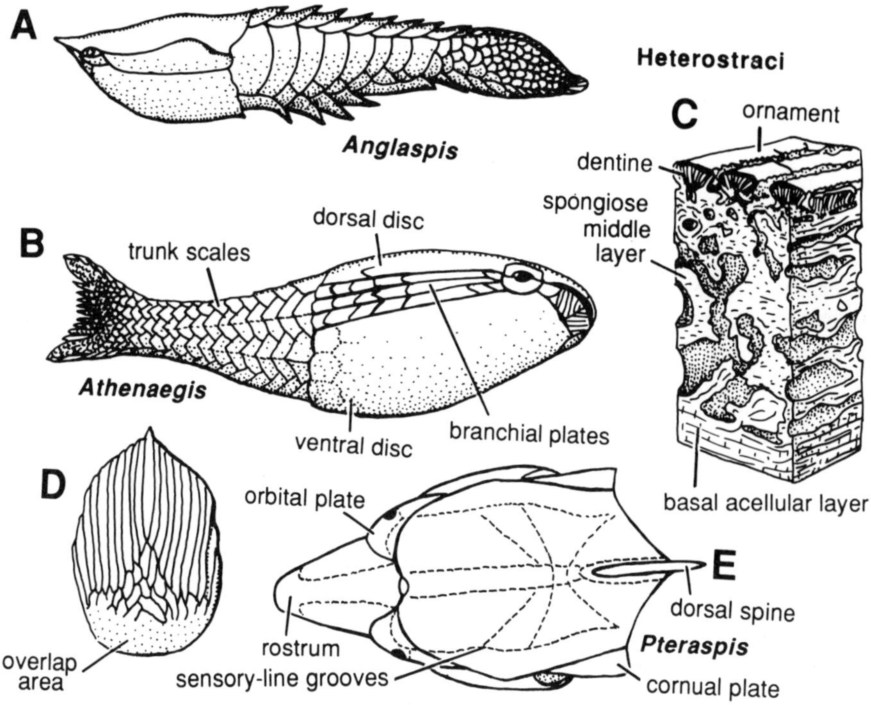

Figure 1.3. Heterostraci. A, B, body shapes of typical heterostracans; C, 'aspidin' bone structure; D, body scale, E, morphological features of a heterostracan armour, dorsal view; not to scale (after Moy-Thomas and Miles, 1971; Soehn and Wilson, 1990).

Heterostraci (Middle Ordovician–Late Devonian)

Heterostracans are agnathans having a bony shield composed of several plates, usually a dorsal shield, a ventral shield, branchial plate or plate series, plus a number of smaller plates around the areas of the mouth (oral plates) and eyes (suborbital, lateral plates etc.)—Figure 1.3A, B, E. Their dermal skeleton has elaborate ornamental patterns on the outer surface, a thick spongiose middle layer and a basal acellular layer, and is often called 'aspidin' (Figure 1.3C). The body is covered by many scales, each of which also has ornament matching the type seen on the larger dermal bones (Figure 1.3D). There are many diverse groups of heterostracans ranging from primitive proto-heterostracans from the Ordovician of Australia (Arandaspidiformes) and South America (*Saccabambaspis*) to the well-known pteraspids, cyathaspids, amphiaspids, and many other groups, of the Silurian and Devonian. Heterostracans are, therefore, of limited use biostratigraphically in the Ordovician Period, but become quite useful during the Late Silurian–Early Devonian, especially in the biogeographic provinces of Euramerica, Siberia and Tuva. They are recognised largely from macrofossil remains of dermal plates, but also from microremains of body scales.

Anaspida (Silurian–?Upper Devonian)

These are agnathan fishes which lack extensive bony plates, but have a laterally-compressed scale-covered body. Perhaps one of the best known forms is *Jaymoytius*, from the Silurian of Scotland. Anaspids range from the Middle Silurian until the Early Devonian (forms previously thought to have lived in the Late Devonian have now been referred to lamprey-like groups; Arsenault and Janvier, 1991). Currently they have little biostratigraphic significance, although their restricted distribution (Euramerican Province) makes them useful in a zoogeographic context.

Thelodonts (Early Silurian–Late Devonian)

Like anaspids, thelodonts lack extensive bony plates but have a complete scale cover over the body. The body is known in some forms (Figure 1.4A) showing well-developed pectoral fins, dorsal and anal fins, and a reverse heterocercal tail. The scales of thelodonts (Figure 1.4B–E) are small (usually 0.5–1.5 mm) and robust unlike the large flat scales of anaspids. Thelodont scales are conspicuous in having a dentine-covered crown, sometimes elaborately ornamented (Figure 1.4C–E), and a bony base with a large central pulp cavity (Figure 1.4B, D). The scales vary in form and size from the trunk, head, gill-covers, fins and mucous membrane areas. Thelodonts are widespread in the Late Silurian and Early Devonian, known from all biogeographic provinces, but by the Middle and Late Devonian they appear to be restricted to Gondwana (Australia, Antarctica, Iran), with only known Late Devonian (Frasnian) thelodonts coming from Western Australia. They are an important group biostratigraphically (e.g. Karatajute-Talimaa, 1978; Turner, 1973), known principally from isolated scales, especially during the Early–Middle Devonian. Thelodonts occur in a range of facies from marine, to marginal marine and continental facies.

Galeaspida (Early Silurian–?Late Devonian)

The galeaspids are a peculiar group of armoured agnathans having a one-piece bony shield with a large median dorsal orifice situated in front of the eyes (Figure 1.5A, C). They have a wide-ranging variety of armour shapes, from simple half-moon shaped types to bizarre forms with long tubular rostra and widely flairing wings (Figure 1.5C, D). Where preserved in rare cases, the body is covered by numerous ornamented scales. The bone

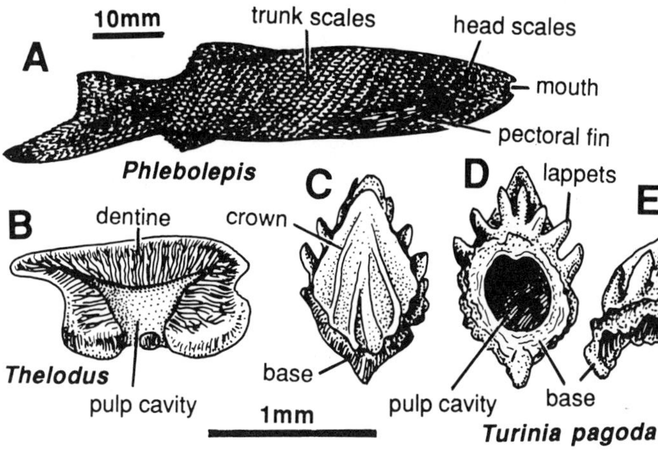

Figure 1.4. Thelodonti. A, body form; B–E, scale morphology and structure, not to scale (after Moy-Thomas and Miles, 1971; Wang *et al.*, 1986).

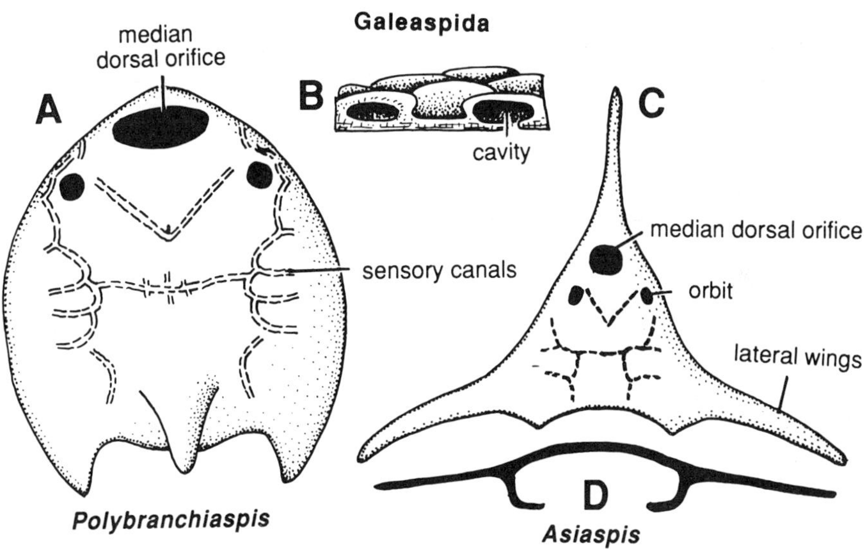

Figure 1.5. Galeaspida. A, C, cephalic shields in dorsal view; B, section of galeaspid dermal bone; D, cross-section through posterior of *Asiaspis* shield, not to scale (after Pan, 1984; and Janvier, 1984).

structure of galeaspids shows hollow cavities within ornamental tubercles or blisters (Figure 1.5B). The group is useful biostratigraphically during the Silurian–Early Devonian in China and neighbouring Vietnam. So far the group is restricted only to the South China terrane (see Chapters 10, 11), although Pan *et al.* (1987) has recorded a probable polybranchiaspid from Ningxia, in Northern China, also the only Late Devonian occurrence of the group. They are recognised mostly from macrofossils of cephalic shields.

Pituriaspids (Devonian)

The Class Pituriaspida was erected in 1991 based on armours of two genera of rostrate agnathan fishes from Early–Middle Devonian sandstones of central Australia (Young, 1991). As they are of restricted sites and age, they are only of local use biostratigraphically within Australia. Young (Chapter 12) gives further details about the group.

GNATHOSTOMATA

Chondrichthyes (Silurian–Recent)

The Chondrichthyes include all sharks (Figure 1.6A), rays and holocephalans. They are characterised by having a micromeric squamation composed of placoid scales (e.g. Figure 1.6B, C), teeth that grow continuously from a dental lamina, or hardened toothplates (e.g. holocephalans, batoids), and the skeleton features mostly cartilage with the only ossified tissues being dentinous types (teeth, fin-spines, scale crowns) and calcified globular cartilage. Their fossils are usually isolated teeth (Figure 1.6D–F, H, I), spines (Figure 1.6G) or scales (Figure 1.6B, C), although complete fishes and isolated braincases are occasionally found. Biostratigraphically they are most useful as microvertebrate remains, known principally from teeth and scales. Shark teeth date back to the Early Devonian (*Leonodus*, Mader, 1986), but shark scales may go back as well into the Silurian (Karatajute-Talimaa, 1973) perhaps as far as Early Silurian (Turner, pers. comm.), indicating that perhaps the earliest sharks were toothless forms. Palaeozoic shark teeth may occur in a great range of morphotypes. Each tooth has a crown (biting area) and a root (bony base). The crown may have one or more pointed cusps, and each of these may have smooth or striated ornamentation. The roots may have distinctive features like nutritive formaina, swellings or buttons and ridges, grooves or canals on the undersurface. Some sharks did not have predatory type teeth but instead show durophagous dentitions, having flat or gently rounded crushing teeth (some are shown in Figure 11.3F, G, I, p. 281). Symphysial teeth may be present in differing morphologies from the biting dentition, and in many cases these are developed as small symphysial tooth whorls. Zangerl (1981) gives a good outline of Palaeozoic shark groups and their respective tooth and scale morphologies and histologies.

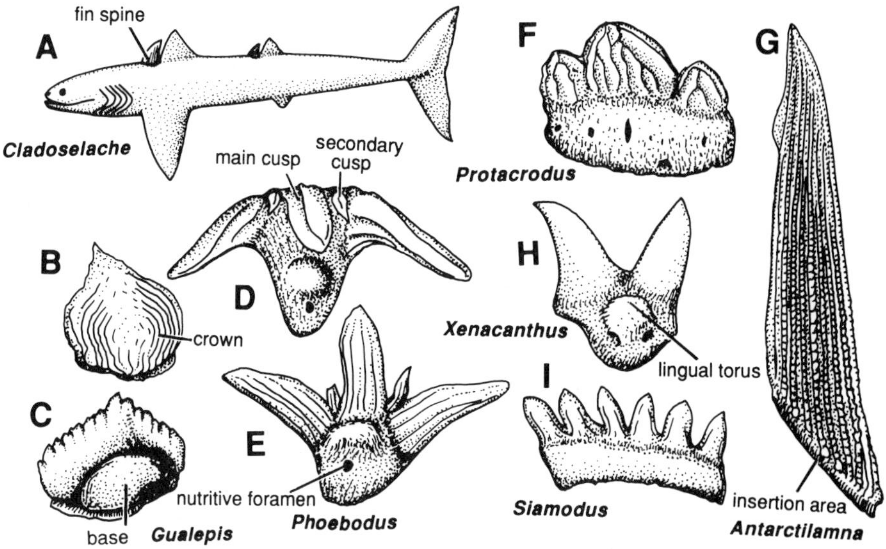

Figure 1.6. Chondrichthyes. A, body form of a Late Devonian cladodont shark; B, C, simple placoid shark scale in external (B) and basal (C) views; D–F, H, I variation and morphology in some Palaeozoic shark teeth. G, fin-spine of a primitive shark, not to scale (A, after Moy-Thomas and Miles, 1971; B, C, from Wang, 1984; D, E, from Ginter, 1990; F, H, from Turner, 1982; G, from Young, 1982; I, from Long, 1990).

Scale types also vary enormously, both within the individual and between taxa. In general the earliest shark scales appear to be simple placoid types such as *Elegestolepis* (Karatajute-Talimaa, 1973), but by the Late Devonian there are many complex scale types, including some with several episodes of ornamental growth on the crown. Later sharks have specialised scales present as cranial denticles, some of which may look remarkably tooth-like in form (e.g. hybodontids).

Acanthodians (Silurian–Middle Permian)

Acanthodians are primitive gnathostomes characterised by having paired fin-spines preceding all fins (Figure 1.7A), and having a micromeric squamation featuring scales with usually thick bulbous bases and flat crowns (Figure 1.7B–I). The scales grow like an onion, having concentric layerings (Figure 1.7D, E). Acanthodians are commonly found as microfossils, represented by scales, spines, tooth whorls (in climatiids and ischnacanthids), isolated dentigerous bones (ischnacanthids only; Figure 1.7J), and in rare cases as whole fishes. They were abundant in the

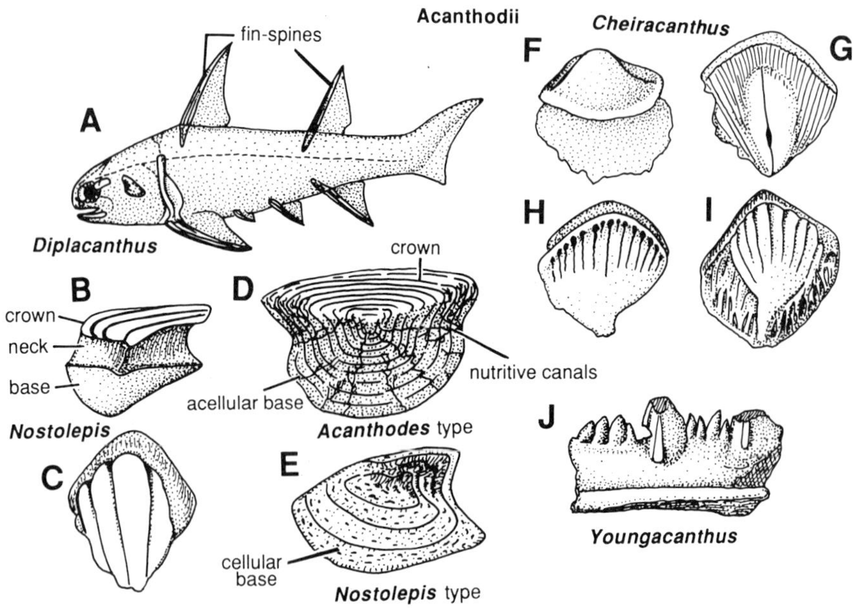

Figure 1.7. Acanthodii. A, body form of a Middle Devonian acanthodian; B–I, scale structure and morphology; J, ischnacanthid dentigerous bone fragment, not to scale (A, D, E, from Moy-Thomas and Miles, 1971; B, C, F–H, from Valiukevicius, 1985; I, from Vieth-Schreiner, 1983; J, from Wang, 1984).

Devonian Period, after which only acanthodiform species survived—a group of filter feeders having rhombic scales with flat, featureless crowns, and spines featuring only one thick rib (Denison, 1979).

As biostratigraphic tools they are most commonly studied as isolated scales or fin-spines. The scales have a wide range of morphologies and histologies. They fit closely but do not imbricate. The crown may vary from a featureless flat surface as in *Acanthodes* to tumid types with complex ornamentation (e.g. *Wetteldorfia*, Figure 1.7I). The crown is separated from the bony base by a well-defined constriction called the neck, and neck canals may or may not be visible on some species. The base varies from being a cellular bone to layered bone with osteocytes present, and from almost flat to markedly tumid. The two major histological types of scale are the *Nostolepis* type (Figure 1.7E) having a crown of mesodentine, with vascular canals and cell spaces, and a base of cellular bone; and the *Acanthodes*-type (Figure 1.7D) having a crown of true dentine, lacking cell spaces, with a base of acellular bone traversed by thin non-vascular canals. Acanthodian spines also occur as microfossils (mostly as fragmentary pieces) and macrofossils (when complete). They are characterised by the nature of ornamentation and curvature, and the ratio of external ornamented spine to insertion area (finely striated). It may be difficult to

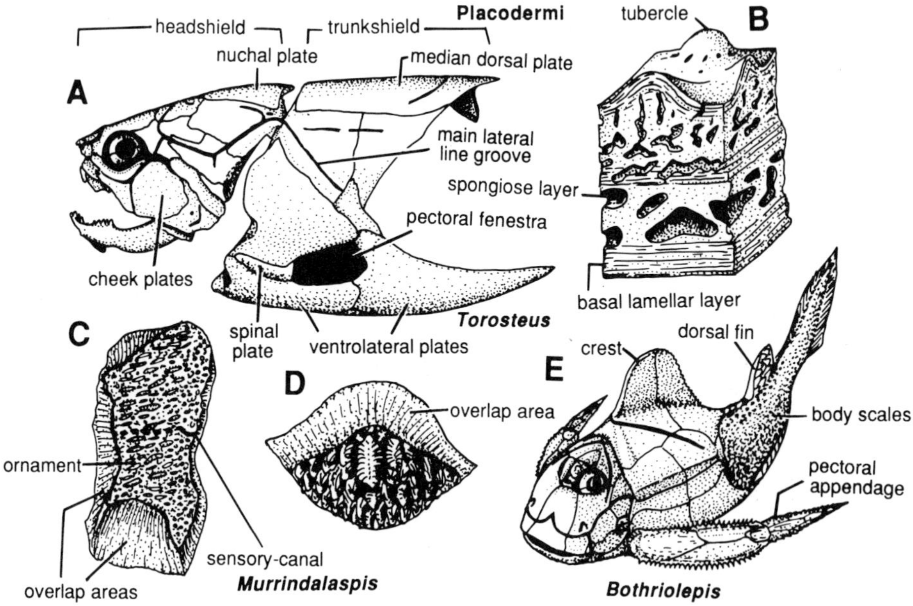

Figure 1.8. Placodermi. A, plourdosteid arthrodire; B, arthrodire bone structure; C, acanthothoracid dermal plate; D, acanthothoracid trunk scale; E, bothriolepid antiarch, not to scale (A, after Gardiner and Miles, 1990; B, after Moy-Thomas and Miles, 1971; C, D, from Long and Young, 1988; E, after Long, 1991).

distinguish shark spines from acanthodian spines in some cases (e.g. xenacanth sharks have simple non-ornamented spines with thick ribs).

Placoderms (Late Silurian–Devonian)

Placoderms are characterised by having a headshield and trunkshield composed of overlapping bony plates which articulate at a neck joint (Figure 1.8A). As fossils they are commonly found in Devonian rocks as isolated plates or articulated armours. The outer surface of placoderm bone has simple dentine-capped tubercles on most forms with smooth bone in others; a spongiose middle layer and basal laminated layer (Figure 1.8B). Those having body scales (most primitive forms of each respective placoderm group) leave abundant microvertebrate remains in marine sediments, typically in the Early Devonian (e.g. Figure 1.8D). Many groups of placoderms have been recognised, the most abundant being the arthrodires (Figure 1.8A), which feature paired upper jaw toothplates, and the antiarchs (Figure 1.8E), which have bony pectoral appendages, segmented in most groups. The more primitive placoderms tend to have heavily ossified dermal skeletons and well-ossified braincases, and their cranial anatomy is well-known (e.g. *Brindabellaspis*, Young, 1980). There are seven main families of placoderms (arthrodires, antiarchs, phyllolepids, acanthothoracids, petalichthyids, ptyctodontids and stensioellids; see Denison, 1978 for specific details).

The dominant groups most useful for biostratigraphic correlations are the antiarchs and the arthrodires, which together comprise more than 80 per cent of known placoderms, especially those forms ranging from Middle to Late Devonian such as *Bothriolepis* (Figure 1.8E), although antiarchs are also most useful in the Early Devonian of China (yunnanolepidoids; see Chapters 10, 11). Placoderms may be related to either chondrichthyans or to bony fishes, their exact phylogenetic affinities is still much debated (Stensiö, 1963, 1969; Gardiner, 1984a; Young, 1986; Maisey, 1986).

As biostratigraphic indicators placoderms are generally studied from macrofossil remains of isolated dermal plates or articulated armours (headshields, trunkshields). The characteristics of each plate in the placoderm armour (e.g. Figure 1.8C) make it easily identifiable to an expert, based on its outline morphology, nature of overlap surfaces, presence of sensory-line canals or body ridges (e.g. dorsolateral ridge), and the type of dermal ornamentation developed (tubercles, ridges, anastomising or linear patterns, smooth surfaces). A single placoderm plate, if well preserved, can give specific identification, or if fragmentary can often give generic level identification.

The study of placoderm body scales as microfossils is still largely in its infancy, but form types such as *Ohiolepis* are now recognised in many Early–Middle Devonian assemblages around the world. The main features of placoderm body scales (e.g. Figure 1.8D) are that they are simple rectangular to polygonal platelets with similar dermal ornamentation as the larger dermal plates, but generally with simple stellate tubercles present. Body scales have been described for a number of placoderm families, such as arthrodires (Denison, 1978), antiarchs (Hemmings, 1978; Young, 1984; Long and Werdelin, 1986), rhenanids (Stensiö, 1969), acanthothoracids (Ørvig, 1975; Long and Young, 1988). Recent discoveries from the Gogo Formation, Western Australia, show that some ptyctodontids, like *Campbellodus*, also had large overlapping body scales.

Osteichthyes (Latest Silurian–Recent)

The Osteichthyes, or 'true bony fishes' as they are often called, represent the largest living group of vertebrates. They first appear as minute scales in microresidues of Late Silurian age, and by the Early Devonian had differentiated into several major clades, particularly the ray-finned fishes (Actinopterygii) and the lobe-finned fishes (Sarcopterygii). Figure 1.9 shows the diversity of osteichthyan groups, including their derived subgroup, the tetrapods. Biostratigraphically they are most useful from the Late Silurian, where remains of primitive actinopterygians (*Ligulalepis*, Figure 1.10A, B) and lophosteiforms (*Lophosteus*, *Andreolepis*) have been recorded in marine sediments of China, Europe and Australia. Early Devonian

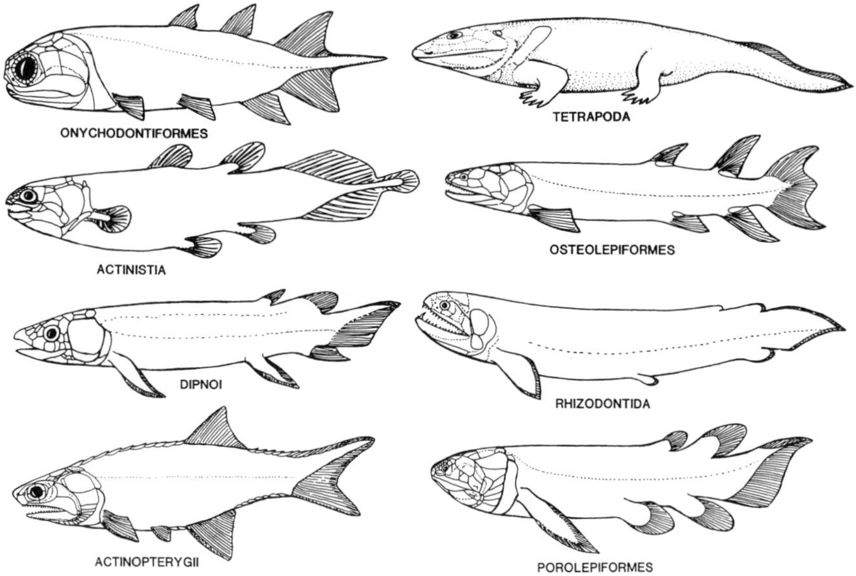

Figure 1.9. Major groups of Palaeozoic osteichthyans, including tetrapods.

sarcopterygian remains are also useful biostratigraphically, as some families, like the Porolepidae, are restricted to this age.

Lophosteiformes (Late Silurian)
These are known only from isolated scales, teeth and fragmentary bones, but are important as being the possible sister group to all other osetichthyans (Janvier, 1978). Two genera are known from the Late Silurian of Gotland, *Lophosteus superbus* and *Andreolepis heddei* (Gross 1969, 1971). The scales are small (about 1–2 mm) and are characterised by lacking ganoine, instead having simple ornament of dentine ridges on a rectangular bony base. Unlike actinopterygians they lack the characteristic peg and socket articulation, and the isolated conical teeth lack acrodin caps.

Actinopterygians (Latest Silurian–Recent)
These are the ray-finned fishes which comprise the largest known group of living fishes. They show a great range of body shapes in the Palaeozoic, but primitively are fusiform with large gaping mouths and forwardly situated orbits. The skulls have many small bones, and feature long toothed mandibular bones (dentaries) and maxillaries, and a snout with a median rostral bone (called 'postrostral' by some workers). Some primitive actinopterygians have micromeric square to rhombic scales (e.g. *Cheirolepis*, Pearson and Westoll, 1979), but the majority of Palaeozoic actinopterygians have rhombic scales with a well-defined dorsal peg and/or a dorsal process, as these scales overlap and fit neatly into the scales around them (Figure 1.10A–C).

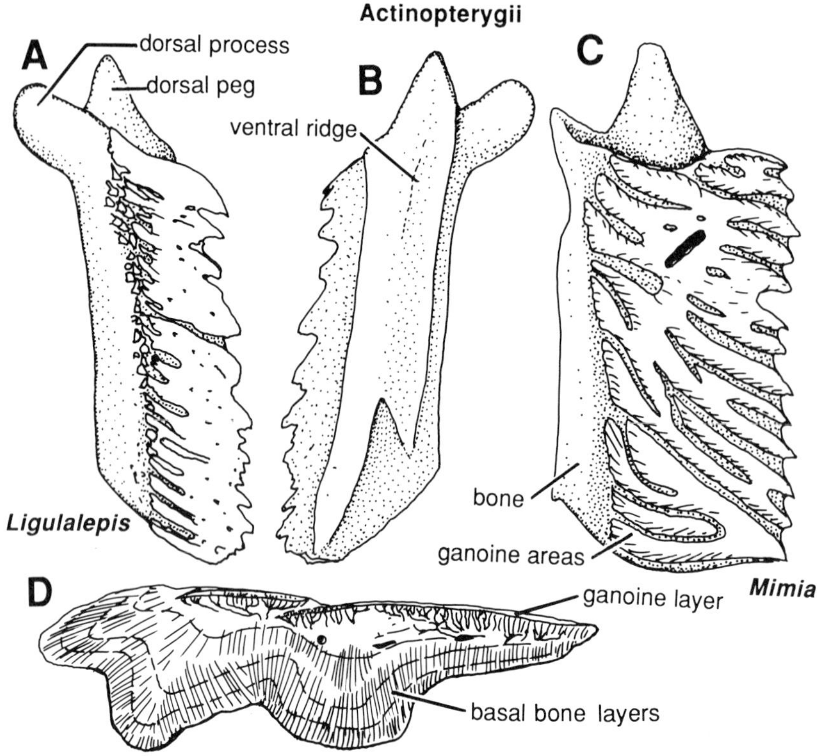

Figure 1.10. Actinopterygian scale morphology and structure. A, B, Early Devonian palaeoniscoid; C, D, Late Devonian palaeoniscoid, D showing scale cross-section, not to scale (A, B, after Schultze, 1968; C, D, after Gardiner, 1984a).

Actinopterygians first appear as isolated scales from the late Silurian (Wang and Dong, 1989) and Early Devonian (*Ligulalepis toombsi*, *Dialipina orvigi*, Schultze, 1968). By the Middle Devonian about three genera have emerged, and these include freshwater (e.g. *Cheirolepis*) and marine forms (*Stegotrachelus*, *Moythomasia*). By the late Devonian some six or more genera had evolved, and well-preserved specimens from the Gogo Formation, Western Australia reveal every aspect of their osteology in great detail (Gardiner, 1984b). Although complete, whole palaeoniscoids are scarce in Devonian sediments, by the late Palaeozoic their remains occur all around the world.

The scales of actinopterygians are of two types: most commonly they are of rhombic or cycloid shape with peg and socket articulations, or in primitive forms like cheirolepids or tegeolepids, they may be micromeric with large bases. The form of the scale is essentially similar in both types, having a rectangular bony base with external ornament comprising layers of dentine forming ridges (ganoine). This ganoid type of scale grows by accretion from both surfaces. The base consists of horizontal layers with contained cell spaces (Figure 1.10D), and may contain canals of Williamson and canals for fibres of Sharpey.

The other common microscopic remains from actinopterygians are teeth. These are invariably simple conical teeth in early palaeoniscoids, characterised by a cap of dense dentine called acrodin. Only the cheirolepids lack acrodin caps on teeth. The outer surfaces of the acrodin caps may be finely striated with ridges. Parts of the fin skeleton, such as segments of fin lepidotrichia are also commonly found as microfossils, but as they are difficult to characterise, they presently offer little information as to the identity of the genus.

Sarcopterygians (Devonian–Recent)

The Sarcopterygii are generally robust osteichthyan fishes with fleshy lobe-fins. Their dermal skeleton is characterised by the primitive presence of the shiny dermal tissue cosmine, a feature secondarily lost in all advanced lineages, and by the presence of advanced tissues in the dentition, such as enameloid on the teeth of predators, and mineralised dentines, such as petrodentine, in the toothplates and denticles of lungfishes (Smith, 1977, 1984). The earliest sarcopterygians include forms whose affinities are still debated (e.g. *Youngolepis*, Chang and Yu, 1981; Chang 1982; *Powichthys*, Jessen, 1980), and are here included as 'stem-group sarcopterygians'. The only living sarcopterygians are the few extant lungfishes (Dipnoi) and the sole crossopterygian, the coelacanth *Latimeria*.

Dipnoans (Devonian–Recent)

Dipnoans (lungfishes) are one of the most characteristic of all early fossil osteichthyans. The braincase is ossified as a single unit with palatal bones fused to the base of neurocranium. Dentition is either by rasping denticles over the palate and lower jaws and gill arch toothplates (e.g. *Uranolophus*, *Griphognathus*), or by toothplates formed from coalescing cusps or tuberosities (e.g. *Dipterus*, *Chirodipterus*). The skull roof is formed of many small bones that bear no resemblance to the skull-roof patterns of other osteichthyans, apart from the 'B' bones which may be homologues to the parietals in stem-group sarcopterygians. Schultze and Campbell (1987) list some 21 synapomorphies that define the Dipnoi.

Biostratigraphically the dipnoans are of interest during Early Devonian as macrofossil remains (skulls, jaws, toothplates) as the rapid evolution of the group saw many changes to dentition and tooth tissues during this time. By the Middle and Late Devonian many genera had appeared with a variety of feeding mechanisms, from heavy-jawed crushers to suctorial or tritorial feeders. Dipnoans began to invade freshwater environments by the Middle Devonian, resulting in modifications of body shape leading to the lineage of ceratodontids and lepidosirenids still extant today. Since the Devonian all lungfishes have been exclusively freshwater dwellers, and their ability to aestivate has been traced to fossils of Permian age.

Biostratigraphically the dipnoans are best studied from isolated skulls and toothplates, although some use is made of isolated scales (e.g. Long *et al.*, 1988). The characteristic dental plates of chirodipterids, dipterids, sagenodids and other families of lungfishes may prove useful in biostratigraphic schemes as more isolated material becomes described. Westoll (1949), in a classic early study of lungfish evolution, devised a point-scoring system which allows one to place the age of a lungfish by its tally of morphological features. Perhaps future refinement of this system may allow new fields to determine an approximate relative age date in the absence of other fossil or radiometric dating information.

Actinistians (Middle Devonian–Recent)
The actinistians or coelacanths are represented by several Palaeozoic forms, dating back to the late Middle Devonian (Stensiö, 1937). There is one living member, *Latimeria*, so much is known of the anatomy and physiology of coelacanths. The main features of the group are the skull which lacks maxillary bones, has a loose arrangement of cheek bones (in most forms) and the lower jaw has marginal biting teeth only at the anterior end, and a double tandem jaw articulation (braced by the large ceratohyal). The tail in most genera is developed as a tassel-shaped caudal fin. The scales are rounded with small rows of radiating ridges and tubercles (Figure 1.11). Coelacanths have limited use as biostratigraphic indicators at present because remains are quite rare in Palaeozoic rocks, although their distribution at this time makes them relevant to biogeographic discussions.

Onychodontiformes (Devonian)
These are still a poorly known group, based largely on a study of the little *Strunius* specimens from the Bergisch-Gladbach of Germany (Middle–Late Devonian), and isolated fragmentary remains of the larger *Onychodus* (Jessen, 1966). Three-dimensional specimens of *Onychodus* from Gogo, Western Australia, are currently being studied and will reveal much new anatomical information about the group. *Onychodus* was a large predator, up to 2 metres estimated length, with a row of symphysial teeth on the lower jaw. Microscopic remains of small onychodonts are occasionally found in Early Devonian limestones (e.g. Ørvig, 1969). The teeth of onychodonts have linear striations which show up as a herring-bone pattern under high magnification (Smith, 1979). Scales of onychodonts are rounded with tubercles on the exposed areas.

Rhipidistians
This term is used commonly to denote the lobe-finned predatory fishes, such as osteolepiforms, porolepiforms and rhizodontiforms.

Porolepiformes (Devonian)
Porolepiformes were large predators with long feathery pectoral fins, broad heads and very small orbits. The group is largely marine until the Middle and Late Devonian when they are found in continental deposits (e.g. Jarvik, 1972). Like onychodonts they have a lower jaw symphysial tooth

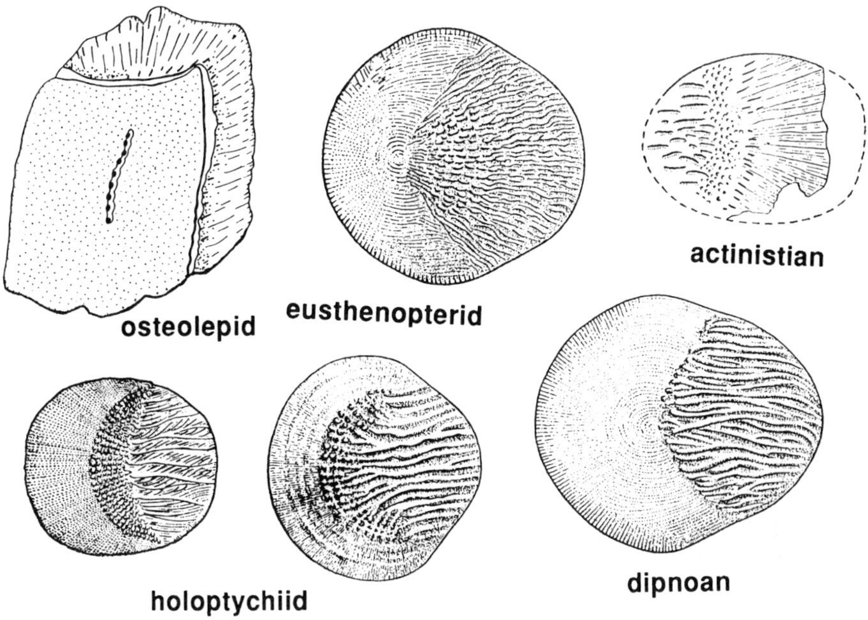

Figure 1.11. Sarcopterygian scale types, not to scale (after Jarvik, 1950).

whorl, but may have several rows of large teeth on the whorl, unlike the single row of onychodonts. The scales vary from thick rhombic cosmine types in primitive forms, to large, coarsely ornamented rounded scales in holoptychioids (Figure 1.11). The name is derived from the characteristic rows of pores visible on the scales of some genera (e.g. *Porolepis*, Ørvig, 1957). The main biostratigraphic use of porolepiforms is in the Latest Devonian, where scales of *Holoptychius*, found ubiquitously, indicate Famennian age.

Youngolepids (Early Devonian)
Youngolepids are here used to include two Early Devonian genera, *Youngolepis* (from China; Chang and Yu, 1981; Chang 1982) and *Powichthys* (from Arctic Canada; Jessen, 1980). They are a poorly known group, and have limited use in biostratigraphy at this stage. Isolated rhombic cosmoid scales of youngolepids cannot be confidently distinguished from those of other early sarcopterygians, so their main use is as macrofossils of skulls or jaws. They do appear to be restricted biogeographically to Euramerica and South China.

Rhizodontiformes (Middle Devonian–Carboniferous)
Rhizodontiformes were moderate to large fishes that appear in the uppermost Middle Devonian but are more commonly found in Carboniferous deposits.

They reached enormous sizes, up to 6 metres estimated length (*Rhizodus*), and are found commonly as isolated macroremains of jaws, teeth, scales and shoulder girdle bones. Only a few specimens exist which show details of body plan (Andrews, 1986) or skull anatomy (Long, 1989). The principle specialisation of the group is to have paddle-like fins with long, stiff unbranched lepidotrichia. Rhizodontiforms are of biostratigraphic use in the Late Palaeozoic, where their remains may indicate Carboniferous age in the lack of other fossil evidence. They are largely freshwater dwellers, although some remains have been found in marine limestones, such as in the Canning Basin of Western Australia.

Osteolepiformes (Middle Devonian–Permian)
Osteolepiformes are a diverse group of lobe-finned predatory fishes, characterised by their cheek pattern of seven bones, large basal scute scales on fins, single pair of external nares in the skull, and well-formed choana in the palate. Many early forms show thick rhombic cosmoid scales (osteolepid, Figure 1.11), but within several lineages they later developed rounded cycloid scales with a distinctive median boss on the basal surface (e.g. eusthenopterids, Figure 1.11). Perhaps the best known Palaeozoic fish is the osteolepiform *Eusthenopteron*, the subject of much study by the Swedish school of workers (see Jarvik, 1980, for summary). Most families of osteolepiforms died out at the end of the Devonian Period, although the cosmine-covered megalichthyinids are a common group in Carboniferous sediments around the world. Biostratigraphic usefulness of osteolepiforms lies principally in recognition of macroremains, such as skulls or whole fish, with limited use of isolated scales, though more so in the Late Devonian where a diverse range of scale types are found.

Tetrapoda (?Middle Devonian–Recent)
The tetrapods, or four-legged animals, are represented in the Devonian by the first amphibians, the ichthyostegalids, and in the Late Palaeozoic by a diverse array of amphibians and early reptiles (see Milner, Chapter 13). Their morphology varies a great deal, but all share the presence of limbs with digits, the lack of opercular-gular breathing, and open sensory-line canals on the skull and dermal bones. The earliest amphibians may have dated back as far as Early Devonian if footprints are considered (Warren *et al.*, 1986), although the oldest confirmed skeletal remains are late Frasnian to early Famennian (Ahlberg, 1991; Campbell and Bell, 1977). Recent discoveries of well-preserved Late Devonian forms from East Greenland reveal much about the anatomy of the first tetrapods, such as the presence of polydactylous digits and gill arch breathing in *Acanthostega* (Coates and Clack, 1990, 1991). The group is useful as biostratigraphic indicators in certain terrestrial deposits of late Palaeozoic age, and also for the wealth of palaeogeographic data provided by their essentially land-living distributions (see Chapter 13).

REFERENCES

Ahlberg, P.E., 1991, Tetrapod or near tetrapod fossils from the Upper Devonian of Scotland, *Nature*, **354**: 298–301.

Andrews, S.M., 1986, Rhizodont fishes from the Dinantian of Fouldenshire, *Transactions of the Royal Society of Edinburgh (Earth Sciences)*, **76**: 67–95.

Arsenault, M. and Janvier, P., 1991, The anaspid-like craniates of the Escuminac Formation (Upper Devonian) from Miguasha (Québec, Canada), with remarks on anaspid-petromyzontid relationships. In M.M. Chang, Y.H. Liu and G.R. Zhang (eds), *Early Vertebrates and Related Problems of Evolutionary Biology*, pp. 19–40. Science Press, Beijing.

Campbell, K.S.W. and Bell, M.W., 1977, A primitive amphibian from the Late Devonian of New South Wales, *Alcheringa*, **1**: 369–81.

Chang, M.-M., 1982, The braincase of *Youngolepis*, a Lower Devonian crossopterygian from Yunnan, south-western China, *Papers in the Department of Geology, University of Stockholm*, 1982, 1–111.

Chang, M.-M. and Yu, X.-B., 1981, A new crossopterygian, *Youngolepis precursor*, gen. et sp. nov., from the Lower Devonian of E. Yunnan, China, *Scientia Sinica*, **24**: 89–97.

Coates, M. and Clack, J., 1990, Polydactyly in the earliest known tetrapod limbs, *Nature*, **347**: 66–9.

Coates, M. and Clack, J., 1991, Fish-like gills and breathing in the earliest known tetrapod. *Nature*, **352**: 234–6.

Denison, R.H., 1978, Placodermi. In *Handbook of Paleoichthyology. Pt 3*, H.-P. Schultze (ed.), Gustav Fischer Verlag, Stuttgart, New York, 124pp.

Denison, R.H., 1979, Acanthodii. In *Handbook of Paleoichthyology. Pt 5*, H.-P. Schultze (ed.), Gustav Fischer Verlag, Stuttgart, New York, 62pp.

Gardiner, B.G., 1984a, Relationships of the palaeoniscoid fishes, a review based on new specimens of *Mimia* and *Moythomasia* from the Upper Devonian of Western Australia, *Bulletin of the British Museum of Natural History (Geology)*, **37**: 173–428.

Gardiner, B.G., 1984b, The relationship of placoderms, *Journal of Vertebrate Paleontology*, **4**: 379–95.

Gardiner, B.G. and Miles, R.S., 1990, A new genus of eubrachythoracid arthrodire from Gogo, Western Australia, *Zoological Journal of the Linnean Society*, **99**: 159–204.

Ginter, M., 1990, Late Famennian shark teeth from the Holy Cross Mts, Central Poland, *Acta Geologica Polonica*, **40**: 69–81.

Gross, W., 1969, *Lophosteus superbus* Pander, ein Teleostome aus dem Silur Oesels, *Lethaia*, **2**: 15–47.

Gross, W., 1971, *Lophosteus superbus* Pander: Zähne. Zahnknocken und besondere Schuppenformen, *Lethaia*, **4**: 131–52.

Hemmings, S.K., 1978, The Old Red Sandstone antiarchs of Scotland: *Pterichthyodes* and *Microbrachius*, *Palaeontological Society Monographs*, **131**: 1–64.

Janvier, P., 1978, On the oldest known teleostome fish *Andreolepis hedei* Gross (Ludlow of Gotland), and the systematic position of the lophosteids, *Esti NSV Teaduste Akadeemia Toimetised Koide Geologiski*, **27**: 88–95.

Janvier, P., 1984, The relationships of the Osteostraci and the Galeaspida, *Journal of Vertebrate Paleontology*, **4**: 344–58.

Jarvik, E., 1950, Middle Devonian vertebrates from Canning Land and Wegeners Halvö (East Greenland). II. Crossopterygii, *Meddelelser om Grønland*, **96, 4**: 1–132.

Jarvik, E., 1972, Middle and Upper Devonian Porolepiformes from East Greenland with special reference to *Glyptolepis groenlandica* n.sp., *Meddelelser om Grønland*, **187**, **2**: 1–295.
Jarvik, E., 1980, *Basic Structure and Evolution of Vertebrates*, Volume 2, Academic Press, London, New York, Toronto, Sydney, San Francisco.
Jessen, H., 1966, Die Crossopterygier des Oberen Plattenkalkes (Devon) der Bergisch-Gladbach-Paffrather Mulde (Rheinisches Schiefergebirge) unter Berücksichtigung von amerikanischem und europäischem *Onychodus*-material, *Arkivs für Zoologie*, **18**: 305–89.
Jessen, H., 1980, Lower Devonian Porolepiformes from the Canadian Arctic with special reference to *Powichthys thorsteinssoni* Jessen, *Palaeontographica*, **167A**: 180–214.
Karatajute-Talimaa, V.N., 1973, *Elegestolepis grossi* gen. et sp. nov., ein neuer typ der Placoidschuppe aus dem oberen Silur der Tuwa, *Palaeontographica*, **143A**: 35–50.
Karatajute-Talimaa, V.N., 1978, *Telodonti Silura i Devona S.S.S.R. i iSpitsbergena, Litovskiy naucho-isseldovatel'skiy geologorazvedochnyy Institut*, Mosklas, Vilnius, 334pp.
Long, J.A., 1989, A new rhizodontiform fish from the Early Carboniferous of Victoria, Australia, and comments on the relationships of the group, *Journal of Vertebrate Paleontology*, **9**: 1–17.
Long, J.A., 1990, Late Devonian chondrichthyans and other microvertebrate remains from Northern Thailand, *Journal of Vertebrate Paleontology*, **10**: 59–71.
Long, J.A., 1991, The long history of Australian fossil fishes. In P. Vickers-Rich, J.N. Monaghan, R.F. Baird and T.H. Rich (eds), *Vertebrate Palaeontology of Australasia*, Pioneer Design Studios with Monash University Publications Committee, Melbourne: 336–428.
Long, J.A. and Werdelin, L., 1986, A new species of bothriolepid antiarch from Victoria, Australia, with descriptions of other species from the state, *Alcheringa*, **10**: 345–99.
Long, J.A., Turner, S. and Young, G.C., 1988, A Devonian fish fauna from subsurface sediments in the eastern Officer Basin, South Australia, *Alcheringa*, **12**: 61–78.
Long, J.A. and Young, G.C., 1988, Acanthothoracid remains from the Early Devonian of New South Wales, including a complete sclerotic capsule and pelvic girdle, *Memoirs of the Association of Australasian Palaeontologists*, **7**: 65–80.
Mader, H., 1986, Schuppen and Zahne von Acanthodiern und Elasmobranchiern aus dem Unter-Devon Spaniens (Pisces), *Göttinger Arbeiten zur Geologie und Paläontologie*, **28**: 1–59.
Maisey, J., 1986, Heads or tails: a chordate phylogeny, *Cladistics*, **2**: 201–56.
Moy-Thomas, J.A. and Miles, R.S., 1971, *Palaeozoic Fishes*, Chapman and Hall, London.
Ørvig, T., 1957, Notes on some Palaeozoic Lower Vertebrates from Spitsbergen and North America, *Norsk geologische tiddskrifter*, **37**: 285–351.
Ørvig, T., 1969, Vertebrates from the Wood Bay Group and the position of the Emsian–Eifelian boundary in the Devonian of Vestspitsbergen, *Lethaia*, **2**: 273–328.
Ørvig, T., 1975, A new radotinid arthrodire from Arctic Canada, In *Problèmes actuels de paléontologie (évolution des vertébrés)*. Editions du CNRS, Paris, 1975: 41–72.
Pan Jiang, 1984, The phylogenetic position of the Eugaleaspida in China. *Proceedings of the Linnean Society of New South Wales*, **107**: 309–19.

Pan Jiang, Hou Fuchen, Cao Jingxuan, Gu Qichang, Liu ShiYu, Wang Junqing, Gao Linada and Liu Chun, 1987, *Continental Devonian of Ningxia and its biotas*, Geological Publishing House, Beijing, 237pp.

Pearson, D.M. and Westoll, T.S., 1979, The Devonian actinopterygian *Cheirolepis* Agassiz, *Transactions of the Royal Society of Edinburgh*, **70**: 337–99.

Schultze, H.-P., 1968, Palaeoniscoidea-Schuppen aus dem Unterdevon Australiens und Kanadas und aus Mittelsdevons Spitzbergens, *Bulletin of the British Museum of Natural History (Geology)*, **16**: 343–68.

Schultze, H.-P. and Campbell, K.S.W., 1987, Characterization of the Dipnoi: a monophyletic group, *Journal of Morphology, Supplement*, **1**: 93–131.

Smith, M.M., 1977, The microstructure of the dentition and dermal ornament of three dipnoans from the Devonian of Western Australia: a contribution towards dipnoan interrelations, and morphogenesis, growth and adaptation of skeletal tissues, *Philosophical Transactions of the Royal Society of London*, **281B**: 29–72.

Smith, M.M., 1979, SEM of the enamel layer in oral teeth of fossil and extant crossopterygian and dipnoan fishes, *Scanning Electron Microscopy*, **2**: 483–90.

Smith, M.M., 1984, Petrodentine in extant and fossil dipnoan dentition: microstructure histogenesis and growth, *Proceedings of the Linnean Society of New South Wales*, **107**: 367–407.

Soehn, K.L. and Wilson, M.H.V., 1990, A complete articulated heterostracan from the Wenlockian (Silurian) beds of the Delorme Group, Mackenzie Mountains, Northwest Territories, Canada, *Journal of Vertebrate Paleontology*, **10**: 405–19.

Stensiö, E., 1937, On the Devonian coelacanthids of Germany with special reference to the dermal skeleton, *Kungligar Svenska Vetenskap Akadamiens Handlingar, 3rd ser.*, **16**: 1–56.

Stensiö, E.A., 1963, Anatomical studies on the arthrodiran head. Part 1. Preface, geological and geographical distribution, the organization of the arthrodires, the anatomy of the head in the Dolichothoraci, Coccosteomorphi and Pachyosteomorphi. Taxonomic appendix, *Kungligar Svenska Vetenskap Akadamiens Handlingar, 4th ser.*, **9**: 1–419.

Stensiö, E., 1969, Arthrodires. In J. Piveteau (ed.), *Traite de Paleontologie, vol. IV, part 1, Poissons*, Mason and Co., Paris.

Turner, S., 1973, Siluro–Devonian thelodonts from the Welsh Borderland, *Journal of the Geological Society of London*, **29**: 557–84.

Turner, S., 1982, Middle Palaeozoic elasmobranch remains from Australia, *Journal of Vertebrate Paleontology*, **2**: 117–31.

Valiukevicius, J.J., 1985, *Acanthodians from the Narva Regional Stage of the Main Devonian Field*, Mosklas, Vilnius, 143pp. [In Russian with English summary.]

Vieth-Schreiner, J., 1983, Fisch-Schuppen und -Zahne aus der Eifeler Kalkmulden-Zone (Emsium, Eifelium), *Senckenbergiana Lethaea*, **64**: 129–77.

Wang, N., 1984, Thelodont, acanthodian, and chondrichthyan fossils from the Lower Devonian of southwest China, *Proceedings of the Linnean Society of New South Wales*, **107**: 419–42.

Wang, N. and Dong, Z., 1989, Discovery of Late Silurian microfossils of Agnatha and fishes from Yunnan, China, *Acta Palaeontologica Sinica*, **8**: 192–206.

Warren, A., Jupp, R. and Bolton, B., 1986, Earliest Tetrapod trackway, *Alcheringa*, **10**: 183–6.

Westoll, T.S., 1949, On the evolution of the Dipnoi. In G.L. Jepson, G.G. Simpson and E. Mayr (eds), *Genetics, palaeontology and evolution*, University Press, Princeton: 121–84.

Young, G.C., 1980, A new Early Devonian placoderm from New South Wales,

Australia, with a discussion of placoderm phylogeny, *Palaeontographica*, **167A**: 10–76.

Young, G.C., 1982, Devonian sharks from south-eastern Australia and Antarctica, *Palaeontology*, **25**: 817–43.

Young, G.C., 1984, An asterolepidoid antiarch (placoderm fish) from the Early Devonian of the Georgina Basin, central Australia, *Alcheringa*, **8**: 65–79.

Young, G.C., 1991, The first armoured agnathan vertebrates from the Devonian of Australia. In Chang, M.M., Liu, Y.H. and Zhang, G.R. (eds), *Early Vertebrates and Related Problems of Evolutionary Biology*, pp. 67–85. Science Press, Beijing, China.

Zangerl, R., 1981, *Paleozoic Chondrichthyes*. In *Handbook of Paleoichthyology. Pt 3A*, H.-P. Schultze (ed.), Gustav Fischer Verlag, Stuttgart, New York, 114pp.

Chapter 2

PALAEOZOIC GLOBAL RECONSTRUCTIONS

Li, Z.-X., Christopher McA. Powell and Allan Trench

INTRODUCTION

Palaeomagnetism offers perhaps the most important constraints on the pre-Mesozoic movement of the continents. Given high-quality palaeomagnetic data, the orientation and palaeolatitude of a continent can be established to within 10° at the 95 per cent confidence level: a precision which is unattainable using either biogeographic or palaeoclimate indicators. Nevertheless, palaeomagnetic studies carry significant limitations and important assumptions which should be clearly outlined at the onset. First, the palaeomagnetic method cannot generally constrain the palaeolongitude of continental blocks. Second, Palaeozoic studies assume the premise of a geocentric axial dipole magnetic field and that true polar wander in Palaeozoic times was either absent or negligible. Neither of these assumptions has been rigorously tested to date and could lead to systematic errors in the order of 10° (e.g. Andrews, 1985; Idnurm, 1985, 1986). Third, and probably of greatest importance, there remains a deficiency of reliable Palaeozoic palaeomagnetic data, in space and time, leading to a degradation of overall precision. A good example of this is the Palaeozoic palaeomagnetism of Gondwanaland, discussed below. Whereas future high quality work will improve the situation, current hypotheses based on the available data must fully appreciate these limitations.

Constraints on Palaeozoic palaeogeography can also be deduced from the distribution of latitude-dependent sedimentary facies (e.g. Caputo and Crowell, 1985; Witzke, 1990; Scotese and Barrett, 1990; Nie, 1991), and latitudinal and/or provincial controls on the distribution of certain faunas. Good examples of the latter are presented within the present volume and the volume edited by Scotese and McKerrow (1990). However, palaeoclimatic and biogeographic considerations are limited by their essentially

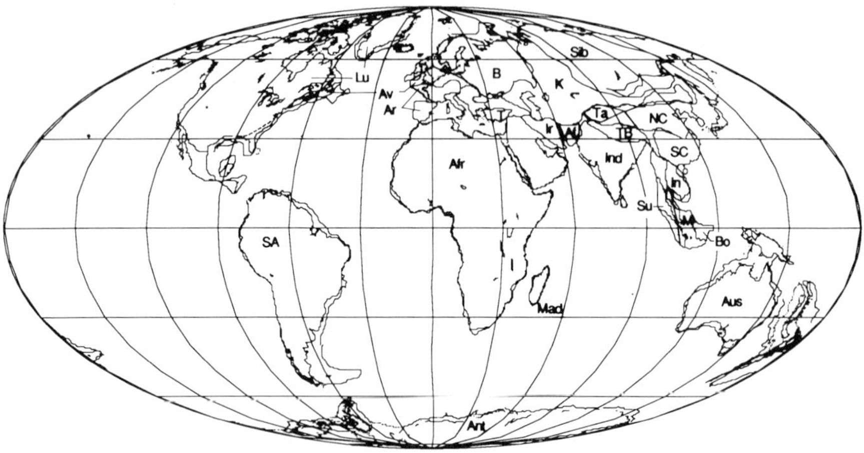

Figure 2.1. Subdivision of continental blocks mentioned in this study. Mollweide equal area projection. SA - South America; Lu - Laurentia; Av - Avalonia; B - Baltica; K - Kazakhstan; Sib - Siberia; Afr - Africa; Ar - Arabia; T - Turkey; Ir - Iran; Af - Afghanistan; Ind - India; TB - Tibet; Ta - Tarim; NC - North China; SC - South China; In - Indochina; Su - Sibumasu; M - E Malaya; Bo - SW Borneo; Aus - Australia; Mad - Madagascar; Ant - East Antarctica.

qualitative nature, the assumption of a latitudinally zonal climatic pattern throughout Palaeozoic time, and by our present understanding of the living habits and habitat of Palaeozoic biotas.

In addition to the above methods, comparisons of the tectonostratigraphic histories of different continents and terranes may also be used to evaluate their movement history (e.g. Hoffman, 1991; Dalziel, 1991a, 1991b; Bluck and Dempster, 1991). Once again, this technique is limited by its qualitative nature, and perhaps by the non-uniqueness of deformation patterns.

When used in isolation, each of these reconstruction techniques is likely to yield an inconclusive palaeogeography. When integrated, however, the combined dataset allows a significant reduction in overall ambiguity. In this study, our strategy has been to construct global palaeogeographic maps by positioning continents using palaeomagnetic evidence where reliable data exist. To this end, palaeomagnetic data from each continent were assessed using published reliability criteria (Van der Voo, 1990; Li and Powell, 1992). Where palaeomagnetic data are either inconsistent or absent, our recourse has been to palaeoclimatic and biogeographic data. Relative palaeolongitudes have been assigned using a combination of tectonostratigraphic and biogeographic information. Using this methodology, we present a series of global palaeogeographic maps corresponding to approximately 510, 420, 400, 365, 310 and 265 million years ago (Ma).

Parts of the reconstructions for all of the time segments are well constrained by palaeomagnetic data and independently supported by non-palaeomagnetic methods. Other parts of the reconstructions that are less well constrained by palaeomagnetic data are perhaps more controversial, and may be significantly modified by future work. Subdivision of the continents into their Palaeozoic constituents is shown in Figure 2.1

CONSTRAINTS ON PALAEOZOIC MOVEMENTS

Gondwanaland

The movement history of Gondwanaland, a supercontinent equivalent in size to approximately 50 per cent of the present-day land surface, forms the dominant factor in any reconstruction of Palaeozoic plate tectonics. Unfortunately, the past movements of Gondwanaland based upon palaeomagnetic, biogeographic and palaeoclimatic data are both equivocal and contentious.

Since two alternative loci for the Australian Palaeozoic apparent polar wander path (APWP), and therefore the Gondwana APWP, were outlined by Embleton *et al.* (1974) and McElhinny and Embleton (1974), several other models have been proposed and revised (Schmidt and Morris, 1977; Morel and Irving, 1978; Goleby, 1980; Li *et al.*, 1990; Bachtadse and Briden, 1990, 1991). Each of these models implies a significantly different Palaeozoic movement history for Gondwanaland.

During the 1970s, this somewhat chaotic situation was attributed to a lack of reliable Gondwanan palaeomagnetic data for the Silurian–Middle Devonian interval (e.g. Schmidt and Morris, 1977). Despite the addition of new data from Africa and Australia, however (i.e. Schmidt *et al.*, 1986; Hargraves *et al.*, 1987; Schmidt *et al.*, 1987; Schmidt and Embleton, 1990; Bachtadse and Briden, 1990, 1991), the situation has not significantly improved.

Mid-Cambrian to Permian palaeomagnetic poles from Gondwanaland, which satisfy either A or B class quality criteria (Li and Powell, 1992), are plotted in Figure 2.2. The collective dataset established a number of points regarding the movement of the South Pole over Gondwana as follows:

- Cambro–Ordovician poles from cratonic Australia, Africa and South America are grouped around northwest Africa.
- Late Devonian poles, exclusively determined from Australia, are located around eastern South America and central Africa (in a Gondwana reassembly).
- Poles of Permo–Carboniferous age, from all the Gondwanan continents, cluster around Antarctica.

Significant inconsistencies exist between Silurian–Devonian poles however, contrasting interpretations of which have yielded the various APWP models. For example, the mid-Devonian pole from the Gilif Hills ring

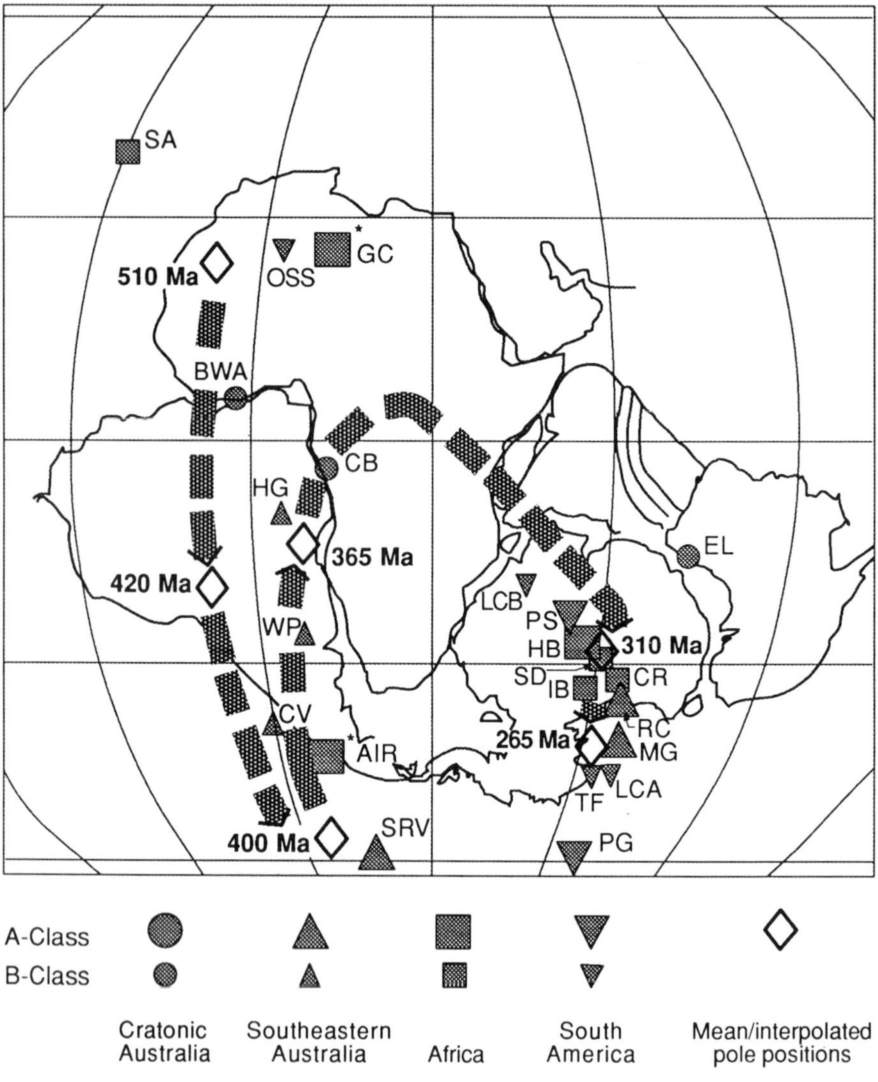

Figure 2.2. Distribution of Cambro-Ordovician to Permian Key-palaeopoles (solid symbols) for Gondwana, the poorly defined apparent polar wander path (APWP), and the interpolated pole positions (open diamonds).

complex (GC, Bachtadse and Briden, 1991) lies close to Cambro–Ordovician results in northern Africa. Conversely, poles from ?Early Silurian Aïr ring complexes (AIR) of Africa (Hargraves *et al.*, 1987), the Siluro–Devonian Snowy River Volcanics (SRV) of SE Australia (Schmidt *et al.*, 1987), and the mid-Devonian Comerong Volcanics (CV) of SE Australia (Schmidt *et al.*, 1986), plot more than 65°–85° away, close to the SW coast of South America. In an attempt to resolve this discrepancy, Bachtadse and Briden (1991) suggested that the Aïr pole records an interval of Carboniferous remagnetisation. In their interpretation, the SE Australian data are interpreted as allochthonous to Gondwana. Whereas the Bachtadse and Briden APWP negates the requirement for rapid drift of Gondwana over the South Pole in Silurian–Devonian times (in the order of 20–30 cm/yr, calculated from APWPs of Morel and Irving, 1978; Bachtadse and Briden, 1990; Li *et al.*, 1990), it introduces additional complexity into the Carboniferous path. In the model followed here, the South Pole traverses central Africa to the tip of South America and then returns to 'Eastern' Antarctica during the Permo–Carboniferous time. Clearly, additional palaeomagnetic studies of Silurian–Devonian rocks are required to resolve these controversies.

In view of the itinerant palaeomagnetic interpretations, Gondwanan APWPs have been constructed using palaeoclimatic indicators by a number of workers (Caputo and Crowell, 1985; Scotese and Barrett, 1990). Once

---

Figure 2.2 – cont.
Palaeomagnetic poles on the diagram are: BWA - mid-Cambrian sedimentary rocks of central Australia (Klootwijk, 1980); SA - Middle Ordovician Salala ring complex (Bachtadse and Briden, 1989); OSS - Ordovician sediments at Salta and Jujuy (Creer, 1970); SRV - Siluro-Devonian Snowy River Volcanics (Schmidt *et al.*, 1987); AIR - Lower Silurian (?) Aïr ring complex (Hargraves *et al.*, 1987); GC - mid-Devonian (?) Gilif Hills ring complex (Bachtadse and Briden, 1991); CV - mid-Devonian Comerong Volcanics (Schmidt *et al.*, 1986); WP - Upper Devonian Worange Point Formation (Thrupp *et al.*, 1991); HG - Siluro-Devonian Carboniferous Hervey Group (Li *et al.*, 1988); CB - Upper Devonian Canning Basin limestones (Hurley and Van der Voo, 1987); EL - overprint in Carboniferous Mount Eclipse Sandstone (Li *et al.*, 1989); HB - mid-Carboniferous Hassi Bachir Formation (Daly and Irving, 1983); MG - Upper Carboniferous Main Glacial Stage (Irving, 1966); RC - Upper Carboniferous Rocky Creek Conglomerate (Irving, 1966); LCB - Upper Carboniferous La Colina Formation (Sinito *et al.*, 1979); PS - Upper Carboniferous Pular and Cas formations (Jesinkey *et al.*, 1987); IB - Permo-Carboniferous El Adeb Larache Formation at Illizi Basin (recalculated from Henry *et al.*, 1992, combining the 15 sites, each of which has at least 3 samples with a characteristic remanence); SD - Lower Permian (?) Upper Serie d'Abadla (Morel *et al.*, 1981); CR - Lower Permian Chougrane Red Beds (Daly and Pozzi, 1976); PG - mid-Permian middle Paganzo Group (Valencio *et al.*, 1977); LCA - Permian La Colina Formation (Thompson, 1972); TF - mid-Permian Tambillos Formation (Rapalini and Vilas, 1991). Poles marked with a star were not used in defining the APWP.

Table 2.1 *Selected palaeopoles used for palaeolatitude control in the reconstructions*

Age of reconstruction (Ma)	PLAT (°N)	PLONG (°E)	a_{95} (DP, DM)	Sources
Gondwanaland (S Pole, in present Africa coordinates)				
Early to mid-P (265 Ma)	–42	051		A synthetic pole within the error of the mean palaeomagnetic pole position below.
	(–43	064	12	Mean of poles SD, CR, PG, LCA & TF in Figure 2.2.
Late Carb. (310 Ma)	–24	051		A synthetic pole within the error of the mean palaeomagnetic pole below.
	(–28	060	10	Mean of poles MG, RC, EL, HB, LCB, PS and IB in Figure 2.2.
Late Devonian (365 Ma)	–14	008	19	Mean of poles WP, HG & CB in Figure 2.2.
Early Devonian (400 Ma)	–56	008		Interpolated from poles SRV & CV in Figure 2.2.
Mid-Silurian (420 Ma)	–20	352		Interpolated from poles O_3/S_1 and S_2/D_1 of Scotese and Barrett (1990).
Cambro-Ordovician (510 Ma)	24	352	37	Mean of poles BWA, SA and OSS (Figure 2.2).
Laurentia and late Palaeozoic Europe (N Pole, in (or interpolated from) table 7 of Van der Voo (1990). All in American coordinates)				
Early to mid-P	46	125	2.1	Pl pole in table 7 of Van der Voo (1990).
Late Carb.	40	129	1.7	Pole for Cu/P and Cu in table 7 of Van der Voo (1990).
Late Devonian	29	111	8.6	Pole for Du and Dm/Du in table 7 of Van der Voo (1990).
Early Devonian	13	103		Interpolated from poles for Su/Dl and Dm, Dl in table 7 of Van der Voo (1990).
Mid-Silurian	19	124	5.8	Pole for Su,m in table 7 of Van der Voo (1990).
Cambro-Ordovician	12	161		Mean of poles for ϵu, ϵm, ϵ and for Ol/m, Ol in table 7 of Van der Voo (1990).
East Avalonia (S Pole)				
Mid-Silurian	15°S			after Torsvik and Trench (1992); and Channell *et al.* (1992b).
Cambro-Ordovician	56	306	9 10	O_1 pole of Trench *et al.* (1992): palaeolatitude only; see discussion in Trench *et al.* (1992).

Table 2.1 *cont.*

Age of reconstruction (Ma)	PLAT (°N)	PLONG (°E)	a_{95} (DP, DM)	Sources
Baltica (S Pole)				
Cambro-Ordovician	31	86	11 15	O_1 pole of Torsvik *et al.* (1990) (C-class pole due to its poor age constraint).
Armorica (mostly from the pre-Late Carboniferous S Poles with Q⩾3, in table 1 of Kent and Van der Voo (1990))				
Late Devonian	−30	009		Bachtadse *et al.* (1983).
Early Devonian	−22	319	10	Perroud and Bonhommet (1984): position slightly adjusted in the reconstruction.
Mid-Silurian				Palaeolatitude ~20°S for Spain indicated by an Early Silurian pole (Perroud *et al.*, 1991).
Cambro-Ordovician	29	357		Mean of the two O_1 poles by Duff (1979) and Lewandowski (1987).
Siberia (N Poles, all from Table 3 of Khramov *et al.* (1981))				
Late Carb.	30	150	14	CM-CL pole.
Late Devonian	20	140	12	DU pole.
Early Devonian				Low palaeolatitude as suggested by palaeoclimatic indicators (see text discussion).
Mid-Silurian	−7	103	-	SL pole.
Cambro-Ordovician	−39	127	-	Mean of poles CU and OL.
Tarim (N Poles)				
Early to mid-P	62	181	8	An average of five Permian studies (i.e. Bai *et al.*, 1987; Li and McWilliams *et al.*, 1988; McFadden *et al.*, 1988; Sharps *et al.*, 1989 & Fang *et al.*, 1990).
Late Carb.	47	170	4 6	Fang *et al.* (1990).
Early Devonian and Late Devonian	16	165	4	Li and McWilliams *et al.* (1990).
Cambro-Ordovician and mid-Silurian				Low palaeolatitudes (see text).
North China (N Poles)				
Early to mid-P	47	359	9	The average of 6 studies (Zhao *et al.*, 1990).
Cambro-Ordovician	−29	131	12 21	ϵ_3-O_2 pole of Zhao *et al.* (1992).

Table 2.1 *cont.*

Age of reconstruction (Ma)	PLAT (°N)	PLONG (°E)	a_{95} (DP,	DM)	Sources
South China (N Poles)					
Early to mid-P Camb.-Carb.	49	245		8	Average of nine studies (Li, 1989). (Equatorial palaeolatitudes (see text).)
Sibumasu (N Poles)					
Late Carb.	17	258		9	Huang and Opdyke (1991). (The palaeolatitude of the Sibumasu Block has been slightly adjusted in the reconstruction).
Central Iran (N Pole)					
Early Devonian	51	196	5	10	Wensink (1983). (Palaeolatitude of Central Iran has been slightly adjusted within the a_{95} limit, in the reconstruction).

again, however, the results are not always consistent. Whereas Caputo and Crowell's model suggests that the South Pole moved from offshore of western South America to central Africa during the Devonian–Early Carboniferous interval, Scotese and Barrett's (1990) analysis suggests that the pole moved directly from southern South America towards the Antarctica Peninsula over this time. Despite these discrepancies, both of the palaeoclimatic analyses agree with the palaeomagnetic positioning of the pole in NW Africa during Cambro–Ordovician times and over 'Eastern' Antarctica during Permo–Carboniferous times.

For the Gondwanan reconstructions in this study, we adopt palaeomagnetic poles for the Cambro–Ordovician and Permo–Carboniferous intervals, which are in agreement with palaeoclimatic data. For the Late Devonian interval, we choose a pole in eastern South America/western-central Africa as indicated by palaeomagnetic results from cratonic and SE Australia, and independently supported by the palaeoclimatic APWP of Caputo and Crowell (1985). We regard the Late Devonian–Early Carboniferous pole in southern South America proposed by Scotese and Barrett (1990) as a possible artefact of sparse palaeoclimatic data. For the Silurian–Early Devonian interval, when Gondwanan palaeomagnetic data show a marked inconsistency, an interpolated mid-Silurian pole from the palaeoclimatic poles of Scotese and Barrett (1990), and an interpolated Early Devonian pole between the Snowy River Volcanic pole (SRV) and the Comerong Volcanics pole (CV), have been used. Additional palaeomagnetic studies are urgently required in order to validate these poles. All Gondwanan poles used in the plate reconstructions are listed in Table 2.1.

North America and Northern Europe (Laurentia)

The Palaeozoic palaeomagnetism of North America and Northern Europe has recently been reviewed by Van der Voo (1990). He demonstrated that when rotated into North American co-ordinates using the rotation parameters of Bullard *et al.* (1965), the mean Palaeozoic poles from Northern Europe agree closely with those from North America. In this analysis, 'European' poles comprised:

1. Palaeozoic poles from the British Isles north of the Iapetus Ocean suture;
2. Middle Silurian–Permian poles from the Caledonian Belt south of the Iapetus suture, the Baltic shield and Russian Platform;
3. Late Carboniferous–Permian poles from Hercynian Europe.

The large number of moderate to high reliability poles (Quality Factor, Q, of Van der Voo (1990) > 3) and the high consistency between these data, suggest that the Palaeozoic APWP of North America and Northern Europe is now reasonably well defined. In particular, there is no longer a disparity between the North American and European Palaeozoic datasets requiring large-scale relative movements between the two (e.g. Morris, 1976; Van der Voo and Scotese, 1981). In this study, the mean Palaeozoic poles listed by Van der Voo (1990), or interpolated poles, are used to constrain the position of North America throughout the Palaeozoic, and of Northern Europe from Middle Silurian times.

In Cambro–Ordovician times, palaeomagnetic data indicate an East–West orientation for the southern Laurentian margin at low southerly palaeolatitudes of 15°–20°S. This palaeogeography is consistent with the widespread occurrence of Cambro–Ordovician warm-water carbonates across North America, Greenland and NW Britain (McKerrow *et al.*, 1991). From Ordovician to Siluro–Devonian times, palaeomagnetic data indicate that Laurentia gradually rotated counterclockwise. By the Early Devonian, the southwest tip of Laurentia had reached approximately 45°S. Since then, Laurentia gradually drifted towards the northern hemisphere. This is also consistent with the palaeolatitudes implied by palaeoclimatic data (Witzke, 1990).

Baltica

The Cambro–Ordovician position of Baltica is constrained by palaeomagnetic data from the North Scandinavian Caledonides (Torsvik and Olesen *et al.*, 1990). Although this data set is not itself of high reliability (C-class pole), the pole lies close to a time-sequence of results from the Arenig–Caradoc limestones of the Swedish platform (Torsvik and Trench, 1991; Perroud *et al.*, 1992). The fidelity of the palaeomagnetic record in the limestones is confirmed by the presence of bedding-parallel polarity reversals. New mid-Silurian data from Gotland, SE Sweden

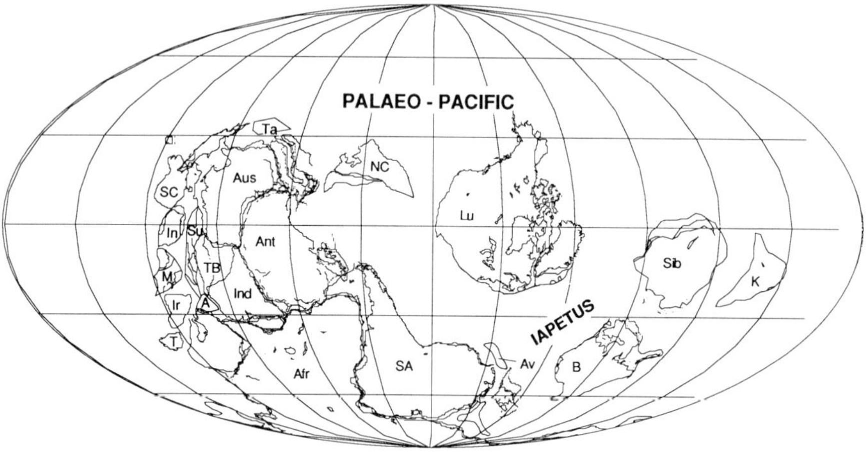

Figure 2.3. Cambro–Ordovician reconstruction. For mnemonics see Figure 2.1.

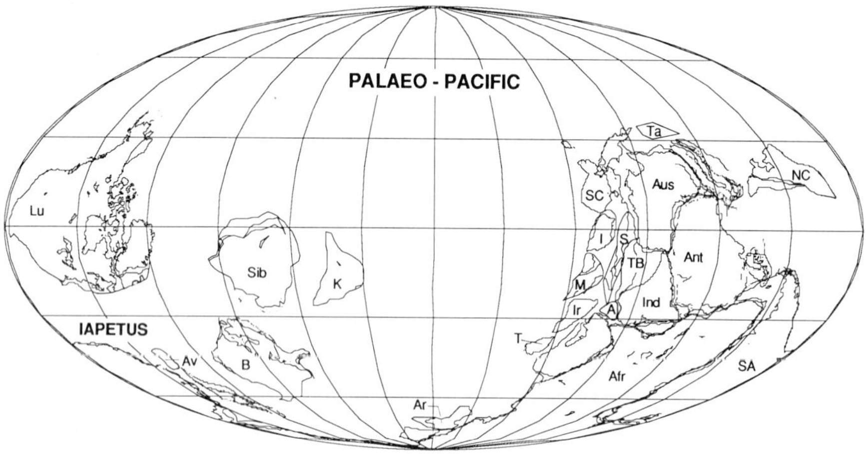

Figure 2.4. Cambro–Ordovician reconstruction from a different view-point.

(Trench and Torsvik, 1991) and Siluro–Devonian data from the Russian platform (Smethurst and Khramov, 1992) are consistent with the synthesis of European poles complied by Van der Voo (1990). The combined data indicate a convergence of the Baltica APWP with the Laurentian path by mid-Silurian times (Torsvik and Smethurst *et al.*, 1990).

The palaeomagnetic data indicate mid-southerly latitudes (30°–50°) for Baltica during Cambro–Ordovician times with an orientation significantly rotated from its present-day geography (Figures 2.3 and 2.4). Northward drift and counterclockwise rotation then occurred during Ordovician–Silurian times. Faunal regimes in and around Baltica are consistent with such a 'rotated' position for the continent in Early Ordovician times (Torsvik *et al.*, 1991). The transition from temperate to warm-water carbonates on the Baltic platform in Late Ordovician times (Webby, 1980) independently supports the northward drift indicated by palaeomagnetic data.

Avalonia

The Cambro–Ordovician palaeolatitude of Avalonia has been constrained using palaeomagnetic data from the Tremadoc–Arenig volcanics of southern Britain (Trench *et al.*, 1992). Avalonia then drifted northward throughout Ordovician times as recorded by palaeomagnetic poles from the Welsh Basin (Channell *et al.*, 1992a) and Nova Scotia (Van der Voo and Johnson, 1985; Johnson and Van der Voo, 1990). Recently, Early Wenlock volcanics (Hancock, 1982) of the Mendip Hills in SW England have yielded data indicating that Eastern Avalonia had reached 10°–15°S by this time (Torsvik and Trench, 1992) implying suturing of Avalonia with Laurentia and Baltica and the closure of the Iapetus Ocean by that time. A palaeolatitude of 15°–20°S is confirmed by results from the Siluro–Devonian Old Red Sandstone of the Anglo–Welsh cuvette (Channell *et al.*, 1992b). Mid-Late Palaeozoic poles from Avalonia are consistent with those of North America compiled by Van der Voo (1990).

Armorica and Central Europe

Lower Palaeozoic palaeomagnetic data from Armorica are sparse due to the prevalence of Late Palaeozoic overprinting associated with the Hercynian orogeny (e.g. Bachtadse, 1988). In a number of cases, however, pre-Hercynian magnetisations are resolved by detailed demagnetisation experiments. The pre-Carboniferous palaeopoles of central and southern Europe selected by Kent and Van der Voo (1990, Table 1), and the recently reported results from the Silurian volcanics at Almaden, southern Spain (Perroud *et al.*, 1991), are used to constrain the palaeolatitude of this region (Table 2.1). Independent evidence bearing on the palaeoposition of Armorica is afforded through the sedimentary facies and biogeography, both methods indicating close affinity between Armorica and north Africa during the Cambrian–Devonian interval (e.g. Noblet and Lefort, 1990; Robardet *et al.*, 1990; McKerrow *et al.*, 1992).

Early and Late Ordovician palaeomagnetic data from Armorica indicate high southerly latitudes (>60°S, Duff, 1979; Perroud *et al.*, 1984, 1986; Perroud and Van der Voo, 1985; Cogné, 1988). Similarly, Early Ordovician

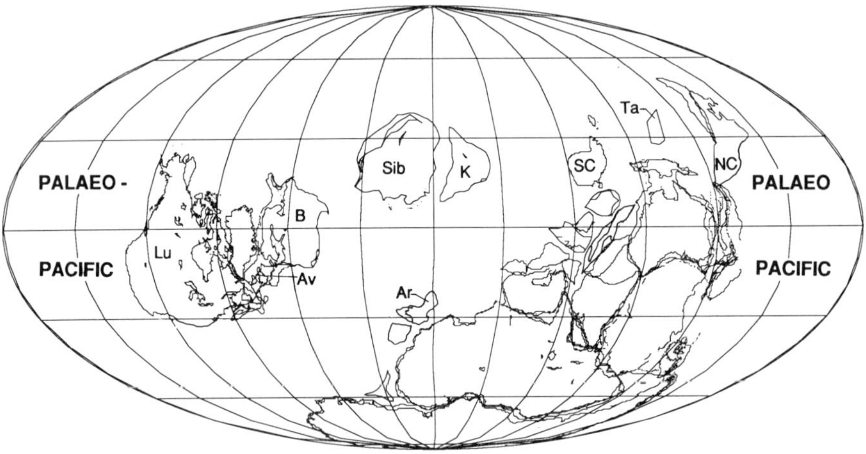

Figure 2.5. Mid-Silurian reconstruction.

results from the Holy Cross Mountains in Poland yield palaeolatitudes in the order of 60°S (Lewandowski, 1987). The collective data are interpreted to represent a palaeoposition close to the Gondwanan margin for Armorica during Cambro–Ordovician times (Figures 2.3, 2.4). Sedimentary facies evidence also indicates the proximity of Armorica to Africa in Early Ordovician times when the two share a common shallow-marine sandstone facies, the Armorican Quartzite (Noblet and Lefort, 1990).

Recent Early Silurian data from southern Spain (Perroud *et al.*, 1991), of uncertain declination but well-controlled inclination, yields a palaeolatitude of 20°. Although the original authors inferred a northern hemisphere position, we favour a southerly palaeolatitude adjacent to northern Africa (Figure 2.5). Devonian palaeolatitudes are constrained by high quality palaeomagnetic data from the San Pedro redbeds of Spain (Perroud and Bonhommet, 1984) and sedimentary and volcanic rocks of Germany (Bachtadse *et al.*, 1983).

Siberia and Kazakhstan

Given the lack of recent palaeomagnetic results from the Siberian Craton, palaeopoles listed by Khramov *et al.* (1981) (Table 2.1), are used to control its palaeolatitude. The Early Devonian pole is not used because its implied middle palaeolatitude for Siberia is in conflict with palaeoclimatic indicators such as the development of carbonate and evaporite sequences (Heckel and Witzke, 1979), and the presence of tropical faunas (e.g. Stock, 1990; Blodgett *et al.*, 1990).

The Kazakhstan Plate is now believed to consist of a number of Precambrian massifs, and Palaeozoic accretionary complexes and island-arcs (Zonenshain *et al.*, 1990, Chapter V). As there is no systematic palaeomagnetic investigation of its constituent terranes, Kazakhstan is shown as a single plate which is arbitrarily positioned close to the Siberian Plate.

East and Southeast Asian blocks

China and Southeast Asia also comprise of a number of microplates or terranes, which sutured together during Latest Palaeozoic to Early and Middle Mesozoic time (e.g. Sengör *et al.*, 1988). Among them are: the Tarim Block, the North China Block, the South China Block, the Indochina Block, the Sibumasu (or Shan-Thai) Block, and the Tibetan blocks.

Whether the South China Block itself consists of a number of minor terranes is also a matter of current debate (e.g. Hsü *et al.*, 1988, 1989; Rowley *et al.*, 1989; Gupta, 1989; Rodgers, 1989; Hsü *et al.*, 1990; Chen *et al.*, 1991; Nie, 1991). Without any Palaeozoic palaeomagnetic results from the proposed Hunan Terrane, and considering that the Hunan Terrane and the Yangtze Block share the same vertebrate fauna in the Devonian (Young, 1990; Wang, this volume), we have treated the whole of South China as a single tectonic entity in this study, and refer to it as the South China Block.

Tarim

None of the Chinese and Southeast Asian blocks has good palaeomagnetic control for the whole Palaeozoic. For the Tarim Block, key-poles are available only for the Late Palaeozoic (Table 2.1). However, as both a reliable Sinian (i.e. late Precambrian) (Li and Li *et al.*, 1991) and Devonian results indicate low to equatorial palaeolatitudes for the Tarim Block, we interpolated equatorial palaeolatitudes for it for the whole early Palaeozoic interval. Such an interpolation is consistent with the limited palaeoclimatic indicators available (Nie, 1991).

North China

For the North China Block, key-poles are available for the latest Precambrian to the Middle Ordovician interval (Zhao *et al.*, 1992) and for the Late Permian (Table 2.1). The lack of mid-Palaeozoic data is largely due to a Late Ordovician to Early Carboniferous hiatus (Zhang *et al.*, 1984). This has caused equivocal interpretation about the polarity of the pre-Late Ordovician data (Zhao *et al.*, 1992). The preferred polarity of Zhao *et al.* (1992) is adopted here because the implied low-latitude in the northern hemisphere is consistent with the close biogeographic relationship of the North China Block with Australia during the Cambrian (Palmer, 1974; Jell, 1974; Burrett and Richardson, 1980). For the Silurian–Early Permian interval, where neither palaeomagnetic data nor other palaeoclimate indicators are available, low palaeolatitudes are applied in the palaeogeographic reconstructions. These are derived by simple interpolation of the low palaeolatitude before and after that interval.

South China

Although much palaeomagnetic work has been done in the past few years in the Yangtze part of the South China Block, the Palaeozoic APWP for the block remains controversial. Lin *et al.* (1985a) first proposed an APWP based on their preliminary results from Cambrian, Carboniferous and Permian strata. However, due to the lack of mid-Palaeozoic results, and the poor quality of the Carboniferous results, the polarity for their Cambrian pole was regarded as equivocal by Li (1989). Later work on Ordovician (Fang *et al.*, 1990), Silurian (Opdyke *et al.*, 1987), Early to Middle Devonian (new and reinterpreted Lin's (1984) data in Li, 1989; Fang *et al.*, 1989; Wu, 1990), and Early Carboniferous (Li, 1989) rocks still leave the problem unresolved. Interpretation remains equivocal because the Cambrian, Silurian, and Fang *et al.*'s (1989) Early to Middle Devonian poles all fall closely together, whereas the Ordovician pole, and other Early to Middle Devonian poles fall over 60° away. Therefore, more careful investigations are needed to establish the Palaeozoic APWP for this block. In the present synthesis, we adopt the low palaeolatitude for the Palaeozoic Yangtze Block, as indicated by most of the palaeomagnetic data, as well as palaeoclimatic indicators (e.g. Nie, 1991). The established Late Permian pole from the Emeishan Basalt (summarised in Li, 1989) is used to control the palaeolatitude of this block in Figure 2.9.

Indochina

There are no Palaeozoic palaeomagnetic data from the Indochina Block. Therefore, palaeoreconstructions depend solely on stratigraphic and biogeographic information (see later discussions).

Sibumasu (Shan-Thai)

Our division of the Sibumasu (also known as Shan-Thai), East Malaya, and Southwest Borneo blocks (Figure 2.1) follows that of Metcalfe (1988). Palaeozoic palaeomagnetic results of the Sibumasu Block exist from Ordovician–Silurian limestones in the northwest of Peninsular Malaysia (Haile, 1980), Devonian limestones in western Yunnan (Fang *et al.*, 1989), and the Late Carboniferous basalts in western Yunnan (Huang and Opdyke, 1991). Preliminary results from Late Palaeozoic rocks of Peninsular Malaysia were reported by McElhinny *et al.* (1974). Although all but the last data set are pre-folding in origin, as the folding occurred no earlier than the Triassic, they are not necessarily all primary. In fact, the moderate palaeolatitude implied by the Ordovician–Silurian data (±43°) and the Devonian data (±42°) for the studied localities are contrary to the warm palaeoclimate of these regions (Metcalfe, 1988). On the other hand, such moderate palaeolatitudes are in good agreement with the Late Carboniferous palaeomagnetic results (Huang and Opdyke, 1991), which are supported by the occurrence of Permo–Carboniferous glacial-marine sediments and cool-water faunas in the Sibumasu Block (Metcalfe, 1988). Furthermore, the single reversed polarity of the data from the Ordovician–Silurian rocks (Haile, 1980) is consistent with its being an overprint formed during the Permo–Carboniferous Kiaman reversed interval. The data with mixed polarities from the Devonian strata (Fang *et al.*, 1989) could represent an overprint acquired either just before, or after, that reversed interval.

Therefore, we regard only the Late Carboniferous palaeomagnetic data as a reliable set of results (Table 2.1). However, considering the possible significant relative rotations around the vertical axis within this block, as shown by Huang and Opdyke (1991), the declination of that result may not be representative for the whole block.

East Malaya and Southwest Borneo

The Palaeozoic tectonic history of East Malaya is different from that of the Sibumasu Block in that it has a large proportion of volcaniclastics, and has developed warm-water faunas during the Permo–Carboniferous (Metcalfe, 1988).

Palaeomagnetic results from this block are scarce. McElhinny *et al.* (1974) reported a preliminary result from five samples from the Bentong Group on the Raub–Bentong Suture between the West and East Malaya blocks. However, the age of the rock formation is highly uncertain (Tjia, 1989), and the magnetic remanence, without any field test, may represent a Mesozoic overprint (Metcalfe, pers. comm., 1992).

There is no Palaeozoic palaeomagnetic data for the Southwest Borneo Block, and little is known about its origin apart from the development of warm-water sediments and faunas during the Late Palaeozoic (Metcalfe, 1988). In the palaeoreconstructions of this study, this block is plotted together with the East Malaya Block, with their palaeolatitudes constrained by palaeoclimatic indicators and their biogeographic relationships with other blocks.

Tibet

There is no palaeomagnetic control on the palaeolatitude of the Tibet Block. The palaeoreconstructions are based on other evidence such as biogeography, palaeoclimatic information and tectonic analysis (e.g. Mitchell, 1981). For simplicity, the tectonic subdivision within the Tibet Block (i.e. the Lhasa and Qiangtang Blocks) is not shown.

Cimmeria

Few Palaeozoic palaeomagnetic works on the microplates composing the western Cimmerian Continent, i.e. the Turkey Block, the Iran Block and the Afghanistan Block (Figure 2.1) have been published in detail. The only reliable results are those reported by Wensink (1983) on Early Devonian rocks from Central Iran. The results, when compared with the 'Y' model of Gondwanan APWPs (Morel and Irving, 1978), permit one to put the Iran Block against the Arabian margin of Gondwanaland (Wensink, 1983). In the palaeoreconstructions of this study, we adopt the hypothesis that the western Cimmerian Continent formed a part of the Gondwanaland adjacent to Arabia until the Triassic (cf. Sengör, 1985).

WORLD PALAEOGEOGRAPHY DURING THE PALAEOZOIC

The palaeogeography at six time-intervals is depicted in Figures 2.3–2.9. Whereas the palaeomagnetic and palaeoclimatic data discussed above (also see Table 2.1) have been used to constrain the palaeolatitude of each continent (or block), published biogeographic and other geological information has been utilised to constrain their relative longitudinal relationships. The main features of these reconstructions are briefly described below.

Cambro–Ordovician

The first reconstruction given is for the Cambro–Ordovician interval (Figures 2.3 and 2.4). The reason we did not go to the Early to Middle Cambrian was that the latest Precambrian to the Middle Cambrian period is believed to be tectonically an extremely active period (e.g. Unrug, 1992), and Gondwanaland probably was not formed until Cambro–Ordovician times (McWilliams, 1981; Dalziel, 1991b; Li and Powell, 1992). Following the hypothesis that Laurentia formed the core of a Proterozoic supercontinent (Moores, 1991; Hoffman, 1991; Dalziel, 1991a, b), we kept Laurentia surrounded by the other rifting-away continents.

The close biological and geological affinity between the South China Block, the South and Southeastern Asian blocks, and East Gondwanaland (Australia in particular) during much of the Palaeozoic has been well documented (e.g. Ridd, 1971; Audley-Charles, 1983; Stait and Burrett, 1987; Metcalfe, 1988; Long and Burrett, 1989; Cocks and Fortey, 1990; Burrett *et al.*, 1990; Young, 1990). Although varying in detail, these reconstructions are similar to that shown in Figures 2.3 and 2.4.

However, there has been a controversy about the position of the North China Block. In most of the previous Cambrian or Ordovician reconstructions (e.g. Burrett and Richardson, 1980; Burrett and Stait, 1985; Lin *et al.*, 1985b; Scotese and McKerrow, 1990), the North China Block was put close to the Southeast Asian blocks, against either the northwestern margin of Australia, or Arabia. As pointed out by Cocks and Fortey (1990), these configurations failed to explain why the Early Ordovician tropical bathyurid trilobites extend from Laurentia, Siberia to North China, but do not appear in the adjacent Southeast Asian blocks and Australia. These earlier reconstructions also have difficulty in explaining why the close biological affinity of the North China Block and South China Block in the Early and Middle Cambrian time became broken in the Cambro–Ordovician time, when there was still strong biological affinity between North China, Central Australia and Tasmania (Burrett *et al.*, 1990; Stait and Burrett, 1987).

Our reconstruction (Figures 2.3, 2.4) could be the answer to this controversy. We propose that the North China Block was attached to the Proterozoic supercontinent in a position opposite northeastern Australia. When East Gondwanaland gradually drifted away from Laurentia during the Cambrian (Moores, 1991; Hoffman, 1991; Dalziel, 1991a, 1991b,), the

North China Block was left in its wake (Figures 2.3 and 2.4). This reconstruction explains why the biological link with the South China Block and other Southeast Asian blocks on the other side of Australia became broken by the Cambro–Ordovician time, while it still shared common faunas with both Laurentia and parts of Australia.

Mid-Silurian

The mid-Silurian palaeoreconstruction (Figure 2.5) is less reliable than that of Cambro–Ordovician due to the contradictory palaeomagnetic results from Gondwanaland (see discussion earlier). The main feature in the mid-Silurian palaeogeography is the closure of the Iapetus Ocean. Armorica was still along the margin of Africa, as indicated by its biological affinity with Gondwanaland which lasted until the Early Devonian (Young, 1990). In the palaeogeography reconstruction, its palaeolatitude is slightly higher than what is indicated solely by palaeomagnetic data.

The South China Block probably had started drifting away from northwest Australia from that time, as indicated by the strong provincial fauna development (Wang *et al.*, 1984).

Early Devonian

Figure 2.6 gives a possible Early Devonian reconstruction. The palaeolatitudes of Laurussia (combined Laurentia and Baltica) and Gondwanaland permit them to join together to form a 'Pangaea' (e.g. Scotese and McKerrow, 1990; Kent and Van der Voo, 1990). However, if the development of

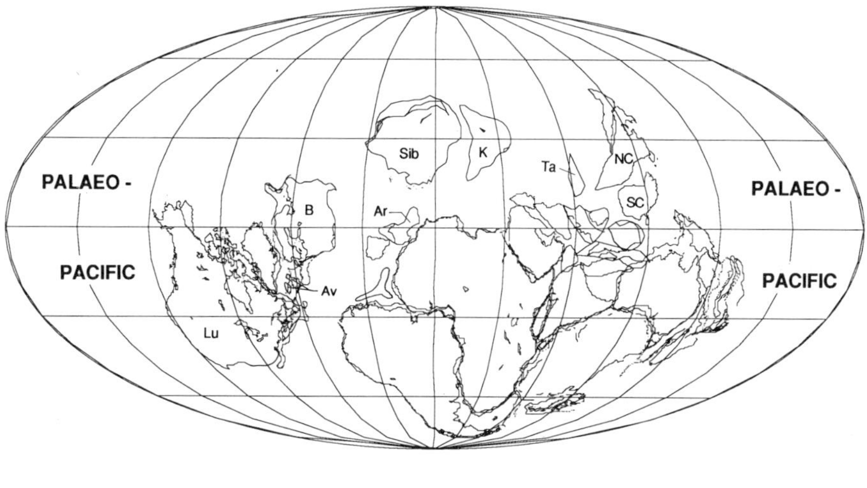

Figure 2.6. Early Devonian reconstruction.

distinctive vertebrate faunas in the two land masses (Young, 1990) is to be accommodated, one has to propose some sort of barrier, such as a large ocean, between them (Figure 2.6). Armorica was still adjacent to northern Africa, which is consistent with biogeography.

The North China and South China blocks were close to each other at that time, but separated from East Gondwanaland during the Early Devonian, as indicated by the development of a highly endemic vertebrate on these two blocks (Burrett *et al.*, 1990; Young, 1990).

Siberia would have been located between 30° to 60°N according to palaeomagnetic data of Khramov *et al.* (1981). However, it has been argued by numerous workers (e.g. Heckel and Witzke, 1979; Young, 1990; Stock, 1990; Blodgett *et al.*, 1990) that such a middle palaeolatitude is inconsistent with the tropical palaeolatitude suggested by palaeoclimatic indicators. Being unable to assess the quality of the palaeomagnetic data, we follow Scotese and McKerrow (1990) in moving Siberia and Kazakhstan about 15° towards the palaeoequator.

Late Devonian

The palaeoreconstruction for the Late Devonian (Figure 2.7) is also controversial. According to palaeomagnetic data, there was a wide ocean between the margins of West Gondwanaland and Laurussia (Figure 2.7). The existence of such an ocean is supported by some biogeographic studies (e.g. phytoplankton study of Colbath, 1990), but opposed by others (e.g. vertebrate study of Young, 1990, and miospore distribution of Streel *et al.*, 1990). Scotese and McKerrow's (1990) reconstruction supported the near-closure of the palaeo-ocean between the two supercontinents.

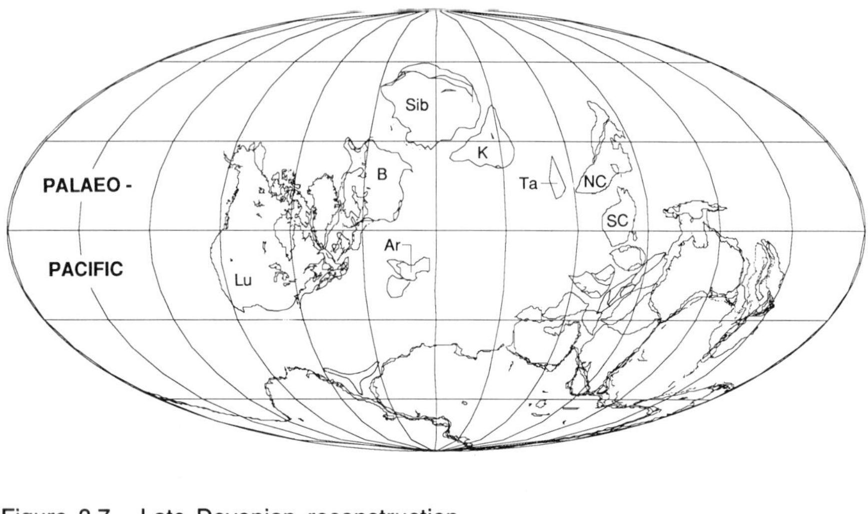

Figure 2.7. Late Devonian reconstruction.

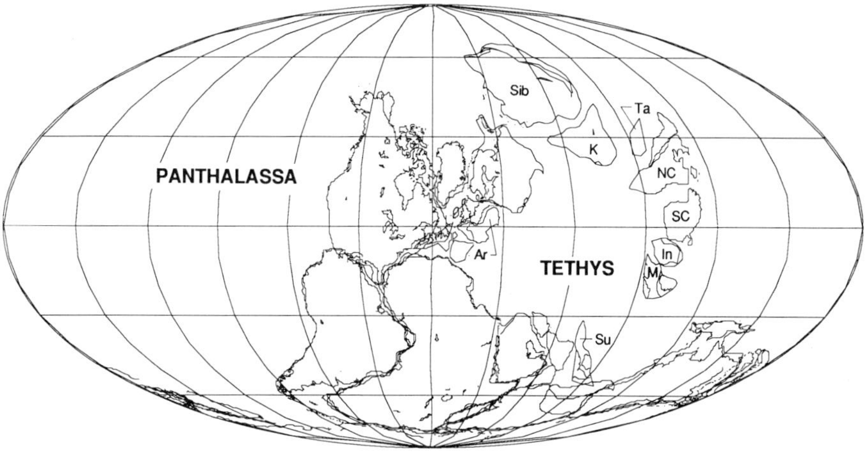

Figure 2.8. Late Carboniferous reconstruction.

In contrast to the suggestion of Young (1990) that the Late Devonian palaeomagnetic data are probably seriously in error, we believe that those data, including the pole from the Canning Basin limestones (Hurley and Van der Voo, 1987), and poles from the Hervey Group (Li *et al.*, 1988) and the Worange Point Formation (Thrupp *et al.*, 1991) of southeast Australia, are reliable. The central African location of the South Pole during latest Devonian–Early Carboniferous is further confirmed by recent work in Central Australia (Chen *et al.*, 1991). There is no reason to suspect the quality of the selected Laurussia results either (Van der Voo, 1990). Therefore, alternative interpretations, such as a land connection between the two supercontinents, either between the nearest margins of the two land masses (e.g. Hurley and Van der Voo, 1987), or, probably more likely, through the Asian blocks, Kazakhstan, and Siberia (Figure 2.7), have to be invoked to explain some of the biogeographic relationships.

The biological barrier between the North China and South China blocks in vertebrate distribution has disappeared since the Middle Devonian, and both blocks, as well as the Sibumasu Block (or Shan-Thai Block), have close links to Australia in the Late Devonian (Burrett *et al.*, 1990).

Late Carboniferous

The main feature of the Late Carboniferous reconstruction (Figure 2.8) is that by the Late Carboniferous Pangaea had formed (Veevers, 1988). According to palaeomagnetic results, East Gondwanaland had started fast drifting towards the South Pole since late Early Carboniferous (Chen *et al.*, 1991). The collision of Gondwanaland with Laurussia, which resulted in the

Hercynian/Alleghenian orogens, could have started during mid-Carboniferous (Li *et al.*, 1989).

There is a range of possible late Palaeozoic Pangaea configurations. Following the suggestion given by Smith and Livermore (1991) in a comprehensive review, we adopted the Pangaea 2 model of Van der Voo and French (1974) in Figure 2.8. To avoid a ~10° continental overlap between Gondwanaland and Laurussia resulting from strictly applying the palaeomagnetic data, we adjusted the position of Gondwanaland within the 95 per cent confidence allowance of its mean palaeomagnetic pole.

Other features in that reconstruction include the rifting away of Indochina and East Malaya–Southwest Borneo blocks from Gondwanaland, and the development of the warm-temperature Chinese biogeographic province (e.g. Chaloner and Lacey, 1973; Metcalfe, 1988) in the eastern part of the Tethys Sea. The Sibumasu Block may have started to rift away, but was still at moderate southern latitude. Armorica had probably accreted to Laurussia by then.

Early to mid-Permian

As for the Late Carboniferous reconstruction (Figure 2.9), an adjustment of ~10° was also made in positioning Gondwanaland in this reconstruction to avoid an overlap with Laurussia. During the Early to mid-Permian, the North China and South China blocks still occupied tropical positions. A warm-temperature Cathaysian Flora Province had developed over them and the adjacent Southeast Asian blocks. Although these blocks may have been close to being accreted by then, their relative orientations and positions were still subject to minor adjustment during subsequent northward drift.

Siberia and Kazakhstan accreted to Laurussia by the Early Permian (Zonenshain *et al.*, 1990). Tarim may have started accreting to Kazakhstan, but had not yet sutured with it.

SUMMARY

Palaeomagnetic data for the Palaeozoic have been assessed critically, and incorporated with palaeoclimatic and biogeographic information to derive global palaeogeography for six Palaeozoic stages.

Palaeomagnetic data for North America and north Europe are regarded as reasonably adequate for palaeolatitude control. However, data for Gondwanaland, Siberia, Kazakhstan, and most of the East and Southeast Asian blocks remains patchy. Precise palaeolatitude controls for each geological interval for these blocks await future studies.

Our reconstructions start from the Cambro–Ordovician interval, when Gondwanaland is believed to have just formed. During that time, most of the continents were in the southern hemisphere. There was a large ocean between Laurentia and the India side of East Gondwanaland.

According to our reconstructions, Gondwanaland underwent a large

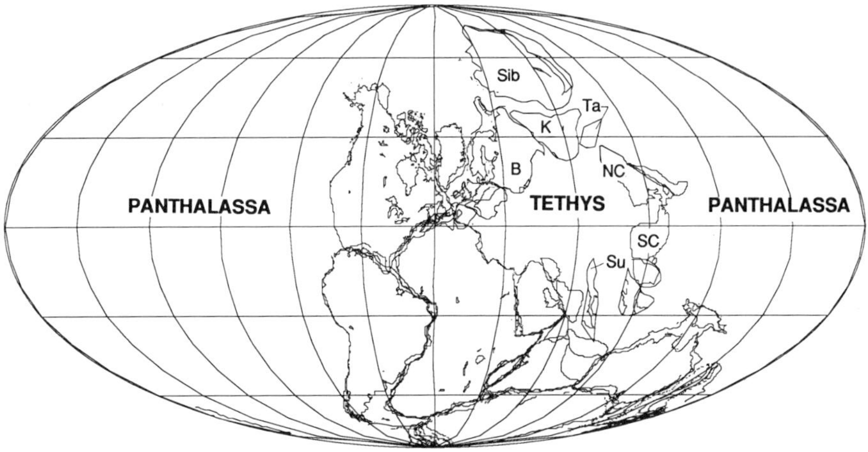

Figure 2.9. Early to mid-Permian reconstruction.

clockwise rotation around an axis close to Australia during the Early Palaeozoic. Laurentia underwent a large eastward movement, as well as a slight southward drift. Baltica, which joined with Laurentia during the Silurian, drifted from a moderate southern hemisphere position in Cambro–Ordovician time to an equatorial position in Silurian–Devonian time. Siberia, and possibly the Kazakhstan terranes, drifted towards the northeast across the equator. All the East and Southeast Asian terranes were around the India–Australia margin of Gondwanaland during the Early Palaeozoic.

During the Devonian, the whole of Laurussia started to drift northward, whereas Gondwanaland underwent a counterclockwise rotation around the Australian axis. Some of the Chinese blocks and Armorica have started to rift away from the Gondwanan margin. Siberia and the Kazakhstan terranes continued to drift northward.

In late Early Carboniferous, East Gondwanaland, for the first time since the Early Palaeozoic, began to drift toward the South Pole. By early Late Carboniferous, the South American–North African margin of Gondwanaland had collided with Laurussia, and thereby formed the Late Palaeozoic Pangaea. While East Gondwanaland drifted toward the South Pole, most of the East and Southeast Asian terranes were left in equatorial positions, forming a chain of continental terranes at the eastern edge of the Tethys Ocean. By Permian time, Siberia and the Kazakhstan terranes were sutured to Laurussia and the Chinese blocks started accreting to them.

REFERENCES

Andrews, J.A., 1985, True polar wander: an analysis of Cenozoic and Mesozoic paleomagnetic poles, *Journal of Geophysical Research*, **90**: 7737–50.

Audley-Charles, M.G., 1983, Reconstruction of eastern Gondwanaland, *Nature*, **306**: 48–50.

Bachtadse, V., 1988, Hercynian palaeomagnetism of Europe: arguments for large scale complex thin-skinned and thick-skinned rotations, *Cuidernos de Geologie Iberica, special publication*, No. **12**: 147–62.

Bachtadse, V. and Briden, J.C., 1989, Palaeomagnetism of the Early to mid-Ordovician Salala igneous ring complex, Red Sea Hills, Sudan, *Geophysical Journal International*, **99**: 677–85.

Bachtadse, V. and Briden, J.C., 1990, Palaeomagnetism constraints on the position of Gondwana during Ordovician to Devonian times. In W.S. McKerrow and C.R. Scotese (eds), *Palaeozoic Palaeogeography and Biogeography. The Geological Society Memoir*, No. 12, The Geological Society of London: 43–8.

Bachtadse, V. and Briden, J.C., 1991, Palaeomagnetism of Devonian ring complexes from the Bayuda Desert, Sudan—new constraints on the apparent polar wander path for Gondwanaland, *Geophysical Journal International*, **104**: 635–46.

Bachtadse, V., Heller, F. and Kroener, A., 1983, Palaeomagnetism investigations in the Hercynian mountain belt of western Europe, *Tectonophysics*, **91**: 285–99.

Bai, Y.H., Cheng, G.L., Sun, Q.G., Sun, Y.H., Li, Y.A., Dong, Y.J. and Sun, D.J., 1987, Late Paleozoic polar wander path for the Tarim platform and its tectonic significance, *Tectonophysics*, **139**: 145–53.

Blodgett, R.B., Rohr, D.M. and Boucot, A.J., 1990, Early and Middle Devonian gastropod biogeography. In W.S. McKerrow and C.R. Scotese (eds), *Palaeozoic Palaeogeography and Biogeography. The Geological Society Memoir*, No. 12, The Geological Society of London: 277–84.

Bluck, B.J. and Dempster, T.J., 1991, Exotic metamorphic terranes in the Caledonides: Tectonic history of the Dalradian block, Scotland, *Geology*, **19**: 1133–6.

Bullard, E.C., Everett, J.E. and Smith, A.G., 1965, A symposium on continental drift, IV, The fit of the continents around the Atlantic, *Philosophical Transactions of Royal Society of London*, **A258**: 41–51.

Burrett, C.F. and Richardson, R., 1980, Tribolite biogeography and Cambrian tectonic models, *Tectonophysics*, **63**: 155–192.

Burrett, C.F. and Stait, B., 1985, South East Asia as a part of an Ordovician Gondwanaland—a palaeogeographic test of a tectonic hypothesis, *Earth Planetary Science Letters*, **75**: 184–90.

Burrett, C.F., Long, J. and Stait, B., 1990, Palaeozoic biogeography of Asian terranes derived from Gondwana. In W.S. McKerrow and C.R. Scotese (eds), *Palaeozoic Palaeogeography and Biogeography. The Geological Society Memoir*, No. 12, The Geological Society of London: 163–74.

Caputo, M.V. and Crowell, J.C., 1985, Migration of glacial centers across Gondwana during Paleozoic Era, *Geological Society of America Bulletin*, **96**: 1020–36.

Chaloner, W.G. and Lacey, W.S., 1973, The distribution of Late Palaeozoic floras. In N.F. Hughes (ed.), *Organisms and continents through time. Special Papers in Palaeontology*, **12**: 271–89.

Channell, J.E.T., McCabe, C., Torsvik, T.H., Trench, A. and Woodcock, N.H., 1992a, Palaeozoic palaeomagnetic studies in the Welsh Basin—Recent Advances, *Geological Magazine* (in press).

Channell, J.E.T., McCabe, C. and Woodcock, N.H., 1992b, Early Devonian (pre-Acadian) magnetization directions in Lower Old Red Sandstone of South Wales (UK), *Geophysical Journal International*, **108**: 883–94.
Chen, J., Foland, K.A., Xing, F., Xu, X. and Zhou, T., 1991, Magmatism along the southeast margin of the Yangtze block: Precambrian collision of the Yangtze and Cathysia block of China, *Geology*, **19**: 815–18.
Chen, Z., Li, Z.X. and Powell, C.McA., 1991, New latest Devonian to Early Carboniferous palaeomagnetic poles from Central Australia and their tectonic implications, *Abstracts of the Eighth International Symposium on Gondwana*, Hobart: 17–18.
Cocks, L.R.M. and Fortey, R.A., 1990, Biogeography of Ordovician and Silurian faunas. In W.S. McKerrow and C.R. Scotese (eds), *Palaeozoic Palaeogeography and Biogeography. The Geological Society Memoir*, No. 12, The Geological Society of London: 97–104.
Cogné, J.P., 1988, Strain, magnetic fabric and palaeomagnetism of the Pont Rean Formation, Brittany, France, *Journal of Geophysical Research*, **93**: 13673–87.
Colbath, G.K., 1990, Palaeogeography of Middle Palaeozoic organic-walled phytoplankton. In W.S. McKerrow and C.R. Scotese (eds), *Palaeozoic Palaeogeography and Biogeography. The Geological Society Memoir*, No. 12, The Geological Society of London: 207–13.
Creer, K.M., 1970, A palaeomagnetic survey of South American rock formations, *Philosophical Transactions of the Royal Society*, **A267**: 457–558.
Daly, L. and Irving, E., 1983, Paleomagnetisme des roches carboniferes du Sahara central; analyse des aimantations juxtaposees; configuration de la Pangee, *Annales Geophysicae*, **1(3)**: 207–16.
Daly, L. and Pozzi, J.-P., 1976, Resultats paleomagnetiques du Permien inferieur et du Trias marocain: comparaison avec les donnees africaines et sud americaines, *Earth and Planetary Science Letters*, **29**: 71–80.
Dalziel, I.W.D., 1991a, Pacific margins of Laurentia and East Antarctica–Australia as a conjugate rift pair: Evidence and implications for an Eocambrian supercontinent, *Geology*, **19**: 598-601.
Dalziel, I.W.D, 1991b, Gondwana as the offspring of an Eocambrian supercontinent (abstract), *Eighth International Symposium on Gondwana*, Hobart: 20–1.
Duff, B.A., 1979, The paleomagnetism of Cambro–Ordovician redbeds, the Erquy Spilite Series, and the Tregastel-Ploumanac'h granite complex, Armorican Massif (France and the Channel Islands), *Geophysical Journal of the Royal Astronomical Society*, **59**: 345–65.
Embleton, B.J.J., McElhinny, M.W., Crawford, A.R. and Luck, G.R., 1974, Palaeomagnetism and the tectonic evolution of the Tasman orogenic zone, *Journal of Geological Society of Australia*, **21**: 187–93.
Fang, D., Chen, H., Jin, G., Guo, Y., Wang, Z., Tan, X. and Yin, S., 1990, Late Paleozoic and Mesozoic paleomagnetism and tectonic evolution of the Tarim terrane. In T.J. Wiley, D.G. Howell and F.L. Wong (eds), *Terrane analysis of China and the Pacific rim*, Circum-Pacific Council for Energy and Mineral Resources Earth Science, Houston, **13**: 251–55.
Fang, W., Van der Voo, R. and Liang, Q., 1989, Devonian paleomagnetism of Yunnan Province across the Shan Thai–South China suture, *Tectonics*, **8**: 939–52.
Fang, W., Van der Voo, R. and Liang, Q., 1990, Ordovician paleomagnetism of eastern Yunnan, China, *Geophysical Research Letters*, **17**: 953–6.
Goleby, B.R., 1980, Early Palaeozoic palaeomagnetism in South East Australia, *Journal of Geomagnetism and Geoelectricity*, **32**: 11–21.
Gupta, S., 1989, In Comments and reply on 'Mesozoic overthrust tectonics in South China', *Geology*, **17**: 669–71.

Haile, N.S., 1980, Palaeomagnetic evidence from the Ordovician and Silurian of northwest Peninsular Malaysia, *Earth and Planetary Science Letters*, **48**: 233–6.

Hancock, N.J., 1982, Stratigraphy, palaeogeography and structure of the East Mendips Silurian Inlier, *Proceedings of the Geologists Association*, **93(3)**: 247–61.

Hargraves, R.B., Dawson, E.M. and Van Houten, F.B., 1987, Palaeomagnetism and age of mid-Palaeozoic ring complexes in Niger, West Africa, and tectonic implications, *Geophysical Journal of the Royal Astronomical Society*, **90**: 705–29.

Heckel, P.H. and Witzke, B.J., 1979, Devonian world palaeogeography determined from distribution of carbonates and related lithic palaeoclimatic indicators, *Special Paper of the Palaeontological Association*, **23**: 99–123.

Henry, B., Merabet, N., Yelles, A., Derder, M.M. and Daly, L., 1992, Geodynamical implications of a Moscovian paleomagnetic pole from the stable Saharan craton (Illozo basin, Algeria), *Tectonophysics*, **201**: 83–96.

Hoffman, P.F., 1991, Did the breakout of Laurentia turn Gondwanaland inside-out? *Science*, **252**: 1409–12.

Hsü, K.J., Li, J., Chen, H., Wang, Q., Sun, S. and Sengör, A.M.C., 1990, Tectonics of South China: Key to understanding West Pacific geology, *Tectonophysics*, **183**: 9–39.

Hsü, K.J., Sun, S., Li, J., Chen, H., Pen, H. and Sengör, A.M.C., 1988, Mesozoic overthrust tectonics in South China, *Geology*, **16**: 418–27.

Hsü, K.J., Sun, S., Li, J., Chen, H., Pen, H. and Sengör, A.M.C., 1989, In Comments and reply on 'Mesozoic overthrust tectonics in South China', *Geology*, **17**: 672–3.

Huang, K. and Opdyke, N.D., 1991, Paleomagnetic results from the Upper Carboniferous of the Shan-Thai-Malay block of western Yunnan, China, *Tectonophysics*, **192**: 333–44.

Hurley, N.F. and Van der Voo, R., 1987, Paleomagnetism of Upper Devonian reefal limestones, Canning Basin, Western Australia, *Geological Society of America Bulletin*, **98**: 138–46.

Idnurm, M., 1985, Late Mesozoic palaeomagnetism of Australia–II. Implications for geomagnetism and true polar wander, *Geophysical Journal of the Royal Astronomical Society*, **83**: 419–33.

Idnurm, M., 1986, Late Mesozoic palaeomagnetism of Australia–III. Bias-corrected pole paths for Australia, Antarctica and Inda, *Geophysical Journal of the Royal Astronomical Society*, **86**: 277–87.

Irving, E., 1966, Paleomagnetism of some Carboniferous rocks from New South Wales and its relation to geological events, *Journal of Geophysical Research*, **71**: 6025–51.

Jell, P.A., 1974, Faunal provinces and possible planetary reconstruction of the Middle Cambrian, *Journal of Geology*, **82**: 319–50.

Jesinkey, C., Forsythe, R.D., Mpodozis, C. and Davidson, J., 1987, Concordant late Paleozoic paleomagnetization from the Atacama Desert: implications from tectonic models of the Chilean Andes, *Earth and Planetary Science Letters*, **85**: 461–72.

Johnson, R.J.E. and Van der Voo, R., 1990, Pre-folding magnetisation reconfirmed for the Late Ordovician–Early Silurian Dunn Point volcanics, Nova Scotia, *Tectonophysics*, **178**: 193–205.

Kent, D.V. and Van der Voo, R., 1990, Palaeozoic palaeogeography from palaeomagnetism of the Atlantic-bordering continents. In W.S. McKerrow and C.R. Scotese (eds), *Palaeozoic Palaeogeography and Biogeography. The Geological Society Memoir*, No. 12, The Geological Society of London: 49–56.

Khramov, A.N., Petrova, G.N. and Pechersky, D.M., 1981, Paleomagnetism of Soviet Union. In *Paleoreconstruction of the continents, AGU Geodynamics Series*, **2**: 177–94.

Klootwijk, C.T., 1980, Early Palaeozoic palaeomagnetism in Australia, *Tectonophysics*, **64**: 249–332.

Lewandowski, M., 1987, Results of the preliminary paleomagnetic investigations of some Lower Paleozoic rocks from the Holy Cross Mountains (Poland), *Kwartalnik Geologicne*, **31**: 543–56.

Li, Y., Li, Y., Sharps, R., McWilliams, M. and Gao, Zhengjia, 1991, Sinian paleomagnetic results from the Tarim block, western China, *Precambrian Research*, **49**: 61–71.

Li, Y., McWilliams, M., Cox, A., Sharps, R., Li, R., Gao, Z., Zhang, Z. and Zhai, Y., 1988, Late Permian paleomagnetic pole from dykes of the Tarim craton, China, *Geology*, **16**: 275–8.

Li, Y., McWilliams, M., Sharps, R., Cox, A., Li, Y., Li, Q., Gao, Z., Zhang, Z. and Zhai, Y., 1990, A Devonian paleomagnetic pole from red beds of the Tarim Block, China, *Journal of Geophysical Research*, **95**: 19,185–98.

Li, Z.X., 1989, Some preliminary Paleozoic paleomagnetic results from Yangtze block: I, the polarity problem for the Early Paleozoic APWP, *Abstracts of the 4th National Conference on Paleomagnetism*, Hangzhou: 10–12.

Li, Z.X. and Powell, C.McA., 1992, Late Proterozoic to early Paleozoic paleomagnetism and the formation of Gondwanaland. In R.H. Findlay *et al.* (eds), *Gondwana 8: Assembly, Evolution and Dispersal* (in press).

Li, Z.X., Powell, C.McA., Schmidt, P.W., 1989, Syn-deformational remanent magnetization of the Mount Eclipse Sandstone, central Australia, *Geophysical Journal International*, **99**: 205–22.

Li, Z.X., Powell, C.McA., Thrupp, G.A. and Schmidt, P.W., 1990, Australian Palaeozoic palaeomagnetism and tectonics—II: a revised apparent polar wander path and palaeogeography, *Journal of Structural Geology*, **12**: 567–75.

Li, Z.X., Schmidt, P.W. and Embleton, B.J.J., 1988, Paleomagnetism of the Hervey Group, central New South Wales and its tectonic implications, *Tectonics*, **7**: 351–67.

Lin, J.-L., 1984, *The apparent polar wander paths for the North and South China blocks*, PhD thesis, University of California, Santa Barbara: 248pp.

Lin, J.-L., Fuller, M. and Zhang, W.-Y., 1985a, Preliminary Phanerozoic polar wander paths for the North and South China blocks, *Nature*, **313**: 444–9.

Lin, J.-L., Fuller, M. and Zhang, W.-Y., 1985b, Paleogeography of the North and South China blocks during the Cambrian, *Journal of Geodynamics*, **2**: 91–114.

Long, J. and Burrett, C., 1989, Fish from the Upper Devonian of the Shan-Thai terrane indicate proximity to east Gondwana and south China terranes, *Geology*, **17**: 811–13.

McElhinny, M.W. and Embleton, B.J.J., 1974, Australian paleomagnetism and the Phanerozoic plate tectonics of Eastern Gondwanaland, *Tectonophysics*, **22**: 1–29.

McElhinny, M.W., Haile, N.S. and Crawford, A.R., 1974, Palaeomagnetic evidence shows Malay Peninsula was not part of Gondwana, *Nature*, **252**: 641–5.

McFadden, P.L., Ma, X.H., McElhinny, M.W. and Zhang, Z.K., 1988, Permo–Triassic magnetostratigraphy in China: northern Tarim, *Earth and Planetary Science Letters*, **87**: 152–60.

McKerrow, W.S., Dewey, J.F. and Scotese, C.R., 1991, The Ordovician and Silurian development of the Iapetus Ocean, *Special Papers in Palaeontology*, **44**: 165–78.

McKerrow, W.S., Scotese, C.R. and Brasier, M.D., 1992, Early Cambrian continental reconstructions, *Journal of Geology Society*, London, **149**: 599–606.

McWilliams, M.O., 1981, Palaeomagnetism and Precambrian tectonic evolution of Gondwana. In A. Kroner (ed.), *Precambrian Plate Tectonics Developments in Precambrian Geology*, **4**: 649–87.

Metcalfe, I., 1988, Origin and assembly of south-east Asia continental terranes. In M.G. Audley-Charles and A. Hallam (eds), *Gondwana and Tethys, Geological Society Publication*, No. 37, Oxford University Press: 101–18.

Mitchell, A.H.G., 1981, Phanerozoic plate boundaries in mainland SE Asia, the Himalayas and Tibet, *Journal of Geology Society, London*, **138**: 109–22.

Moores, E.M., 1991, Southwest US–East Antarctic (SWEAT) connection: a hypothesis, *Geology*, **19**: 425–8.

Morel, P. and Irving, E., 1978, Tentative paleocontinental maps for the Early Phanerozoic and Proterozoic, *Journal of Geology*, **86**: 535–61.

Morel, P., Irving, E., Daly, L. and Moussine-Pouchkine, A., 1981, Paleomagnetic results from Permian rocks of the northern Saharan craton and motions of the Moroccan Meseta and Pangea, *Earth and Planetary Science Letters*, **55**: 65–74.

Morris, W.A., 1976, Transcurrent motion determined palaeomagnetically in the North Appalachians and Caledonides and the Acadian Orogeny, *Canadian Journal of Earth Sciences*, **13**: 1236–43.

Nie, S.-Y., 1991, Paleoclimatic and paleomagnetic constraints on the Paleozoic reconstructions of south China, north China and Tarim, *Tectonophysics*, **196**: 279–308.

Noblet, Ch. and Lefort, J.P., 1990, Sedimentological evidence for a limited separation between Armorica and Gondwana during the Early Ordovician, *Geology*, **18**: 303–6.

Opdyke, N.D., Huang, K., Xu, G., Zhang, W.Y. and Kent, D.V., 1987, Paleomagnetic results from the Silurian of the Yangtze paraplatform, *Tectonophysics*, **139**: 123–32.

Palmer, A.R., 1974, Search for the Cambrian world, *American Scientist*, **62**: 216–24.

Perroud, H. and Bonhommet, N., 1984, A Devonian pole for Armorica, *Geophysical Journal of the Royal Astronomical Society*, **77**: 839–45.

Perroud, H., Bonhommet, N. and Thebault, J.P., 1986, Palaeomagnetism of the Ordovician Moulin de Chateaupanne formation, Vendee, western France, *Geophysical Journal of the Royal Astronomical Society*, **85**: 573–82.

Perroud, H., Calza, F. and Khattach, D., 1991, Paleomagnetism of the Silurian volcanism at Almaden, southern Spain, *Journal of Geophysical Research*, **96**: 1949–62.

Perroud, H., Robardet, M. and Bruton, D.L., 1992, Palaeomagnetic constraints upon the palaeogeographic position of the Baltic Shield in the Ordovician, *Tectonophysics*, **201**: 97–120.

Perroud, H. and Van der Voo, R., 1985, Paleomagnetism of the late Ordovician Thouars Massif, Vendee Province, France, *Journal of Geophysical Research*, **90**: 4611–25.

Perroud, H., Van der Voo, R. and Bonhommet, N., 1984, Palaeozoic evolution of the Armorica plate on the basis of palaeomagnetic data, *Geology*, **12**: 579–82.

Rapalini, A.E. and Vilas, J.F., 1991, Tectonic rotations in the Late Palaeozoic continental margin of southern South America determined and dated by palaeomagnetism, *Geophysical Journal International*, **107**: 333–51.

Ridd, M.F., 1971, South-East Asia as a part of Gondwanaland, *Nature*, **234**: 531–3.

Robardet, M., Paris, F. and Racheboeuf, P.R., 1990, Palaeogeographic evolution of south-western Europe during Early Palaeozoic times. In W.S. McKerrow and C.R. Scotese (eds), *Palaeozoic Palaeogeography and Biogeography. The Geological Society Memoir*, No. 12, The Geological Society of London: 411–19.

Rodgers, J., 1989, In Comments and reply on 'Mesozoic overthrust tectonics in south China', *Geology*, **17**: 671–2.

Rowley, D.B., Ziegler, A.M. and Nie, S.-Y., 1989, Comment on 'Mesozoic overthrust tectonics in south China', *Geology*, **17**: 384–6.

Schmidt, P.W. and Embleton, B.J.J., 1990, The palaeomagnetism of the Tumblagooda Sandstone, Western Australia: Gondwana Palaeozoic apparent polar wandering, *Physics of the Earth and Planetary Interiors*, **64**: 303–13.

Schmidt, P.W. and Morris, W.A., 1977, An alternative view of Gondwana Palaeozoic apparent polar wander path, *Canadian Journal of Earth Sciences*, **14**: 2674–8.

Schmidt, P.W., Embleton, B.J.J., Cudahy, T.J. and Powell, C.McA., 1986, Prefolding and premegakinking magnetizations from the Devonian Comerong Volcanics, New South Wales, Australia, and their bearing on the Gondwana Pole Path, *Tectonics*, **5**: 135–50.

Schmidt, P.W., Embleton, B.J.J. and Palmer, H.C., 1987, Pre- and post-folding magnetization from the Devonian Snowy River Volcanics and Buchan Caves Limestone, Victoria, *Geophysical Journal of the Royal Astronomical Society*, **91**: 155–70.

Scotese, C.R. and Barrett, S.F., 1990, Gondwana's movement over the South Pole during the Palaeozoic: evidence from lithological indicators of climate. In W.S. McKerrow and C.R. Scotese (eds), *Palaeozoic Palaeogeography and Biogeography. The Geological Society Memoir*, No. 12, The Geological Society of London: 75–85.

Scotese, C.R. and McKerrow, W.S., 1990, Revised World maps and introduction. In W.S. McKerrow and C.R. Scotese (eds), *Palaeozoic Palaeogeography and Biogeography. The Geological Society Memoir*, No. 12, The Geological Society of London: 1–21.

Sengör, A.M.C., 1985, The story of Tethys: How many wives did Okeanos have? *Episodes*, **8**: 3–12.

Sengör, A.M.C., Altiner, D., Cin, A., Ustaomer, T. and Hsü, K.J., 1988, Origin and assembly of the Tethyside orogenic collage at the expense of Gondwana Land. In M.G. Audley-Charles and A. Hallam (eds), *Gondwana and Tethys Geological Society Special Publication*, No. 37, Oxford University Press: 119–81.

Sharps, R., McWilliams, M., Li, Y., Cox, A., Zhang, Z., Zhai, Y., Gao, Z., Li, Y. and Li, Q., 1989, Lower Permian paleomagnetism of the Tarim block, northwestern China, *Earth and Planetary Science Letters*, **92**: 275–91.

Sinito, A.M., Valencio, D.A. and Vilas, J.F., 1979, Palaeomagnetism of a sequence of Upper Palaeozoic–Lower Mesozoic red beds from Argentina, *Geophysical Journal of the Royal Astronomical Society*, **58**: 237–47.

Smethurst, M.A. and Khramov, A.N., 1992, A new Devonian palaeomagnetic pole for the Russian platform and Baltica, and related apparent polar wander, *Geophysical Journal International*, **108**: 179–92.

Smith, A.G. and Livermore, R.A., 1991, Pangea in Permian to Jurassic time, *Tectonophysics*, **187**: 135–79.

Stait, B. and Burrett, C., 1987, Biogeography of Australian and Southeast Asian Ordovician nautiloids. In C.D. McKenzie (ed.), *Gondwana Six: Stratigraphy, Sedimentology and Paleontology*, AGU, Washington, D.C.: 21–8.

Stock, C.W., 1990, Biogeography of Devonian stromatoporoids. In W.S. McKerrow and C.R. Scotese (eds), *Palaeozoic Palaeogeography and Biogeography. The Geological Society Memoir*, No. 12, The Geological Society of London: 257–65.

Streel, M., Fairon-Demaret, M. and Loboziak, S., 1990, Givetian–Frasnian phytogeography of Euramerica and western Gondwana based on miospore distribution. In W.S. McKerrow and C.R. Scotese (eds), *Palaeozoic Palaeogeography and Biogeography. The Geological Society Memoir*, No. 12, The Geological Society of London: 291–6.

Thompson, R., 1972, South American Palaeozoic palaeomagnetic results and the welding of Pangaea, *Earth and Planetary Science Letters*, **18**: 266–78.

Thrupp, G.A, Kent, D.V., Schmidt, P.W. and Powell, C.McA., 1991, Palaeomagnetism of the red beds of the Late Devonian Worange Point Formation, SE Australia, *Geophysical Journal International*, **104**: 179–201.

Tjia, H.D., 1989, The Bentong Suture. In B. Situmorang (ed.), *Proceedings of the 6th Regional Conference on Geology, Mining and Hydrological Research in SE Asia*, Jakarta, 1987. Indonesian Association of Geologists.

Torsvik, T.H. and Trench, A., 1991, The Lower-Middle Ordovician palaeofield of Scandinavia: southern Sweden 'revisited', *Physics of the Earth and Planetary Interiors*, **65**: 283–91.

Torsvik, T.H., Olesen, O., Ryan, P.D. and Trench, A., 1990, On the palaeogeography of Baltica during the Palaeozoic: new palaeomagnetic data from Scandinavian Caledonides, *Geophysical Journal International*, **103**: 261–79.

Torsvik, T.H., Ryan, P.D., Trench, A. and Harper, D.A.T., 1991, Cambro–Ordovician palaeogeography of Baltica, *Geology*, **19**: 7–10.

Torsvik, T.H., Smethurst, M.A., Briden, J.C. and Sturt, B.A., 1990, A review of palaeomagnetic data from Europe and their palaeogeographic implications. In W.S. McKerrow and C.R. Scotese (eds), *Palaeozoic Palaeogeography and Biogeography. The Geological Society Memoir*, No. 12, The Geological Society of London: 25–41.

Torsvik, T.H. and Trench, A., 1992, Palaeozoic results from the East Mendip Silurian inlier, southern Britain (E. Avalonia): palaeogeography implications, *Annales Geophysicae*, Supplement I to Volume **10**: 24.

Trench, A. and Torsvik, T.H., 1991, The Lower Palaeozoic apparent polar wander path for Baltica: palaeomagnetic data from Silurian limestones of Gotland, Sweden, *Geophysical Journal International*, **107**: 373–9.

Trench, A., Torsvik, T.H., Dentith, M.C., Walderhaug, H. and Traynor, J.-J., 1992, A high southerly palaeolatitude for Southern Britain in Early Ordovician times; palaeomagnetic data from the Treffgarne Volcanic Formation SW Wales, *Geophysical Journal International*, **108**: 89–100.

Unrug, R., 1992, The Gondwana Supercontinent: Middle Proterozoic crustal fragments, Late Proterozoic assembly, and unsolved problems. In R.H. Findlay *et al.* (eds), *Gondwana 8: Assembly, Evolution and Dispersal* (in press).

Valencio, D., Vilas, J.F.A. and Mendía, J.E., 1977, Palaeomagnetism of a sequence of red beds of the Middle and Upper Section of Paganzo Group (Argentina) and the correlation of Upper Palaeozoic–Lower Mesozoic rocks, *Geophysical Journal of the Royal Astronomical Society*, **51**: 59–74.

Van der Voo, R., 1990, Phanerozoic paleomagnetic poles from Europe and North America and comparisons with continental reconstructions, *Reviews of Geophysics*, **28**: 167–206.

Van der Voo, R. and French, R.B., 1974, Apparent polar wandering for the Atlantic-bordering continent: Late Carboniferous to Eocene, *Earth-Science Reviews*, **10**: 99–119.

Van der Voo, R. and Johnson, R.J.E., 1985, Palaeomagnetism of the Dunn Point Formation (Nova Scotia): High palaeolatitudes for the Avalon Terrane in the Late Ordovician, *Geophysical Research Letters*, **12**: 337–40.

Van der Voo, R. and Scotese, C.R., 1981, Palaeomagnetism evidence for a large (2000 km) sinistral offset along the Great Glen Fault during Carboniferous time, *Geology*, **9**: 583–9.

Veevers, J.J., 1988, Gondwana facies started when Gondwanaland merged in Pangea, *Geology*, **16**: 732–4.

Wang, Y., Boucot, A.J., Rong, J.Y. and Yang, X.C., 1984, Silurian and Devonian biogeography of China, *Geological Society of America Bulletin*, **95**: 265–79.

Webby, B.D., 1980, Biogeography of Ordovician stromatoporoids, *Palaeogeography, Palaeoclimatology and Palaeoecology*, **32**: 1–19.

Wensink, H., 1983, Paleomagnetism of red beds of Early Devonian age from Central Iran, *Earth and Planetary Science Letters*, **63**: 325–34.

Witzke, B.J., 1990, Palaeoclimatic constraints for Palaeozoic palaeolatitudes of Laurentia and Euramerica. In W.S. McKerrow and C.R. Scotese (eds), *Palaeozoic Palaeogeography and Biogeography. The Geological Society Memoir*, No. 12, The Geological Society of London: 57–73.

Wu, N., 1990, *Mid-Palaeozoic palaeomagnetism of western Zhejiang Province and southern Anhui Province, and tectonic significance*, MSc. thesis, Zhejiang University: 129pp.

Young, G.C., 1990, Devonian vertebrate distribution patterns and cladistic analysis of palaeogeographic hypotheses. In W.S. McKerrow and C.R. Scotese (eds), *Palaeozoic Palaeogeography and Biogeography. The Geological Society Memoir*, No. 12, The Geological Society of London: 243–55.

Zhang, Xh.M., Liou, J.G. and Coleman, R.G., 1984, An outline of the plate tectonics of China, *Geological Society of America Bulletin*, **95**: 295–312.

Zhao, X., Coe, R., Zhou, Y., Wu, H. and Wang, J., 1990, New paleomagnetic results from northern China: collision and suturing with Siberia and Kazakhstan, *Tectonophysics*, **181**: 43–81.

Zhao, X., Coe, R.S., Liu, C. and Zhao, Y., 1992, New Cambrian and Ordovician paleomagnetic poles for the North China Block and their paleogeographic implications, *Journal of Geophysical Research*, **97**: 1767–88.

Zonenshain, L.P., Kuzmin, M.I. and Natapov, L.M., edited by B.M. Page, 1990, *Geology of the USSR: A plate tectonic synthesis. Geodynamic Series*, Vol. 21, AGU, Washington, D.C.: 242pp.

Chapter 3

EARLY–MIDDLE PALAEOZOIC VERTEBRATE EXTINCTION EVENTS

J.A. Long

INTRODUCTION

Throughout the Palaeozoic Era several major extinction events have been recognised. One of these, the Frasnian–Famennian boundary, is often cited as a time of major invertebrate extinctions (e.g. McLaren, 1970, 1982; McGhee, 1982, 1989) but vertebrate extinctions have not been discussed in much detail (although see McGhee, 1982). The following analysis takes into account all known higher taxonomic level groups of vertebrates for the Early and Middle Palaeozoic, incorporating new data from Australia, Antarctica and Asia. As some groups are of dubious phylogenetic position (as to the status of family or superfamily, order etc.) the level of family has been used throughout this work to represent natural taxonomic higher grouping. In many cases the age range of a family, or its status as a separate family, is based on new or recent data from Australia, Antarctica or China, e.g. Phyllolepidae (Long, 1984), Culmacanthidae (Long, 1983; Young, 1989) Turiniidae (Turner, 1991), polybranchiaspid and galeaspid agnathans (Pan, 1984; Pan *et al.*, 1987). The age ranges cited by Tarlo and Andrews *et al.* in Harland *et al.* (1967) have thus been significantly modified in view of recent works as have classifications commonly used. As time ranges are not precisely known for some families, the extinction of a family is taken at the end of the particular stage during which it made its last recorded presence. This study excludes the Carboniferous and Permian faunas due to the vast numbers of actinopterygian families, some of which are in dubious taxonomic status, and because many of the major groups of fishes died out before the Late Palaeozoic. Also there are greater taxonomic problems associated with late Palaeozoic fishes as much of the chondrichthyan record is based upon isolated teeth, scales or fin-spines of uncertain taxonomic affinities (e.g. Tway and Zidek, 1982, 1983). Permian–Recent fish extinctions have been analyzed (Patterson and Smith, 1987)

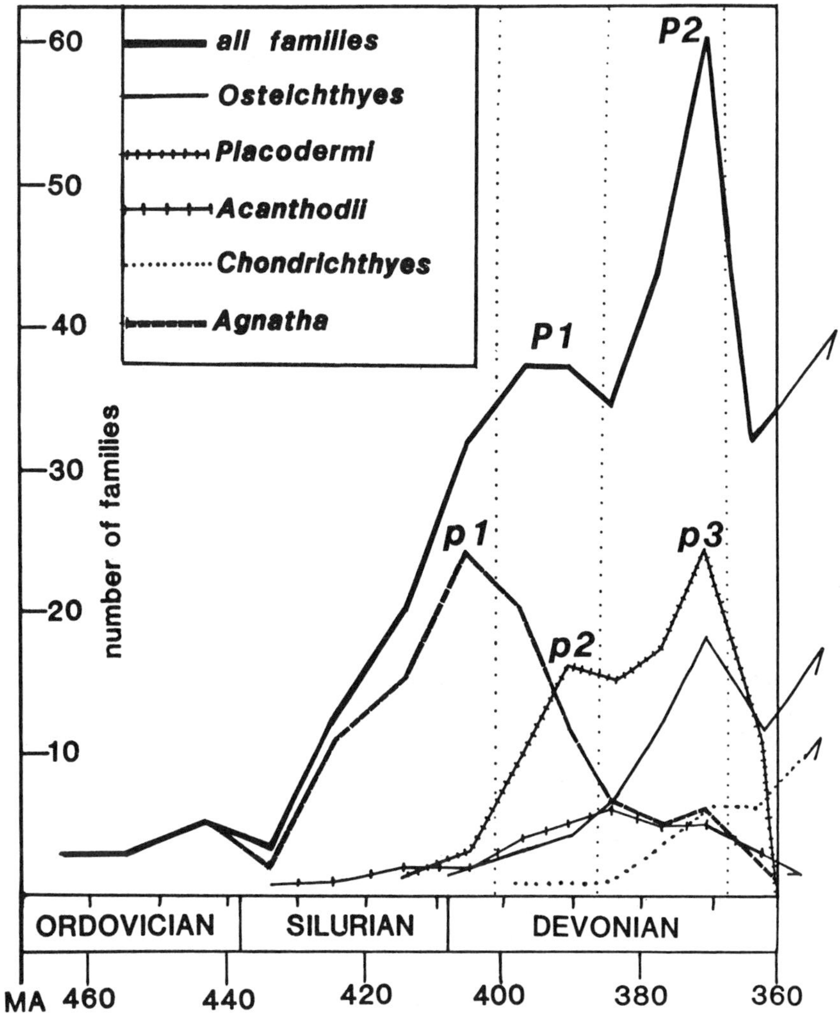

Figure 3.1. Diversity of fish families in the Early and Middle Palaeozoic. Data for Figures 3.1 and 3.2 modified from Harland *et al.* (1967), Denison (1978, 1979) and Zangerl (1981) with input of new data from Australia and China. P1, P2, p1–3 are diversity peaks discussed in text. Arrows indicate continuing diversity trends in groups surviving the end of the Devonian.

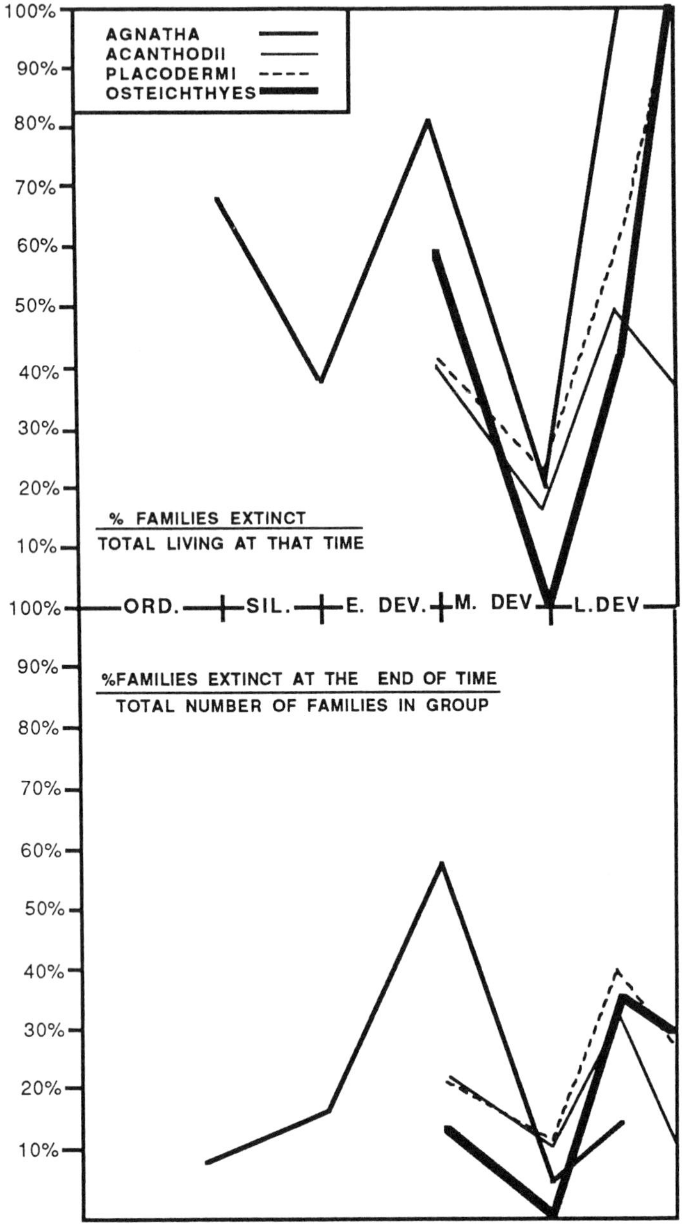

Figure 3.2. Origination (top) and extinction (below) graphs for Early–Middle Palaeozoic vertebrate families.

as have terrestrial vertebrate extinctions for the Late Palaeozoic (Benton, 1985).

EARLY–MIDDLE PALAEOZOIC VERTEBRATE EXTINCTIONS

Figure 3.1 shows early vertebrate families plotted against geological time (stages), and the breakdown for the major groups—Agnatha, Acanthodii, Placodermi, Chondrichthyes and Osteichthyes. There are two significant peaks in diversity: P1 during the Early Devonian comprises the sum of smaller peaks 'p1' (agnathan diversity peak in the Lochkovian) and 'p2' (early placoderm diversity peak in the Pragian). 'P2' represents the maximum peak of vertebrate diversity for the Early and Middle Palaeozoic, the major group forming this peak being the Placodermi (p3), with significant numbers of osteichthyans also present at this time.

Figure 3.2 shows family origination and extinctions. Major extinction events are identified (E1–E4) in Figure 3.3. Background extinction ratios (Table 3.1) are calculated using numbers of extinctions of families/total number of families living during that time interval, divided by time (ma). Figure 3.3 summarizes these events, illustrating major higher taxonomic groups which succumbed to their biogeographic ranges.

The Agnatha is the only group known reliably from the Ordovician and, of the few families known, three to four are apparently extinct by the end of the Ordovician (E1, Arandaspididae, Eryptychiidae, Astraspididae and possibly the saccabambaspids, Elliott *et al.*, 1991), although as some of these are known only from the Middle–Late Ordovician it is difficult to correlate these extinctions with invertebrate mass extinctions in the Ashgillian.

Agnathan diversity increased steadily throughout the Silurian, but no major extinction events are recognised. A significantly high number of agnathan families disappeared at the end of the Early Devonian (E2), corresponding with the rise in gnathostome groups (particularly placoderms and osteichthyans). Agnathans and placoderms suffered extinctions at the end of the Early Devonian (E2), including several higher taxonomic groups from South China (e.g. galeaspida, Yunnanolepiformes), and a few of the osteichthyan families known for that time (Dipnorhynchidae, 'youngolepids', Porolepidae).

Figure 3.2 - cont.
Originations are simply the number of new families appearing for each stage; extinctions are calculated as the percentage of families ending in each stage relative to the total number of families living at that time.
Data are plotted for the middle of each stage (av. 6.7×10^6m.a. for the Devonian; three points plotted for the Silurian, av. 12.6×10^6m.a.).

Table 3.1 *Background extinctions, extinction ratios (per stage)*

Stage	A: Extinctions	T (m.a.)	Ratio A/T	Background	
Famennian	21	3	3	L. DEV 3.71	
Frasnian	31	7	4.42		
Givetian	5	6	0.83		DEV.
Eifelian	2	7	0.28	M. DEV 0.54	1.97
Emsian	15	7	2.14		
Pragian	10	7	1.42	E. DEV 1.7	
Lochkovian	11	7	1.57		
Prid.-Ludlow	4	13	0.31		
Wenlock	3	7	0.43	SIL. 0.23	
Llandovery	0	10	0		
Ashgill	3	10	0.3		
Caradoc	?	10	0	app. 0.15	
pre-Caradoc	1	>10	<0.1		

The end of the Frasnian (E3) marks the extinction of several agnathan families, including the last of the armoured forms (osteostracans, psammosteids, obrucheviids, see Chapters 4 and 5), the thelodontids (turiniids, Turner and Dring, 1981), and possibly the polybranchiaspids (Pan *et al.*, 1987). The 'anaspids' from the Late Devonian Escuminac Formation are now known to represent early lamprey-like forms (Arsenault and Janvier, 1991). The only Agnatha to survive this datum are the two extant groups (petromyzontids and myxinoids).

A few placoderm families died out at the end of the Early Devonian, in particular the Acanthothoraci (palaeacanthaspids), the Yunnanolepidae, Buchanosteidae, Williamsaspidae, Gemuendinaspidae, and the enigmatic groups Stensioellidae and Paraplesiobatidae (the last four families are monotypic, Denison, 1978). During, or at the end, of the Frasnian the highest number of placoderm families died out (Table 3.2) with those surviving lasting only until the end of the Devonian (E4), some eight million years later. Table 3.2 shows that most of the families of placoderms which succumbed at the end of the Frasnian were marine forms, with fewer freshwater or brackish groups being affected, a correlation noted earlier by McGhee (1982) for placoderm and acanthodian species. Furthermore, of those that survived about one third appear to have left marine environments for freshwater or brackish habitats.

The osteichthyans were rare as a group during the Early Devonian, and few of those families survived into the Middle Devonian. At the end of the Middle Devonian there are no osteichthyan extinctions at family level, but several groups disappear by the end of the Frasnian. The dipnoan families lost at this time include the dipterids *s.s.* (including *Rhinodipterus*), the chirodipterids, holodipterids and possibly fleurantiids (dependent on which

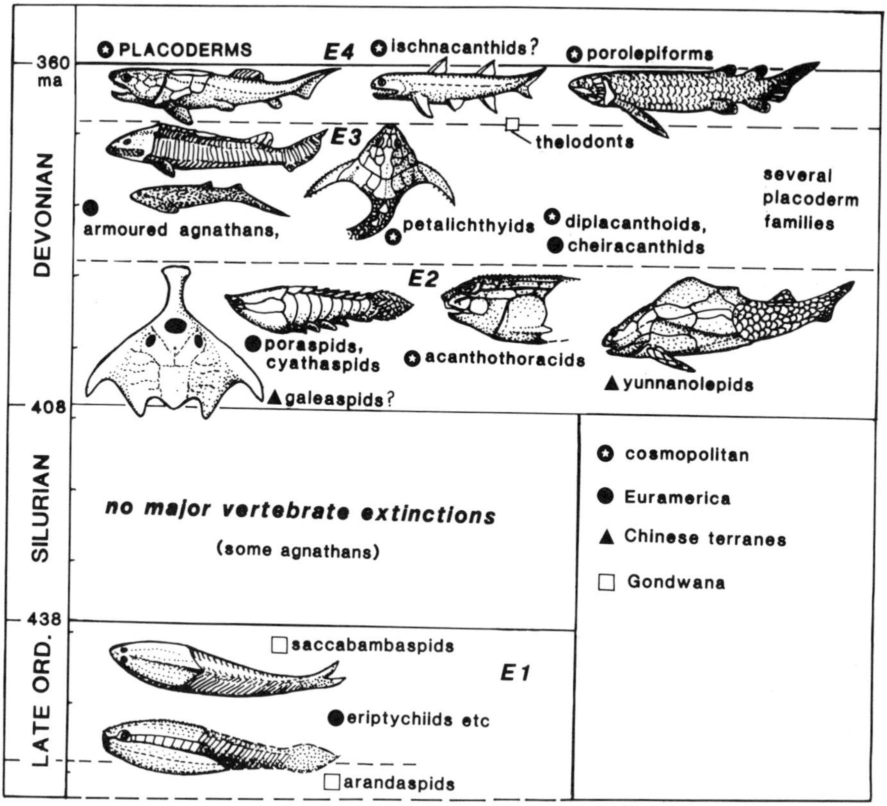

Figure 3.3. Major extinction events of higher taxonomic groups of Early-Middle Palaeozoic vertebrates, with biogeographic ranges indicated.

genera are classified herein). Other osteichthyan groups which become extinct at this time include the cheirolepid actinopterygians (Pearson and Westoll, 1979) and the diplocercid actinistians. No major osteichthyan orders or subclasses were effected at this boundary, but at the end of the Devonian (E4) the orders Porolepiformes and Onychodontiformes disappeared along with many osteolepiform groups (e.g. Eusthenopteridae), the last marine-dwelling dipnoans (e.g. Rhynchodipteridae) and several primitive stem-group actinopterygians.

The chondrichthyans are difficult to assess because of the status of their classification and the nature of their fossils (mostly scales, teeth, spines, Zangerl, 1981). It is noted, however, that no family extinctions are known at, or near, the Frasnian–Famennian boundary. The only chondrichthyan 'family' extinct at the end of the Devonian was the coronodontids. The acanthodians, represented by nine families (Denison, 1979; Long, 1986),

Table 3.2 *Frasnian placoderm extinctions, biogeographic ranges and habitats (data after Denison, 1978; also recent work from China, Australia added)*

Family	Range		Environment	
	Frasnian	Famennian	Frasnian	Famennian
Asterosteidae	EUR	-	M	-
Ptyctodontidae	COS	?EUR	M	M
Macropetalichthyidae	EUR	-	M	-
Antarctaspidae	E.G	-	FR	-
Phyllolepidae	E.G	COS	FR	FR
Sinolepidae	S.C	S.C, E.G	FR	FR
Bothriolepididae	COS	COS	M,FR	FR
Asterolepididae	COS	COS	M,FR	FR
Phlyctaenaspidae	COS	-	FR	-
Holonematidae	COS	-	M,FR	-
Groenlandaspidae	COS	COS	M,FR	FR
Brachydeiridae	EUR	-	M	-
Leptosteidae	EUR	-	M	-
Pholidosteidae	EUR	-	M	-
Rhachiosteidae	EUR	-	M	-
Plourdosteidae*	COS	-	M,FR	-
Coccosteidae†	EUR	-	M,FR	-
Incisoscutidae	E.G	-	M	-
Camuropiscidae	E.G	-	M	-
Dinichthyidae	COS	COS	M	M
Leiosteidae	EUR	-	M	-
Trematosteidae	EUR	-	M	-
Mylostomatidae	COS	EUR?	M	M
Selenosteidae	EUR	EUR	M	M

* Plourdosteidae includes *Panxiosteus*
† Coccosteidae excludes Famennian reports based on inferognathals only
Range: regions defined as in Young
EUR - Euramerica; E.G - East Gondwana; S.C - South China; COS - cosmopolitan; M - marine; FR - freshwater.

show monotypic family extinctions at the end of the Early Devonian (Mesacanthidae, Euthacanthidae) and at the end of the Frasnian (Diplacanthidae, Culmacanthidae), but no major taxa above family level are affected until the end of the Devonian when the Order Ischnacanthida (here excluding *Acanthodopsis*, regarded as an advanced acanthodiiform, Long 1986) died out.

Several causes for mass extinctions at the Frasnian–Famennian boundary have been suggested: global sea-level rises, global cooling (Copper, 1987; Johnston *et al.*, 1985) and catastrophic impact (Playford *et al.*, 1984) although further geochemical evidence for the latter suggestion may be required (McGhee *et al.*, 1984). Young (1988) has commented that apparent vertebrate extinctions at the Frasnian–Famennian boundary are artificial, due to bias of the Euramerican sampling centre (exclusion of much data from Australia, Antarctica and Asia). However, if the ranges and habitats of placoderm families are considered before and after the start of the Famennian (Table 3.2), certain trends are apparent. The correlation between high extinction for marine vertebrates and invertebrates supports the notion of rising sea-levels affecting large areas of shallow epicontinental environments (Schopf, 1979).

However, a major factor which has been overlooked is the build up of high levels of endemism during the Early Devonian and Middle Devonian, with distinct endemic vertebrate provinces recognised (Young, 1981, 1990; see also Chapter 12). A global faunal interchange event occurred sometime between the start of the Middle Devonian and the Frasnian (Young, 1981), resulting in the breakdown of highly endemic faunas in Euramerica, East Gondwana and South China. Such biogeographic events would have profound effects on extinction rates as long standing endemic communities are suddenly invaded by immigrant taxa, thereby increasing competition for niche occupation. Analysis of Frasnian extinctions for the Placodermi, the most diverse group at that time (Table 3.2) shows that of approximately 22 families, only nine are cosmopolitan, and five of these survived until the end of the Devonian. Endemic families, particularly marine groups, mostly suffered extinction. Extinction of benthic invertebrate communities may have secondarily effected benthic feeding vertebrates and thus caused some extinctions (e.g. chirodipterid and holodipterid dipnoans).

The late Frasnian was a time of peak marine transgression, with marine faunas inhabiting large areas of epicontinental sea ('perched faunas', Johnston, 1974). The Famennian regression resulted in relatively quick loss of shallow marine habitats, with high numbers of marine benthic invertebrate extinctions. This does not, however, account for the large number of nektonic marine vertebrate groups also succumbing at the end of the Frasnian. Other factors, such as increased competition from immigrant taxa must have played a significant role in these extinctions.

The extinction of many vertebrates at the end of the Devonian corresponds with a rapid increase in diversity and radiation of actinopterygians and chondrichthyans, as well as a second peak in marine transgression. Extinctions are less dramatic in numbers than at the end of the Frasnian, and are accompanied by high origination rates (Actinopterygii, Chondrichthyes). The demise of the placoderms and some osteichthyan and acanthodian groups were most likely caused by niche occupation by incoming groups (Long, 1990), in both freshwater and marine environments. These extinctions were effective world-wide, without evidence of provincial bias.

It is concluded that vertebrate mass extinction occurred at, or near, the end of the Frasnian, brought about largely through changes in the marine environment and from increased competition following global biotic interchange sometime in the Middle Devonian; the vertebrate extinction event at the end of the Devonian was probably due to increasing competition from the rapidly diversifying actinopterygian and chondrichthyan fishes.

ACKNOWLEDGEMENTS

For reading the manuscript and giving helpful comments I thank Dr Clive Burrett, University of Tasmania, and for helpful discussion I thank Dr Gavin Young, Canberra.

REFERENCES

Arsenault, M. and Janvier, P., 1991, The anaspid-like craniates of the Escuminac Formation (Upper Devonian) from Miguasha (Québec, Canada), with remarks on anaspid-petromyzontid relationships. In M.M. Chang, Y.H. Liu and G.R. Zhang (eds), *Early Vertebrates and Related Problems of Evolutionary Biology*, pp. 19–40. Science Press, Beijing.

Benton, M.J., 1985, Mass extinctions among non-marine tetrapods, *Nature*, **316**: 811–14.

Copper, P., 1987, Frasnian/Famennian mass extinction and cold water oceans, *Geology*, **14**: 835–83.

Denison, R.H., 1978, *Placodermi*. In *Handbook of Paleoichthyology. Pt 2*, H.-P. Schultze (ed.), Gustav Fischer Verlag, Stuttgart, New York, 128pp.

Denison, R.H., 1979, *Acanthodii*. In *Handbook of Paleoichthyology. Pt 5*, H.-P. Schultze (ed.), Gustav Fischer Verlag, Stuttgart, New York, 62pp.

Elliott, D.K., Blieck, A.R.M. and Gagnier, P.-Y., 1991, Ordovician vertebrates. In C. Barnes and S. Williams (eds), *Advances in Ordovician Geology, Geological Survey of Canada Paper*, **90–9**: 93–106.

Harland, W.B., Holland, C.H., House, M.R., Hughes, N.F., Reynolds, A.B., Rudwick, M.J.S., Satterthwaite, G.E., Tarlo, L.B.H. and Willey, E.C., 1967, *The Fossil Record*, Geological Society, London, Burlington House.

Johnston, J.G., 1974, Extinction of perched faunas, *Geology*, **2**: 479–82.

Johnston, J.G., Klapper, G. and Sandberg, C.A., 1985, Devonian eustatic fluctuations in Euramerica, *Geological Society of America Bulletin*, **96**: 567–87.

Long, J.A., 1983, A new diplacanthoid acanthodian from the Late Devonian of Victoria, *Memoirs of the Association of Australasian Palaeontologists*, **1**: 51–65.

Long, J.A., 1984, New phyllolepids from Victoria and the relationships of the group, *Proceedings of the Linnean Society of New South Wales*, **107**: 263–304.

Long, J.A., 1986, New ischnacanthid acanthodians from the Early Devonian of Australia, with a discussion of acanthodian interrelationships, *Zoological Journal of the Linnean Society*, **87**: 321–39.

Long, J.A., 1990, Chapter 11: Fishes. In K.J. McNamara (ed.), *Evolutionary Trends*, Belhaven Press, London, pp. 255–278.

McLaren, D.J., 1970, Time, life, and boundaries, *Journal of Paleontology*, **44**: 801–15.

McLaren, D.J., 1982, Frasnian–Famennian extinctions. In L.T. Silver and P.H. Schultz (eds), *Geological implications of impacts of large asteroids and comets on the Earth, Special Paper of the Geological Society of America*, **190**: 477–84.

McGhee, G.R., Jr, 1982, The Frasnian–Famennian extinction event: a preliminary analysis of Appalachian marine ecosystems. In L.T. Silver and P.H. Schultz (eds), *Geological implications of impacts of large asteroids and comets on the Earth, Special Paper of the Geological Society of America*, **190**: 491–500.

McGhee, G.R., Jr, 1989, The Frasnian–Famennian extinction event. In S.K. Donovan (ed.), *Mass Extinctions*, Belhaven Press, London: 133–51.

McGhee, G.R., Jr, Gilmore, J.S., Orth, C.J. and Olsen, E., 1984, No geochemical evidence for an asteroid impact at Late Devonian mass extinction horizon, *Nature*, **308**: 629–31.

Pan, J., 1984, The phylogenetic position of the Eugaleaspida in China, *Proceedings of the Linnean Society of New South Wales*, **107**: 309–20.

Pan Jiang, Hou Fuchen, Cao Jingxuan, Gu Qichang, Liu ShiYu, Wang Junqing, Gao Linada and Liu Chun, 1987, *Continental Devonian of Ningxia and its biotas*, Geological Publishing House, Beijing, 237pp.

Patterson, C. and Smith, A.B., 1987, Is the periodicity of extinctions a taxonomic artefact? *Nature*, **330**: 248–51.

Pearson, D.M. and Westoll, T.S., 1979, The Devonian actinopterygian *Cheirolepis* Agassiz, *Transactions of the Royal Society of Edinburgh*, **70**: 337–99.

Playford, P.E., McLaren, D.J., Orth, C.J., Gilmore, J.S. and Goodfellow, W.D., 1984, Iridium anomaly in the Upper Devonian of the Canning Basin, Western Australia, *Science*, **226**: 437–9.

Schopf, T.J.M., 1979, The role of biogeographic provinces in regulating marine faunal diversity through geologic time. In J. Gray and A.J. Boucot (eds), *Historical biogeography, plate tectonics and the changing environment*, Oregon State University Press, Corvallis: 449–57.

Turner, S., 1991, Palaeozoic Vertebrate Microfossils in Australia. In P. Vickers-Rich, J.N. Monaghan, R.F. Baird and T.H. Rich (eds), *Vertebrate Palaeontology in Australasia*, Pioneer Design Studios with Monash University Publications Committee, Melbourne: 429–64.

Turner, S. and Dring, R.S., 1981, Late Devonian thelodonts (Agnatha) from the Gneudna Formation, Carnarvon Basin, Western Australia, *Alcheringa*, **5**: 39–48.

Tway, L.E. and Zidek, J., 1982, Catalog of Late Pennsylvanian ichthyoliths, Part 1, *Journal of Vertebrate Paleontology*, **2**: 328–61.

Tway, L.E. and Zidek, J., 1983, Catalog of Late Pennsylvanian ichthyoliths, Part 2, *Journal of Vertebrate Paleontology*, **3**: 414–38.

Young, G.C., 1981, Biogeography of Devonian vertebrates, *Alcheringa*, **5**: 225–43.

Young, G.C., 1988, Frasnian–Famennian vertebrate extinctions, *Geological Society of Australia, Abstracts*, **21**: 436.

Young, G.C., 1989, New occurrences of culmacanthid acanthodians (Pisces, Devonian) from Antarctica and southeastern Australia, *Proceedings of the Linnean Society of New South Wales*, **111**: 11–24.

Young, G.C., 1990, Devonian vertebrate distribution patterns and cladistic analysis of palaeogeographic hypotheses. In W.S. McKerrow and C.R. Scotese (eds), *Palaeozoic Palaeogeography and biogeography, The Geological Society of London, Memoir*, **12**: 243–55.

Zangerl, R., 1981, *Chondrichthyes*. In *Handbook of Paleoichthyology. Pt 3A*, H.-P. Schultze (ed.), Gustav Fischer Verlag, Stuttgart, New York, 128pp.

PART TWO

PALAEOZOIC VERTEBRATE BIOSTRATIGRAPHY

Chapter 4

THE SILURIAN–DEVONIAN AGNATHAN BIOSTRATIGRAPHY OF THE OLD RED CONTINENT*

P. Janvier and A. Blieck

INTRODUCTION

When L. Agassiz published his monograph 'Recherches sur les Poissons Fossiles' (1835), he uncovered to the palaeontological community of his time a number of remarkable fossil fish taxa from the Devonian Old Red Sandstone of Britain which displayed totally new anatomical structures. His first reaction, as one of Cuvier's students, was to compare, and even refer, these new forms to some extant species with heavily ossified dermal skeleton, such as the siluriform or the 'ganoid' fishes (e.g. sturgeons). Of course, he could hardly have imagined that some of these massive Old Red Sandstone fishes may have been related to the naked and slender-bodied lampreys. His *Cephalaspis lyelli*, for instance, looked more like extant catostomid fishes than lampreys. However, in his later 'Monographie des Poissons fossiles des Vieux Grès Rouges' (1844), he admitted that 'Mais d'un autre côté, la structure et la position de la bouche, et le développement des nageoires paires et impaires, sont si différents dans ces deux familles que, tout en reconnaissant leur affinité, il sera toujours impossible de les réunir'. The jawless vertebrates from the Old Red Sandstone of Britain and the Silurian limestones of Estonia (Pander, 1856) were for half a century referred to as the 'siluroids', 'acipenseroids' or to groups of their own, until Cope (1889) gathered them under the name 'Ostracoderm' which he placed in the new taxon Agnatha, that is 'jawless'. He was later followed by Woodward (1898), Dean (1895) and Dollo (1903) but his views were strongly objected to by Lankester (1897), Traquair (1899) and Goodrich (1909) (see for review Stensiö, 1927).

After World War I, very thorough studies of these fossil jawless vertebrates, or agnathans, began to appear, first with Kiaer's (1924) work

* A contribution to IGCP Project 328.

on the Anaspida and, above all, with Stensiö's (1927) monograph on the Spitsbergen 'cephalaspids' where, for the first time, a well argumented account of their affinities is presented. Stensiö gave an incredibly detailed account of the internal anatomy of the Osteostraci and his conclusions, that the Recent hagfishes and lampreys were descended independently from the heavily armoured Siluro–Devonian 'ostracoderms' was a surprise to most zoologists. There has since been vivid debate about the relationships of these fossil groups, and there is now a tendency toward considering them as a stem-group for lampreys and gnathostomes (see for review Forey, 1984; Janvier, 1985; Maisey, 1986).

These armoured jawless vertebrates have long been referred to as the 'Old Red Sandstone agnathans'. This is in fact true for the rich material from Scotland, the Welsh Borderland, Spitsbergen and Podolia, which occur in this particular kind of sediment—a massive reddish to greenish sandstone widely spread in the Devonian of Europe. This is also true for the faunas from the still undeformed, unmetamorphosed, red, green or yellow sands of the Baltic area. However, the 'Old Red Sandstone facies' is now known to occur anywhere in the Devonian of the world (China, Vietnam, Australia, etc.), and conversely, the armoured agnathans occur in other types of facies, yet they remain most abundant in detritic, shallow-water, marginal facies. It is now well-known that the Old Red Sandstone facies is not associated to any particular geographical area. In fact, it is a mollassic deposit which, in Early Devonian times, expanded largely on flood plains and along continental margins in near-shore environments.

The term 'Old Red Continent' (ORC) is still used by some authors for a palaeogeographical unit including North America, Greenland, northern Europe (to the Urals, including Novaya Zemlya) and Spitsbergen. Modern geological and geophysical studies show the ORC to correspond to separate continental blocks in Lower Palaeozoic times, being connected in Mid-Palaeozoic times by mainly the Caledonian–Taconian orogenic events (see, e.g., McKerrow and Scotese, 1990; Pickering *et al.*, 1988; Soper and Woodcock, 1990; Zonenshain *et al.*, 1985). Some allochthonous (exotic) terranes are not considered within the ORC, viz., westernmost North America and Florida, the latter having Gondwanan affinities. On the contrary, other separate microblocks as Acadia or 'England' (sometimes united under Avalonia) are connected to the ORC during the time-span under consideration here. All along this century, ORC has received different names: North Atlantis, Laureuropa, Laurrussia, Euramerica. It does correspond to the modern concept of Laurentia (North America p.p. + Greenland), Baltica (northern Europe) and Avalonia (northeastern North America + 'England' or Midlands-Brabant) to which exotic Arctic terranes are sometimes added [northern slope of Alaska, extreme Soviet Far East (Kolyma, Chukotka), Spitsbergen, Severnaya Zemlya–Tajmyr] (Figure 4.1).

Siberia and certain domains of Central Asia (Tuva, Chakass) were almost certainly separated from the ORC in Devonian times, but the affinities of their Siluro–Devonian agnathans show that they had closer biogeographic relationships to the latter than to, say, China or Australia. As far as these animals are concerned, the Old Red Continent, Siberia (including western

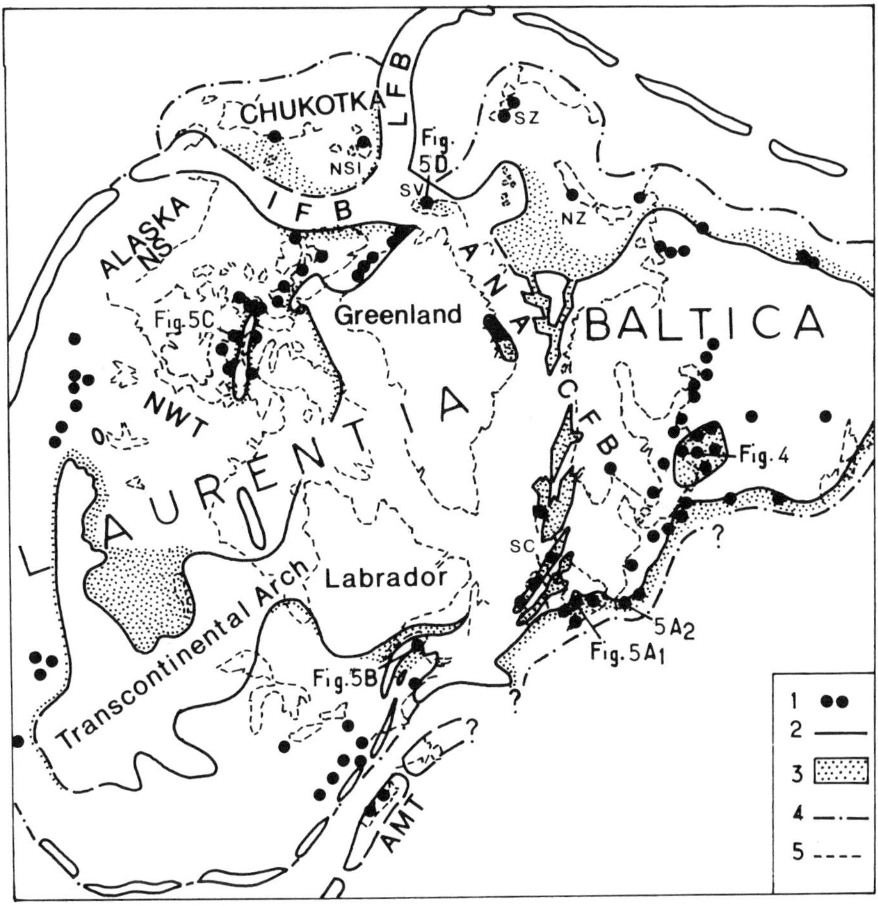

Figure 4.1. Silurian–Devonian agnathan-bearing localities of the Old Red Continent, plotted on the Emsian (Lower Devonian, ca. −390 My) reconstruction of P.A. Ziegler (1988, Figure 4.4, simplified). SW Europe (Armorica-Iberia) and the Alpine terranes are excluded (see, e.g., Paris and Robardet, 1990; Robardet *et al.*, 1990). Beware not to wrongly interpret the palaeoenvironments of localities, as the shore lines have changed throughout Silurian–Devonian times (on this reconstruction, the Emsian shore lines only are represented). Note also that the Silurian localities of the Netherlands, N. Germany and N. Poland are erratic boulders, originally more northerly located in the Baltic area.
1 - agnathan localities, 2 - boundaries of positive areas (cratons and fold belts), 3 - Old Red Sandstone distribution (both mainly continental clastics and deltaic-shallow marine sandy facies), 4 - boundaries between continents and ocean crusts, 5 - present day shore lines.
Abbreviations: AMT - Avalon-Meguma Terranes (Nova Scotia + S. Newfoundland); ANACFB - Arctic-North Atlantic Caledonian Fold Belt; IFB - Innuitian Fold Belt; LFB - Lomonosov Fold Belt; NS - North Slope; NSI - New Siberian Islands; NWT - North West Territories, Canada; NZ - Novaya Zemlya; SC - Scotland; SV - Svalbard; SZ - Severnaya Zemlya.

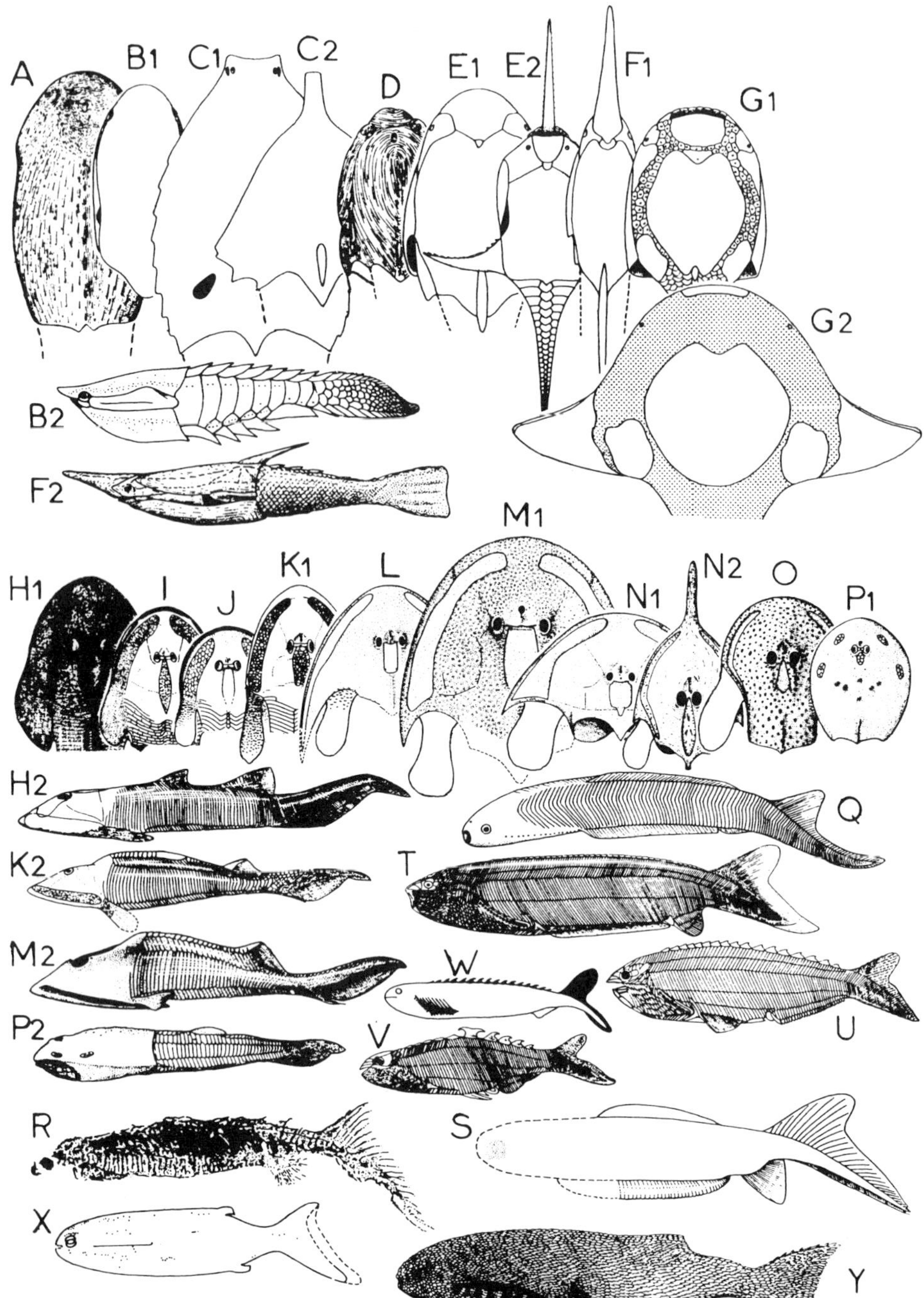
A
B1
C1
C2
D
E1
E2
F1
G1
G2
B2
F2
M1
H1
I
J
K1
L
N1
N2
O
P1
H2
Q
K2
T
M2
W
U
P2
V
R
S
X
Y

Siberia and Severnaya Zemlya), and the Tuva and Chakass regions are characterized, in Siluro–Devonian times, by the association of the Osteostraci and Heterostraci. They are, however, treated in this book as separate chapters: the Old Red Continent as defined here (Euramerica cephalaspid province *sensu* Young, 1981, 1990), and the Siberia and related terranes (Siberia amphiaspid province plus Tuva tannuaspid province; see Chapter 5).

All the major groups of Siluro–Devonian agnathans of these areas appear in a radiation-like manner in the Early Silurian (latest Llandovery–Wenlock), and disappear before the Frasnian–Famennian boundary. Whether extant lampreys are living members of one of these groups, the Anaspida, is still uncertain, and there is no evidence that any of these groups is ancestral to either hagfishes or the Gnathostomes.

THE TAXA AND THEIR STRATIGRAPHICAL DISTRIBUTION

The Siluro–Devonian agnathans of the Old Red Continent, are represented by four groups, the Heterostraci (often known as 'pteraspids'), Osteostraci (often known as 'cephalaspids'), Anaspida and Thelodonti, the latter being possibly a paraphyletic assemblage of primitive vertebrates. In addition, there are some minor taxa of uncertain affinities, classically referred to the Anaspida, but which may be stem-group lampreys (*Jamoytius*, *Euphanerops*, *Endeiolepis*).

Figure 4.2 – cont.

Figure 4.2. Silurian-Devonian jawless vertebrates from the Old Red Continent (not to scale). A–G, Heterostraci, head shields in dorsal view and reconstructions in lateral view (B2, F2); H–P, Osteostraci, head shields and paired fins in dorsal view, and reconstructions in lateral view (H2, K2, M2, P2); Q–W, Anaspida and related forms, reconstructions in lateral view; X, Y, Thelodonti, reconstructions in lateral view. A – Traquairaspidiformes (*Traquairaspis*); B – Cyathaspidiformes (B1, *Poraspis*; B2, *Anglaspis*); C – Amphiaspidiformes (C1, *Kureykaspis*; C2, *Eglonaspis*); D – Anchipteraspididae (*Anchipteraspis*); E – Protopteraspididae (E1, *Protopteraspis*; E2, *Doryaspis*); F – Pteraspididae (F1, *Rhinopteraspis*; F2, *Errivaspis*); G – Psammosteidae (G1, *Drepanaspis*; G2, *Pycnosteus*); H – *Ateleaspis*; I – *Aceraspis*; J – *Hirella*; K – *Hemicyclaspis*; L – Cephalaspidida (*Parameteoraspis*); M – Zenaspidida (M1, *Diademaspis*; M2, *Zenaspis*); N – Benneviaspidida (N1, *Benneviaspis*; N2, *Belonaspis*); O – Kiaeraspidida (*Kiaeraspis*); P – Thyestiida (*Tremataspis*); Q – *Jamoytius*; R – *Legendrelepis*; S – *Endeiolepis*; T – *Pharyngolepis*; U – *Pterygolepis*; V – *Birkenia*; W – *Lasanius*; X – *Phlebolepis*; Y – *Loganellia*. Data after Dineley and Loeffler (1976), Blieck and Heintz (1983), Novitskaya (1986), Elliott (1984), Blieck (1984), Obruchev and Mark-Kurik (1965), Ritchie (1964, 1967, 1968a, 1980), Heintz (1939, 1968), Stensiö (1932), Janvier (1985), Arsenault and Janvier (1991), Parrington (1964), Märss (1979), Turner (1973).

Heterostraci

The Heterostraci possess a large dermal armour of dentine and acellular bone (aspidin), which covers the head region and is pierced laterally by a pair of common external branchial openings. The armour consists of a variable number of plates, which may sometimes fuse into a rigid box (amphiaspids of Siberia), or conversely, break down into numerous platelets or tesserae. Heterostracans *sensu stricto* are restricted to the Silurian and Devonian (Llandovery–Frasnian), but Ordovician vertebrates, the Arandaspida, Astraspida and Eriptychiida, were formerly referred to as Heterostraci *sensu lato*. In fact, these Ordovician forms share some unique features with heterostracans, such as the presence of large median dorsal and ventral plates, or 'discs', but do not possess common external branchial openings. Astraspids and arandaspids are now regarded as the primary and secondary sister-groups of the Heterostraci (Gagnier, 1989).

Heterostracans comprise two major monophyletic groups, the Pteraspidiformes and Cyathaspidiformes, and a number of minor groups (traquairaspids, corvaspids, *Lepidaspis*, etc.), some of which may just be primitive heterostracans (Figure 4.2). The stratigraphical distribution of the major heterostracan taxa is given in Figure 4.3. The Cyathaspidiformes diversify rapidly in the Late Silurian but become rare by the end of the Early Devonian, when they seem to be replaced by the Pteraspidiformes. The Pteraspidiformes occur first in the Late Silurian but become much diversified during the Early Devonian. By the Middle Devonian they survive almost only in the form of their most derived family, the Psammosteidae, which became extinct by the Late Frasnian. The Psammosteidae comprises the largest heterostracans, such as *Pycnosteus* or *Tartuosteus*, which reaches about one metre in breadth (Figure 4.2 G2).

Osteostraci

The Osteostraci are characterized by a generalized horseshoe-shaped endo- and exoskeletal head shield which bears a pair of peculiar lateral and one median 'field'. These are depressions covered with dermal platelets and connected by canals to the otic capsule. These 'fields' are of unknown nature, having been interpreted as either dynamosensory or electric organs. Contrary to heterostracans, osteostracans possessed paired and median dorsal fins, an epicercal tail, and both their endo- and exoskeleton are made up of true bone with cell spaces. Since they possessed a dorsally placed naso-hypophyseal opening the Osteostraci have long been regarded as closely related to lampreys (and the Anaspida); however, they are also known to share a number of synapomorphies with the gnathostomes (true bone, epicercal tail, sclerotic ring, endolymphatic openings, dorsal jugular vein, concentrated paired fins).

Osteostracans are known from the Late Wenlock to the Frasnian. The detail of their stratigraphical distribution is given in Figure 4.3. The most remarkable feature of osteostracan distribution in time is that, contrary to

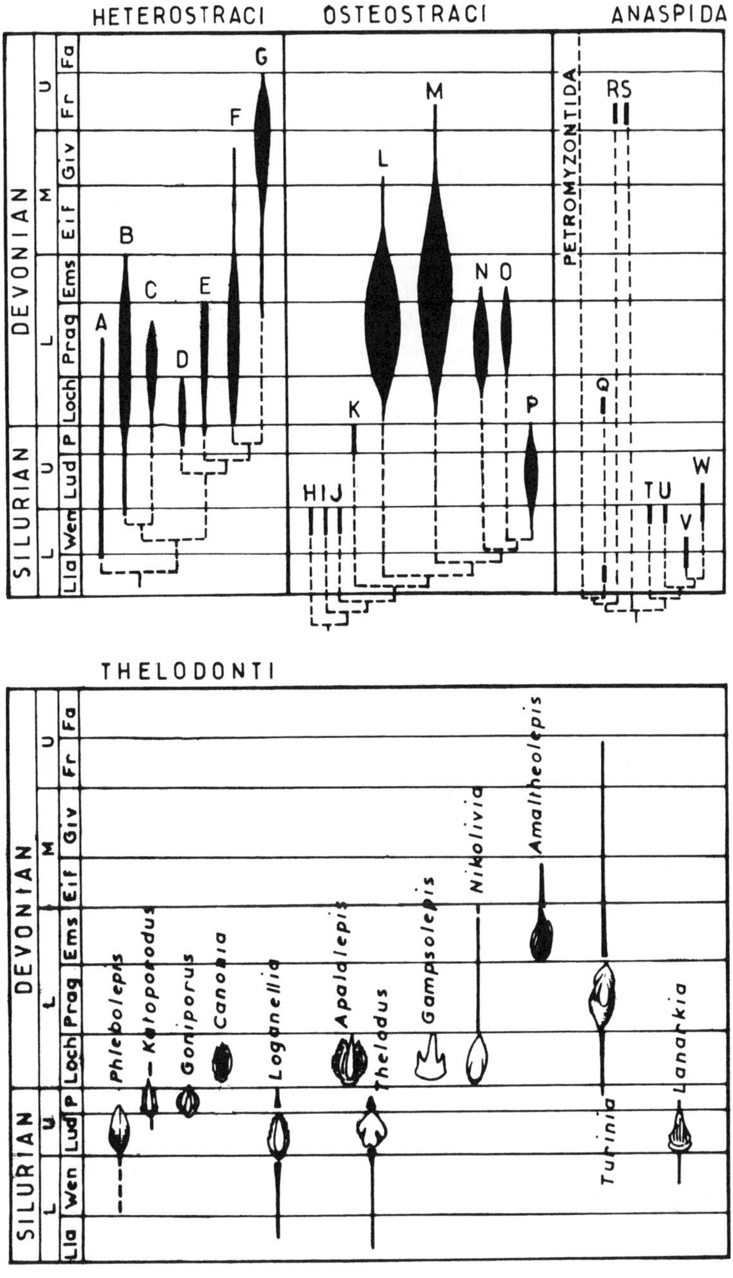

Figure 4.3. Stratigraphical distribution of the Silurian–Devonian agnathans of the Old Red Continent. Abbreviations: A–W, main monophyletic taxa as on Figure 4.2; L–M–U (in stratigraphy), Lower, Middle, Upper; Lla, Llandovery; Wen, Wenlock; Lud, Ludlow; P, Pridoli; Loch, Lochkovian; Prag, Pragian; Ems, Emsian; Eif, Eifelian; Giv, Givetian; Fr, Frasnian; Fa, Famennian.

heterostracans, their earliest (Wenlock) occurrences are represented by the most primitive (*Ateleaspis*, *Aceraspis*) and the most derived (Tremataspididae) members of the group. This pattern clearly indicates that the diversification of the group took place long before the earliest known occurrences.

Most osteostracans belong to a large monophyletic group, the Cornuata, characterized by pointed cornual processes in front of paired fins (Figure 4.2). The Cornuata comprise five major monophyletic groups, the Cephalaspidida (*Cephalaspis*, *Parameteoraspis*, *Mimetaspis*, *Pattenaspis*), the Zenaspidida (*Zenaspis*, *Diademaspis*, *Tegaspis*, *Machairaspis*, *Scolenaspis*), the Benneviaspidida (*Benneviaspis*, *Hoelaspis*, Boreaspididae), the Kiaeraspidida (*Kiaeraspis*, *Norselaspis*, *Axinaspis*, *Acrotomaspis*, *Gustavaspis*, *Nectaspis*) and the Thyestiida (*Procephalaspis*, *Witaaspis*, *Thyestes*, Tremataspididae), and a few minor groups of uncertain affinities (Tannuaspidida, *Didymaspis*, *Waengsjoeaspis*, *Securiaspis*). In addition, there are a few taxa of non-cornuate osteostracans (primitively devoid of cornual processes) which are likely to represent the primitive osteostracan condition (*Ateleaspis*, *Aceraspis*, *Hirella*, *Hemicyclaspis*). The Thyestiida became diversified very early in the Silurian lagunas of the Baltic area and may have later reached Great Britain, Timan and the Urals but seem to have disappeared by the Early Devonian. Their most primitive members (*Procephalaspis*, *Auchenaspis*) had the typical cornuate shield shape whereas the most derived (Tremataspididae) had lost paired and unpaired fins (except the caudal). The Kiaeraspidida are small forms with reduced cornual processes and are restricted to the Lower Devonian of Spitsbergen. The Benneviaspidida have a depressed head shield with short abdominal division, and they include a group of small forms with long rostral process (Boreaspididae). They are known from Britain, Podolia, Spitsbergen and Severnaya Zemlya, but the Boreaspididae are restricted to the latter two areas. The Zenaspidida are usually large, massive osteostracans, with a coarsely tuberculated ornamentation and a large hypophyseal foramen. They are abundant in the Lower Devonian of Britain, Podolia and Spitsbergen. Finally, the Cephalaspidida are generally broad-shielded forms with flattened cornual processes. *Cephalaspis* occurs in Britain and Spitsbergen, *Parameteoraspis*, reaching a very large size (about 35 cm in shield breadth), is particularly abundant in Spitsbergen but is lacking in Britain and Podolia, and smaller cephalaspidids, such as *Pattenaspis* and *Mimetaspis* occur in Britain, Spitsbergen, Podolia, Rhineland and France.

Beside these five major groups, some minor groups of uncertain affinities (often a single species) occur locally as probable endemics. For example, *Didymaspis*, which may be related to either the Kiaeraspidida or Thyestiida, seems to be restricted to Britain (with one doubtful record from the Russian Platform). *Waengsjoeaspis* is restricted to Spitsbergen.

Osteostracans are rare in the Devonian of North America and practically none of the North American taxa can be assigned with certainty to any of the major Old World cornuate taxa (all the North American species referred to as '*Cephalaspis*' do not belong to this genus and are named so on the basis of generalized cornuate characteristics), yet the Middle and Upper

Devonian genus *Yvonaspis* may be related to the European *Parameteoraspis* and the genera *Alaspis* and *Escuminaspis* may possibly be derived Zenaspidida. *'Cephalaspis' wyomingensis*, *Camptaspis utahensis*, *'Cephalaspis' gabrielsei*, all from western North America, are cornuates of unknown affinities.

Anaspida

The Anaspida are slender-bodied jawless vertebrates characterized by triradiate postbranchial spines. They possess rod-like body scales (when not naked), large median dorsal scales or scutes, broad-based paired fins, and a strongly hypocercal tail. Their exoskeleton (scales, dermal plates) are made up of acellular bone (aspidin). Anaspids are rare and apparently restricted to the Silurian. The earliest record consists of isolated scales from the Llandovery of Arctic Canada, and the latest from the Upper Silurian of the Baltic area. The Anaspida occur mostly in Norway and Scotland, with some sparse records from eastern Canada, the Baltic area, and possibly China.

One of the possible phylogenies of the Anaspida suggests a tendency towards the reduction of the number of gill openings, enlargement of the median dorsal scutes, reduction of anal fin, but is neither corroborated nor refuted by stratigraphy, since most anaspid taxa are of the same age. The genera *Jamoytius*, from the Lower Silurian of Scotland, *Endeiolepis*, *Euphanerops* and *Legendrelepis*, all three from the Upper Devonian of Québec, are classically referred to as anaspids because of their strongly hypocercal tail, a characteristic which is now known to occur in other vertebrate groups (e.g. thelodonts). In fact, the apparently non-mineralized scale-covering and the possible presence of an annular cartilage in these forms suggest that they may be more closely related to lampreys than to the Anaspida proper and, thus, may be stem-group Petromyzonida.

Thelodonti

The Thelodonti are supposedly jawless vertebrates whose exoskeleton consists exclusively of minute placoid-like scales, having a crown, a root, and a pulp cavity. There is no clear evidence of the monophyly of the thelodonts, as such a micromeric exoskeleton is just what can be expected in primitive, generalized vertebrates. Only the peculiar 'handle' which prolongs anteriorly the scale base in most thelodonts may possibly be considered as a synapomorphy. Thelodont scales display a wide range of diversity in morphology and histology, and their shapes vary within the individual animal. Their histologic structure varies from orthodentine (*Turinia*, *Thelodus*) to mesodentine (*Phlebolepis*), but their base is always made up of acellular bone (aspidin). Only a few thelodonts are known from articulated specimens (*Turinia*, *Loganellia*, *Phlebolepis*, Ritchie, 1968b; Turner, 1982, 1991), some of which allowing relatively detailed reconstructions and evincing the presence of paired flaps, or fins, dorsal and anal fins, and a hypocercal tail. However, most thelodont taxa are based on a series of isolated scales, which have proved to be useful for stratigraphical studies. The distribution of some thelodont species is given in Figure 4.3. The earliest scales are from the Llandovery of the Welsh Borderland, and the Thelodonti reach their greatest diversity in the Late Silurian and Early Devonian. Only a few species of *Turinia* survive until the Frasnian in the East Gondwana province.

BIOSTRATIGRAPHICAL CORRELATIONS OF VERTEBRATE-DOMINATED SEQUENCES

It is a well-known fact that the vertebrate-bearing sequences of the Old Red Sandstones (ORS) arc difficult to correlate between each other, and with the invertebrate-dominated sequences. The problem of the Silurian/Devonian boundary is a good example for illustrating that point (see, e.g., Martinsson, ed., 1977; Rudwick and Ziegler, in House *et al.*, eds, 1979). The agnathans have been important elements in the discussion as, e.g., pteraspid heterostracans have been used for the first biozonation of the Lower ORS of England by E.I. White (refs. in Stubblefield, 1985; see also Ball and Dineley, 1961). However, cooperative works by the IUGS Subcommissions on Silurian and Devonian Stratigraphy have contributed much to solving the problem (Chlupác and Oliver, 1989; Holland, 1985; Holland and Bassett, eds, 1989; House, 1988; Ziegler and Klapper, 1985).

A nearly complete Silurian–Devonian vertebrate-bearing succession based mainly on the East Baltic sequence is correlated to the standard graptolite and conodont biozonations of the Silurian and Devonian respectively (Blieck *et al.*, 1988; Märss, 1989; Rzhonsnitskaya, 1988; Valiukevicius, 1988). This is summarized on Figure 4.4 where the index-species of the agnathan- and gnathostome-based biozones are stratigraphically located from the mid-Wenlock up to the late Famennian, i.e., for a time-span of about 70 million years. This sequence yielded in reality more than 200 vertebrate species, one quarter in the Silurian, three-quarters in the Devonian. Maxima in number of species occur at some levels: Viita in the Upper Wenlock (15 species), Himmiste (Lower Ludlow, 19 spp.), Tilze (Lower Devonian, 18 spp.), Aruküla (Middle Devonian, 21 spp., which can be considered as the acme-zone), Pamusis and Tervete (Upper Devonian, 13 and 16 spp. respectively). A rough statistical analysis (Figure 4.4C) shows that agnathans are dominant groups among vertebrates during Silurian and Early Devonian times although gnathostomes increase greatly in number and variety in Middle Devonian times and later. Some bioevents can be visualized: the radiation of vertebrates in the Early Silurian (Llandovery–Wenlock), a strong decrease in the late Early Devonian (partly due to lithostratigraphical hiatuses in the sequence, thus lack of outcrops and fossiliferous localities, and difficulties in correlating these levels to the international zonations), and a second radiation in the Middle Devonian. All armoured agnathans disappear just before the Frasnian/Famennian

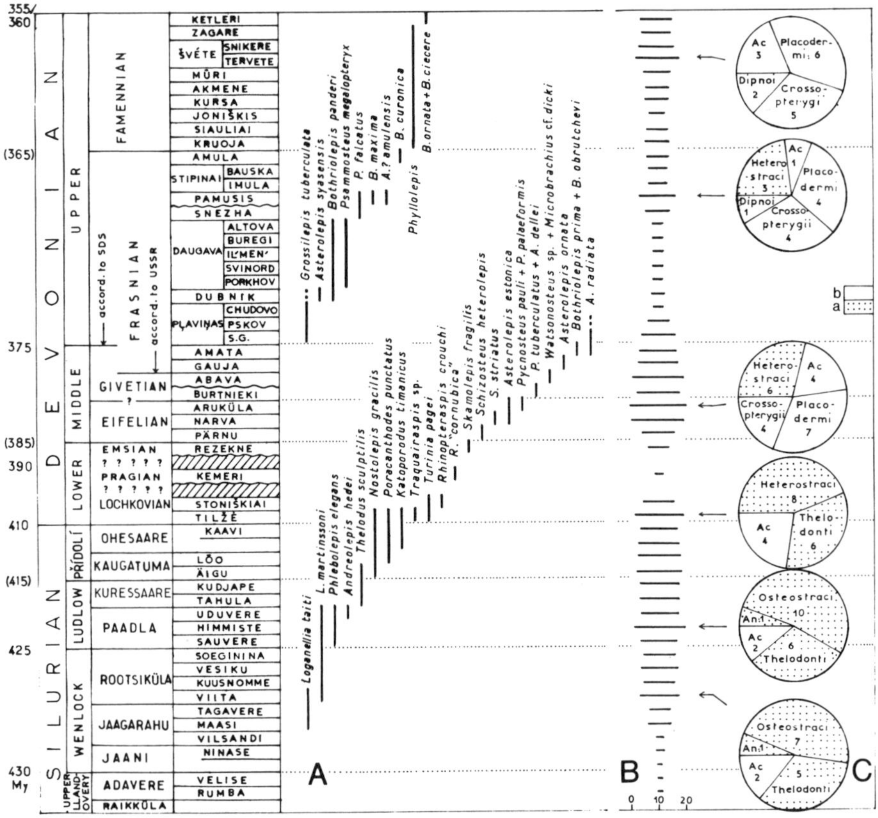

Figure 4.4. The Silurian-Devonian sequence of the East Baltic area with: A - distribution of the vertebrate index-species [the whole ichthyofauna includes ca. 207 species: 156 Devonian, 51 Silurian; see references in Blieck *et al.*, 1988, modified according to Rzhonsnitskaya, 1988, Mark-Kurik, pers. comm., Murchison Symposium, Keele, 1989]; B - number of species for each time-span; C - schematic analysis of the proportions of agnathan (a) and gnathostome (b) higher taxa at the maxima observed on B (the number of species is indicated in each sector). Ac - Acanthodii, An - Anaspida.
For a more detailed analysis of the Silurian, see Märss (1982, 1986). Note 1, that actinopterygians are known in certain levels but not at the maxima of species; 2, that chondrichthyans have not been published; 3, that the Silurian and Devonian scales are correlated to the standard graptolite and conodont zonations respectively, according to Bassett *et al.* (1989) and Rzhonsnitskaya (1988).

boundary (with *Psammosteus* in the Bauska beds) and placoderms ('armoured fishes') in the uppermost Famennian (with *Bothriolepis* in the Ketleri beds).

Nevertheless even richer sequences are known outside the Baltic area. For instance, the Lower Devonian succession of Spitsbergen has revealed about 77 genera of vertebrates, among them 52 genera of agnathans (no. 1–52 on Figure 4.5D) and about twice as many species (a real paradise for palaeoichthyologists in the field!). Some horizons yield about 30 different species (Blieck *et al.*, 1987). One of the important topics for Palaeozoic palaeoichthyologists in the near future will be to correlate more precisely those vertebrate successions with the invertebrate- and plant (spore)-based ones (see, e.g., Turner, ed., 1988–90; recent schemes have also been published by Dineley, 1988; Halstead, 1987).

JAWLESS VERTEBRATES IN SILURIAN–DEVONIAN VERTEBRATE COMMUNITIES

How did Siluro–Devonian agnathans live? Were they freshwater or marine animals, or both? Did the gnathostomes prey on them? And what did they eat? Most of these questions still remain unanswered, yet some local case studies have yielded reliable data.

Since the early nineteenth century, the agnathans—and all Devonian fishes in general—associated to Old Red Sandstone facies were long regarded as freshwater fishes. In fact, they were generally not associated with typically marine invertebrates, they almost never occurred in typically marine facies (marine limestone for example), and were referred by early authors, such as Agassiz, to extant freshwater fish groups ('ganoids', siluriforms, sturgeons). However, detailed sedimentological and palaeontological studies of Siluro–Devonian agnathan-dominated communities (Spitsbergen, Saaremaa) have provided clear evidence that most of them lived in near-shore marginal marine environments. This has been confirmed recently by the study of strontium isotope ratios, made on a number of Silurian and Devonian fish localities (Schmitz *et al.*, 1991), which has yielded a typical marine 'signal' in several instances (Upper Silurian of the Welsh Borderland and Gotland, Lower Devonian of Podolia), although other samples from the Lower Devonian of Wales and from the Frasnian of Latvia yielded a freshwater 'signal'.

During the Silurian, most vertebrate communities are dominated by the agnathans (Figure 4.4), mostly heterostracans and thelodonts in Euramerica, with a few rare gnathostomes such as acanthodians and, by the end of the Silurian, some primitive osteichthyans and chondrichthyans (evidence of Silurian gnathostomes, in particular placoderms, seem to be more abundant in the Chinese realm). Silurian vertebrate-bearing localities are scarce and display local characteristics in faunal diversity that may be linked with environmental conditions. The Late Silurian locality of Ringerike-Rudstangen (Norway) is characterized by the abundance of osteostracans (mostly non-cornuate forms) and anaspids, with a few thelodonts, but no heterostracans or acanthodians at all. In contrast, the Late Wenlockian–Ludlowian localities of Saaremaa (Estonia) are particularly rich in osteostracans, thelodonts and acanthodians, with only rare anaspid and osteichthyan remains and a few heterostracans. The Welsh Borderland and Scottish localities yield abundantly all these groups but osteichthyans. One might easily conclude that the Silurian was the 'golden age' of agnathans, as the acanthodians, osteichthyans and chondrichthyans of that time were mostly small forms and could hardly prey on these comparatively large and heavily armoured jawless vertebrates. One might thus logically expect the latter to disappear as the gnathostomes became larger and more abundant. But this is not at all the case. On the contrary, as soon as the vast sandy shores of the Old Red Continent begin to form, some groups of jawless vertebrates, such as osteostracans and heterostracans, diversify into an incredible variety of forms, almost always in close association with large gnathostomes such as arthrodires, sarcopterygians and acanthodians. In particular, the Wood Bay Formation of Spitsbergen is a good example of this kind of association, given that the thanatocenosis reflects the biocenosis (Blieck and Janvier, 1991), where both large and small osteostracans and heterostracans lived together with their presumed predators (Figure 4.5D).

There is a long-lived theory that the dawn of the gnathostomes caused the fall of their jawless companions. This is wrong. The dawn of the gnathostomes took place much earlier, in the Silurian or even before, and both groups lived together for many tens of million years (this tale can be compared to the story of mammals eliminating dinosaurs!). There is only one piece of evidence that gnathostomes could eventually eat armoured agnathans, in the form of a trace of a bite (possibly of the large sarcopterygian *Glyptolepis*) with subsequent healing in the branchial plate of the large psammosteid *Psammolepis* from the late Middle Devonian of Estonia (Mark-Kurik, 1966). What probably caused the decline of armoured agnathans is the reduction of their favourite type of environment, that is, shallow sandy grounds, close enough to the shore to be constantly enriched in small food particles by the streams and deltas which brought in debris from the incipient land biotopes (spores, plant fragments, small arthropods, etc.). By Emsian–Eifelian times, this type of environment became rare worldwide and vertebrates (mostly gnathostomes) lived in more frankly marine environments, where they are now found in association with crinoids, brachiopods, corals, etc. Consequently, most agnathan groups underwent a considerable reduction in diversity, with the exception of the thelodonts, which may have been partly pelagic, when gnathostomes became dominant in the palaeocommunities (Figure 4.4). The osteostracans remained abundant in the detritic facies of Gaspé (Québec) and survived in this area until the Frasnian (Figure 4.5B: *Alaspis*). Similarly, the large psammosteid heterostracans survived in sandy facies of Central and Eastern Europe until the Late Frasnian. If any process of ecological elimination of the jawless vertebrates has ever taken place, it may have been by virtue of the success of a peculiar group of placoderms, the antiarchs, which invaded the sandy grounds of the entire world in early Middle Devonian times. The

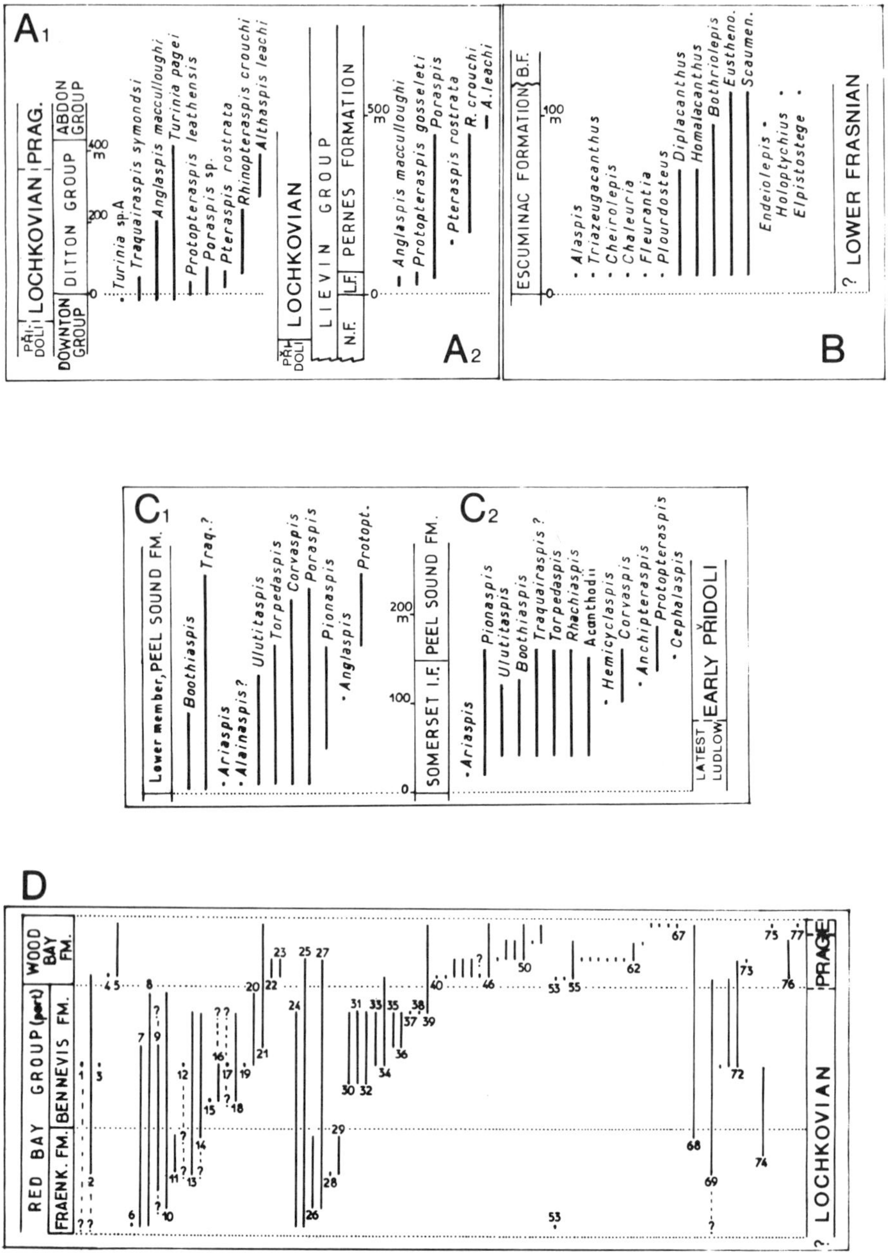
A1
PŘI-DOLI
LOCHKOVIAN
PRAG.
DOWNTON GROUP
DITTON GROUP
ABDON GROUP
Turinia sp.A
Traquairaspis symondsi
Anglaspis maccullougbi
Turinia pagei
Protopteraspis leathensis
Poraspis sp.
Pteraspis rostrata
Rhinopteraspis crouchi
Althaspis leachi
A2
PŘI-DOLI
LOCHKOVIAN
LIEVIN GROUP
N.F.
L.F.
PERNES FORMATION
Anglaspis maccullougbi
Protopteraspis gosseleti
Poraspis
Pteraspis rostrata
R. crouchi
A. leachi
B
ESCUMINAC FORMATION
B.F.
Alaspis
Triazeugacanthus
Cheirolepis
Chaleuria
Fleurantia
Plourdosteus
Diplacanthus
Homalacanthus
Bothriolepis
Eustheno.
Scaumen.
Endeiolepis
Holoptychius
Elpistostege
? LOWER FRASNIAN
C1
Lower member, PEEL SOUND FM.
Boothiaspis
Traq.?
Ariaspis
Alainaspis?
Ulutitaspis
Torpedaspis
Corvaspis
Poraspis
Pionaspis
Anglaspis
Protopt.
C2
SOMERSET I.F.
PEEL SOUND FM.
Ariaspis
Pionaspis
Ulutitaspis
Boothiaspis
Traquairaspis ?
Torpedaspis
Rhachiaspis
Acanthodii
Hemicyclaspis
Corvaspis
Anchipteraspis
Protopteraspis
Cephalaspis
LATEST LUDLOW
EARLY PŘIDOLI
D
RED BAY GROUP (part)
FRAENK. FM.
BENNEVIS FM.
WOOD BAY FM.
LOCHKOVIAN
PRAGE

same group may also have contributed to the elimination of the Galeaspida in the Chinese realm. In fact, these benthic detrital feeders and possible burrowers with appendage-like pectoral fins may well have fed on very much the same type of particle as most agnathans, and their great abundance in detritic facies shows that they could pullulate in such a way that most competitors had no chance to save their niche.

There is only one piece of evidence of the diet of Siluro–Devonian agnathans. This is the intestinal content of the anaspid-like stem-group petromyzonid *Endeiolepis*, from the Frasnian of Québec (Figure 4.5B), but this phosphatized mass has not yet been studied in detail. Nevertheless, it does not show any trace of large food particles, contrary to stomacal contents of acanthodians in the same locality, which show abundant crustacean remains.

Figure 4.5 - cont.

Figure 4.5. Simplified stratigraphical distributions of (A) vertebrate index-species, and (B–D) vertebrate genera, in the Silurian and/or Devonian of (A1) the Welsh Borderland, England, at Brown Clee Hill; (A2) Artois, France; (B) Eastern Canada, Québec, at Miguasha; (C) the Canadian Arctic Archipelago; (C1) at Transition Bay, Prince of Wales Island; (C2) at Pressure Point, West Creswell, and Cape Anne-Cunningham Inlet, Somerset Island; (D) Svalbard, Spitsbergen, at Raudfjorden and Woodfjorden. After Blieck *et al.* (1987), Blieck and Janvier (1989), Elliott (1984), Gagnier (1989), Hesse and Sawh (1982). Location of sequences is indicated on Figure 4.1. On (D), no metric scale is given; total thickness of the whole sequence is estimated at ca. 4000 m (Friend, 1961). For a complete Silurian-Devonian sequence, see Figure 4.4.
Abbreviations: B.F. - Bonaventure Formation, E - Emsian, *Eustheno.* - *Eusthenopteron*, FM - Formation, Fraenk. Fm. - Fraenkelryggen Formation, L.F. - Liévin Formation, N.F. - Noulette Formation, Prag. - Pragian, *Protopt* - *Protopteraspis*, *Scaumen.* - *Scaumenacia*, Somerset I.F. - Somerset Island Formation, *Traq.?* - *Traquairaspis?*.
1- *Nikolivia*, 2- *Turinia*, 3- *Apalolepis*, 4- *Sigurdia*, 5- *Amaltheolepis*, 6- *Davelaspis*, 7- *Traquairaspis*, 8- *Corvaspis*, 9- *Protopteraspis*, 10- *Poraspis*, 11- *Dinaspidella*, 12- *Lepidaspis*, 13- *Anglaspis*, 14- *Homalaspidella*, 15- *Canadapteraspis?*, 16- *Miltaspis*, 17- *Larnovaspis*, 18- *Ctenaspis*, 19- *Weigeltaspis*, 20- *Irregulareaspis*, 21- *Doryaspis*, 22- *Zascinaspis*, 23- *Gigantaspis*, 24- *Cephalaspis*, 25- *Machairaspis*, 26- *Pattenaspis*, 27- '*Cephalaspis*', 28- *Ectinaspis*, 29- *Waengsjoeaspis*, 30- *Mimetaspis*, 31- '*Securiaspis*', 32- *Tegaspis*, 33- '*Benneviaspis*', 34- *Benneviaspis*, 35- *Kiaeraspis*, 36- *Hoelaspis*, 37- *Scolenaspis*, 38- *Zenaspis*, 39- *Parameteoraspis*, 40- *Norselaspis*, 41- '*Boreaspis*', 42- *Cephalaspis?*, 43- *Axinaspis*, 44- *Boreaspis*, 45- *Belonaspis*, 46- *Diademaspis*, 47- *Acrotomaspis*, 48- *Spatulaspis*, 49- *Dicranaspis*, 50- *Nectaspis*, 51- *Hildenaspis*, 52- *Gustavaspis*, 53- Palaeacanthaspids?, 54- *Sigaspis*, 55- *Arctaspis*, 56- *Lataspis*, 57- '*Svalbardaspis*', 58- *Lehmanosteus*, 59- *Dicksonosteus*, 60- *Elegantaspis*, 61- *Heintzosteus*, 62- Euleptaspids, 63- *Arctolepis*, 64- *Actinolepis*, 65- *Heterogaspis*, 66- *Herasmius*, 67- *Homostius*, 68- Elasmobranchii, 69- '*Nostolepis*', 70- '*Protodus*', 71- '*Gomphonchus*', 72- '*Onchus*', 73- *Xylacanthus*, 74- actinopterygians?, 75- *Orvikuina*, 76- '*Porolepis*', 77- *Heimenia*.

CONCLUSIONS

It is now a well-known fact that vertebrate endemism was well marked in the Late Silurian and Early Devonian, and decreased by the Middle and Late Devonian. This has been explained by either changes in geographical conditions, or increasing dispersion ability of organisms, or both. Analytic historical biogeography of exclusively fossil groups is faced with two major problems. One is the difficulty to obtain more than two sources of data of nearly the same age and equivalent quality in information, the other is that endemism in palaeontology is in principle untestable, since one can never expect to have a complete record of the biotas for a given time. Thus, endemism in the fossil record is merely a theory based on the postulate that a taxon actually did not occur where it is not found as fossil. The thyestidian osteostracans, for example, were first known exclusively from Estonia and Wales and, then, discovered in northern Timan, Norway and the Urals. They may now be regarded as endemic to this group of areas, until they are found elsewhere. In contrast, the endemism of the Galeaspida in China and northern Vietnam, i.e., the South and North Chinese blocks, remains unchallenged.

A number of areas of apparent endemism can thus be defined on the basis of mainly thelodont, heterostracan, and osteostracan faunas (Blieck and Janvier, 1991 and in press; Turner and Tarling, 1982). Anaspids are too rare to provide reliable data. This endemism is, of course, a handicap to correlate ichthyofaunas of separate continental blocks where only few species or genera are found in common. Correlations are better made on a single block such as the Old Red Continent, where agnathans have been studied for more than a century and a half. Faunas of the same age from not too far distant localities (on Devonian geography) as Welsh Borders–Artois, or Canadian Arctic Archipelago–Spitsbergen, may be very similar in taxonomic composition. For a global analysis of vertebrate provinces in the Middle Palaeozoic, we refer to G.C. Young (Chapter 12).

REFERENCES

Agassiz, L., 1835, *Recherches sur les poissons fossiles: 2. contenant l'histoire des Ganoïdes*. Neufchâtel, Imprimerie du Petit-Pierre, **1833–44**: 85–200.

Agassiz, L., 1844–1845, *Monographie des Poissons fossiles du Vieux Grès Rouge ou système dévonien (Old Red Sandstone) des Iles Britanniques et de Russie*, Editions Jent and Gassmann, Neufchâtel, 171pp.

Arsenault, M. and Janvier, P., 1991, The anaspid-like craniates of the Escuminac Formation (Upper Devonian) from Miguasha (Québec, Canada), with remarks on anaspid-petromyzontid relationships. In M.M. Chang, Y.H. Liu and G.R. Zhang (eds), *Early Vertebrates and Related Problems of Evolutionary Biology*, pp. 19–40, Science Press, Beijing.

Ball, H.W. and Dineley, D.L., 1961, The Old Red Sandstone of Brown Clee Hill and the adjacent area, I: Stratigraphy, *Bulletin of the British Museum of Natural History (Geology)*, **5**: 177–242.

Bassett, M.G., Kaljo, D. and Teller, L., 1989, The Baltic Region. In C.H. Holland

and M.G. Bassett (eds), *A global standard for the Silurian System*, National Museum of Wales, Geological Series, Cardiff, **9**: 158–70.

Blieck, A., 1984, Les Hétérostracés Ptéraspidiformes, Agnathes du Silurien–Dévonien du continent Nord-Atlantique et des blocs avoisinants. *Cahiers de Paléontologie (Vertébrés)*, 199pp.

Blieck, A., Goujet D. and Janvier P., 1987, The vertebrate stratigraphy of the Lower Devonian (Red Bay Group and Wood Bay Formation) of Spitsbergen, *Modern Geology*, **11**: 197–217.

Blieck, A. and Heintz, N., 1983, The cyathaspids of the Red Bay Group (Lower Devonian) of Spitsbergen. The Downtonian and Devonian Vertebrates of Spitsbergen XIII, *Polar Research*, **1**: 49–74.

Blieck, A. and Janvier, P., 1991, Silurian Vertebrates. In M.G. Bassett *et al.* (eds), *The Murchison Symposium* (International Symposium on the Silurian System, Keele, 1989), *Palaeontology, Special Papers*, **44**: 345–89.

Blieck, A. and Janvier, P., in press, Silurian–Devonian vertebrate-dominated communities, with particular reference to agnathans. In A.J. Boucot and J. Lawson (eds), *IGCP Project Ecostratigraphy Final Report*, Cambridge University Press.

Blieck, A., Mark-Kurik, E. and Märss, T., 1988, Biostratigraphical correlation between Siluro–Devonian invertebrate-dominated and vertebrate-dominated sequences: the East Baltic example. In N.J. McMillan, A.F. Embry and D.J. Glass (eds), *Devonian of the World* (Proceedings IInd International Symposium on the Devonian System, Calgary, 1987), *Canadian Society of Petroleum Geologists Memoir*, **14, III**: 579–87.

Chlupác, I. and Oliver, Jr, W.A., 1989, Decision on the Lochkovian–Pragian boundary stratotype (Lower Devonian), *Episodes*, **12**: 109–13.

Cope, E.D., 1889, Synopsis of the families of Vertebrata, *The American Naturalist*, **23**: 849–77.

Dean, B., 1895, *Fishes, living and fossil. An outline of their forms and probable relationships*. McMillan *et al.* (eds).

Dineley, D.L., 1988, The radiation and dispersal of Agnatha in Early Devonian time. In N.J. McMillan, A.F. Embry and D.J. Glass (eds), *Devonian of the World* (Proceedings IInd International Symposium on the Devonian System, Calgary, 1987), *Canadian Society of Petroleum Geologists Memoir*, **14, III**: 567–77.

Dineley, D.L. and Loeffler E.J., 1976, Ostracoderm faunas of the Delorme and associated Siluro–Devonian Formations, North West Territories, Canada, *Palaeontology, Special Papers*, **18**: 1–214.

Dollo, L., 1903, Le *Pteraspis* dans l'Ardenne, *Comptes-rendus de l'Académie des Sciences de Paris*, **136**: 699–701.

Elliott, D.K., 1984, Siluro–Devonian fish biostratigraphy of the Canadian Arctic Islands, In *Symposium on Evolution and Biogeography of Early Vertebrates* (Sydney–Canberra, 1983), *Proceedings of the Linnean Society of NSW*, **107**: 197–209.

Forey, P.L., 1984, Yet more reflection on agnathan-gnathostome relationships, *Journal of Vertebrate Paleontology*, **4**: 330–43.

Friend, P.F., 1961, The Devonian Stratigraphy of North and Central Vestspitsbergen, *Proceedings of the Yorkshire Geological Society*, **33**, **1**: 77–118.

Gagnier, P.Y., 1989, *Analyses anatomique et phylogénétique de quelques Vertébrés paléozoïques américains, Volume 1: Les Vertébrés ordoviciens de Bolivie. Volume 2: Les Poissons Acanthodiens de la Formation d'Escuminac, Province du Québec, Canada*, Thèse Doctorat en Sciences Naturelles du l'Université de Paris 7 (unpublished).

Goodrich, M.A., 1909, Cyclostomes and Fishes. In R. Lankester, *A treatise on zoology, Part IX: Vertebrata Craniata*, fascicule 1. A. and C. Black (eds), London.

Halstead, L.B., 1987, Agnathan extinctions in the Devonian. In E. Buffetaut, J.J. Jaeger and J.M. Mazin (eds), *Les extinctions dans l'histoire des Vertébrés* (Paris, 1985). *Mémoires de la Société géologique de France*, Nouvelle Série, **150**: 7–11.

Heintz, A., 1939, Cephalaspida from the Downtonian of Norway, *Norske Videnskaps Akademiens Skrifter (Matematiske- naturvidenskapslige Klasse)*, **1939**, **5**: 1–119.

Heintz, A., 1968, New agnathans from Ringerike Sandstone, *Norske Videnskaps Akademiens Skrifter (Matematiske- naturvidenskapslige Klasse)*, **1969**: 1–28.

Hesse, R. and Sawh, H., 1982, Escuminac Formation. In R. Hesse, G.V. Middleton and B.R. Rust (eds), *Paleozoic continental margin sedimentation in the Quebec Appalachians*, International Association of Sedimentology, 11th International Congress (Hamilton). Excursion Guidebook **7B**: 72–80.

Holland, C.H., 1985, Series and stages of the Silurian System, *Episodes*, **8**: 101–3.

Holland, C.H. and Bassett, M.G. (eds), 1989, *A global standard for the Silurian System*, National Museum of Wales, Geological Series, Cardiff, **9**.

House, M.R., 1988, International definition of Devonian System boundaries, *Proceedings of the Ussher Society*, **7**: 41–6.

House, M.R., Scrutton, C.T. and Bassett, M.G. (eds), 1979, *The Devonian System, Special Papers Palaeontology*, **23**.

Janvier, P., 1985, Les Céphalaspides du Spitsberg: anatomie, phylogénie et systématique des Ostéostracés siluro-dévoniens; révision des Ostéostracés de la Formation de Wood Bay (Dévonien inférieur du Spitsberg), *Cahiers de Paléontologie*, CNRS édit., Paris, 244pp.

Kiaer, J., 1924, The Downtonian fauna of Norway. I: Anaspida with a geological introduction, *Videnskappsselskapets Skrifter (Matematiske- naturvidenskapslige Klasse)*, **6**: 1–139.

Lankester, E.R., 1897, The taxonomic position of the Pteraspidae, Cephalaspidae and Asterolepidae, *Natural Science*, **11**: 45.

Maisey, J.G., 1986, Heads and tails: a chordate phylogeny, *Cladistics*, **2**: 201–56.

Mark-Kurik, E., 1966, On some alteration of the exoskeleton in the Psammosteidae (Agnatha). In [*The Organism and its Environment in the Past*], 'Nauka', Moskva: 55–60 (in Russian).

Märss, T., 1979, [Lateral-line sensory system of the Ludlovian thelodont *Phlebolepis elegans* Pander], *Eesti NSV Teaduste Akadeemia Toimetised*, **28**: 108–11 (in Russian).

Märss, T., 1982, [Vertebrate biozones (Silurian of northern Baltic States)]. In D. Kaljo and E. Klaamann (eds), [*Assemblages and biozones of the Silurian of the Baltic States*]. 'Valgus', Tallinn, **7**: 97–116 (in Russian).

Märss, T., 1986, [Silurian vertebrates of Estonia and West Latvia], *Fossilia Baltica*, Akademia Nauk Estonia SSR, Instituut Geologii, 'Valgus', Tallinn, 104pp. (in Russian).

Märss, T., 1989, Vertebrates. In C.H. Holland and M.G. Bassett (eds), *A global standard for the Silurian System, National Museum of Wales, Geological Series, Cardiff*, **9**: 284–9.

Martinsson, A. (ed.), 1977, *The Silurian-Devonian boundary*, International Union of Geological Sciences, Series A, **5**, Stuttgart.

McKerrow, W.S. and Scotese, C.R. (eds), 1990, *Palaeozoic Palaeogeography and Biogeography. Geological Society of London, Memoir*, **12**.

Novitskaya, L.I., 1986, [Fossil agnathans of USSR—Heterostracans: cyathaspids, amphiaspids, pteraspids], Akademia Nauk SSSR, *Trudy Paleontologicheskogo Instituta*, **219**: 1–159, Nauka, Moscow. (In Russian.)

Obruchev, D.V. and Mark-Kurik, E., 1965, [Devonian Psammosteids (Agnatha, Psammosteidae) of the USSR], *Eesti NSV Teaduste Akadeemia Geoloogia Instituut*, 304pp. (in Russian).
Pander, C.H., 1856, *Monographie der fossilen Fische des silurischen Systems der russisch-baltischen Gouvernements*. St Petersburg.
Paris, F. and Robardet, M., 1990, Early Palaeozoic palaeobiogeography of the Variscan regions. In P. Matte (ed.), *Terranes in the Variscan belt of Europe and Circum-Atlantic Paleozoic Orogens* (IGCP Project 233). *Tectonophysics*, **177**: 193–213.
Parrington, R., 1964, On the nature of the Anaspida. In T.S. Westoll (ed.), *Studies of Fossil Vertebrates*, pp. 108–28, The Athlone Press, London.
Pickering, K.T., Bassett, M.G. and Siveter, D.J., 1988, Late Ordovician–Early Silurian destruction of the Iapetus Ocean: Newfoundland, British Isles, and Scandinavia—a discussion, *Transactions of the Royal Society of Edinburgh*, **79**: 361–382.
Ritchie, A., 1964, New lights on the Norwegian Anaspida, *Norske Videnskaps Akademiens Skrifter (Matematiske- naturvidenskapslige Klasse)*, **14**: 1–35.
Ritchie, A., 1967, *Ateleaspis tessellata* Traquair, a non-cornuate cephalaspid from the Upper Silurian of Scotland, *Zoological Journal of the Linnean Society*, London, **47**: 69–81.
Ritchie, A., 1968a, New evidence on *Jamoytius kerwoodi*, an important ostracoderm from the Silurian of Lanarkshire, Scotland, *Palaeontology*, **11**: 21–39.
Ritchie, A., 1968b, *Phlebolepis elegans* Pander, an Upper Silurian thelodont of Oesel, with remarks on the morphology of the thelodonts. In T. Ørvig (ed.), *Current Problems in Lower Vertebrate Phylogeny*, Nobel Symposium **4**: 81–8, Almqvist and Wiksell, Stockholm.
Ritchie, A., 1980, The Late Silurian anaspid genus *Rhyncholepis* from Oesel, Estonia, and Ringerike, Norway, *American Museum Novitates*, **2699**: 1–18.
Robardet, M., Paris, F. and Racheboeuf, P.R., 1990, Palaeogeographic evolution of southwestern Europe during Early Palaeozoic times. In W.S. McKerrow and C.R. Scotese (eds), *Palaeozoic Palaeogeography and Biogeography. Geological Society of London, Memoir*, **12**: 411–19.
Rzhonsnitskaya, M.A., 1988, Biostratigraphic scheme of the Devonian of the Russian Platform. In N.J. McMillan, A.F. Embry and D.J. Glass (eds), *Devonian of the World* (Proceedings IInd International Symposium on the Devonian System, Calgary, 1987), *Canadian Society of Petroleum Geologists Memoir*, **14, III**: 691–701.
Schmitz, B., Åberg, G., Werdelin, L., Forey, P. and Bendix-Almgreen, S.E., 1991, 87Sr/86Sr, Na, F, Sr, and La in skeletal fish debris as a measure of the paleosalinity of fossil-fish habitats, *Geological Society of America Bulletin*, **103**: 786–94.
Soper, N.J. and Woodcock, N.H., 1990, Silurian collision and sediment dispersal patterns in southern Britain, *Geological Magazine*, **127**: 527–42.
Stensiö, E.A., 1927, The Downtonian and Devonian vertebrates of Spitsbergen. Part I. Family Cephalaspidae, *Skrifter om Svalbard og Ishavet*, **12**: 1–391.
Stensiö, E.A., 1932, *The Cephalaspids of Great Britain*, British Museum (Natural History), London.
Stubblefield, J., 1985, Errol Ivor White 1901–1985, elected FRS 1956, *Biographical Memoir of Fellows of the Royal Society*, **31**: 635–51.
Traquair, R.H., 1899, Report on Fossil Fishes collected by the Geological Survey of Scotland in the Silurian Rocks of the South of England, *Transactions of the Royal Society of Edinburgh*, **XXXIX**: 827–64.
Turner, S., 1970, Fish help to trace continental movements, *Spectrum*, **79**: 8–10.

Turner, S., 1973, Siluro–Devonian thelodonts from the Welsh Borderland, *Journal of the Geological Society of London*, **129**: 557–84.

Turner, S., 1982, A new articulated thelodont (Agnatha) from the Early Devonian of Britain, *Palaeontology*, **25**: 879–89.

Turner, S. (ed.), 1988–1990, *Ichthyolith Issues: news and views on Palaeozoic vertebrate microfossils*. No. 1–5 (mimeographed).

Turner, S., 1991, Monophyly and interrelationships of the Thelodonti. In M.M. Chang, Y.H. Liu and G.R. Zhang (eds), *Early Vertebrates and Related Problems of Evolutionary Biology*, pp. 87–119, Science Press, Beijing.

Turner, S. and Tarling, D.H., 1982, Thelodont and other agnathan distributions as tests of Lower Palaeozoic continental reconstructions, *Palaeogeography, Palaeoclimatology and Palaeoecology*, **39**: 295–311.

Valiukevicius, J., 1988, Correlation of Lower and Middle Devonian deposits of the USSR with acanthodian assemblages. In N.J. McMillan, A.F. Embry and D.J. Glass (eds), *Devonian of the World* (Proceedings IInd International Symposium on the Devonian System, Calgary, 1987), *Canadian Society of Petroleum Geologists Memoir*, **14, III**: 601–7.

Woodward, A.S., 1898, *Outlines of vertebrate palaeontology*, Cambridge Biological Series, Cambridge.

Young, G.C., 1981, Biogeography of Devonian vertebrates, *Alcheringa*, **5**: 225–43.

Young, G.C., 1990, Devonian vertebrate distribution patterns and cladistic analysis of palaeogeographic hypotheses. In W.S. McKerrow and C.R. Scotese (eds), *Palaeozoic Palaeogeography and Biogeography. Geological Society of London, Memoir*, **12**: 243–55.

Ziegler, P.A., 1988, Laurussia—the Old Red Continent. In N.J. McMillan, A.F. Embry and D.J. Glass (eds), *Devonian of the World* (Proceedings IInd International Symposium on the Devonian System, Calgary, 1987), *Canadian Society of Petroleum Geologists Memoir*, **14, I**: 15–48.

Ziegler, W. and Klapper, G., 1985, Stages of the Devonian System. In *Global Standards in Stratigraphy, Episodes*, **8**: 104–9.

Zonenshain, L.P., Kuzmin, M.I. and Kononov, M.V., 1985, Absolute reconstructions of the Palaeozoic oceans, *Earth and Planetary Science Letters*, **74**: 103–16.

Chapter 5

SILURIAN–DEVONIAN VERTEBRATE BIOSTRATIGRAPHY OF SIBERIA AND NEIGHBOURING TERRANES

*A. Blieck and P. Janvier**

INTRODUCTION

Compared to the other northern landmasses, the continental blocks of CIS, east of the Ural mountains, and Mongolia have been much more recently prospected for Palaeozoic vertebrates. Few localities have been discovered over a very wide area spanning from the Urals to the Pacific ocean (Figure 5.1). The Siberian block and its neighbouring terranes have agnathan and gnathostome taxa in common with Laureuropa (Euramerica) although endemic taxa, e.g., amphiaspids and tannuaspids, are known in rather restricted areas. The oldest vertebrate localities in this region are Ordovician in age, with probable thelodonts and astraspids (an heterostracan-related taxon) which have not yet been described. However, Silurian–Devonian vertebrates are better documented. Recent biostratigraphical correlations have been based on thelodonts, acanthodians and placoderms, mainly in the Silurian and Lower Devonian.

This part of the world is classically included within a large northern landmass called Laurasia, facing a southern one called Gondwana, but recent re-evaluations of the palaeogeography of its Silurian–Devonian continental components have been processed by Talent *et al.* (1987a and b) and Talent and Yolkin (1987). Siberia itself (or Angara) does correspond to the Central Siberian Uplands south of the Tajmyr Peninsula (Figure 5.1). The Siberian block includes most of Mongolia, and is attached to the Jiamusi-Bureinsk block (i.e., northern Manchuria), each on one side of the Mongolo-Okhotsk (Zabaikal) Rift which is, however, probably composed of several displaced terranes. The West Siberian Lowlands are excluded from the Siberian block and are, hypothetically, related to Kazakhstan.

The northern part of Tajmyr Peninsula, together with Severnaya Zemlya

* A contribution to IGCP Project 328.

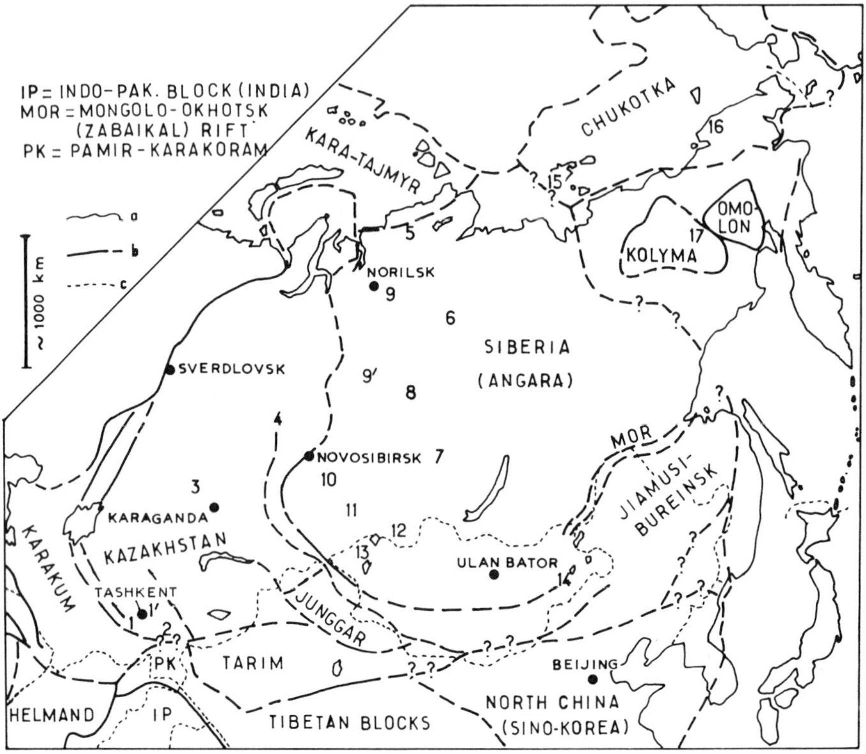

Figure 5.1. Post-Silurian continental blocks, east of the Ural mountains, after Talent *et al.* (1987 a–b, slightly simplified), with the Tajmyr suture modified according to Churkin *et al.* (1981). Severnaya Zemlya, on the Kara-Tajmyr block, is related to Europe. The West Pacific terranes (Japan, Sikhote Alin, Kamchatka, etc.) are not represented.
a - extant coast lines; b - ophiolite belts and hypothesized sutures; c - political boundaries; 1–17 - Ordovician to Devonian vertebrate localities (all in CIS [ex USSR] except 13–14 in Mongolia).

and the Kara Sea, is considered as a separate block, related to Europe (see Janvier and Blieck, Chapter 4). This block is named Kara-Tajmyr by Talent *et al.* (1987b) or Barentsia by Ziegler P.A. (1988; but Barentsia *sensu* Scotese and McKerrow, 1990, does not include northern Tajmyr). The Soviet Far East is divided into separate smaller elements, named Kolyma, Omolon and Chukotka blocks (the last one including the North Slope of Alaska). All these blocks are separated by known or supposed sutures which have closed in Upper Devonian (between Kazakhstan and Siberia), Permo–Carboniferous (Kara-Tajmyr/Siberia) or Jurassic–Cretaceous times (Soviet Far East). They are thus considered as separate elements in Silurian–Devonian times (see a partially different view by Scotese and McKerrow, 1990, Figures 11–16). The question has been further extended by Young

(1981, 1990) to include whether Tuva (Tuvinskaya A.S.S.R., west of Lake Bajkal) has to be considered as separated or not from Siberia (Figures 5.1 and 12.2).

VERTEBRATE TAXA

Agnathans and gnathostomes are present in the Silurian–Devonian of Siberia and related terranes. Agnathans are represented by heterostracans, osteostracans and thelodonts; all higher gnathostome taxa have been collected, viz., chondrichthyans, placoderms, acanthodians, actinopterygians and sarcopterygians.

Among agnathans, heterostracans and osteostracans are mainly represented by derived groups, namely the amphiaspids and tannuaspids respectively. Amphiaspids are a monophyletic group of cyathaspidiform heterostracans with a box-like bony head carapace composed of either a dorsal and a ventral shield or a completely fused 'muff'. In amphiaspidids (Amphiaspidoidei *sensu* Novitskaya, 1971), this carapace bears a peculiar, paired adorbital opening which might be an inhalant device. The armour of amphiaspids has a generally flattened shape suggesting benthic habits comparable to those of extant rays. Some highly derived amphiaspids are even blind forms (*Eglonaspis* and *Lecaniaspis* among hibernaspidids) with a tube-like forward projection ending with the oral aperture (*Eglonaspis*, *Lecaniaspis* and other hibernaspidids, as well as some amphiaspidids but with a shorter oral tube) (Figure 5.2). After revision of the biostratigraphical correlations of the amphiaspid-bearing sedimentary sequences of Tajmyr and the NW Siberian Platform (see next paragraph), it appears that the group is mainly restricted to the Lochkovian, with few species in the Lower Pragian (Lower Devonian) [*Boothiaspis* Broad (1973) from the Canadian Arctic, originally thought to be an amphiaspid, is reinterpreted as a poraspidid cyathaspid by Elliott and Dineley (1985)]. Novitskaya (1976) pointed out some difference between the amphiaspid faunas of the Tajmyr and Siberian Platform respectively. In fact very few genera (*Gunaspis* and *Argyriaspis*) seem to be common to both areas (Figure 5.4). This condition may be due to the high degree of endemism of the group.

Tannuaspids are a small group of osteostracans characterized by a considerable reduction of the lateral and median fields. Like the Tremataspididae, they are devoid of paired fins (Figure 5.2). They come from hilly areas south of Siberia, in Central Asia: the Tuva (or Tuvin) A.S.S.R. and Chakass (or Khakass) autonomous region, where they are known from Upper Silurian to Lower (*Tannuaspis* and *Tuvaspis*) and Middle Devonian (*Ilemoraspis*).

Thelodonts as well have been sorted from various Silurian and Devonian facies of Siberia and around. They are particularly represented by Silurian endemic species (of *Loganellia*) or genera (*Helenolepis*), in Western Siberia and Tuva (Karatajute-Talimaa, 1978, 1990) (Figure 5.2).

Abundant Silurian–Devonian gnathostome remains have been collected from Siberia and neighbouring areas, yet their study has not been as

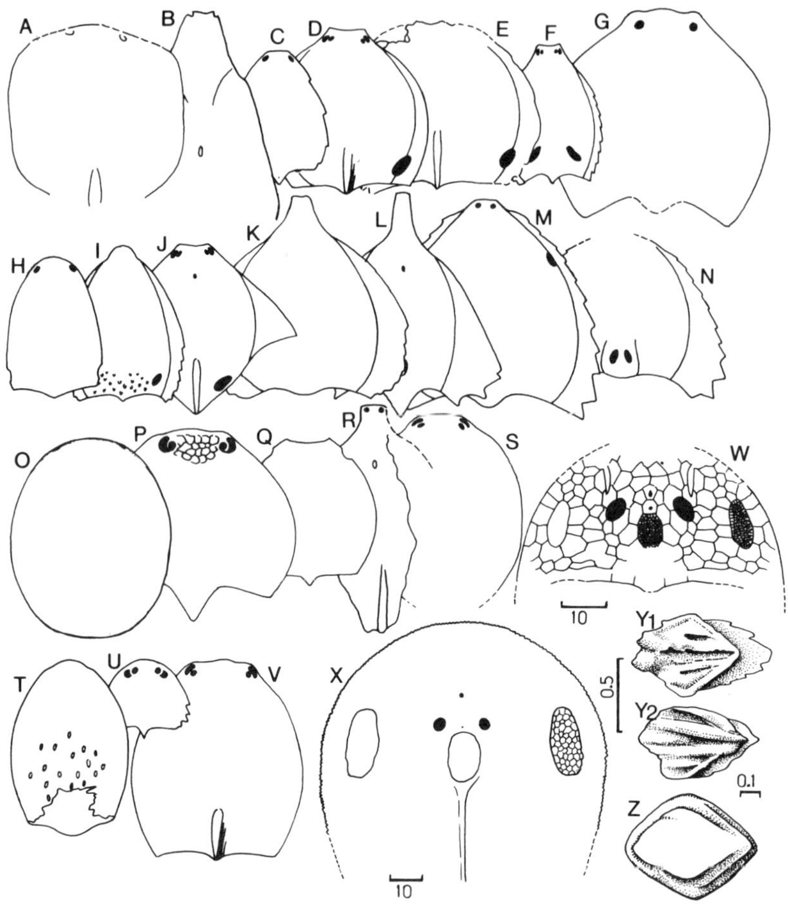

Figure 5.2. Silurian and Devonian jawless vertebrates from Siberia and adjacent terranes.

A–V, outline of the head shield of some amphiaspids in dorsal view (not to scale): A–O, genera recorded from the Lower Devonian of the Siberian Platform (I occurs also in Kotelny island); P–V, genera recorded from Tajmyr only. A: *Amphiaspis*, B: *Pelurgaspis*, C: *Aphataspis*, D–E: *Olbiaspis*, F: *Kureykaspis*, G: *Dotaspis*, H: *Putoranaspis*, I: *Gerronaspis*, J: *Angaraspis*, K: *Lecaniaspis*, L: *Eglonaspis*, M: *Hibernaspis*, N: *Edaphaspis*, O: *Amphoraspis*, P: *Gabreyaspis*, Q: *Siberiaspis*, R: *Empedaspis*, S: *Pelaspis*, T: *Argyriaspis*, U: *Prosarctaspis*, V: *Tareyaspis*, W: *Ilemoraspis kirkinskayae* Obruchev, Middle Devonian, Khakassia, head shield in dorsal view, X: *Tannuaspis levenkoi* Obruchev, Upper Silurian or Lower Devonian, Tuva, head shield in dorsal view, Y: *Helenolepis multicostata* Karatajute-Talimaa, Upper Silurian, Tuva, scale in basal (Y1) and external (Y2) view, Z: *Loganellia tuvaensis* (Karatajute-Talimaa), Silurian, Tuva, scale in external view.
(A–V from Novitskaya, 1986 and Novitskaya and Karatajute-Talimaa, 1989; W, X, from Afanassieva and Janvier, 1985; Y, Z, from Karatajute-Talimaa, 1978). Scale-bars in mm.

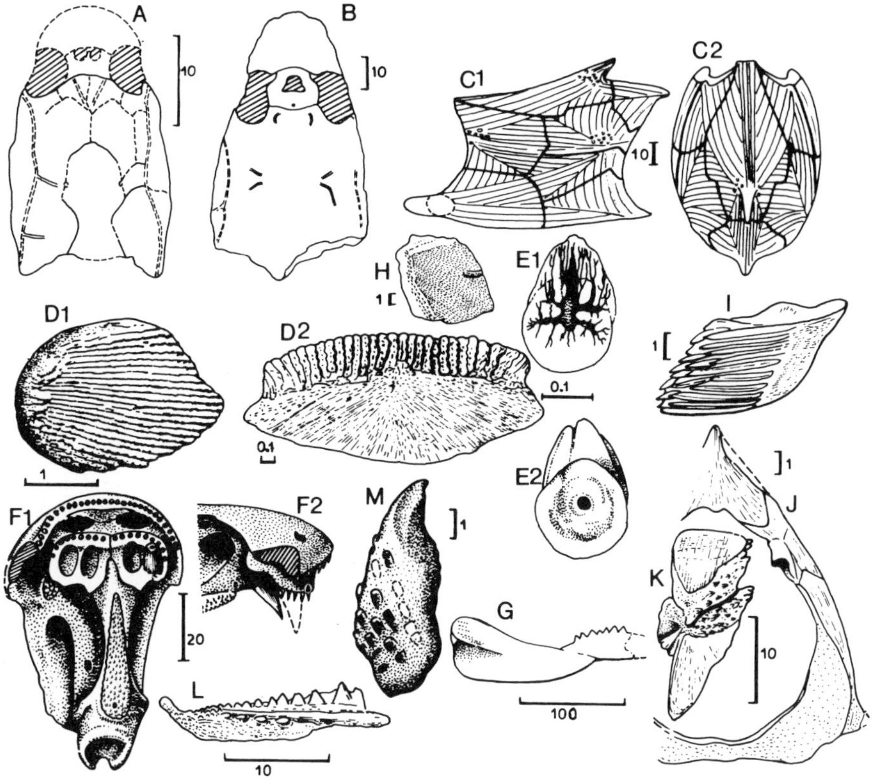

Figure 5.3. Silurian and Devonian gnathostomes from Siberia and adjacent terranes. A: *Kimaspis tienshanica* Mark-Kurik, Lower Devonian of Tienshan, skull in dorsal view; B: *Kolymaspis sibirica* Bystrow, Lower Devonian of the Kolyma Basin, skull in dorsal view; C: *Stegolepis jugata* Malinovskaya, Upper Devonian of Khazakhstan, trunk shield in lateral (C1) and dorsal (C2) view; D: *Mongolepis rozmanae* Karatajute-Talimaa and Novitskaya, Lower Silurian (Llandovery) of Mongolia, isolated scale in external view (D1) and transverse section (D2); E: *Elegestolepis conica* Karatajute-Talimaa, Lower Silurian, Bratsk region, isolated scale in horizontal section (E1) and basal view (E2); F: *Megistolepis klementzi* Obruchev, Upper Devonian of the Minusinsk Basin, orbito-ethmoidian part of skull in ventral (F1) and lateral (F2) view; G: *Angarichthys hyperboreus* Obruchev, Middle Devonian, Bakhta River, Siberia, inferognathal; H–M: specimens from the Lower Devonian of Kotelny Island—H: *Porolepis* sp., Rybnaya, isolated scale; I: *Dialipina markae* Schultze, Rybnaya, isolated scale; J: Ptyctodontida, Sokolova, trunkshield in anterior view; K: Arctolepidae?, Sokolova, anterior dorso-lateral plate; L: Ischnacanthida, Rybnaya, dentigerous bone; M: undetermined (possibly lungfish or bradyodont) tooth plate, Sokolova. Scale-bars in mm.
(A from Mark-Kurik, 1973, and Denison, 1979; B from Denison, 1979; C from Malinovskaya, 1973; D from Karatajute-Talimaa *et al.*, 1990; E from Karatajute-Talimaa, 1990; F from Vorobjeva, 1977; G from Heintz, 1933; H–M from Mark-Kurik, 1974, 1975).

extensive as for agnathans. Contrary to the latter, endemic—or probably endemic—taxa are comparatively rare, and there is no clear example of a highly diversified group restricted to an area, as in the case of amphiaspids. Isolated scales from the Silurian of Mongolia (*Mongolepis*, Karatajute-Talimaa *et al.*, 1990), Central Siberia and Tuva (*Elegestolepis*) are referred to the earliest known chondrichthyans, and are hitherto unknown elsewhere (Figure 5.3). However, this may be due to the relative scarcity of Silurian vertebrate localities all over the world.

The early Devonian gnathostomes, in particular those associated with amphiaspids in the Kotelny Island (Mark-Kurik, 1974) and in Tajmyr (Novitskaya and Karatajute-Talimaa, 1989), are not fundamentally different from those of Euramerica and consist mostly of acanthodians, placoderms ('palaeacanthaspids', ptyctodonts, arctolepids, ?'coccosteomorphs'), actinopterygians (*Dialipina* sp.) and sarcopterygians (*Porolepis kureikensis, P. taimyrica*). Vorobyeva (1963) mentions *Porolepis* scales 'of Siberian type', that is with a strip of dentine tubercles anterior to the cosmine-covered surface, a feature which is often met with in the scales of the Sino-Vietnamese genus *Youngolepis*. The 'palaeacanthaspids' *Kimaspis* from southern Tien Shan and *Kolymaspsis* from the Kolyma, are difficult to interpret palaeobiogeographically, since they belong to a probably paraphyletic taxon (Figure 5.3). Moreover, even if *Kimaspis* seems quite similar to the Czechoslovachian genus *Radotina*, these forms are too rare to permit acceptable biogeographic conclusions.

In Middle and Late Devonian times, the gnathostome fauna displays an even lower degree of endemism. Only Kazakhstan seems to have yielded an endemic antiarchan genus, *Stegolepis* (Figure 5.3), most probably an asterolepiform closely related to *Pterichthyodes* (Young, 1988). The various *Bothriolepis* species described from the Minusinsk basin (Obruchev, 1955) and Kazakhstan (Malinovskaya, 1989) are based on isolated plates and their relationships to other *Bothriolepis* species in the world are not yet elucidated, yet this would be a prerequisite to a biogeographical analysis of this genus. Three osteolepiform genera, *Shirolepis*, *Megistolepis* and *Thaumatolepis* (Figure 5.3), are known only from the Minusinsk basin and Tuva and may possibly be endemic to this region (Obruchev, 1955; Vorobyeva, 1977). Most of the Late Devonian placoderm occurrences from marine facies are widespread taxa, and only well preserved material allowing comparisons at the specific level may yield data of biogeographic interest. Holonematid and coccosteid dermal plates are recorded from the Minusinsk basin, Mongolia, Kotelny, but mean nothing at this level of determination. Isolated scales of acanthodians occur in various localities of the Soviet Arctic and Central Asia but, there again, any conclusion of biogeographic significance would be spurious when based on such parataxa.

GEOGRAPHICAL AND STRATIGRAPHICAL DISTRIBUTION

Since only a few fossiliferous localities are known from this part of the world, and several of them have been recently published, it seems that a (hopefully) exhaustive review may be useful. Some localities are known from a single species or genus of vertebrate which thus looks endemic. Others, on the contrary, have been more completely studied and yielded rather abundant ichthyofaunas. The localities are plotted on Figure 5.1 with the same numbering as follows.

1 – Uzbekistan, South Tien Shan, North Nuratau Ridge, 'Gedinnian'
Kimaspis tienshanica Mark-Kurik (1973) is a small 'palaeacanthaspid' placoderm, which seems phylogenetically related to *Radotina* from Czechoslovachia (Young, 1980).

1' – Uzbekistan?, near Tashkent, western Tien Shan, Frasnian
Bothriolepis turanica (Obruchev, 1955; Denison, 1978; Young, 1990).

2 – Tadjikistan, South Tien Shan, Isfara river, Upper Silurian
Märss (1989, p. 284) mentions thelodonts.

3 – Kazakhstan, SW of lake Tengiz, Givetian to Famennian
Antiarchs (*Bothriolepis*, *Turanolepis*, *Asterolepis*, *Tenizolepis*, *Stegolepis*) and sarcopterygians (*Dipterus*, *Bogdanovia*, 'crossopterygii') are known mainly from the Givetian and Frasnian; antiarchs and arthrodires from the Famennian (Obruchev, 1964; Senkevich, 1967; Malinovskaya and Jurina, 1975; Denison, 1978; Young, 1990). This fauna is mainly composed of endemic genera. It thus seems that the 'pteraspids' mentioned by Bandaletov and Bublitchenko (1977, p. 222; cited by Blieck, 1984, p. 125 and Figure 74) do in fact correspond to one of the above-mentioned genera, and are probably not pteraspids.

4 – Russia (ex RSFSR) in CIS (ex USSR), Western Siberia, Vas'uganje region, Upper Silurian
The thelodont *'Logania' incompta* Karatajute-Talimaa (1990) comes from Pridoli? levels of the drill core Maizasskaya-I, north of River Tara, in the Nyurol basin (Zapivalov and Trofimuk, 1988).

5 – Russia, Central Tajmyr, Tareya river area, Lochkovian to Emsian
A rather complete fauna is known in the Lower Devonian, south of the Tajmyr suture (Figure 5.1). It is particularly well-known for its amphiaspid heterostracans from the Lochkovian Bely Kamen and Uryum beds (Figure 5.4A). The occurrence of a small *Rhinopteraspis* rostrum (Obruchev, 1964, Heterostraci, pl. II:5; *R. crouchi* in Blieck, 1980, p. 42) reinforces the correlation to the Lochkovian (Blieck, 1984). Acanthodian assemblages have been collected from the whole sequence and give precise correlations to the Lower Devonian of the Baltic area and the Russian Platform (Valiukevicius, 1988). The Emsian placoderm assemblage is correlated to records from, at least, Spitsbergen and Severnaya Zemlya on the Barentsia-Kara-Tajmyr block (Mark-Kurik, 1991).

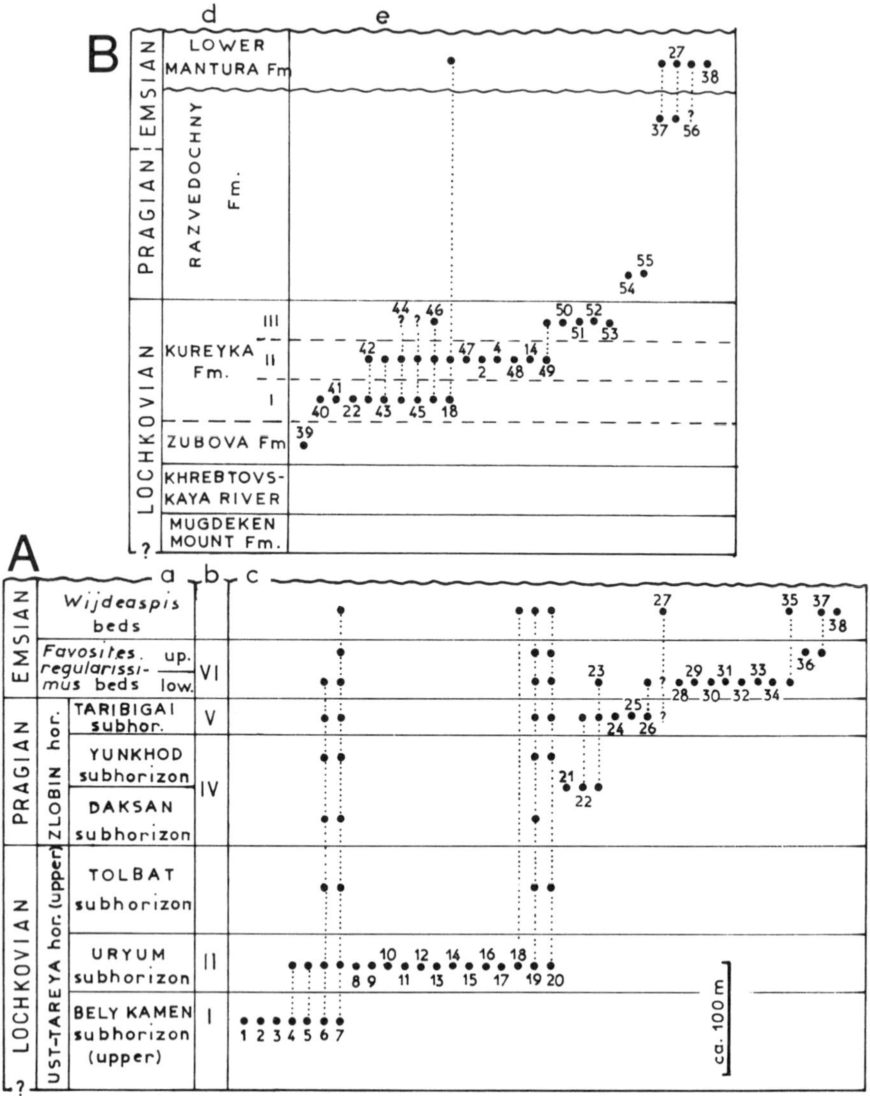

Figure 5.4. Stratigraphical distribution of the vertebrate genera in the Lower Devonian of (A) Central Tajmyr (loc. 5 on Figure 5.1), and (B) north-western Siberian Platform (Kureyka basin, Norilsk area, and Kotuy river; loc. 9).

6 – Russia, Central Siberia, Mojero river, Middle Ordovician
Ørvig (1989, p. 248) considers that some of Moskalenko's (1970, pl. 2: 1–3; pl. 3: 4–7; pl. 4: 1–2; pl. 5: 3–7) conodonts are 'genuine scales of early vertebrates, possibly including thelodontids' (perhaps cf. *Lanarkia*).

7 – Russia, Central Siberia, Bratsk region, Upper Ordovician – Lower Silurian
Karatajute-Talimaa (in Turner, 1985) mentions thelodonts (loganiids, associated to astraspids: Karatajute-Talimaa and Mark-Kurik, oral. comm., Tallinn, 1989) in the Upper Ordovician. She also published the chondrichthyan *Elegestolepis* from the middle? Llandovery of two bore-holes (*E. conica* Karatajute-Talimaa in Novitskaya and Karatajute-Talimaa, 1986).

8 – Russia, Central Siberia, Nizhnaya Chunku river, NE of Bajkit, Llandovery
The small tesserae of *Tesakoviaspis* Karatajute-Talimaa (1978), from the Lower Llandovery, are reinterpreted as astraspids (Karatajute-Talimaa, pers. comm., Tallinn, 1989; Talimaa, 1990; see Moskalenko, 1968; Obruchev, 1973); endemic species of *'Logania'* (*L. sibirica*, *L. moskalenkoae*, *L.* cf. *cruciformis*: Karatajute-Talimaa, 1978) have been collected in the Upper Llandovery.

Figure 5.4 – cont.

a and d - chrono- and litho-stratigraphical schemes after Cherkesova (1988); b - acanthodian assemblages of Valiukevicius (1988); c and e - distribution after Bystrow (1959), Obruchev (1964), Novitskaya (1971, 1976), Novitskaya and German (1983), Valiukevicius (1988), Sobolev *et al.* (1988), Novitskaya and Karatajute-Talimaa (1989), Mark-Kurik (1991); see also Markov *et al.* (1967), Cherkesova (1973).
Each black spot means that the corresponding genus has been collected in the (sub)horizon, but does not give its precise extension within the (sub)horizon; for more details see Sobolev *et al.* (1988).
Abbreviations. Fm. - formation; hor. - horizon.
1 - *Prosarctaspis*, 2 - *Gunaspis*, 3 - *Onchus*, 4 - *Argyriaspis*, 5 - *Tareyaspis*, 6 - *Gomphonchus*, 7 - *Cheiracanthoides*, 8 - *Gabreyaspis*, 9 - *Litotaspis*, 10 - *Pelaspis*, 11 - *Empedaspis*, 12 - *Siberiaspis?*, 13 - *Amphoraspis*, 14 - *Rhinopteraspis*, 15 - *Canadalepis*, 16 - *Machaeracanthus?*, 17 - *Poracanthodes*, 18 - *Porolepis*, 19 - *Nostolepis*, 20 - *Acanthodes?*, 21 - arctolepidids, 22 - radotinids, 23 - *Onychodus*, 24 - palaeoniscoid, 25 - dipnoans, 26 - elasmobranch, 27 - *Luetkeichthys*, 28 - heterostracans?, 29 - *Ohioaspis?*, 30 - *Glyptolepis?*, 31 - Osteolepididae, 32 - *Ganorhynchus*, 33 - cf. *Moythomasia*, 34 - *Atopacanthus?*, 35 - *Watsonacanthus*, 36 - thelodont, 37 - *Wijdeaspis*, 38 - *Dipterus*, 39 - *Steinaspis*, 40 - *Kureykaspis*, 41 - *Dotaspis*, 42 - *Aphataspis*, 43 - *Olbiaspis*, 44 - *Putoranaspis*, 45 - *Gerronaspis*, 46 - *Angaraspis*, 47 - *Siberiaspis*, 48 - *'Drepanaspis'*, 49 - *Lecaniaspis*, 50 - *Eglonaspis*, 51 - *Hibernaspis*, 52 - *Edaphaspis*, 53 - *Norilaspis*, 54 - *Pelurgaspis*, 55 - *Amphiaspis*, 56 - heterostiid.

9 – Russia, NW Siberian Platform, Kureyka basin, Norilsk area and Kotuy river, Lochkovian to Emsian
This region has yielded a Lower Devonian vertebrate succession, very similar to that of Tajmyr (Figure 5.4B). The Lochkovian assemblages are dominated by amphiaspids, with few species in the Lower Pragian. A small *Rhinopteraspis* rostrum (Obruchev, 1964, Heterostraci, pl. II: 4; *R. crouchi* in Blieck, 1980, p. 42) also correlates the equivalent of the Kureyka Formation of the Norilsk area to the Lochkovian. In the upper part of the sequence, the Emsian *Wijdeaspis-Luetkeichthys* placoderm assemblage correlates to other localities of the circum-Arctic area (Mark-Kurik, 1991).

9′ – Russia, River Bakhta, tributary of the Ienissei, Middle Devonian
Obruchev (1964, Placodermi, pl. III: 7) figured the homosteid arthrodire *Angarichthys* (see also Denison, 1978, p. 70) from this site.

10 – Russia, NE slope of Salair, Kuznetsk basin, Emsian to Frasnian
The type-localities of the Emsian Telengit sequence have yielded chondrichthyans, placoderms, acanthodians and osteichthyans; a local zonation is based on acanthodians (Figure 5.5; Yolkin *et al.*, 1988) with few species in common to other localities throughout the CIS (Valiukevicius, 1988). Some placoderms have been collected from the Eifelian (*Deirosteus*, a genus also known from Belgium and the Baltic area) and the Frasnian (*Bothriolepis*, a cosmopolitan genus) (Obruchev, 1964; Denison, 1978).

11 – Russia, Chakass autonomous region, Minusinsk depression, Givetian to Famennian
From the Givetian comes the tannuaspid osteostracan *Ilemoraspis*; from the Frasnian is known a fish assemblage with antiarch (*Bothriolepis*) and sarcopterygians (*Dipterus*, *Onychodus*, *Megistolepis*); and the Famennian has yielded the rhizodontid sarcopterygian *Thaumatolepis* (Vorobjeva, 1977). The genera *Ilemoraspis*, *Megistolepis*, and *Thaumatolepis* are endemic to this part of the world (Obruchev, 1964; Denison, 1978; Afanassieva and Janvier, 1985; Young, 1990). A pentadactyl footprint has been mentioned by Martyanov (1960) from the Middle Devonian Askiz formation near Verkh-Askiz, but the specimen seems to have been lost.

12 – Russia, Tuva (Tuvinskaya A.S.S.R.), Llandovery to Frasnian
It seems that the whole Silurian–Devonian sequence of the Tuva depression has yielded vertebrates. Karatajute-Talimaa (1973, 1978) has published Lower and Upper Silurian endemic species of *'Logania'*, and Pridoli endemic genera of thelodonts (*Helenolepis*) and chondrichthyans (*Elegestolepis*). The tannuaspids are represented here by the Upper Silurian and Lower Devonian, endemic genera *Tannuaspis* and *Tuvaspis* (Obruchev, 1964; Afanassieva and Janvier, 1985; Lower Devonian placoderms are mentioned by Young, 1990). Frasnian antiarchs (*Bothriolepis*) have also been published (Denison, 1978).

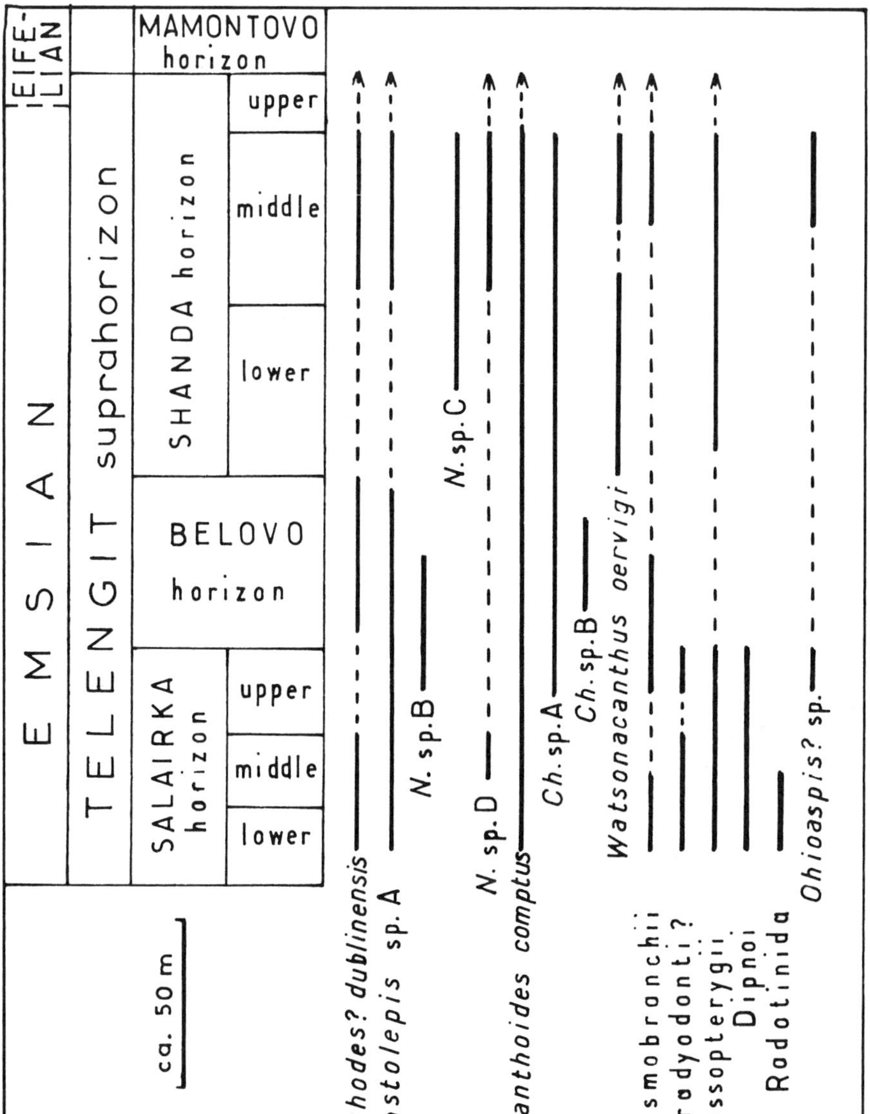

Figure 5.5. Stratigraphical distribution of the vertebrate species in the Emsian (Lower Devonian) of NE Salair, after Yolkin *et al.* (1988; loc. 10 on Figure 5.1; the series is correlated to the standard conodont zonation).

13 – Western Mongolia, N of Khara-Us-Nur lake, Llandovery
Mongolepis Karatajute-Talimaa and Novitskaya (in Karatajute-Talimaa *et al.*, 1990) is the oldest known chondrichthyan, associated with a rather rich microfauna.

14 – Eastern Mongolia, Oushan Obo- Oulan Obo area, Eifelian
Vorobjeva and Janvier (1984) have shortly described fragments of a brachythoracid arthrodire, collected in association with large arthrodire plates, acanthodian fin spines, onychodontid and osteolepid teeth and bones.

15 – Russia, New Siberian Archipelago, Kotelny island, Lochkovian to Frasnian
Several localities have been found along rivers of the western part of the island. Two Early Lower Devonian ('Gedinnian') localities have yielded rather rich assemblages of agnathans and gnathostomes; among them, the amphiaspid *Gerronaspis* is suggestive of the Lower Devonian fauna of the NW Siberian Platform (loc. 9), whereas the other elements are mainly known from the Old Red Continent (Laureuropa), viz., palaeacanthaspid, ptyctodont, arctolepid and 'coccosteomorph' placoderms, acanthodian jaw bones of the *Atopacanthus* type, porolepid and palaeoniscid scales (cf. *Dialipina*), dipnoan-like tooth plates and psammosteid-like heterostracan fragments. An Emsian outcrop gave indeterminate arthrodire remains, although a richer Frasnian assemblage is known (*Bothriolepis*, holoptychiid, dipnoan) (Mark-Kurik, 1974, 1975).

16 – Russia, De Long Strait (Proliv Longa), Emsian
Vertebrates have been mentioned by Cherkesova (1973: 'Heterostraci Pteraspidae' and 'Arctolepidida') and Mark-Kurik (1974).

17 – Russia, Kolyma river, Lochkovian to Emsian
An endemic 'palaeacanthaspid' placoderm, *Kolymaspis*, comes from the Lochkovian (Obruchev, 1964; Denison, 1978); it has been considered the possible sister-group of either *Brindabellaspis* from the Lower Devonian of New South Wales, Australia, or of *Romundina* + *Kimaspis* (loc. 1) (Young, 1980). Acanthodian scale-assemblages have been collected throughout the whole Lower Devonian, the Late Emsian one being completely unknown from other regions (Valiukevicius, 1988).

PALAEOBIOGEOGRAPHICAL REMARKS AND CONCLUSIONS

As already said, this wide part of the Old World is still rather poor in Silurian–Devonian vertebrate localities. So comparisons between the different continental blocks east of the Urals and with the other major palaeobiogeographical provinces (Old Red Continent, W-Gondwana, E-Gondwana, South China, North China, etc.) are difficult to make.

Several endemic taxa are known in Siberia and neighbouring terranes.

The amphiaspid heterostracans have been collected in the Lochkovian–Lower Pragian of Tajmyr and the NW Siberian Platform (Figure 5.1, loc. 5 and 9), but *Gerronaspis* is known from the Lochkovian of Kotelny island (loc. 15). The latter, however, has mainly North Atlantic (Old Red Continent) biogeographic relationships with, e.g., psammosteid-like heterostracan fragments. The tannuaspid osteostracans are endemic to the Upper Silurian to Givetian of Tuva and Chakass (loc. 11–12). Endemic Silurian species or genera of thelodonts occur in Central Siberia and Tuva (loc. 8 and 12). Possibly endemic Silurian chondrichthyans as well are known in S. Central Siberia, Tuva and W-Mongolia (loc. 7, 12, 13). Lower Devonian endemic 'palaeacanthaspid' placoderms come from Uzbekistan (loc. 1) and Kolyma (loc. 17), thus on both opposite sides of the considered area, but this might have little or no significance. Some Middle and Upper Devonian placoderms and sarcopterygians are also geographically restricted (loc. 3, 9′, 11).

On the contrary, the Emsian *Wijdeaspis-Luetkeichthys* placoderm assemblage (with dipnoans, porolepids) is widely distributed in the circum-Arctic area (Tajmyr and NW Siberia, loc. 5 and 9; *Wijdeaspis* being also known from Severnaya Zemlya and Spitsbergen). A Givetian–Upper Devonian antiarch fauna is distributed from Central Asia (loc. 1′, 3, 11, 12) to the Arctic (loc. 15), with endemics in Kazakhstan (*Stegolepis*).

How do we interpret these data? Some taxa, such as heterostracans (*Rhinopteraspis*, psammosteid), placoderms (*Bothriolepis*, *Asterolepis*), acanthodians (*Nostolepis*, *Gomphonchus*, *Cheiracanthoides*) are widespread and provide no biogeographic information. In this context, endemic forms (amphiaspids, tannuaspids, thelodonts) may be due to local palaeoecological conditions, or support the hypothesis that their blocks were isolated in some way. Some widespread taxa may be good nektonic, migrating species (discussion in Young, 1981, 1990; Blieck and Janvier, 1991, and in press). It however remains that the Silurian and Lower Devonian ichthyofaunas of Siberia and neighbouring terranes have enough endemic elements to at least distinguish separate areas: Kazakhstan (loc. 3), NW Siberia (loc. 5 and 9), W Central and SW of Siberia (loc. 8, 11–13). This is not in disagreement with separate continental blocks (ref. on Figure 5.1). Kotelny Island (Chukotka block) should be related to the Old Red Continent as is Severnaya Zemlya (Kara-Tajmyr block). The Chukotka block is indeed connected to 'Laurussia' by Ziegler (1988, Figures 3–6; see Janvier and Blieck, Chapter 4).

REFERENCES

Afanassieva, O. and Janvier, P., 1985, *Tannuaspis*, *Tuvaspis* and *Ilemoraspis*, endemic osteostracan genera from the Silurian and Devonian of Tuva and Khakassia (USSR), *Geobios*, **18**: 493–506.

Bandaletov, S.M. and Bublitchenko, N.L., 1977, Kazakhstan. In A. Martinsson, *The Silurian–Devonian Boundary*, IUGS Series A, Stuttgart, **5**: 219–26.

Blieck, A., 1980, Le genre *Rhinopteraspis* Jaekel (Vertébrés, Hétérostracés) du Dévonien inférieur: systématique, morphologie, répartition, *Bulletin du Muséum National d'Histoire Naturelle, 4e série*, 2, section C, **1**: 25–47.

Blieck, A., 1984, Les Hétérostracés Pteraspidiformes, Agnathes du Silurien-Dévonien du Continent nord-Atlantique et des blocs avoisinants: révision systématique, phylogénie, biostratigraphie, biogéographie, *Cahiers de Paléontologie, CNRS édit.*, Paris, 200pp.

Blieck, A. and Janvier, P., in press, Silurian–Devonian vertebrate-dominated communities, with particular reference to agnathans. In A.J. Boucot and J. Lawson (eds), *IGCP Project Ecostratigraphy Final Report*, Cambridge University Press.

Blieck, A. and Janvier, P., 1991, Silurian Vertebrates. In M.G. Bassett, P.D. Lane and D. Edwards (eds), The Murchison Symposium (International Symposium on the Silurian System, Keele, 1989), *Special Papers in Palaeontology*, **44**: 345–89.

Broad, D.S., 1973, Amphiaspidiformes (Heterostraci) from the Silurian of the Canadian Arctic Archipelago, *Geological Survey of Canada Bulletin*, **222**: 35–51.

Bystrow, A.P., 1959, The microstructure of skeleton elements in some vertebrates from Lower Devonian deposits of the USSR, *Acta Zoologica*, **40**: 59–83.

Cherkesova, S.V., 1973, [Lower Devonian deposits of the Soviet Arctic]. In B.V. Nalivkin (ed.), *Third International Symposium of the Silurian–Devonian Boundary and the Stratigraphy of the Lower and Middle Devonian* (Leningrad, 1968), **II**: 276–9.

Cherkesova, S.V., 1988, Lower and Middle Devonian marine deposits of the Soviet Arctic and the correlation with Arctic Canada. In N.J. McMillan, A.F. Embry and D.J. Glass (eds), *Devonian of the World* (Proceedings IInd International Symposium on the Devonian System, Calgary, 1987), *Canadian Society of Petroleum Geologists, Memoir*, **14, III**: 669–79.

Churkin, Jr., M., Soleimani, G., Carter, C. and Robinson, R., 1981, Geology of the Soviet Arctic: Kola peninsula to Lena river. In A.E.M. Nairn, M. Churkin, Jr. and F.G. Stehli (eds), *The Ocean basins and margins*, 5. Plenum Press, New York and London: 331–75.

Denison, R., 1978, *Placodermi*. In H.-P. Schultze (ed.), *Handbook of Paleoichthyology*, volume 2. Gustav Fischer Verlag, Stuttgart, New York.

Elliott, D.K. and Dineley, D.L., 1985, A new Heterostracan from the Upper Silurian of Northwest Territories, Canada, *Journal of Vertebrate Paleontology*, **5**: 103–10.

Heintz, A., 1933, Revision of the Estonian Arthrodira. 1. Family Homosteidae Jaekel. *University of Tartu, Geological Institute Publication*, 35.

Karatajute-Talimaa, V.N., 1973, *Elegestolepis grossi* gen. et sp. nov., ein neuer Typ der Placoidschuppe aus dem oberen Silur der Tuwa, *Palaeontographica A*, **143**: 35–50.

Karatajute-Talimaa, V.N., 1978, [Silurian and Devonian Thelodonts of the USSR and Spitsbergen]. 'Mosklas', Vilnius, (in Russian).

Karatajute-Talimaa, V.N., 1990, [Vertebrates]. In N.P. Kul'Kov and V.N. Dubatolov (eds), Silur i Devon jugovostoka Zapadno-Sibirskoj plity, Skvajiny Majzasskaya-1 i Maloichskaya-7. Sibiriskoe Otdelevie Institut Geologii i Geofiziki Akademia Nauk SSSR, Moskva, 'Nauka': 59–62, (in Russian).

Karatajute-Talimaa, V.N., Novitskaya, L.I., Rozman, Ch. S. and Sodov, J., 1990, [*Mongolepis*—a new elasmobranch genus from the Lower Silurian of Mongolia], *Paleontologischeskii Zhurnal*, **1**: 76–96 (in Russian).

Malinovskaja, S.P., 1973, [*Stegolepis* (Antiarchi, Placodermi), new genus from the Middle Devonian of Central Kazakhstan]. *Paleontologischeskii Zhurnal*, 1973(2): 71–82, (in Russian).

Malinovskaja, S.P., 1989, Devonian Bothriolepidae (Placodermi) of Central Kazakhstan, *Ispitatelei otd. Geologie 1988*, **63**: 56–70.

Malinovskaja, S.P. and Jurina, A.L., 1975, Obosnovanie vozrasta kontinental'nych

otlozenij devona central'nogo kazakhstana po flore i rybam, *Vestnik Moskovskogo Universiteta, Serie 4, Geologija*, **6**: 32–7, (in Russian).
Mark-Kurik, E., 1973, *Kimaspis*, a new Palaeacanthaspid from the Early Devonian of Central Asia, *Eesti NSV Teaduste Akadeemia Toimetised*, **22**: 322–30.
Mark-Kurik, E., 1974, Discovery of new Devonian fish localities in the Soviet Arctic, *Eesti NSV Teaduste Akadeemia Toimetised*, **23**: 332–5.
Mark-Kurik, E., 1975, A tooth-plate from the Lower Devonian of Kotelny Island, *Eesti NSV Teaduste Akadeemia Toimetised*, **24**: 307–9.
Mark-Kurik, E., 1991, Contribution to the correlation of the Emsian (Lower Devonian) on the basis of placoderm fishes. *Newsletter in Stratigraphy*, **25**: 11–23.
Markov, F.G., Cherkesova, S.V. and Gramberg, I.S., 1967, Devonian of the Soviet Arctic. In D.H. Oswald (ed.), *International Symposium on the Devonian System (Calgary, 1967)*, Calgary, **I**: 503–15.
Märss, T., 1989, Vertebrates. In C.H. Holland and M.G. Bassett (eds), A global standard for the Silurian System, *National Museum of Wales, Geological Series*, Cardiff, **9**: 284–9.
Martyanov, N.Ye., 1960, [A Pentadactyl Footprint], *Priroda, Akademia Nauk SSSR, Moskva*, **49**: 115, (in Russian).
Moskalenko, T.A., 1968, [Llandovery Agnathan remains in Siberia]. In V.N. Yakovlev, D.V. Obruchev *et al.*, Ocherkii po filogenii i sistematikye iskopayemykh ryb i byeschyelyustnikh. *'Nauka', Moskva*: 29–32, (in Russian).
Moskalenko, T.A., 1970, [Conodonts of the Krivaya Luka Stage (Middle Ordovician) from the Siberian Platform], Trudy Institut Geologii i Geofiziki, Sibiriskoe Otdelevie, *Akademia Nauk SSSR*, **61**: 1–116, (in Russian).
Novitskaya, L.I., 1971, Les Amphiaspides (Heterostraci) du Dévonien de la Sibérie. *Cahiers de Paléontologie, CNRS édit*, Paris, 130pp.
Novitskaya, L.I., 1976, [Subdivision of the Kureyka Suite based on amphiaspid (Agnathan) assemblages]. *Akademia Nauk SSSR Izvestia, series geologiski* **1**: 47–55, (in Russian).
Novitskaya, L.I., 1986, Drevneishie beschelyustnye SSSR. Geterostraki: tsiataspidy, amphiaspidy, pteraspidy. [The earliest Agnatha of the USSR. Heterostraci: cyathaspids, amphiaspids, pteraspids]. *Akademia Nauk SSSR, Trudy Paleontologiceskogo Instituta*, **219**, 'Nauka', Moskva, 160pp. (in Russian).
Novitskaya, L.I. and German, A.B., 1983, A new amphiaspid (Heterostraci) from the Lower Devonian deposits of the Siberian Platform, *Paleontological Journal*, **3**: 83–9.
Novitskaya, L.I. and Karatajute-Talimaa, V.N., 1986, [Remarks on the cladistic analysis of the relations of myopterygians, and on the problem of origin of gnathostomes]. In Vorobjeva, E.I. and Lebedkina, N.S., eds [Morphology and evolution of animals], '*Nauka*', Moscva, pp. 102–105, (in Russian).
Novitskaya, L.I. and Karatajute-Talimaa, V.N., 1989, [A new amphiaspid (Heterostraci) from the Lower Devonian deposits of Taimyr], *Paleontologischeskii Zhurnal*, **3**: 112–16, (in Russian).
Obruchev, D.V., 1955, Devonian fishes from the Minusinsk Basin, *V.S.E.G.E. I.*: 45–7.
Obruchev, D.V., 1964, Besceljustnye, Ryby [Agnathan, Fishes]. In J.A. Orlov (ed.), *Osnovy paleontologii* [Fundamentals of Paleontology], 'Nauka', Moskva, IA (11), 552pp. (in Russian).
Obruchev, D.V., 1973, [The importance of the vertebrates for the correlation of the Silurian, Lower and Middle Devonian of the USSR]. In B.V. Nalivkin (ed.), *Third International Symposium of the Silurian–Devonian Boundary and the Stratigraphy of the Lower and Middle Devonian* (Leningrad, 1968), **II**: 189–97, (in Russian).
Ørvig, T., 1989, Histologic studies of ostracoderms, placoderms, and fossil

elasmobranchs. 6. Hard tissues of Ordovician vertebrates, *Zoologica Scripta*, **18**: 427–46.

Scotese, C.R. and McKerrow, W.S., 1990, Revised world maps and introduction. In W.S. McKerrow and C.R. Scotese (eds), *Palaeozoic Palaeogeography and Biogeography*. Geological Society, London, Memoir **12**: 1–21.

Senkevich, M.A., 1967, Devonian continental deposits of Central Kazakhstan. In D.H. Oswald (ed.), *International Symposium on the Devonian System* (Calgary, 1967), Calgary, **II**: 1117–27.

Sobolev, N.N., Karatajute-Talimaa, V.N. and Valiukevicius, Y.Y., 1988, Conodonts and vertebrates from the Zlobin horizon and the *Favosites regularissimus* zone (Lower Devonian) of Taimyr, Siberia, USSR. In N.J. McMillan, A.F. Embry and D.J. Glass (eds), *Devonian of the World* (Proceedings IInd International Symposium on the Devonian System, Calgary, 1987), *Canadian Society of Petroleum Geologists, Memoir* **14, III**: 535–9.

Talent, J.A., Gratsianova, R.T., Shishkina, G.R. and Yolkin, E.A., 1987a, Devonian faunas in relation to crustal blocks: Kazakhstan, Mongolia, Northern China. In R. Werner und W. Ziegler (eds), *Neue Daten zur Biostratigraphie, Paläogeographie und Paläontologie des Devons in Eurasien und Australien. Courier Forschungs-Institut Senckenberg*, **92**: 225–33.

Talent, J.A., Gratsianova, R.T. and Yolkin, E.A., 1987b, Prototethys: fact or phantom? Paleobiogeography in relation to the crustal mosaic for the Asia–Australia hemisphere in Devonian–Early Carboniferous times. In K.G. McKenzie (ed.), *Shallow Tethys 2*. Balkema Publishers, Rotterdam: 87–111.

Talent, J.A. and Yolkin, E.A., 1987, Transgression-regression patterns for the Devonian of Australia and Southern West Siberia. In R. Werner und W. Ziegler (eds), *Neue Daten zur Biostratigraphie, Paläogeographie und Paläontologie des Devons in Eurasien und Australien. Courier Forschungs-Institut Senckenberg*, **92**: 235–49.

Talimaa, V.N., 1990, [Early steps of the evolution of Vertebrates and the problem of stratigraphy of deposits of various facies]. In V.N. Dubatolov and A.T. Moskalenko (eds), *Princip razvitija i istorizma v geologii i paleobiologii* [Principles of development and historical method in geology and paleobiology]. 'Nauka', Sibiriskoe Otdelevie, Novosibirsk: 212–20, (in Russian).

Turner, S., 1985, Remarks on the early history of chondrichthyans, thelodonts, and some 'higher elasmobranchs'. In R. Cooper (ed.), Hornibrook Symposium (Christchurch, 1985). *New Zealand Geological Survey, Record*, **9**: 93–5.

Valiukevicius, J., 1988, Correlation of Lower and Middle Devonian deposits of the USSR with acanthodian assemblages. In N.J. McMillan, A.F. Embry and D.J. Glass (eds), *Devonian of the World* (Proceedings IInd International Symposium on the Devonian System, Calgary, 1987), *Canadian Society of Petroleum Geologists, Memoir* **14, III**: 601–7.

Vorobjeva, E.I., 1963, [On the fish genus *Porolepis* in the Devonian of USSR], *Paleontologischeskii Zhurnal*, **2**: 45–8, (in Russian).

Vorobjeva, E.I., 1977, [Morphology and evolution of crossopterygian fishes], *Akademia Nauk, SSSR*, 163, (in Russian).

Vorobjeva, E.I. and Janvier, P., 1984, Observations to the note by E. Vorobjeva: 'First discovery of Palaeozoic fishes on the territory of the Mongolian Republic'. In *Palaeontology and Biostratigraphy of Mongolia*, Moscow, **3**: 134–6, (in Russian). *Geobios*, **17**: 379–380.

Yolkin, E.A., Gratsianova, R.T., Bakharev, N.K., Izokh, N.E., Yazikov, A.Yu., V'yushkova, L.V., Zheltonogova, V.A. and Petrosyan, N.M., 1988, Facies and faunal associations of the Telengitian (Emsian) in its type locality. In N.J. McMillan, A.F. Embry and D.J. Glass (eds), *Devonian of the World* (Proceedings IInd International Symposium on the Devonian System, Calgary, 1987), *Canadian Society of*

Petroleum Geologists, Memoir **14, III**: 193–207.

Young, G.C., 1980, A new Early Devonian Placoderm from New South Wales, Australia, with a discussion of Placoderm phylogeny. *Palaeontographica A*, **167**: 10–76.

Young, G.C., 1981, Biogeography of Devonian vertebrates, *Alcheringa*, **5**: 225–43.

Young, G.C., 1988, Antiarchs (placoderm fishes) from the Devonian Aztec Siltstone, southern Victoria land, Antarctica, *Palaeontographica A*, **202**: 1–125.

Young, G.C., 1990, Devonian vertebrate distribution patterns and cladistic analysis of palaeogeographic hypotheses. In W.S. McKerrow and C.R. Scotese (eds), *Palaeozoic Palaeogeography and Biogeography, Geological Society, London, Memoir* **12**: 243–55.

Zapivalov, N.P. and Trofimuk, A.A., 1988, Distributions of oil and gas in Devonian rocks of West Siberia. In N.J. McMillan, A.F. Embry and D.J. Glass (eds), *Devonian of the World* (Proceedings IInd International Symposium on the Devonian System, Calgary, 1987), *Canadian Society of Petroleum Geologists, Memoir* **14, I**: 553–6.

Ziegler, P.A., 1988, Laurussia—the Old Red Continent. In N.J. McMillan, A.F. Embry and D.J. Glass (eds), *Devonian of the World* (Proceedings IInd International Symposium on the Devonian System, Calgary, 1987), *Canadian Society of Petroleum Geologists, Memoir* **14, I**: 15–48.

Chapter 6

BIOSTRATIGRAPHY OF THE SILURIAN AND DEVONIAN GNATHOSTOMES OF THE EURAMERICA PROVINCE

D.L. Dineley and E.J. Loeffler

INTRODUCTION

Biostratigraphy records the succession of fossils in sedimentary sequences and it provides a means of relative dating of these rocks. It is most successful where the fossils are numerous, well preserved and constitute a record of unbroken evolution. In consequence, marine invertebrate fossils have been given the most attention. Palynomorphs, both in marine and non-marine successions, have also been employed as stratigraphic index fossils with great success. In continental facies the vertebrates provide similar biostratigraphic data. The earliest known vertebrates are marine agnathous Ordovician forms. By the late Silurian, gnathostome fishes had begun to enter non-marine habitats where, during the Devonian period, they underwent a remarkable radiation and evolution. Their bony armour, scales and spines are locally common in marine formations, but are more abundant in the continental facies known as the Old Red Sandstone.

Early vertebrate evolution was profoundly influenced both by physical environment and by the changes taking place in the plant kingdom, especially the colonisation of freshwaters and land surfaces by vascular plants. New ecosystems arose and were geographically very widespread in the Devonian period. The Old Red Sandstone facies first appeared in late Silurian time but during the Devonian it developed on all the continents, especially in active orogenic and post-orogenic basins. Thus vertebrate biostratigraphy has new and strong regional and facies controls in the late Silurian and thereafter, and this causes major problems in correlating the marine with the non-marine facies. Few of the non-marine vertebrates are known also from the marine facies, and vice versa. Progress is, however, being made; miospores are common to both facies types (Richardson and McGregor, 1986).

The emergence of distinct jawed forms probably took place at a time of active radiation. The earliest remains of (acanthodian) gnathostome affinities are of late Ordovician age. Others appear in the Middle Silurian.

Westoll (1979) has outlined the use of Devonian fish in biostratigraphy in terms of major taxonomic groups. Since then new data have appeared, with new taxa being added to faunal lists, especially from the Euramerica, South China and East Gondwana provinces (Young, 1981). In particular, many new species of placoderms are known, and the geographical and stratigraphical ranges of others have been extended. Nevertheless, the difficulties of constructing a continuous succession of vertebrate zones even within one province, based either upon lineages or assemblages, remain much as before (Dineley, 1982, 1984).

The most detailed study of vertebrate stratigraphy in Silurian–Devonian mixed marine/non-marine successions concerns the eastern Baltic (Blieck *et al.*, 1988). A sequence of vertebrate faunas or assemblages can be recognised, each of which has a relatively restricted stratigraphic range. In due course it may be possible to distinguish such a sequence elsewhere in the marginal basins around the Old Red Sandstone continent of Euramerica. Other, internal, basins reveal more or less similar successions, but with conspicuous gaps between the fish-yielding strata and without the marine intervals. Correlation between these basins is more difficult but is being achieved gradually nonetheless.

This paper offers a broad survey of gnathostome distribution throughout the major outcrops of the Euramerican Old Red Sandstone, and draws some broad conclusions.

THE EURAMERICA PROVINCE

In the debate about global palaeogeographies the configuration of the major continental masses is continually subject to revision and reinterpretation as agreement between palaeobiological, palaeoclimatological and geophysical/tectonic evidence is sought. The recent Oxford conference on Palaeozoic palaeogeography (McKerrow and Scotese, 1990) and the second international Calgary conference on the Devonian system (McMillan, Embry and Glass, 1988) have provided several such syntheses for the Silurian and Devonian.

Among the offerings, Young (1981, 1990) has contributed a palaeogeography of Devonian vertebrate provinces, distinguishing five, viz. Euramerica (or cephalaspid), Siberian (or amphiaspid), Tuva, South China (or yunnanolepid) and East Gondwana (or wuttagoonaspid). Each is centred upon a continental mass, the characteristic vertebrates being in the continental (or internal) and peripheral (or external) basins, i.e. in Old Red Sandstone and/or intermediate and marine facies. Each province was distinct and discrete in early Devonian time with strong endemic elements in its faunas. Later these elements were reduced and the faunas took on

Figure 6.1. The Late Silurian-Devonian continent, known as Euramerica (= Laurussia, North Atlantis, the Old Red Sandstone Continent) was formed by the collision of the Baltica craton and the Laurentia craton. The Caledonian orogen marks the suture between the two; the peripheral mobile belts are named. Continental and post-orogenic deposition occurred in all of these regions. Tropical, epeiric seas flooded the cratons themselves (after Ziegler, 1988).

more cosmopolitan aspects. Young illustrated these features using maps by Scotese (1986); the Scotese and McKerrow map (1990) shows only minor differences. The model of the Lochkovian world shows Euramerica centred upon latitude 10°S, with the Siberian/Tuva provinces close by. Within the Euramerican province are the major cratonic blocks of Laurentia, Baltica, Barentia and Kara with accreted minor terranes. Novaya Zemlya, Podolia and the Rhineland–Artois are also included.

These nuclear blocks constitute the essential Old Red Sandstone continent or Euramerica (also known as Laurussia; Ziegler, 1988) crossed by the Caledonian/Arctic megasuture. Around it a complete circle of active orogenic belts persisted from the Pridoli to late Carboniferous. One of the more active orogenic and accretionary episodes, the Acadian, occurred in the Emsian–Givetian interval, a time of sudden change in the vertebrate faunas.

Most of the Old Red Sandstone basins underwent deformation before the end of the Devonian. They had sharp tectonic and physiographic boundaries; palaeoenvironments were sharply delineated, somewhat migratory and relatively short-lived. In the external basins continental facies interfinger with the marine as a result of extensive transgressional–regressional movements of sea level over a period of about 50 million years (Allen, Dineley and Friend, 1968).

In outcrop many of these basins are isolated but it is convenient to group their principal vertebrate localities as shown on Figure 6.3. Young (1990) referred to some 34 vertebrate-yielding locales in his cladistic analysis of the evolution of the province. These and others are referred to below.

SILURIAN AND DEVONIAN GNATHOSTOMES

Little is known of pre-Devonian jawed fishes, although some occurred as early as the late Ordovician and others existed in the Silurian in several parts of the world. One may assume that the habitats in which Agnatha flourished gave gnathostome fishes similar opportunities for evolutionary radiation. Nevertheless, jawed forms are fewer than the Agnatha in pre-Emsian strata. At the beginning of the Eifelian a sudden radiation took place, emphatically reversing the case. Widespread changes also occurred at the beginning of the Frasnian. A mass extinction event took place at the end of the Famennian, with several groups of vertebrates disappearing with many of the marine invertebrates (see Figure 6.2).

It will be useful at this point to survey briefly the major gnathostome groups extant during the Silurian and Devonian. See Long (Chapter 1) for more detailed information on the morphology of each group. The literature is voluminous but there are useful summaries and overviews of the different groups and their phylogenies. Young (1990) lists several of these and notes the great increase in knowledge over the last two decades. In evidence during the Devonian period were four major groups of gnathostome fishes—acanthodians, placoderms, osteichthyes and chondrichthyes. By the end of the period there were also at least five genera of tetrapods.

The Acanthodii

Two views of acanthodian affinities have been put forward and the classification within the group is still somewhat uncertain (Denison, 1979). Acanthodii are present in both marine and non-marine facies and in the former they range from the late Ordovician to the Permian. Over 60 genera are recorded. The Climatiformes are abundant and varied in the Lochkovian and Pragian, the Ischnacanthiformes are abundant throughout the

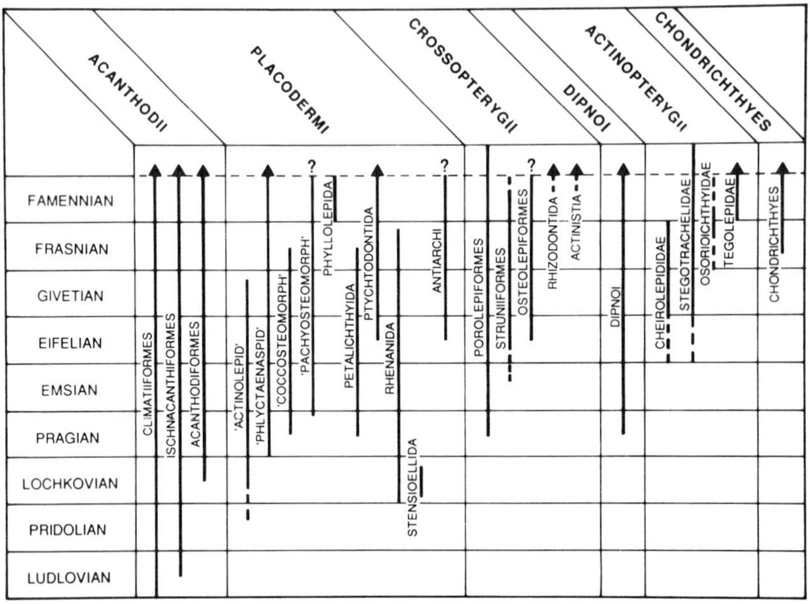

Figure 6.2. The stratigraphical ranges of Devonian gnathostomes in Euramerica. Based on information from several sources, principally Andrews, Gardiner, Miles and Patterson, *The Fossil Record*, Geological Society of London, 1967, 637–83.

early Devonian but persist in the late, and the Acanthodiformes range is early Devonian to Permian. The family Diplacanthidae are characteristically Devonian in ranging from the Lochkovian to the Frasnian. Although Acanthodii have rarely been used in biostratigraphic schemes, Valiukevicius (1988) has been able to recognise 13 successive acanthodian assemblages in the Lower and early Upper Devonian of the USSR. Undoubtedly a major obstacle to the more extensive use of acanthodian remains in biostratigraphy is the small size of the animal, coupled with the minute size and uniformity of the scales. Articulated remains are rare and most records are based upon isolated spines or scales.

The Actinopterygii

Problematic scales from the Silurian of the Baltic, China and elsewhere have been referred to this group (Gross, 1968; Mark-Kurik, 1969). Devonian actinopterygii—the palaeoniscoids—range from the Emsian onwards, but they are not common. The best known genus, *Cheirolepis*, occurs in the upper Eifelian of Scotland and the lower Frasnian of eastern Canada. *Orvikulina* is Eifelian–Givetian in range; *Stegotrachelus* is Givetian;

Moythomasia is Frasnian–Famennian and *Tegeolepis* appears to be the last Devonian actinopterygian of the Famennian.

The Crossopterygii

This large group includes the 'orders' Holoptychiida (Porolepiformes), Actinistia (Coelacanthiformes), Osteolepida (Osteolepiformes), Rhizodontida and Onychodontida (Struniiformes). They are an important element in the early fish faunas, especially in the mid and late Devonian. The earliest Holoptychiida occur in Lochkovian strata and the latest in the uppermost Famennian of East Greenland. The Osteolepida are highly conspicuous with over 30 species in the middle and late Devonian; the Actinistia, Rhizodontiida and Onychodontida are mostly late Devonian although rare species occur earlier elsewhere.

The Dipnoi

This specialised group of osteichthyes produced up to 25 genera (49 species) (Marshall, 1987) during the Devonian period in a unique radiation. The early forms occur in the Lower Devonian within several vertebrate provinces and many are associated with marine facies. The majority of genera are from Middle and Upper Devonian rocks.

The Placodermi

Including over 250 genera, this is the largest and most important category of Silurian–Devonian gnathostomes in terms of both taxonomic diversity and abundance. The Antiarchi and the Arthrodira are well known in all provinces and throughout the Devonian. Other 'orders' of placoderms are the phyllolepids (freshwater) and the more commonly occurring marine forms the ptyctodonts, the petalichthyids and the acanthothoracids.

The antiarchs are very numerous, their origins being perhaps indicated by Silurian forms in South China (Pan, 1981). The family Bothriolepididae is unique in its worldwide distribution during the late Devonian and in possessing some 80 species, of which over 70 are of *Bothriolepis*. The major radiation of the Antiarchi seems to have occurred at the beginning of the Middle Devonian. The last antiarchs are known from Devonian–Carboniferous beds in East Greenland. The majority are recovered from non-marine strata. Long (1983) and Blieck *et al.* (1988) offer antiarch correlation of mid and late Devonian successions in different parts of the world which illustrates the 'cosmopolitan' nature of the group. Young (1984) illustrated his discussion of the use of fossils in biostratigraphy by reference to the 24 antiarch genera.

The arthrodires comprise more than 70 per cent of all known placoderms and occurred in all provinces during the Devonian period. At least 70 genera are recorded in the Euramerican Old Red Sandstone and adjacent outcrops of marine Devonian. Silurian forms are rare and very small. The late Devonian dinichthyids were gigantic, reaching estimated lengths of 6 m. Classification and evolution within the group have been much debated (Denison, 1984). Westoll (1979) distinguished four evolutionary 'grades' of arthrodires in his stratigraphy, viz. actinolepids ranging from Lochkovian to Givetian, phlyctaenaspids ranging from Lochkovian to ?basal Carboniferous, coccosteomorphs ranging from Pragian to Frasnian and the pachyosteomorphs ranging from Lochkovian to Famennian. The last group underwent a great radiation in the late Devonian (Moy-Thomas and Miles, 1971).

The Chondrichthyes

Pre-Devonian Chondrichthyes are practically unknown (Zangerl, 1981). Thorsteinsson (1973), however, has described scales and spines from Wenlockian strata in arctic Canada. In the seas Devonian 'sharks' evolved rapidly, fossil remains being generally limited to teeth, dermal denticles and fin spines (Wells, 1944). Long (1990) notes the ranges of some 13 Devonian taxa occurring in the Euramerican province. Remarkable preservation of complete cadaver outlines of *Cladoselache* occurs in the late Devonian Cleveland Shales (Newberry, 1889; Dean, 1990). As yet, chondrichthyan remains are of uncertain stratigraphical value in pre-Carboniferous rocks.

GNATHOSTOME BIOSTRATIGRAPHY

Early gnathostome habitats ranged from continental freshwaters to neritic marine and are preserved in both the internal and the external basins of Euramerica (Laurussia) (Figure 6.3). In the red-bed sequences, the vertebrates are accompanied by a few invertebrates and vascular plants which afford little biostratigraphic aid. Palynomorphs are of irregular distribution but are of stratigraphic value where present. The early Devonian freshwater vertebrate faunas were apparently agnathan-dominated worldwide. Much of the highly varied lithostratigraphy of Devonian Euramerica reflects the development of major tectonic systems. Caledonian, Rhenohercynian, Uralian, Franklinian, Appalachian and Cordilleran. In addition, there are deposits of the Baltic–Russian platform and the Laurentian interior. Major outcrops are distributed between these as shown in Table 6.1.

In the text that follows (and Table 6.1), units that are distinct in terms of their vertebrate content may be referred to as 'horizons', 'fish beds' or even 'series'. They are essentially assemblage-defined entities, which may or may not possess common boundaries with the next such unit above or below. Intervals of barren strata may occur between the successive vertebrate-yielding beds. Thus degrees of overlap between successive

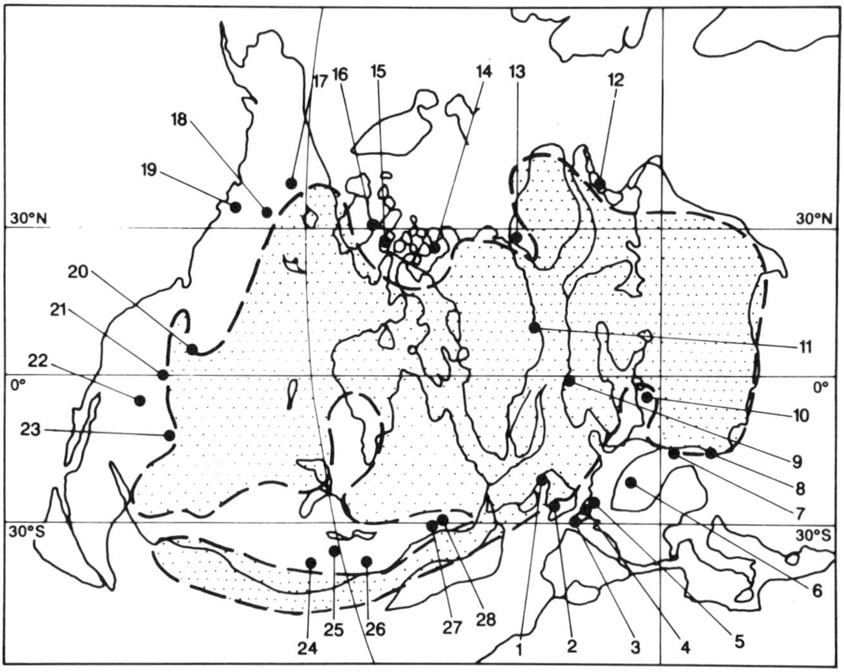

Figure 6.3. Localities for Devonian gnathostomes referred to in the text. The palaeogeographical base is after Scotese, 1986. Approximate limit of the land mass as in Lochkovian-Emsian time. Localities: 1, S. Scotland and N. England; 2, Anglo-Welsh Basin; 3, Brittany; 4, N. France and Belgium; 5, Rheinische Schiefergebirge, Germany; 6, Czechoslovakia; 7, Holy Cross Mts, Poland; 8, Podolia, USSR; 9, W. Norway; 10, Baltic States; 11, East Greenland; 12, Novaya Zemlya, USSR; 13, Spitsbergen; 14, Ellesmere I., NWT, Canada; 15, Cornwallis, Somerset and Prince of Wales Islands, NWT, Canada; 16, Bathurst and Melville Islands, NWT, Canada; 17, Anderson River, NWT, Canada; 18, Mackenzie Mts, NWT, Canada; 19, Bear River, Yukon, Canada; 20, Idaho; 21, Utah and Wyoming; 22, Nevada and N. Arizona; 23, Colorado; 24, Cincinnati, Ohio; 25, Cleveland, Ohio; 26, Catskill Mts, New York and Pennsylvania; 27, S. Gaspé, Québec; 28, E. Gaspé, Québec.
Stippled area = land area/non-marine deposits.

vertebrate 'zones' may be difficult to identify or delineate, and biostratigraphic gaps may be difficult to measure in terms of vertebrate evolution. Interregional correlation is still on a rather broad general basis, although many taxa appear to characterise specific short spans of geologic time throughout this faunal province.

Table 6.1 *Siluro-Devonian and Devonian depositional sites of the Euramerican Vertebrate Province (Old Red Sandstone Continent)*

The Baltic–Russian Platform	[Craton]
Caledonian Basins	[Intercratonic belt]
Anglo-Welsh Basin and Southern England	
Scotland	
Western Norway	
East Greenland	
Spitsbergen	
Rhenohercynian Basins	[Active continental margin]
Brittany	
Northern France–Belgium	
The Rhineland	
Holy Cross Mountains, Poland	
Podolia	[Part of Ukrainian platformal sequence]
Uralian Basins	[Active continental margin]
Ural Mountains	
Novaya Zemlya	
Appalachian Basins	[Active continental margin]
Eastern Canada	
Appalachian USA	
Franklinian Basins	[Active continental margin]
Boothia Uplift Region	
Bathurst, Melville and Ellesmere Islands	
Cordilleran Basins	[Active continental margin]
Mackenzie Mountains	
Rocky Mountains - SW USA	
Laurentian Interior	[Craton]

The Baltic–Russian Platform

Blieck, Mark-Kurik and Märss (1988) have complied data for the correlation of the vertebrate-dominated and non-vertebrate-dominated communities in the platformal Silurian to Carboniferous of the south Baltic area (Figure 6.4). It is a very complete succession, as summarised below, and is offered as a global standard for mid-Palaeozoic correlation. In brief, the gnathostome vertebrates figure in zones together with Agnatha, but are dominant in those of the middle and upper parts of the Devonian. Distinct acanthodian assemblages (19 species in all) are most in evidence in the Llandovery, then are sparingly represented as far as the Famennian (Valiukevicius, 1988). Four species of Actinopterygii are present in the Emsian–Eifelian. The placoderms are conspicuous with over 45 species in the Middle and Upper Devonian; of these seven are of *Asterolepis* and 14 of *Bothriolepis*. The Crossopterygii total 24 species and the Dipnoi five. The vertical distribution of these species is used to erect a set of zones which is correlated to the local palynomorph and brachiopod zones. At the genus level this region has affinities with the other 'Caledonide' basins such as in Scotland (see also Figure 6.9). Species differences, however, emphasise the local nature of the communities.

Caledonian Basins

The British Isles

Gnathostomes occur in each of the principal Old Red Sandstone basins of the British Isles. In the Silurian they are represented solely by acanthodian scales and spines from the Anglo-Welsh area (Miles, 1973a) and the Midland Valley of Scotland. By the Devonian, acanthodians had been joined by other gnathostomes, and their remains occur in nearly all major outcrop areas of the Old Red Sandstone in Britain. Rare plates and bones are also found in the predominantly marine facies of the Devonian in Southwest England and from the subsurface in Southern England (Allen, in House *et al.*, 1977).

The Anglo-Welsh Basin and Southern England

Silurian acanthodian scales and bones occur in the (marine) Ludlovian strata, the greatest concentrations being in the Ludlow Bone Bed and similar winnowed deposits. In the Lower Old Red Sandstone red-beds, agnathans are of stratigraphic importance but acanthodians and small arthrodires such as *Kujdanowiaspis* also occur (White, 1950). Upper Old Red Sandstone vertebrate faunas are rare but include species of *Ctenacanthus*, *Bothriolepis*, ?*Coccosteus*, *Holoptychius*, *Prescottaspis* and cf. *Sauripterus* (Allen *et al.*, 1977; Ball and Dineley 1961; White, 1961). In northern England the Cheviot Hills contain late Devonian genera *Bothriolepis* and *Holoptychius* and a Famennian fauna like that of Dura Den in Scotland.

Scotland

Some of the earliest studies of Devonian vertebrates were made on Scottish material, but the difficulties in resolving the biostratigraphy persist. It is customary to speak of two separate depositional basins for the Devonian of Scotland—the Caledonian Basin occupies the Midland Valley and has Lower and Upper Old Red Sandstone strata within it; the Orcadian Basin extends over much of the northern North Sea and coastal areas of northeastern Scotland.

The Orcadian Basin contains thick Middle and Upper Old Red Sandstone. Gnathostomes do not figure in Scottish Old Red Sandstone biostratigraphy (see Figure 6.5) until the Middle Devonian, even though acanthodian remains exist in the Lower Old Red Sandstone of the Midland Valley. On the mainland northeastern seaboard, Orkney and Shetland, gnathostome fishes occur most abundantly in local 'fish beds', between

CHRONOSTR. UNITS				BALTIC STRATIGRAPHICAL SCALE		BIOSTRATIGRAPHICAL CORRELATIONS: SPORES	BRACHIOPODS	VERTEBRATES	
DEVONIAN	UPPER	FAMENNIAN	UPPER	KETLERI		S. lepidophytus	Adolfia talasica -		
				ŽAGARE		Cornispora	Cyrtospirifer kurban		
				ŠVETÉ	SNIKERE	varicornata			
					TĒRVETE				
				MŪRI			Ptychomoloetechia turanica		
				AKMENE					
			LOWER	KURSA		? — — ?			
				JONIŠKIS			?rtospirifer		
				ŠIAULIAI		Hymenozonotriletes	ex. gr. C. archiaci		
				KRUOJA					
		FRASNIAN	UPPER	AMULA		speciosus -	Theodossia	Asterolepid. ? gen. indet.	
				STIPINAI	BAUSKA	H. radiatus	ex. gr. T. anossofi		
					IMULA	? — — ?			
				PAMŪŠIS				Asterolepis ? amulensis	Psammosteus falcatus
			MIDDLE	SNEZHA		Archaeoperisaccus	Tenticospirifer		
				DAUGAVA	IŞTRA — ALTOVA	mirus -	tenticulum -		
					BUREGI				
					IL'MEN'		Cyrtospirifer		
					SVINORD		ex. gr. C. disjunctus		
					PORKHOV	A. menneri			
			LOWER	DUBNIK			Hypothiridina	A. syasiensis	
				PLAVINAS	CHUDOVO		calva -		
					ATZELE		Ladogia meyendorfii -		
					SĒLIJA		Lamellispirifer		
					SNETNAYA GORA	? — — ?	muralis -	Asterolepis radiata	
				ŠVENTOJI	AMATA	Archaeozonotriletes	L. novosibiricus		B. obrutschevi - B. prima
					GAUJA	micromanifestus		A. ornata - sp. 3 - essica	A. ornata
	MIDDLE	GIVET.		BURTNIEKI		A chaeozonotriletes	?	A. dellei	P. tuberculatus
				ARUKÜLA		extensus	Bicarinatina uqulana - livica - etc	Asterolepis estonica	P. pauli - P. palaeformis
		EIFELIAN		NARVA	upper	H. polymorphus	Bicarin. borchi		Sch. striatus
					lower	Periplecotriletes tortus	Bicarinatina acuta - scaphulata Laima seliana		
				PĀRNU			?		Sch. heterolepis
	LOWER	EMSIAN		RĒZEKNE		H. inassuetus	Laima latgalica	Asterolepid. indet.	Sk. fragilis
		"SIEGEN."		ĶEMERI		R. clandestinus ?	?		? — — ? R. "cornubica"
		"GEDINN."	UPPER	STONIŠKIAI			Lingula aff. lata L. cf. cornea etc.		? — — ? R. crouchi
			LOWER	TILŽE		GRAPTOLITES ↓			P. leathensis - Traquairaspis
SILURIAN	PRIDOLI	UN-NAMED		OHESAARE	KAAVI	? ultimus	? S. stonishkensis	J. ovalis - M. orbignyi - D. magna	K. timanicus; Por. punctatus
				KAUGATUMA	LÕO				N. gracilis
					ÄIGU				
	LUDLOW	LUDFORDIAN		KURESSAARE	KUDJAPE	?	D. navicula - P. striatellus - G. rotunda	J. canaliculata - etc.	Th. sculptilis
					TAHULA				
				PAADLA	UDUVERE	balticus		B. costatula - G. magna	A. hedei
					HIMMISTE	tauragensis	P. minimus - D. navicula	C. pseudoknighti	P. elegans
		GORSTIAN			SAUVERE	scanicus; progenitor/nilssoni	P. aknistensis - D. navicula	J. crassa	L. martinssoni
	WENLOCK	HOMERIAN		ROOTSIKÜLA	SOEGININA	ludensis	L. minima - G. lithuanica		
					VESIKU	nassa			
					KUUSNÕMME	testis			
					VIITA	radians			L. taiti
				JAAGARAHU	TAGAVERE	perneri			
		SHEIN-WOODIAN			MAASI		D. biloba - S. cingulatus	R. elegantula	↓ ?
					VILSANDI	flexilis			
				JAANI	NINASE	antennularius	D. paralata - S. cingulatus	R. sabrinae - sawddensis	
						riccartonensis			
						murchisoni-bohemicus			
	LLANDOVERY	TELYCHIAN		ADAVERE	VELISE	spiralis; griestonensis; crispus; turriculatus	D. baltica - A. grayi - Plectodonta sp.	V. visbyensis ?	
		AERONIAN			RUMBA	linnaei			
				RAIKKÜLA	RAIKKÜLA	sedgwicki; convolutus; triangulatus	Plectodonta sp. "Chonetes" sp.		
		RHUDDA-NIAN			SAARDE	cyphus			
				JUURU		?	?		

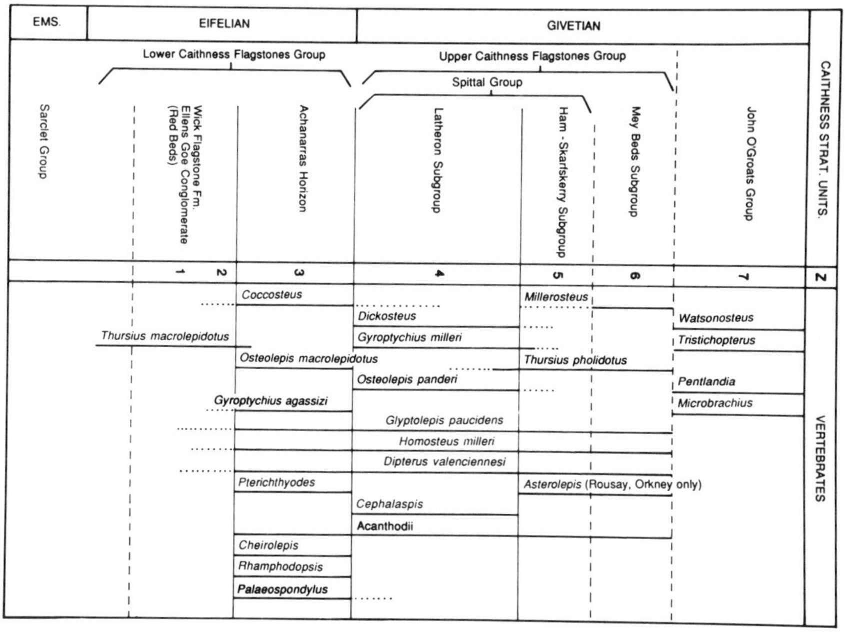

Figure 6.5. Vertebrate succession in the Middle Old Red Sandstone of the Orcadian Basin (largely after Donovan, Foster & Westoll, 1974). 1–7 Faunal divisions recognised by Donovan *et al.*

Figure 6.4. The correlation of vertebrate and invertebrate zonations of the Silurian and Devonian of the Baltic area has progressed to a remarkable degree, as witnessed by this table from Blieck *et al.* (1988). The intercalation of marine and non-marine fossil assemblages is extensive, but breaks in the succession occur at several levels. Spores are important in their occurrence in both the marine and the non-marine facies.

CAITHNESS, ORKNEY, SHETLAND	MORAY AND NAIRN	VERTEBRATES
	Rosebrae Beds	*Bothriolepis alvesensis, B. cristata, Phyllolepis* sp., *Phaneropleuron* sp., *Rhynchodipterus elginensis, Eusthenopteron traquairi, Holoptychius* spp., *Glyptopomus elginensis*
	Scaat Craig Beds	*Psammosteus falcatus, Cosmacanthus malcolmsoni, Bothriolepis* sp., *Conchodus ostreiformis, Panderichthys* sp., *Holoptychius* spp.
	Alves Beds	*Bothriolepis alvesensis, B. gigantea, Psammosteus megalopteryx*
	Boghole and Whitemire Beds	*Asterolepis* sp., *Bothriolepis* sp., *Cosmacanthus* sp., *Bothriolepis gigantea, Psammosteus megalopteryx, Conchodus* sp., *Eusthenopteron traquairi, Holoptychius* spp., *'Coccosteus' magnus*
Dunnet Head Sandstone	Nairn Sandstones	*Asterolepis* sp., *Psammolepis undulata, Panderichthys leptognathus, 'Coccosteus' magnus, Holoptychius* spp.
Bressay Flags		*Asterolepis* sp., *Holonema ornatum, Holoptychius* sp.
Lerwick Sdsts.		*Stegotrachelus finlayi, Tristichopterus? , Microbrachius?*
Rora Head M.O.R.S.		

Figure 6.6. Faunal succession of gnathostomes in the Upper Old Red Sandstone of the Orcadian Basin (from various sources).

which they are generally rare. The literature on Scottish Old Red Sandstone vertebrate biostratigraphy is sparse, the most useful accounts being by Westoll (1977, pp. 66–93; 1979), and Donovan *et al.* (1974). Mark-Kurik (1978) has suggested that a series of arthrodire 'zones' in the Scottish Middle Old Red Sandstone may correlate closely with a similar series in the east Baltic Devonian.

In the Midland Valley the Upper Old Red Sandstone rests unconformably on the Lower. Both series are largely coarse intermontane clastics with minor arenites in which occur rare Agnatha and Acanthodii. The Upper Old Red Sandstone was deposited following a long erosive interval. In these horizons vertebrates are very rare but a larger number of species is present (Figure 6.6). The most significant occurrence of gnathostomes is in the Dura Den Formation with *Bothriolepis hydrophila*, *Phyllolepis woodwardi*, *Glyptopomus minor*, *G. kinnairdi*, *Eusthenopteron dalgleishensis*, *Holoptychius flemingi* with the possibly co-specific *H. nobilissimus* and *Phaneropleuron andersoni*. This is a very late Devonian fauna and is perhaps the youngest in Scotland. The Middle and Upper Devonian rocks to the north show much lateral lithological variation from one local basin to another. The situation is summarised in the accompanying table and reveals successions of vertebrate assemblages of five to twenty different taxa.

Some Scottish Middle Old Red Sandstone vertebrates such as *Pterichthyodes*, may be endemic; on the other hand, the small antiarch *Microbrachius*, hitherto known only in the John O'Groats Group, has been reported from China (Pan, 1988). *Remigolepis* may also be present (Andrews, 1978).

System	Stage	Unit	Canning Land	Kap Franklin area (subdivision according to Butler; correlation uncertain)
CARBONIFEROUS			*Gronlandaspis* "series". *Gronlandaspis mirabilis*, indet. Arthrodira, Palaeonisciformes, *Holoptychius* sp., plants	
UPPER UPPER DEVONIAN	Famennian	*Remigolepis* "series" (c. 800 m)	Upper Division (c.520 m) *Remigolepis* spp., *Oervigia nordica, Jarvikia arctica*, *Holoptychius* sp., *Eusthenodon wangsjoi, Ichthyostega* spp., *Ichthyostegopsis wimani, Acanthostega gunnari*	
			Middle Division (c. 200 m) Unfossiliferous	
			Lower Division (c. 80 m) *Remigolepis* spp., *Bothriolepis nielseni, Phyllolepis nielseni, Soederberghia groenlandica, Holoptychius* sp., *Ichthyostega* spp.	
		Phyllolepis "series"	Upper part *Phyllolepis orvini, Bothriolepis gronlandica*, indet. Acanthodii, *Nielsenia nordica, ?Soederberghia, Holoptychius* sp.	
			Lower part *Bothriolepis jarviki*, indet. Selachii, *Cladodus* sp., *Holoptychius* sp.	
Lower Sandstone Complex. Poor in fossils. Indet. Arthrodira, *Holoptychius* sp., Osteolepiform (cf. *Eusthenopteron* sp.)				
UPPER MIDDLE DEVONIAN	Givetian	Upper part	*Gyroptychius groenlandicus* "series" (c. 350 m). Indet. *Porolepiformes, Gyroptychius groenlandicus*	Randbol "series" (c. 800 m). *Asterolepis* cf. *save-soderberghi*, cf. *Clarkeosteus halmodeus*, Osteolepiformes cf. *Thursius macrolepidotus*
			Plant-bearing "series" (c. 500m).	
		Lower part	*Asterolepis save-soderberghi* series (c.400). *Psammolepis groenlandica*, indet. Acanthodii, *Homostius kochi, Asterolepis save-soderberghi*, indet. Dipnoi, *Glyptolepis groenlandica, Gyroptychius* cf. *groenlandicus*.	Unfossiliferous beds (c. 1000 m)
			Heterostius "series" (c50 m). *Heterostius groenlandicus, Homostius kochi*	Viddal "series" (c.1500 m). *Estheria*, indet. Porolepiformes, *Gyroptychius* cf. *groenlandicus*

Figure 6.7. Faunal succession in the Old Red Sandstone formations of East Greenland (after Jarvik, 1961).

Western Norway

The red-bed intramontane basin of the Hornelen Field in western Norway contains osteolepids and porolepids which suggest a direct connection and correlation with Achanarras, Sandwick and Melby fish beds of the Middle Old Red Sandstone Orcadian basin, and with the *Asterolepis* Series in East Greenland (Jarvik, 1949).

East Greenland

The well-preserved Middle and Upper Devonian vertebrate faunas of the region between Steno Land and Kong Oscars Fjord have been intensively studied since the beginning of the century (see Bendix-Almgreen, 1976). Many of the lithostratigraphic units are coarse clastics, reaching in all about 3000 m. Deposition was continental, comparatively rapid and episodic; delta fan and fluvial sediments are dominant. The complete faunal list from these rocks is very extensive (op. cit., p. 542). In addition to the numerous gnathostomes, a single drepanaspid has been found. The antiarchs are the most common of the vertebrates and this has led to the recognition of the different antiarch 'series'. Upper Devonian porolepiform fishes are locally common, while the Middle Devonian osteolepiformes are present in strength in the Kapp Franklin area and in Canning Land. Dipnoans are rare in the Middle Devonian, but are not uncommon in the Upper. Jarvik (1961) correlated the East Greenland faunas (Figure 6.7) with those of Scotland, noting that *Gyroptychius groenlandensis* is very similar to *G. milleri* and suggesting that the *Gyroptychius* Series of Canning Land and Wegener Halvø equates with the Thurso Flagstone Group and the Rousey Bed of northern Scotland. The *Heterostius* Series, with both *Heterostius* and *Homostius*, he correlated to the Stromness Beds of Eifelian–Givetian age in Scotland.

Jarvik (1950) also noted that in the Upper Devonian *Remigolepis* Series the bed with *Phyllolepis orvini* and *Bothriolepis groenlandica* may correlate with the Baltic–Russian beds with *B. ornata* and *Phyllolepis*. He concluded that the Lower Division of the *Remigolepis* Series is probably Famennian; the Upper Division and the overlying *Groenlandaspis* Series may not only be higher Famennian but also include an element of the earliest Carboniferous.

Spitsbergen

Although vertebrates have been known in the Old Red Sandstone of Spitsbergen since 1890, and many taxa have been described in detail, comparatively little has been published on the biostratigraphy. Friend (1961, 1973; Friend *et al.*, 1966) has provided an overall account of the stratigraphy, listing the fossils occurring in each formation. The 6000 m-thick sequence is the product of torrential alluvial fans, river flood plain, brackish lagoon and lake environments within intermontane basins. Much of the early work on the vertebrates was by Kiaer, Heintz and Stensiö (Friend, 1961, refs.). Later contributions were made by Ørvig (1957a). Recent Franco-Norwegian expeditions have extended knowledge of the faunas of the lower stratigraphic units (Blieck, 1984; Blieck, Goujet and Janvier, 1987; Goujet, 1984).

The faunas include numerous Agnatha as well as gnathostomes, over 130 species in all. Six acanthodian genera occur in the (Pridolian–Lochkovian) Red Bay Group, but fewer are in higher units. The arthrodires are conspicuous, if not the dominant, elements of the faunas higher in the Red Bay Group and in the overlying Wood Bay Formation. Blieck (1984) established a series of agnathan biozones (I–V) for the Red Bay Group and the Wood Bay Formation, equivalent to the Lochkovian and Pragian (see Figure 6.8). In the Grey Hoek Formation the arthrodires are common and are generally like those of Lower Devonian formations elsewhere. *Homostius* and *Heterostius*, however, suggest a Middle Devonian age. Crossopterygii are indicated by locally abundant scales and other remains. The Wijde Bay Formation contains a smaller selection of vertebrates, but also molluscs and plants which led Føyn and Heintz (1943) to postulate a Givetian age for it. The Mimer Valley Formation, which is up to 1000 m thick, includes the youngest Devonian strata in Spitsbergen and might have originated in a freshwater lake and delta setting. This unit has a small number of psammosteids plus arthrodires and the antiarch *Asterolepis*. The crossopterygii are scantily represented in the Wijde Bay, but a larger assemblage has been recovered from the Mimer Valley Formation. Both formations were regarded as Givetian by Føyn and Heintz (op. cit.), but Friend (1961) correlated the Mimer Valley Formation to the Nairn and Boghole formations of Scotland, indicating a Frasnian age. Spore

		SILURIAN	DEVONIAN							
		Pridoli	Lochkov -Emsian					Eifelian		
	I - V = Proposed zonation of Blieck et al., 1987 VI - VIII = Possible further zonation	RED BAY GROUP	Sigurdfjellet Member (WOOD BAY FORMATION)	Kapp Kjeldsen Member (WOOD BAY FORMATION)	Keltiefjellet Member (WOOD BAY FORMATION)	Stjordalen Member (WOOD BAY FORMATION)	GREY HOEK FORMATION	WIJDE BAY FORMATION	MIMER VALLEY FORMATION
PLACODERMI	Palaeacanthaspids	●	●						
	Arctaspis			●		●			
	Lataspis			●					
	Actinolepis			●					
	Dicksonosteus			●					
	Arctolepis						●		
	Angarichthys				●				
	Mediaspis		●	●		●	●		
	Sigaspis		●						
	"Svalbardaspis"		●	●	●				
	Heintzosteus			●					
	Lehmanosteus			●					
	Plataspis			●	●				
	Elegantaspis			●					
	Euleptaspids			●	●				
	Heterostius					●		●	
	Homostius				●	●	●	●	
	Huginaspis						●		
	Monaspis					●			
	Holonema							●	
	Lunaspis						●		
	Asterolepis								●
OSTEICHTHYES	*Dictyonosteus*							●	
	Glyptolepis							●	
	Heimenia							●	
	Onychodus							●	
	"Porolepis"	●	●	●	●	●	●	●	●
	Actinopterygians?	●							
	Orvikuina					●			
	I, II	III	IV		V		VI	VII	VIII

Figure 6.8. Faunal succession in the Silurian–Devonian of Spitsbergen (after Blieck *et al.*, 1987).

assemblages have confirmed the age of the Wood Bay Formation as Pragian–Emsian and the Mimer Valley Formation as Eifelian to lower Frasnian (Allen, 1967).

The Rhenohercynian Basins

In Western Europe vertebrate localities are scattered in Spain, Brittany and northern France and Belgium, while in the major Devonian outcrops of Germany they occur in stratigraphically well-understood sequences, with large commonly well-preserved faunas. (For general accounts, see McMillan, Embry and Glass, 1988.) They evolved within a broad, continually mobile belt adjacent to the margin of Euramerica, and throughout which existed parallel facies belts of continental, neritic and deeper water sediments. The literature is very large and the principal contributor in recent decades has been W. Gross. (See especially Gross, 1950.) Other localities in Central Europe are minor. In the southeast, Podolia has a significant Silurian–Lower Devonian passage sequence and an Old Red Sandstone facies with vertebrates. Rare gnathostome fragments have been discovered in the marine Devonian of Czechoslovakia.

Brittany
Isolated gnathostome bones and plates, scales and spines come from various levels in the Devonian of the Armorican region. They are mostly associated with shallow neritic marine facies. Acanthodians occur in the Lower Devonian and arthrodires include *Breizosteus armoricensis* in the Pragian and a ?holonematid in the Emsian. In the Frasnian, *Holonema* cf. *radiatum* may indicate a connection with the fish fauna of Iran (Lelièvre, Goujet and Morzadec, 1983).

Northern France and Belgium
Devonian rocks have long been known in the Boulonnais and Artois areas and in the cores of the major anticlines near the Franco-Belgian border (Dutertre, 1930; Goujet and Blieck, 1979). Acanthodians and *Kujdanowiaspis* are present in rocks correlated with the *crouchi* zone of the British Dittonian on the basis of Heterostraci. Marine Middle Devonian strata in scattered small outcrops yield rare arthrodire fragments, including ?*Macropetalichthys*. The more widespread Upper Devonian has a considerable gnathostome fauna; *Aspidichthys*, *Ganorhynchus* and *Ptychodus* occur in the Frasnian. From the marine Famennian come *Dinichthys belgicus*, *Dipterus nelsoni* and *D. fourmarieri*, *Holoptychius flemingi* and Acanthodii. The continental Psammites de Montfort have given a Famennian fauna with *Bothriolepis lohesti*, *B.*sp., *Phyllolepis undulata*, *Holoptychius flemingi*, *H. murchisoni*, *Glyptolepis benedeni*, *G. radians* and *Glyptopomus kinnairdi*. The correlation of these horizons with the (Famennian) Dura Den in Scotland is obvious.

The Rhineland

The Rhineland and adjacent Devonian outcrops in Germany offer exceptional stratigraphic sequences with good biostratigraphical control via conodonts and other micro- and macro-invertebrate fossils. Devonian rocks with local red-bed units rest upon deformed Lower Palaeozoics. Over 50 species of vertebrates are known from the Lower Devonian, around 25 from the Middle and some 50 species from the Upper Devonian. Many are well-preserved specimens; some are in marine strata, others are associated with intermediate and non-marine facies. Acanthodii, placoderms and osteichthyes occur in all three series.

The Lower Devonian has about 15 species of Agnatha and about 10 Acanthodii; arthrodires, principally actinolepids, are numerous but not well known. Rarities are *Radotina*, *Lunaspis*, *Pseudopetalichthys* and *Stensioella*. Recent work on microvertebrates proves them to be increasingly useful in stratigraphy (Veith, 1980, 1983). Porolepids and Dipnoi are few, but large in size and known in Pragian and Emsian strata.

The Middle Devonian vertebrates are much reduced in diversity from the level of the Lower Devonian. A single acanthodian genus, *Onchus*, is widespread. The arthrodires are, in contrast, numerous and many are large. A spectacular fauna has been recorded from the marine late Eifelian Oberer Plattenkalk around Bergisch Galdbach (Jessen, 1966; Ørvig, 1960, 1961)—*Atopacanthus* sp., *Coccosteus*? sp., Coccosteomorphi gen. et sp. indet, Brachithoraci gen. et sp. indet., *Rhachiosteus pterygiatus*, *Ptyctodus* sp., *Ctenurella gladbachensis*, Asterolepiformes indet., *Dipterus* cf. *valenciennesi*, *Dipterus*? sp., *Rhinodipterus ulrichi*, *Ganorhynchus*? sp., *Moythomasia nitida*, *Stegotrachelus*? sp., *Thursius*? sp., *Gyroptychius* sp., Osteolepidae *Latvius decherti*, *L*.n.sp.A, *L*.n.sp.B, *Thursius*? sp., Struniformes *Strunius walteri*, *S. rolandi*, *Onychodus obliquimarginatus*, Coelacanthiform, *Diplocercides heiligenstockiensis*. This horizon must be remarkably near to that of the *Gyroptychius groenlandicus* Series in East Greenland, and of the Cellulosa Marl (Plavinas Beds) in Latvia.

The Upper Devonian is primarily an open-water goniatite facies throughout much of the region. Nevertheless its vertebrate fauna is a large one. The various arthrodires make up the greater part of the fauna; antiarchs are extremely rare. There are few petalichthyids, but selachians are locally common, represented by *Cladodus* denticles and teeth. Crossopterygians and dipnoans are limited to two genera each. *Aldingeria* is the sole palaeoniscoid present. In the succession of Devonian vertebrate faunas the German instances move progressively from great similarity to those of the British Old Red Sandstone facies to faunas of marine aspect, as indeed they do in comparison with the succession in the Baltic–Russian platform area, where the antiarchs are common.

Holy Cross Mountains, Poland

Devonian strata, present in the mountains in the southeastern corner of Poland, have yielded vertebrates from both the Lower and the Upper Devonian. Gnathostomes present in the Upper Devonian include several arthrodires, elasmobranchs, holoptychiids and at least one crossopterygian

(Kulczycki, 1957). The fauna appears to correlate generally to the Lower Frasnian. Ginter (1990) reports a fauna of late Famennian sharks, including new species of *Phoebodus*, and *Thrinacodus ferox* (Turner) which is also known in Australia and China.

Podolia
The succession in this part of the Ukraine is far to the present southeast of the western European Siluro–Devonian but has a succession of late Silurian marine rocks passing upwards into a Pridoli–Lochkovian Old Red Sandstone facies, which bear abundant agnathans. Strictly speaking these rocks belong to the epeiric succession of the Ukrainian platform, rather than to the Rhenohercynian orogenic belt. The faunas bear strong resemblance to those of Britain and Spitsbergen (Nikforova, 1977). *Machaeracanthus*, *Guerichosteus* and *Hariostius* are recorded here. From thc thin Frasnian unit that caps the Devonian sequence '*Dinichthys*' and *Anomalichthys* have been reported.

Uralian Basins

Ural Mountains
This is an enormous region with extensive areas of Silurian and Devonian deposits. The facies are predominantly marine, neritic and deeper-water. Continental strata are rare, and no account of the few vertebrates is given here.

Novaya Zemlya
In the islands of Novaya Zemlya Devonian Old Red Sandstone facies are widely distributed and contain vertebrates similar to, and in a succession of faunas like, that of Podolia and the Anglo-Welsh Basin (Karatajute-Talimaa, 1978).

Appalachian Basins

Continental deposits resulting from orogenic movements in eastern North America lie within two broad domains. Those in a belt from New England to easternmost Quebec span most, if not all, of the Devonian period. South from New York State into Pennsylvania and West Virginia, continental Devonian rocks occur in a single great basin which subsided most extensively in the later half of the period.

Appalachian USA
The lithostratigraphy and sedimentology of the 'Catskill Delta facies' of New York State, northern Pennsylvania and western Maryland are relatively well known (Woodrow and Sevon, 1985). The Silurian and Lower Devonian are thin, but the Middle and Upper Devonian clastic basin infill is up to 3000 m thick. Fossils, especially vertebrates, are rare. Gnathostomes occur

as isolated pieces or as concentrations in conglomerates; articulated remains are very rare. Nevertheless the faunal list includes species typical of the Middle and Upper Old Red Sandstone elsewhere. Spores are also employed in correlation (McGregor, 1977; Richardson and McGregor, 1986). Banks (1980) has recognised in the Catskill facies a series of macroplant assemblages (I to VII) that extends from the Pridoli to the Famennian.

Around 20 species of fish are recorded (Eastman, 1907, 1917), some based on rather questionable material (and of uncertain stratigraphic positions or ranges), viz: Acanthodii (*Gyracanthus sherwoodi, Onchus rectus*), Antiarchi (*Bothriolepis minor, B. nitidus*), Arthrodires (*Coccosteus* sp., *Dinichthys* sp., *Acanthaspis* sp., *Holonema rugosum*), Dipnoi (*Dipterus angustus, D. controversus, D. fleischeri, D. sekwoodi*), Crossopterygii (*Ganorhynchus beecheri, Holoptychius americanus, H. giganteus, H. halli, H. radiatus*). Of these, the Acanthodii are not significant in stratigraphy; the antiarchs indicate the Middle and Upper Devonian. So far it has not been possible to distinguish distinct gnathostome zones, but Thomson (1976) regarded the Upper Devonian rhipidistians *Holoptychius, Eusthenodon* and *Glyptopomus* as common to all major circum-atlantic Old Red Sandstone outcrops while *Hyneria, Sterropterygion* and *Sauripterus* are not known outside the Appalachian basin.

Eastern Canada

The Devonian redbeds of the Gaspé Peninsula, P.Q., contain well-preserved fossil fish and early vascular plant remains, known since the latter half of last century. The eastern Gaspé succession is part of a major clastic wedge, shed northward and northwestward from the rising Acadian mountains. Although stratigraphic correlations have been made between these rocks and those of Chaleur Bay, southern Gaspé, palaeocurrent directions suggest that the latter succession is not part of the same clastic wedge, but accumulated in an intermontane basin (Lawrence and Rust, 1988).

In eastern Gaspé a coarsening-upwards sequence of clastics rests unconformably on Lower Devonian shallow marine limestones. Its lower part (Battery Point Formation) has well-preserved fish and plants. Unfortunately, poor exposure and lateral lithological variation make it difficult to determine relative stratigraphic positions of the fossiliferous localities (Carroll *et al.*, 1972).

The fish fauna of the Battery Point Formation (Pageau, 1969) includes agnathans (*Yvonaspis*), acanthodians (*Climatius, Cheiracanthus, Ctenacanthus, Machaeracanthus, Mesacanthus*) and arthrodires (*Phlyctaenius, Pageauaspis, Gaspeaspis, Cartieraspis, Kolypaspis* and *Laurentaspis*) (see Denison, 1978; Belles-Isles, 1989). Belles-Isles (op. cit.) refers to the base of the Battery Point as coincident with the Lower/Middle Devonian boundary. Abundant spores date the Battery Point Formation as early Emsian to early Eifelian (McGregor, 1973, 1977).

Two distinct and remarkable vertebrate assemblages have been found at different stratigraphic levels in the Devonian clastics of the Restigouche River–Chaleur Bay area of southern Gaspé (Dineley and Williams, 1968). The older of the two, probably of late Lower Devonian age, comes from the Atholville Beds (basal LaGard Formation) near Campbelltown, N.B. It includes agnathans (*Cephalaspis* = *Yvonaspis*), Acanthodii (*Climatius*, *Cheiracanthus*, *Gyracanthus*, '*Homacanthus*', *Mesacanthus*, and *Doliodus*) and arthrodires (*Phlyctaenaspis*). There are strong similarities between this fauna and that of the Battery Point Formation of eastern Gaspé at both genus and species level. The Atholville Beds fauna is, however, the poorer both in numbers of individuals and in diversity. Dineley and Williams (1968) suggested that the Battery Point Formation can be correlated with the LaGard Formation and also (Dineley in Carroll *et al.*, 1972, p. 97) that the Atholville fish occurrence represents the fauna of higher or more isolated parts of the same river system as that in which the eastern Gaspé fish lived. Lawrence and Rust (1988) confirm that these two formations appear to be of similar lithology and age.

The younger of the two fish faunas of southern Gaspé is the Escuminac, described by Whiteaves (1881), Stensiö (1939), Jarvik (1944a, b), and Ørvig (1957a). Its 21 species include Osteostraci (*Alaspis*, *Escuminaspis*, Osteostraci indet.), 'anaspids' (*Endiolepis*, *Euphanerops*), an arthrodire (*Plourdosteus*), an antiarch (*Bothriolepis*), Acanthodii (*Diplacanthus*, *Homalacanthus*, *Traizeugacanthus*), Dipnoi (*Fleurantia*, *Scaumenacia*), a palaeoniscoid (*Cheirolepis*), and the crossopterygians (*Eusthenopteron*, *Glyptolepis*, *Holoptychius*, *Thursius*, *Quebecius*, *Miguashua*). It has been dated as Frasnian (Ørvig, 1957a). Russell's (1939) zonation of the assemblages in the Escuminac is suspect (Carroll *et al.*, 1972). This fauna includes the last of the Osteostraci and of the 'Anaspida'; the dipnoan genera are endemic but the other gnathostomes also occur elsewhere. Especially common in the highest unit of the Escuminac Formation, *Eusthenopteron* is the most advanced crossopterygian.

Franklinian Basins

Boothia Uplift Region

Vertebrate faunas, first discovered on Prince of Wales, Somerset and Cornwallis Islands by the Geological Survey of Canada in 1955 (Thorsteinsson, 1958), occur in the palaeozoic rocks flanking the Boothia Uplift. The Cornwallis Disturbance, responsible for the Uplift, had its maximum effect on sedimentation during early Devonian time—reflected in the development of the red-bed facies of the Somerset Island, Peel Sound and Snowblind Bay Formations adjacent to the Uplift. Farther from this arch, or when the influence of the uplift was slight, marine carbonates were deposited. Miall (1970) has described the transition from continental to marine facies within the Peel Sound Formation from east to west across Prince of Wales Island.

Agnatha are abundant at a number of levels between the (?Ludlovian) Cape Storm Formation and the (Lochkovian) Peel Sound and Drake Bay Formations of Cornwallis, Somerset and Prince of Wales Islands. Dating based upon conodonts and invertebrates suggests that similar taxa occur earlier in this area than in Europe and this may support the suggestion that the Canadian Arctic was an 'evolutionary centre' for Agnatha during the Silurian times (Dineley, 1990; Elliott, 1983, 1984; Blieck and Janvier, 1991). Although acanthodian scales and possible elasmobranch fragments have been reported from the Silurian (Thorsteinsson, 1973), the earliest instances of gnathostome macrofossils from this area are from the Upper Member of the Peel Sound Formation on Prince of Wales Island and its correlatives, the Drake Bay Formation of Prince of Wales Island, and the Snowblind Bay Formation of Cornwallis Island. The actinolepid arthrodires, *Baringaspis* (Miles, 1973b) and *Eskimaspis* (Dineley and Liu, 1984), and the porolepid crossopterygian *Powichthys* (Jessen, 1975) occur together with Acanthodii (Vieth, 1980) and Agnatha (Elliott, 1983). The faunas as a whole suggest correlation with the *crouchi* zone of the Anglo-Welsh vertebrate succession, the *vogti* horizon of Vestspitsbergen and the Czortkov Horizon of Podolia, and can be dated as late Gedinnian to early Siegenian (Lochkovian–Pragian) from associated invertebrates (Elliott, 1984).

Bathurst, Melville and Ellesmere Islands
Acanthodii have been described from the Cape Phillips Formation of Ellesmere Island, where they occur in a limestone band with Gedinnian brachiopods within a graptolitic shale sequence (Spjeldnaes, 1967). Scales attributed to thelodonts, Acanthodii and elasmobranchs (Veith, 1980) are present in the Lower Devonian of Devon, Bathurst, Axel Heiberg and Prince of Wales Islands. Comparable faunas in the USSR and Spitsbergen imply some correlation value for these fossils. Clastic strata of Middle to Upper Devonian age underlie large areas of south Ellesmere, Devon, Bathurst, Melville, Prince Patrick and Banks Islands. They take the form of a large clastic wedge, up to 4000 m thick, deposited in a large foreland basin to the south and east of the Ellesmerian Orogenic Belt (Embry, 1988). Fish remains collected from southwestern Ellesmere Island during the second Norwegian 'Fram' expedition (1898–1902; Schei, 1903) were described by Kiaer (1915). He recorded *Macropetalichthys*, now designated *Ellopetalichthys* by Ørvig (1957b), from Schei's (1903, 1904) Horizon Dh. subsequently assigned by McLaren (in Fortier *et al.*, 1963) to the Bird Fjord Formation. Kiaer's (1915, p. 16) original argument for a Middle rather than a Late Devonian age for the Horizon is now confirmed by the dating of the Bird Fjord Formation as Eifelian (*macromanifestus* zone) on the basis of its spores (Embry and Klovan, 1976; Embry, 1988). Kiaer (1915) described species of *Psammosteus*, *Bothriolepis*, *Holoptychius*, *Glyptolepis* and *Osteolepis* from Schei's Series E which, together with the plants described by Nathorst (1904), indicated an Upper Devonian age. This higher fish horizon is within strata now assigned to the Fram Formation, spores from which give an early to middle Frasnian age (*medius* and *maclarenii* zones) (Embry and Klovan, 1976).

Gnathostome remains have been recorded from other formations within or immediately below this clastic wedge, notably from the Hell Gate Formation which occurs on south Ellesmere and Devon Islands (Embry and Klovan, 1976, p. 552), the Beverley Inlet Formation (Thorsteinsson and Tozer, 1970; Embry and Klovan, 1976, p. 559) of Bathurst and Melville

Islands, and the Blue Fjord Formation of Bathurst Island (Norris in Fortier *et al.*, 1963, p. 607). Some of this material has proved to be of value in dating (see last ref.) but the quality and quantity of fossil gnathostomes so far found is insufficient for the basis of any biostratigraphic scheme, even locally. The time-stratigraphic framework for the Middle–Upper Devonian clastic wedge (Embry, 1988, Figure 2) is based on spores (Chi and Hills, 1976; Richardson and McGregor, 1986).

Cordilleran Basins

Stretching over 2000 kms from the Canadian Arctic to California, this huge area includes the Rocky Mountains and lands to the west. Most of the vertebrate localities here are within carbonate facies but at all levels channel deposits intrude. They represent run-off sediments from local large 'highs' and many of the faunas are regarded as freshwater. Both Silurian and Devonian systems yield vertebrates.

Southwestern United States

Vertebrate remains occur in many Devonian formations in Idaho, Utah, Wyoming, Colorado, Arizona, Nevada and eastern California. These rocks represent a wide variety of freshwater, marginal and neritic marine environments lying to the east of the Cordilleran continental margin and north of the Sonoran Trough; typical Old Red Sandstone facies are absent. Denison (1951, 1958, 1968a, 1968b) described new local Devonian gnathostomes, discussing their habitats, biostratigraphy and correlation. Lower, Middle and Upper Devonian faunas are present with numerous gnathostomes, principally placoderms and osteichthyes. This part of the Euramerica province is also characterised by a distinctive suite of Heterostraci, including *Psephaspis* which ranges up into the Middle Devonian.

Acanthodii are few, but Lower Devonian arthrodires include *Aethaspis*, *Simblaspis*, *Bryantolepis*, *Anarthraspis*, genera not found elsewhere. The Dipnoi *Uranolophus* and *Dipterus* also occur (Denison, 1958, 1968a). Middle Devonian faunas contain indeterminate arthrodires, and also *Holonema* and asterolepid antiarchs. At Red Hill, Nevada (Murphy *et al.*, 1976; Gregory *et al.*, 1977) an extraordinary fauna includes *Asterolepis*, *Eusthenopteron*, dipnoans, osteolepids and acanthodians, including *Machaeracanthus?*.

The late Devonian rocks extend into New Mexico and Montana where placoderms and lungfish are reported. *Bothriolepis* species are widespread (Denison, 1951), as are various arthrodires of obscure identity. Lungfishes are not uncommon in the region but many specimens are of uncertain taxonomy. Holoptychiid crossopterygians occur at many localities and Denison records a large rhizodontid, *Litoptychus bryanti*. Osteolepid scales are locally common. Throughout this vast area, vertebrate faunas had a remarkably uniform aspect at any one time; endemic genera were few. Fairly precise chronology is available from the marine palaeontology in the vertebrate-bearing strata or adjacent beds.

Western Canada

Denison (1963, 1964) and Soehn and Wilson (1990) have described vertebrates from the Silurian and Devonian of the southeastern Yukon and northern British Columbia, but the oldest gnathostomes are Acanthodii from the Delorme and associated formations of the Mackenzie Mountains, N.W.T. (Bernacsek and Dineley, 1977). Of these fossils, *Ischnacanthus gracilis* (Egerton) is otherwise known only from the Lower Old Red Sandstone of Scotland; a possible related form, *?I. scheii*, occurs in the Lower Devonian strata on Ellesmere Island (Spjeldnaes, 1967). Palaeoniscoid scales from the Delorme Formation of the Anderson River area (Schultze, 1970) are probably from its Lower Devonian part. Bernacsek (1977) has described a lungfish skull, *Stomiahykus thalodus*, from the early Eifelian Crainswick Formation of the Snake River area, Mackenzie Mountains, and Jarvik (1967) records the late Lower Devonian lungfish *Melanognathus canadensis*, known only from a lower jaw from the Anderson River, N.W.T.

The Laurentian Interior

West of the Appalachian Basin, the epeiric seas of the continental interior spread around the margins of the Canadian shield and the Transcontinental Arch. The neritic sediments here are fine siliciclastics, carbonates and evaporites, part of the Kaskaskia Sequence. Vertebrate remains are limited to scattered, and commonly isolated bones, plates and scales or to locally abundant microvertebrates, as in Ohio (Wells, 1944), and in numerous conodont residues. The taxonomic list is over 150 species but the details of many taxa are obscure. Many of the macrofossils are marine Acanthodii, placoderms and elasmobranchs. The list of Middle Devonian vertebrates runs to over 50 species but many are known from single localities or a restricted area. The Upper Devonian register of vertebrates is even larger, but the same qualification applies (Gross, 1950). Denison (1968b, 1978, 1979) believed that an appreciable number of these taxa were invalid, but the long list may reflect the diversity of vertebrate communities over the great area involved. Dinichthyids and cladodont sharks are exceptionally numerous in Ohio and New York. They appear to have spread beyond the Transcontinental Arch and also eastward to the Gondwana seas. Several other arthrodires (Miles, 1964, 1966) are also known in Europe.

DISCUSSION

The biostratigraphy of the basins external to the Euramerican continent can be resolved in terms of marine fossils on the one hand and of palynology on the other. The intercalation of vertebrate-dominated strata results in a biostratigraphy that not only records succession of vertebrate assemblages or individual taxa but also relates this new dimension to the palynological standard. In many instances the vertebrates occur in marine strata and the

	BALTIC STRATI-GRAPHIC UNITS	BALTIC INDEX VERTEBRATES	SCOTTISH STRATIGRAPHIC UNITS	SCOTTISH INDEX VERTEBRATES
FRASNIAN			ALVES, SCAAT CRAIG AND BOGHOLE BEDS	B. major B. major P. taylori
FRASNIAN	AMATA GAUJA	Plourdosteus livonicus Asterolepis radiata A.ornata	NAIRN SDSTS	P.? magnus A. maxima
?				
GIVETIAN	ABAVA	Watsonosteus sp. nov. Microbrachius cf. dicki	EDAY AND JOHN O'GROATS GROUP	W. fletti M. dicki
GIVETIAN	BURTNIEKI	Dickosteus? markae Gyroptychius elgae	UPPER CAITHNESS FLAGSTONES GROUP	T. pholidotus O. panderi G. milleri D. threiplandi
EIFELIAN	ARUKÜLA	Dickosteus? grossi Osteolepis baltica Gyroptychius pauli Thursius estonicus		
EIFELIAN	NARVA: KERNAVË LEIVU	Coccosteus cuspidatus? Thursius fischeri Gyroptychius grossi Rhamphodopsis cf. threiplandi	ACHANARRAS LMST.	C. cuspidatus G. agassizi T. moythomasi R. threiplandi
EIFELIAN	NARVA: VADJA	Coccosteus sp.	LOWER CAITHNESS FLAGSTONES GROUP	C. cuspidatus T. macrolepidotus
EIFELIAN	PÄRNU	Gyroptychius latvicus Thursius talsiensis	?	
EMSIAN	RËZEKNE		?	
EMSIAN	ĶEMERI	(with Rhinopteraspis dunensis)	SARCLET GROUP	

Figure 6.9. Ranges of gnathostomes in the more important vertebrate-yielding formations of the Devonian of the Baltic region with a suggested correlation with the Devonian of Scotland (information kindly provided by Dr E. Mark-Kurik, 1991 - personal communication).

correlation of vertebrate-dominated and invertebrate-dominated faunas is automatic. In the internal basins the same miospore succession may be present to serve as a control on the vertebrate stratigraphy. This brief survey of gnathostome distribution shows that each of the major fish groups has potential for biostratigraphical correlation. So far this potential

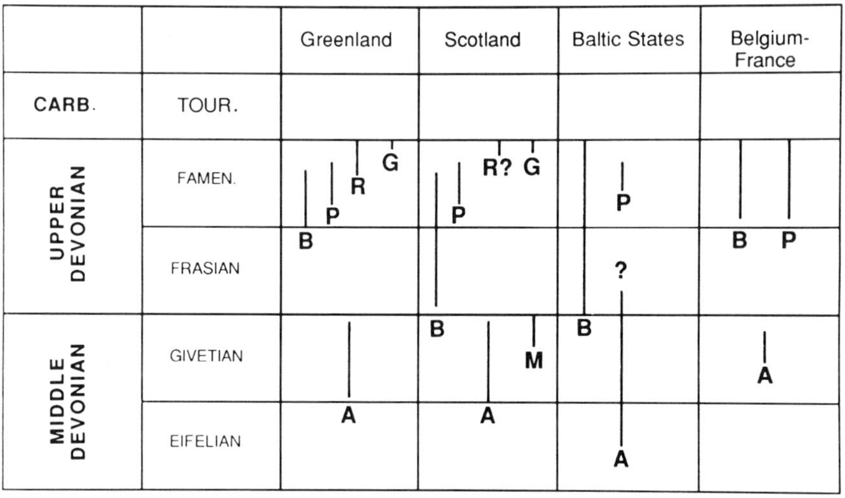

Figure 6.10. Ranges of vertebrate index genera in the Euramerican Old Red Sandstone. A, *Asterolepis*; B, *Bothriolepis*; G, *Groenlandaspis*; M, *Microbrachius*; P, *Phyllolepis*; R, *Remigolepis* (after Blieck, Mark-Kurik and Märss, 1988).

remains somewhat limited by incomplete taxonomic data, which in turn is a result of poor preservation and facies-influenced distribution. Much of the material evidence is of impaired value because of inadequate documentation when the fossils were collected.

The Acanthodii now appear to be of real value locally and even interregionally where they occur in number and are specifically identifiable, both as macro- and microfossils.

The Arthrodira constitute a highly complex and abundant group which so far may be generally regarded as of limited stratigraphic use. The distinctive *Groenlandaspis*, however, seems to be a widespread indicator of the late Famennian (Ritchie, 1975). Arthrodires occur in both marine and non-marine facies and some show evolutionary changes with time which serve to identify particular intervals of time.

The Antiarchi genera are relatively few but are widespread and of limited stratigraphic ranges within the Middle and Upper Devonian (Figure 6.10). Various categories of Osteichthyes appear at different levels in the Devonian System, as shown in Figure 6.2. It is rare that gnathostome faunas contain only one species of bony fish, but of several present one may be the most abundant. It remains to be seen if a pattern or patterns of assemblage or community evolution can be identified in the Middle and Late Devonian. If this can be achieved the stratigraphic value of these faunas will be enhanced.

Chondrichthyan fossils are very restricted in their nature and occurrence

FAUNA	Central Timan: Ustrednaya Beds	Baltic Main Devonian Field: Snetnaya Gora Beds	Poland, Holy Cross Mts. Lower Frasnian	Germany, Bergisch Gladbach U. Plattenkalk	Canada, Escuminac, P.Q.; Escuminac Formation
Holonema.			●	●	
Plourdosteus	●	●	●	●	●
"Ptyctodus"		●	●	●	
Ctenurella	●	●		●	
Rhynchodus		●	●		
Dinichthyidae	●		●		
Asterolepis	●	●		●	●
Bothriolepis	●	●			●
Acanthodii	●	●		●	
Latvius	●	●		●	
Eusthenopteron		●			●
Glyptolepis	●	●			●
Onychodontidae	●	●		●	
Holoptychiidae	●	●	●		●
Coelacanthiformes				●	●
Crossopterygii					●
Griphognathus		●		●	
Rhinodipterus	●	●		●	
Holodipterus			●		
Dipnoi				●	●
Moythomasia	●	●		●	
Palaeonisci				●	●
Elasmobranchi		●	●		

Figure 6.11. Faunal comparison of the Late Devonian gnathostome faunas in the Baltic and other localities on the eastern side of Euramerica (largely after Ivanov, 1990). Of these, only Escuminac is believed to be within an internal basin; the others are associated with marine strata.

and have yet to be credited with great biostratigraphic value. Long (1990) gives the known ranges of Devonian sharks based on teeth.

The outcome of these limitations is that pre-Eifelian gnathostomes in general are neither common enough nor known in sufficient detail to be of service to biochronology. Middle and Upper Devonian vertebrate communities tend to be broadly characteristic of particular stages, but smaller biostratigraphic units such as zones are restricted to local successions rather than wider areas (Figure 6.11). Blieck *et al.* (1988) propose the vertebrate succession of the Baltic area as a standard to which other vertebrate occurrences be compared. Their proposal is backed by a detailed study of the vertebrate faunas of that area and the recognition of vertebrate zones which can be correlated with the invertebrate-dominated facies. Similar reference successions are needed in other parts of the Euramerica Province such as Arctic Canada, SW USA or Appalachia. Eventually, comprehensive vertebrate biostratigraphy may result, with vertebrate evolution at both species and community levels becoming better understood.

In recent years an additional field of research has become significant, i.e. the identification of microvertebrate remains (minute scales, denticles, teeth, etc.) and the use of these fossils in biostratigraphy. This discipline is being followed in both the marine and the non-marine facies with encouraging results (Karatajute-Talimaa, 1978; Mader, 1986; Vieth, 1980, 1983). It has recently received recognition as a promising and productive research subject in the establishment of an IGCP Project. Gnathostomes appear to contribute significantly to the biostratigraphic record and this international endeavour may be expected to enlarge our understanding of their role world-wide.

At the end of the Devonian a large number of organisms, including many gnathostomes, became extinct. Biostratigraphy will assist in an understanding of the nature of these extinctions and, perhaps, also of the timing of the advent of new vertebrates including those descendants of Devonian gnathostomes, the tetrapods.

REFERENCES

Allen, K.C., 1967, Spore assemblages and their stratigraphic application in the Lower and Middle Devonian of North and Central Vestspitsbergen, *Palaeontology*, **10**: 280–97.

Allen, J.R.L., Dineley, D.L. and Friend, P.F., 1968, Old Red Sandstone Basins of North America and Northwest Europe. In D.H. Oswald (ed.), *International Symposium on the Devonian System, Calgary*, Calgary, **I**: 69–98.

Andrews, M.S., 1978, A possible occurrence of *Remigolepis* in the topmost Old Red Sandstone of Berwickshire, *Scottish Journal of Geology*, **14**: 311–15.

Ball, H.W. and Dineley, D.L., 1961, The Old Red Sandstone of Brown Clee Hill and the adjacent area, I: Stratigraphy, *Bulletin of the British Museum (Natural History)*, **A5**: 177–242.

Banks, H., 1980, Floral assemblages in the Silurian–Devonian. In D.L. Dilcher and T.N. Taylor (eds), *Biostratigraphy of fossil plants*, Dowden, Hutchinson and Ross, Stroudsburg, Pa., 1–24.

Belles-Isles, M., 1989, *Yvonaspis*, nouveau genre d'Osteostraci (Vertebrata, Agnatha) de Dévonien (Emsian–Eifelian des Gres de Gaspé) (Québec, Canada), *Canadian Journal of Earth Sciences*, **26**: 2396–401.

Bendix-Almgreen, S.E., 1976, Palaeovertebrate Faunas of Greenland. In A.W. Escher and A. Stuart-Watt (eds), *The Geology of Greenland*, Geological Survey of Greenland, 537–52.

Bernacsek, G.M., 1977, A Lungfish cranium from the Middle Devonian of the Yukon Territory, Canada, *Palaeontographica*, A, **157**: 175–200.

Bernacsek, G.M. and Dineley, D.L., 1977, New acanthodians from the Delorme Formation (Lower Devonian) of Northwest Territories, Canada, *Palaeontographica*, **158**: 1–25.

Blieck, A., 1984, Les hétérostracés pteraspidiformes, agnathes du Silurien–Dévonien du continent Nord-Atlantique et des blocs avoisinants, *Cahiers de Paléontologie; Editions de CNRS.*

Blieck, A., Goujet, D. and Janvier, P., 1987, The vertebrate stratigraphy of the Lower Devonian (Red Bay Group and Wood Bay Formation) of Spitsbergen, *Modern Geology*, **11**: 197–217.

Blieck, A. and Janvier, P., 1991, Silurian Vertebrates. In M.G. Bassett, P.D. Lane and D. Edwards (eds), The Murchison Symposium, Proceedings of an international conference on the Silurian System, *Special Papers in Palaeontology*, Nr. **44**: 345–89.

Blieck, A., Mark-Kurik, E. and Märss, T., 1988, Biostratigraphical correlations between Siluro–Devonian invertebrate-dominated and vertebrate-dominated sequences: the East Baltic example. In N.J. McMillan, A.F. Embry and D.J. Glass (eds), *Devonian of the World. Canadian Society of Petroleum Geologists, Memoir* **14**, **III**: 579–87.

Carroll, R.L., Belt, E.S., Dineley, D.L. and McGregor, D.C., 1972, *Excursion A59–Vertebrate Palaeontology of Eastern Canada*, Guidebook to Fieldtrip A59 for the 24th International Geological Congress, Montreal, Canada, 1972.

Chi, B.I. and Hills, L.V., 1976, Biostratigraphy and taxonomy of Devonian megaspores, Arctic Canada. *Bulletin of Canadian Petroleum Geology*, **24**: 640–818.

Dean, B., 1909, Studies on fossil fishes (sharks, chimaeroids and arthrodires), *Memoirs of the American Museum of Natural History*, **9**: 211–28.

Denison, R.H., 1951, Late Devonian Fresh-water Fishes from the Western United States, *Fieldiana, Geology*, **11**: 221–61.

Denison, R.H., 1958, Early Devonian Fishes from Utah. Part III. Arthrodira, *Fieldiana, Geology*, **11**: 461–551.

Denison, R.H., 1963, New Silurian Heterostraci from Southeastern Yukon, *Fieldiana, Geology*, **14**: 105–41.

Denison, R.H., 1964, The Cyathaspididae, a family of Silurian and Devonian jawless vertebrates, *Fieldiana, Geology*, **13**: 309–473.

Denison, R.H., 1968a, Early Devonian Lungfishes from Wyoming, Utah and Idaho, *Fieldiana, Geology*, **17**: 353–413.

Denison, R.H., 1968b, Middle Devonian Fishes from the Lemhi Range of Idaho, *Fieldiana, Geology*, **16**: 269–88.

Denison, R.H., 1978, Placodermi. Volume 2. In H.-P. Schultze (ed.), *Handbook of Palaeoichthyology*, Gustav Fischer Verlag, Stuttgart, New York, 124pp.

Denison, R.H., 1979, Acanthodii. Volume 5. In H.-P. Schultze (ed.), *Handbook of Palaeoichthyology*, Gustav Fischer Verlag, Stuttgart, New York, 56pp.

Denison, R.H., 1984, Further consideration of the phylogeny and classification of the Order Arthrodira (Pisces: Placodermi), *Journal of Vertebrate Paleontology*, **4**: 396–412.

Dineley, D.L., 1982, Difficulties in the vertebrate biogeography of the Old Red Sandstone of the British Isles. In B.S. Sokolov and M.A. Rzhonsnitskaya (eds), *Biostratigraphy of Lower and Middle Devonian Boundary Deposits*, Akademia NAUK CCCP, 54–57. (In Russian with English abstract.)

Dineley, D.L., 1984, Devonian Vertebrates in Biostratigraphy, *Proceedings of the Linnean Society of New South Wales*, **107** (1983): 185–96.

Dineley, D.L., 1990, Palaeozoic Fishing—the Franklinian Grounds. In C.R. Harington (ed.), *Canada's Missing Dimension*, National Museum of Natural History, Ottawa, 1, 55–80.

Dineley, D.L. and Liu, Y.-H., 1984, A new actinolepid arthrodire from the Lower Devonian of Arctic Canada, *Palaeontology*, **27**: 875–88.

Dineley, D.L. and Williams, B.P.J., 1986, The Devonian continental rocks of the Lower Restigouche River, Québec, *Canadian Journal of Earth Sciences*, **5**: 945–53.

Donovan, R.N., Foster, R.J. and Westoll, T.S., 1974, A Stratigraphical Revision of the Old Red Sandstone of North-eastern Caithness, *Royal Society of Edinburgh Transactions*, **69**: 167–201.

Dutertre, A.-P., 1930, Les Poissons Dévoniens du Boulonnais, *Bulletin, Société géologique de France*, **30**: 571–87.

Eastman, C.R., 1907, Devonic fishes from the New York formations, *New York State Museum*, Memoir 10.

Eastman, C.R., 1917, Fossil fishes in the collection of the United States National Museum, *Bulletin of the United States National Museum*, **52**: 235–304.

Elliott, D.K., 1983, New Pteraspidae (Agnatha, Heterostraci) from the Lower Devonian of Northwest Territories, Canada, *Journal of Paleontology*, **26**: 389–406.

Elliott, D.K., 1984, Siluro-Devonian fish biostratigraphy of the Canadian Arctic Islands, *Proceedings of the Linnean Society of New South Wales*, **107**, (1983): 197–209.

Embry, A.F., 1988, Middle–Upper Devonian Sedimentation in the Canadian Arctic Islands and the Ellesmerian orogeny. In N.J. McMillan, A.F. Embry and D.J. Glass (eds), *Devonian of the World*, Canadian Society of Petroleum Geologists, Memoir **14**, **II**: 15–28.

Embry, A.F. and Klovan, J.E., 1976, The Middle–Upper Devonian clastic wedge of the Franklinian geosyncline, *Canadian Society of Petroleum Geologists, Bulletin*, **24**: 485–639.

Fortier, Y.O. *et al.*, 1963, Geology of the North-Central part of the Arctic Archipelago. Northwest Territories (Operation Franklin), *Geological Survey of Canada*, Memoir **320**: 671pp.

Føyn, S. and Heintz, A., 1943, The Downtonian and Devonian Vertebrates of Spitsbergen. VIII. The English–Norwegian–Swedish Expedition, 1939, Geological Results. *Skrifter om Svalbard og Ishavets*, **No. 85**: 1–51.

Friend, P.F., 1961, The Devonian stratigraphy of North and Central Vestspitsbergen, *Proceedings of the Yorkshire Geological Society*, **33**: 77–118.

Friend, P.F., 1973, Devonian stratigraphy of Greenland and Svalbard, *American Association of Petroleum Geologists*, Memoir 19, 469–70.

Friend, P.F., Heintz, N. and Moody-Stuart, M., 1966, New unit terms for the Devonian of Spitsbergen and a new stratigraphic scheme for the Wood Bay Formation, *Norsk Polarinstitutet Arbok for 1965*, 59–63.

Ginter, M., 1990, Late Famennian sharks' teeth from the Holy Cross Mountains, Central Poland, *Acta Geologica Polonica*, **40**: 69–81.

Goujet, D., 1984, Les poissons placoderms du Spitsberg; Arthrodires Dolichothoraci de la Formation de Wood Bay (Dévonien inferieur), *Cahiers de Paléontologie;*

Editions CNRS, 284pp.

Goujet, D. and Blieck, A., 1979, Les Vertébrés de l'Assise des Schistes et Grès de Pernes (Dévonien de Nord de la France), *Annales de la Société Geologique du Nord*, **98**: 263–78.

Gregory, J.T., Morgan, T.G. and Reed, J.W., 1977, Devonian fishes in Central Nevada. In M.N. Murphy, W.B.N. Berry and C.A. Sandberg (eds), Western North America: Devonian, *University of California Riverside Campus Museum Contributions*, **4**: 112–12.

Gross, W., 1950, Die palaontologische und stratigraphische Bedeutung der Wirbeltierfaunen des Old Reds und der Marine Altpaläozoischen Schichten. *Abhandlungen der deutschen Akademie der Wissenschaften zu Berlin; Mathematische-naturwissenschaftliche Klasse*, 1949, Nr. **1**.

Gross, W., 1968, Fragliche Actinopterygier-Schuppen aus dem Silur Gotlands, *Lethaia*, **1**: 184–218.

Ivanov, A.S., 1990, The Snetnaya Gora fish assemblages of the Main Devonian Field and its biogeographic significance, *Vestnik Leningradskogo Universiteta*, Ser. 7, **1**: 94–8.

House, M.R., Richardson, J.B., Allen, J.R.L., Holland, C.H. and Westoll, T.S., 1977, A correlation of the Devonian rocks of the British Isles, *Geographical Society Special Report*, **No. 7**, London.

Jarvik, E., 1944a, On the dermal bones, sensory canals and pit lines of the skull in *Eusthenopteron foordi* Whiteaves, with some remarks on *E. save-soderberghi* Jarvik, *Kungliga Svenska Vetenskapsakademiens Handlingar*, **3(21)**: 1–48.

Jarvik, E., 1944b, On the exoskeletal shoulder-girdle of teleostomian fishes with special reference to *Eusthenopteron foordi* Whiteaves, *Kungliga Svenska Vetenskapsakademiens Handlingar*, **3(21)**: 1–48.

Jarvik, E., 1948, On the Middle Devonian Crossopterygians from Hornelen Field in Western Norway, *Universitetet Bergen Arbok, Naturvitenskapelig rekke*, **No. 8**: 1–48.

Jarvik, E., 1961, Devonian Vertebrates. In G.O. Raash, (ed.), *Geology of the Arctic*, University of Toronto Press, **1**: 197–20.

Jarvik, E., 1967, On the structure of the lower jaw in dipnoans: with a description of an Early Devonian dipnoan from Canada, *Melanognathus canadensis* gen. et sp. nov., *Zoological Journal of the Linnean Society*, **47**: 155–84.

Jessen, H., 1966, Die Crossopterygier des Oberen Plattenkalkes (Devon) der Bergisch-Gladbach-Paffrather Mulde (Rheinisches Schiefergebirge) unter Berücksichstigung von amerikanischem und europäischem *Onychodus*-Material, *Arkiv für Zoologi*, Serie 2, **18**: 305–89.

Jessen, W., 1975, A new choanate fish, *Powichthys thorsteinsson*, ng.n.sp., from the early Lower Devonian of the Canadian Arctic Archipelago, *Problèmes actuels de paléontologie, Colloques Internationaux du Centre nationale de la Recherche Scientifique*, **218**: 213–22.

Karatajute-Talimaa, V.N., 1978, Thelodonty Silura i Devona SSSR i Spicbergena (Silurian and Devonian Thelodonts of the USSR and Spitsbergen), Mosklas, Vilnius (in Russian).

Kerr, J.W., 1977, Cornwallis Fold Belt and the mechanism of basement uplift, *Canadian Journal of Earth Sciences*, **14**: 1374–1401.

Kiaer, J., 1915, Upper Devonian Fish Remains from Ellesmere Land, with remarks on *Drepanaspis*, Report on the 2nd Norwegian Arctic Expedition in the 'Fram', 1898–1902, **4**, No. 33, *Videnskabs-selskabet i Kristiania*, 72pp.

Kulczycki, J., 1957, Upper Devonian fishes from the Holy Cross Mountains (Poland), *Acta Palaeontologica Polonica*, **2**, (4).

Lawrence, D.A. and Rust, R.B., 1988, The Devonian clastic wedge of eastern Gaspé and the Acadian orogeny. In N.J. McMillan, A.F. Embry and D.J. Glass (eds), *Devonian of the World*, Canadian Society of Petroleum Geologists, Memoir **14**, **III**: 543–64.

Lelièvre, H., Goujet, D. and Morzadec, P., 1983, Les poissons du Frasnien (Dévonien supérieur) de Traonliors en Plougastel-Daoulas (Bretagne occidentale). *Bulletin de la Société Geologique et Mineral Bretagne*, (C) **15**: 75–83.

Long, J., 1983, New bothriolepid fish from the Late Devonian of Victoria, Australia, *Palaeontology*, **26**: 295–320.

Long, L., 1990, Late Devonian chondrichthyan and other microvertebrate remains from northern Thailand, *Journal of Vertebrate Paleontology*, **10**: 59–77.

McKerrow, W.S. and Scotese, C.R., 1990, *Palaeozoic Palaeogeography and Biogeography*, Geological Society, Memoir No. 12, 435pp.

McMillan, N.J., Embry, A.F. and Glass, D.J., 1988, *Devonian of the World, Canadian Society of Petroleum Geologists, Memoir* **14,** 3 vols.

McGregor, D.C., 1973, Lower and Middle Devonian spores of eastern Gaspé, Canada, I, Systematics, *Palaeontographica*, **142B**: 1–77.

McGregor, D.C., 1977, Lower and Middle Devonian spores of eastern Gaspé, Canada, II, Biostratigraphy, *Palaeontographica*, **163B**: 111–42.

Mader, H., 1986, Schuppen und Zahne von Acanthodiern und Elasmobranchiern aus dem Unter-Devon Spaniens (Pisces), *Göttingen Arbeiten Geologie und Palaontologie*, **No. 28**.

Mark-Kurik, E., 1969, Distribution of Vertebrates in the Silurian of Estonia, *Lethaia*, **2**: 145–52.

Mark-Kurik, E., 1978, The correlation of the Middle Devonian of the East Baltic and the north of the British Isles, Abstracts. *Palaeontological Association Symposium on the Devonian System, Bristol, 1978*, 37.

Marshall, C.R., 1987, A List of Fossil and Extant Dipnoans, *Journal of Morphology Supplement*, **1**, (1986), 15–23.

Miall, A.D., 1970, Continental marine transition in the Devonian of Prince of Wales Island, Northwest Territories, *Canadian Journal of Earth Sciences*, **7**: 125–44.

Miles, R.S., 1964, A large arthrodire plate from Chautauqua County, New York, *Arkiv für Zoologi*, Serie 2, Bd.16, 545–50.

Miles, R.S., 1966, *Protitanichthys* and some other coccosteomorph arthrodires from the Devonian of North America, *Kungliga Svenska Vetenskapsakademiens Handlingar*, **4(10)**: 1–49.

Miles, R.S., 1973a, Articulated Acanthodian Fishes from the Old Red Sandstone of England, with a Review of the Structure and Evolution of the Acanthodian Shoulder-Girdle, *Bulletin of the British Museum (Natural History) Geology*, **24**: 112–213.

Miles, R.S., 1973b, An actinolepid arthrodire from the Lower Devonian Peel Sound Formation, *Palaeontographica*, **A143**: 109–18.

Moy-Thomas, J.A. and Miles, R.S., 1971, *Palaeozoic Fishes* (2nd edn.), Chapman and Hall, London, 250pp.

Murphy, M.A., Morgan, T.G. and Dineley, D.L., 1976, *Asterolepis* sp. from the Upper Devonian of Central Nevada, *Journal of Paleontology*, **50**: 467–71.

Nathorst, A.G., 1904, Die oberdevonische Flora des Ellesmerelandes. Report of the 2nd Norwegian Arctic Expedition in the Fram, 1898–1902, *Vidensk.-Selskabet i Kristiania*, **I**, No. 1, 22pp.

Newberry, J., 1889, The Paleozoic Fishes of North America, *United States Geological Survey*, Monograph 16, 225pp.

Nikiforova, O.I., 1977, Podolia. In A. Martinsson (ed.), *The Silurian–Devonian Boundary*, International Union of Geological Sciences, Stuttgart, (A), **5**: 52–64.

Obruchev, D. and Karatajute-Talimaa, V., 1967, Vertebrate faunas and correlation of Ludlovian–Lower Devonian in eastern Europe. In C. Patterson and P.H. Greenwood (eds), *Fossil Vertebrates*, Linnean Society of London and Academic Press, 5–14.

Ørvig, T., 1957a, Remarks on the vertebrate fauna of the lower Upper Devonian of Escuminac Bay, P.Q., Canada, with special reference to the Porolepiform Crossopterygians, *Arkiv für Zoologi*, **2**: 367–426.

Ørvig, T., 1957b, Notes on some Palaeozoic vertebrates from Spitsbergen and North America, *Norsk Geologisk Tidsskrift*, **31**, 285–353.

Ørvig, T., 1960, New Finds of acanthodians, arthrodires, crossopterygians, ganoids and dipnoans in the Upper Middle Devonian calcareous flags (Oberer Plattenkalk) of the Bergisch-Gladbach-Paffrath Trough (Part 1), *Paläontologische Zeitschrift*, **34**: 295–335.

Ørvig, T., 1961, New Finds of acanthodians, arthrodires, crossopterygians, ganoids and dipnoans in the Upper Middle Devonian calcareous flags (Oberer Plattenkalk) of the Bergisch-Gladbach-Paffrath Trough (Part 2), *Paläontologische Zeitschrift*, **35**: 10–27.

Ørvig, T., 1969, Vertebrates of the Wood Bay Group and the position of the Emsian–Eifelian boundary in the Devonian of Spitsbergen, *Lethaia*, **2**: 273–328.

Pageau, Y., 1969, Nouvelle faune ichthyologique de Devonien Moyen dans les Gres de Gaspé (Québec), II Morphologie et systematique, *Le Naturaliste Canadien*, **96**: 399–478, 805–89.

Pan, J., 1981, Devonian Antiarch Biostratigraphy of China, *Geological Magazine*, **118**: 69–75.

Pan, J., 1988, Vertebrates from Devonian facies in China. In N.J. McMillan, A.F. Embry and D.J. Glass (eds), *Devonian of the World*, Canadian Society of Petroleum Geologists, Memoir **14, III**: 609–18.

Richardson, J.B. and McGregor, D.C., 1986, Silurian and Devonian spore zones of the Old Red Sandstone Continent and adjacent regions, *Geological Survey of Canada, Bulletin*, **364**: 79pp.

Ritchie, A., 1975, *Groenlandaspis* in Antarctica, Australia and Europe, *Nature*, **254**: 569–73.

Russell, L.S., 1939, Notes on the occurrence of fossil fishes in the Upper Devonian of Miguasha, Québec, Contributions from the Royal Ontario Museum, *Palaeontology*, **12**: 1–4.

Schei, P., 1903, Summary of Geological results; 2nd Norwegian North Polar Expedition in the Fram, 1898–1902, *Geographical Journal*, **22**: 56–65.

Schei, P., 1904, Preliminary account of the geological investigations made during the second Norwegian North Polar Expedition in the Fram, 1898–1902. In Otto Sverdrup, *New Land*, II, 455–66.

Schultze, H.-P., 1970, Palaeoniscoidea-Schuppen aus dem Unter Devon Australiens und Kanadas und aus dem Mittel Devon Spitsbergens, *Bulletin of the British Museum (Natural History), Geology*, **16**: 343–68.

Scotese, C.R., 1986, Phanerozoic reconstructions: a new look at the assembly of Asia, *University of Texas Institute for Geophysics Technical Report*, **66**: 1–54.

Scotese, C.R. and McKerrow, W.S., 1990, *Palaeozoic Palaeogeography and Biogeography*, Geological Society of London, Memoir 12, 320 pp.

Soehn, K.L. and Wilson, M.V.H., 1990, A complete, articulated heterostracan from Wenlockian (Silurian) beds of the Delorme Group, Mackenzie Mountains, Northwest Territories, Canada, *Journal of Vertebrate Paleontology*, **10**: 405–19.

Spjeldnaes, N., 1967, Acanthodians from the Siluro–Devonian of Ellesmere Island. In D.H. Oswald (ed.), *International Symposium on the Devonian System, Calgary, 1967*, **II**: 807–14, Alberta Society of Petroleum Geologists.

Stensiö, E.A., 1939, A New Anaspid from the Upper Devonian of Scaumenac Bay in Canada, with remarks on the other Anaspids, *Kungliga Svenska Vetenskaps-akademiens Handlingar*, **18**: 125pp.

Thomson, K.S., 1976, The faunal relationships of rhipidistian fishes (Crossopterygii) from the Catskill (Upper Devonian) of Pennsylvania, *Journal of Paleontology*, **50**: 1203–8.

Thorsteinsson, R., 1958, Cornwallis and Little Cornwallis Islands, District of Franklin, Northwest Territories, *Geological Survey of Canada*, Memoir 294, 134pp.

Thorsteinsson, R., 1967, Preliminary note on Silurian and Devonian ostracoderms from Cornwallis and Somerset Islands, Canadian Arctic Archipelago, *Colloques internales de CNRS*, **163**: 45–7.

Thorsteinsson, R., 1973, Dermal elements of a new lower vertebrate from the Middle Silurian (Upper Wenlockian) rocks of the Canadian Arctic Archipelago, *Palaeontographica*, **143, A**: 51–7.

Thorsteinsson, R. and Tozer, E.T., 1970, Geology of the Arctic Archipelago. In R.J.W. Douglas (ed.), Geology and Economic Minerals of Canada, 5th edn., *Geological Survey of Canada Economic Geology Series*, Report Nr. **1**: 547–90.

Valiukevicius, J., 1988, Correlation of Lower and Middle Devonian deposits of the USSR with acanthodian assemblages. In N.J. McMillan, A.F. Embry and D.J. Glass (eds), *Devonian of the World*, Canadian Society of Petroleum Geologists, Memoir **14, III**: 601–7.

Veith, J., 1980, Thelodontier- Acanthodier- und Elasmobranchier-Schuppen aus dem Unter-Devon der Kanadischen Arktis (Agnatha, Pisces), *Göttinger Arbeiten Geologie und Paläontologie*, **23**: 69pp.

Veith-Schreiner, J., 1983, Fisch-Schuppen und -Zahne aus der Eifeler Kalkmulden-Zone (Emsium, Eifelium), *Senckenbergiana lethaea*, **64**: 129–77.

Wells, J.W., 1944, Fish remains from the Middle Devonian bone beds of the Cincinnati Arch Region, *Palaeontographica Americana*, **3**: 1–16.

Westoll, T.S., 1977, Northern Britain. In M.R. House, J.B. Richardson, W.G. Chaloner, J.R.L. Allen, C.H. Holland and T.S. Westoll, *A Correlation of the Devonian Rocks of the British Isles*, Geological Society, Special Report Nr. **8**: 66–93.

Westoll, T.S., 1979, Devonian Fish Biostratigraphy. In M.R. House, C.T. Scrutton and M.G. Bassett (eds), *The Devonian System*, Special Papers in Palaeontology, **23**: 341–53.

White, E.I., 1950, The Vertebrate Faunas of the Lower Old Red Sandstone of the Welsh Borders, *Bulletin of the British Museum (Natural History), Geology*, **A1**: 51–67.

White, E.I., 1961, The Old Red Sandstone of Brown Clee Hill and adjacent area. II. Palaeontology, *Bulletin of the British Museum (Natural History), Geology*, **A5**: 243–310.

Whiteaves, J.F., 1881, On some fossil fishes, crustacea and mollusca from Devonian rocks of Campbellton, N.B., with description of five new species, *Canadian Naturalist*, serie nouvelle, **10**: 93–101.

Woodrow, D.L. and Sevon, W.D., 1985, *The Catskill Delta, Geological Society of America, Special Paper*, Nr. **201**: 1–8.

Woodward, A.S., 1891, The Devonian fish faunas of Spitsbergen, *Annals and Magazine of Natural History*, 6 series, **8**: 1–15.

Young, G.C., 1981, Biogeography of Devonian Vertebrates, *Alcheringa*, **5**: 225–43.

Young, G.C., 1984, Comments on the Phylogeny and Biogeography of Antiarchs (Devonian Placoderm Fishes) and the Use of Fossils in Biogeography, *Proceedings of the Linnean Society of New South Wales*, **107**, (1983): 443–73.

Young, G.C., 1990, Devonian vertebrate distribution patterns and cladistic analysis of palaeogeographic hypotheses. In W.S. McKerrow and C.R. Scotese (eds), *Palaeozoic Palaeogeography and Biogeography*, Geological Society, Memoir 12, 243–55.

Zangerl, R., 1981, Chondrichthyes 1. Paleozoic Elasmobranchs, In H.-P. Schultze (ed.), *Handbook of Paleoichthyology*, 3, Gustav Fischer Verlag, Stuttgart, 114pp.

Ziegler, P.A., 1988, Laurussia–the Old Red Continent. In N.J. McMillan, A.F. Embry and D.J. Glass (eds), *Devonian of the World*, Canadian Society of Petroleum Geologists, Memoir 14, **I**: 15–48.

Chapter 7

SILURIAN–DEVONIAN VERTEBRATE BIOSTRATIGRAPHY OF WESTERN GONDWANA AND RELATED TERRANES (SOUTH AMERICA, AFRICA, ARMORICA-BOHEMIA, MIDDLE EAST)*

H. Lelièvre, P. Janvier and A. Blieck

INTRODUCTION

In Silurian and Devonian times, West Gondwana consisted chiefly of what is now South America, Africa and the Arabic Peninsula. In addition, a number of small terranes are supposed to have extended along the northern margin of this large land mass (Figure 7.1). These were chiefly Armorica (now the Iberic Peninsula and France up to Normandy), Bohemia (down to the alpine area), probably Italy and Greece, and a series of terranes referred to by Sengör (1984) as Cimmeria, comprising western Turkey, southern Caucasus, northern and central Iran, most of Afghanistan, and in its easternmost part, several terranes included now in Tibet (Qiang Tang and Lhasa blocks) and southeast Asia [see, for recent review and current models, McKerrow and Scotese (eds), 1990; also Chapter 2]. According to current plate tectonics models, the fate of these Cimmerian terranes is varied, some becoming later accreted to mainland Gondwana, whereas others formed a zone of contact between the latter and more northerly situated continents, i.e., Baltica, Siberia (Angara) and South China. However, one must be aware that all current large-scale models of Silurian–Devonian geography are probably very far from the truth, and that the worse method would be to study the distribution of Silurian–Devonian organisms (vertebrates or else) in the light of these maps (Figure 7.1). As far as west Gondwanan vertebrate faunas are concerned, the scarcity of the

* A contribution to IGCP Project 328.

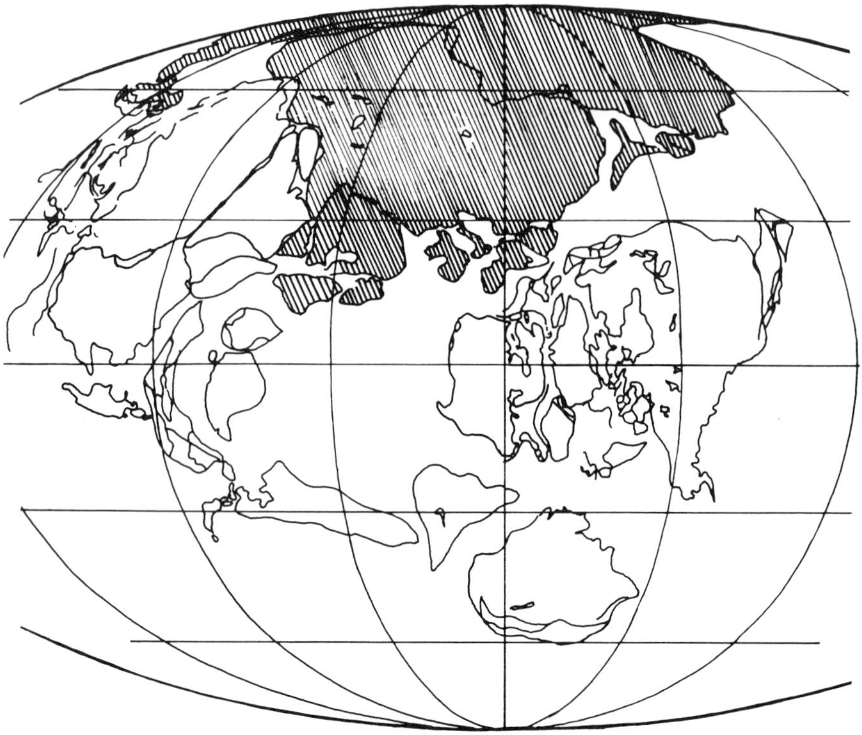

Figure 7.1. Current Middle Devonian palaeogeographic reconstruction, based on palaeomagnetic data (from McKerrow and Scotese, 1990). The areas under consideration in this paper (West Gondwana and related terranes) are hatched.

record is a major obstacle to the discovery of significant palaeobiogeographic patterns. Only a few guesses can be made, some of which are not clearly consistent with the current palaeogeographical base maps (Iranian–Russian or Bolivian–Chinese 'tracks'). Therefore, we have chosen to give here an account of the extensive Silurian–Devonian vertebrate record in these regions, aimed at helping future field investigators or theoreticists. The specimens illustrated in this review have been selected on the basis of various criteria: their relative abundance in the localities under discussion, their restricted distribution (to one area or a locality) or simply their good state of preservation.

First mentions of Early Silurian–Devonian vertebrate records from Gondwanan regions include acanthodian spines from the Bosphorus (Turkey) and from France (Rouault, 1858), acanthodian and 'heterostracan' remains from the Amazon Basin of Brazil (Katzer, 1897), as well as Silurian acanthodians from Portugal (Priem, 1910) and undetermined fish remains from the Early

Devonian of the Falkland Islands (Hallé, 1912, see ref. in Greenway, 1972). Significant records from West Gondwana began to appear after World War II, first in Morocco (Lehman, 1952a, 1956, 1963, 1976, 1977a, b, 1978), then in Iran, Afghanistan and Turkey (Golshani *et al.*, 1972; Schultze, 1973b; Janvier, 1974, 1977a, b, 1978b, 1979, 1980, 1983; Janvier and Marcoux, 1977; Janvier and Martin, 1978, 1979; Janvier and Pan, 1982; Janvier and Ritchie, 1977; Lelièvre *et al.*, 1981; Turner and Janvier, 1979; Hamdi and Janvier, 1981; Janvier *et al.*, 1984; Goujet and Janvier, 1984; Blieck and Goujet, 1977; Blieck *et al.*, 1980, 1982) and finally South Africa (Chaloner *et al.*, 1980) and South America (Janvier, 1976, 1978a, 1987; Goujet *et al.*, 1984, 1985; Janvier and Suarez-Riglos, 1986; Gagnier, 1987; Gagnier *et al.*, 1988, 1989; Janvier and Melo, 1987, 1988, 1992). During the past decade some relatively good vertebrate material has been found in the Armorica, in particular Spain (Schultze, 1973a; Babin *et al.*, 1976; Blieck, 1982; Mader, 1986; Lelièvre and Goujet, 1990), Montagne Noire (France) (Lelièvre *et al.*, 1987) and Brittany (Goujet, 1976, 1980, 1981; Lelièvre *et al.*, 1983). Most of these Silurian and Devonian vertebrate localities are local occurrences (except in Morocco) and do not compare with the vast vertebrate-bearing Devonian fields of Spitsbergen or the Russian Platform. North Indian records have turned out to be flawed with anomalies and misplacings (see Talent *et al.*, 1988).

SOUTH AMERICA

As mentioned above, the earliest record of vertebrate remains from the Devonian of South America are from the Lower Devonian of the Amazon Basin, where Katzer (1897) recorded spines of *Machaeracanthus* and plates of '*Pteraspis*', the latter being certainly an erroneous record. Then, Hallé (1912; see ref. in Greenway, 1972) recorded undetermined fish remains from the Falkland (Malvinas) Islands. This has been confirmed by later findings in the sandstone of the Port Philomel Beds (Figure 7.3Q). The record of '*Asterolepis*' and '*Cephalaspis*' from the Devonian of the Precordillera de San Juan, Argentina (Frenguelli, 1951) is erroneous (possibly arthropod or plant remains).

1. Bolivia

The first extensive Devonian vertebrate collections were made by L. Branisa and the staff of the Bolivian Oil Company YPFB in the late sixties. Part of this material was first described by Janvier (1976, 1978a) who noticed the abundance of peculiar chondrichthyan remains in the concretions of the Belén and Sicasica Formations of the La Paz–Oruro area (Figure 7.2 A1). Later discoveries have yielded placoderm, acanthodian and actinopterygian remains in the same area (Goujet *et al.*, 1984, 1985; Janvier and Suarez-Riglos, 1986; Janvier, 1987, Gagnier *et al.*, 1988, 1989), as well as in the Cochabamba (Figure 7.2 A2) and Sucre areas (Figure 7.2 A3, 4). These

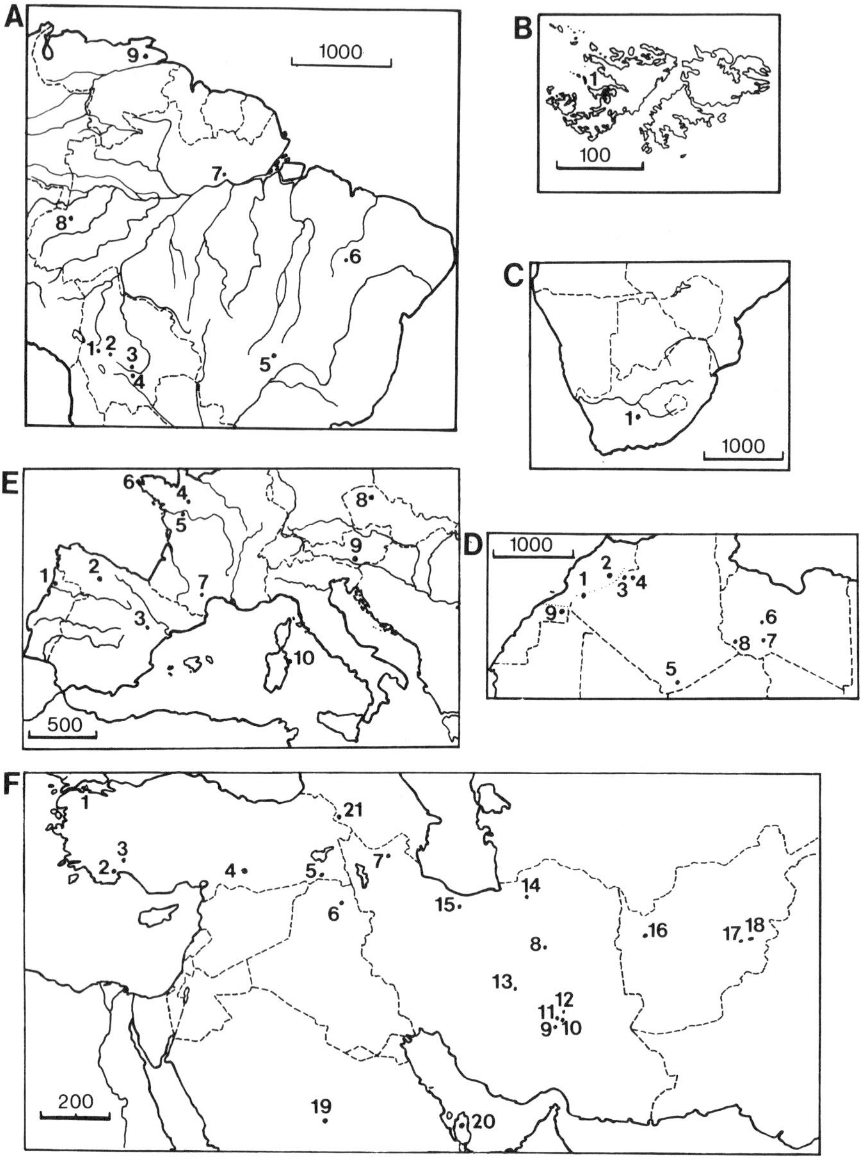

Figure 7.2. Geographic distribution of the Silurian-Devonian vertebrate localities under consideration in this paper.

findings are casual and range from the uppermost Silurian Tarabuco and Catavi Formations, to the Upper Devonian (Famennian) Iquiri Formation (see for details on stratigraphy, Janvier and Suarez-Riglos, 1986; Gagnier *et al.*, 1989).

The Silurian occurrences, mostly from the Tarabuco Formation (Figure 7.2 A4), consist of spines and jaw bones of climatiid (quite similar to *Climatius*) and ischnacanthid acanthodians (Figure 7.3K, L).

The Devonian occurrences are mostly from the shales lying just below and above the two thick quartzite beds referred to as the 'Cuarcita Condoriquina' and the 'Cuarcita Cruz Loma', that is, in the Sicasica and Belén Formations of the La Paz area (Figure 7.2 A1), or their equivalents in the East, the Icla and Huamampampa Formations. The following taxa have been recorded:

Placodermi:
- *Bolivosteus chacomensis* Goujet, Janvier and Suarez-Riglos (Figure 7.3J)

Acanthodii:
- Climatiida gen. et sp. indet. (Figure 7.3L)
- Ischnacanthida gen. et sp. indet. (Figure 7.3K)

Chondrichthyes:
- *Zamponiopteron triangularis* Janvier and Suarez-Riglos (Figure 7.3N)
- *Z. falciformis* Janvier and Suarez-Riglos
- *Z. spinifera* Janvier and Suarez-Riglos
- *Pucapampella rodrigae* Janvier and Suarez-Riglos (Figure 7.3M)
- *Antarctilamna* sp. (Figure 7.3I)

Actinopterygii:
- ?*Moythomasia* sp. (Figure 7.3P)

This fauna is characterized by the abundance of chondrichthyan remains,

---

Figure 7.2 – cont.

A, South America: 1, La Paz–Oruro area (chiefly Chacoma, Belén and Sicasica); 2, Cochabamba area; 3, Seripona and La Higuera; 4, Sucre area (including Tarabuco, Pucapampa, Tabacal–Tomina); 5, Ponta Grossa Formation, Parana; 6, Picos (Parnaiba Basin); 7, Sucuriju Creek (Para) (borehole core and outcrop); 8, Jandiatuba Basin (borehole); 9, Maturin Basin (borehole). B, Falkland (Malvinas) Islands: 1, Dunose Head. C, South Africa: 1, Barrydale. D. North Africa: 1, Dra Valley; 2, Tafilalt and Maider; 3, Béni-Abbès; 4, Ougarta Chain; 5, Idekel Sandstone; 6, Ghadamès Basin and Chatti; 7, Murzuk Basin; 8, Tihemboka; 9, Zemmour. E, Armorica: 1, Sao Felix, Laundos; 2, Cantrabrian Chain (Oviedo area); 3, Zaragossa area; 4, Mayenne; 5, Angers; 6, Brest; 7, Montagne Noire; 8, Prag area; 9, Graz area; 10; Sardinia; F, Middle East: 1, Bosphorus and Marmara Sea area; 2, Armutgözlek Tepe, Kemer; 3, Bademli; 4, Hazro; 5, Cukurca, Hakkari; 6, Pirispiki Sandstone, near Arbol; 7, Zonuz; 8, Niaz, near Tabas; 9, Ab-e-Morad; 10, Chanaruh; 11, Hootk; 12, Tangil-e-ab-Garm (Ravar); 13, Angireh; 14, Khush-Yeilagh; 15, Djeiroud; 16, Robat-e-Pai; 17, 18, Tokya-Dewal anticline; 19, Jawf; 20, Qatar; 21, Nakhitchevan.

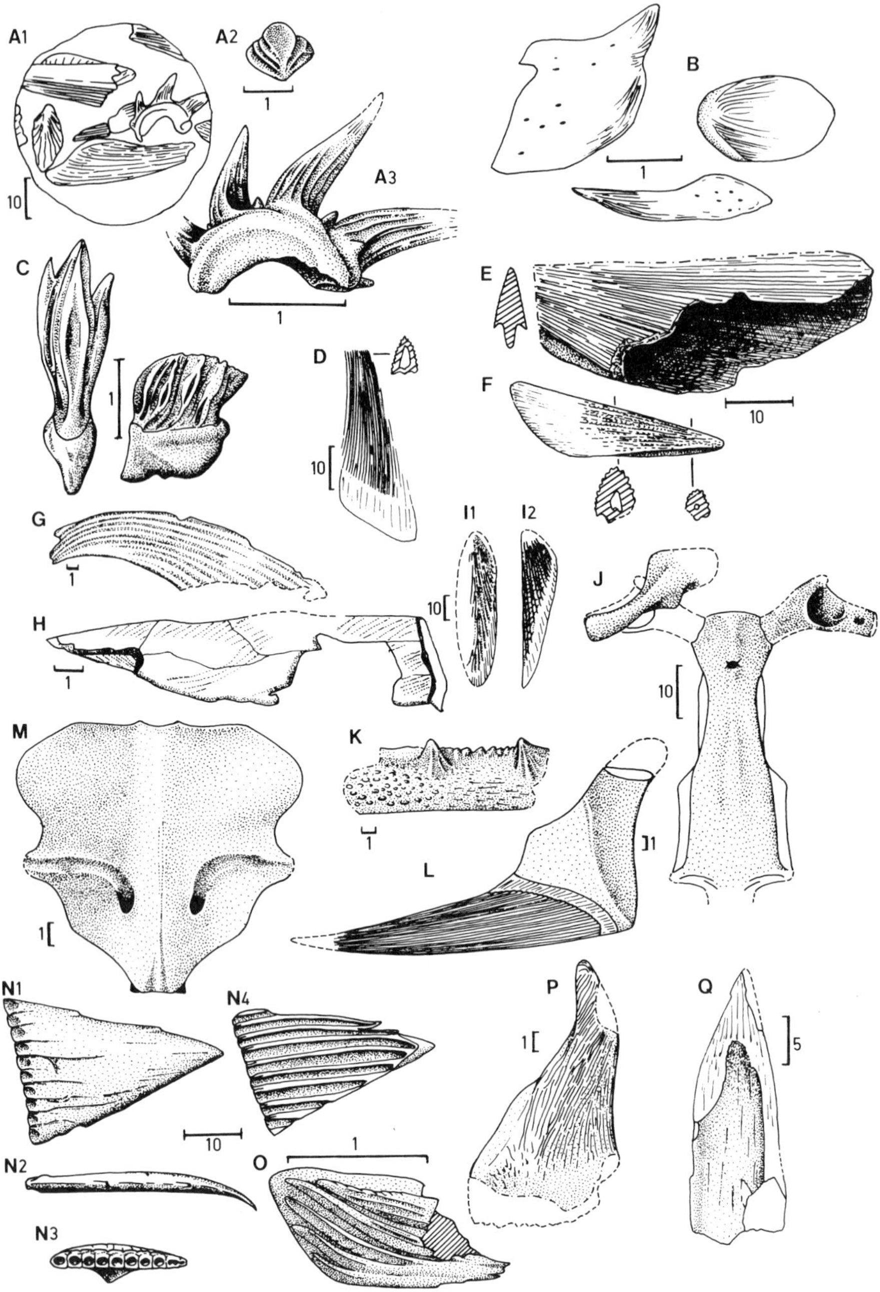
A1
A2
A3
B
C
D
E
F
G
H
I1
I2
J
J1
K
L
M
N1
N2
N3
N4
O
P
Q
10
1
5

most of them being quite different from all previously known Devonian chondrichthyans. The peculiar triangle-shaped elements referred to *Zamponiopteron* (Figure 7.3N) are made up of calcified cartilage (not bone, as suggested by Janvier and Suarez-Riglos, 1986) and clearly belong to the paired fins of some unknown chondrichthyan. The elements referred to as *Pucapampella* (Figure 7.3M) are not of neurocranial origin but are rather synarcual elements, suggesting the presence of holocephalans (bradyodonts or chimaeroids), in fact the earliest known occurrence of the group (Janvier and Dingerkus, 1991).

Most of the material from these formations is not found *in situ* and only a few specimens are well dated. The occurrence of the actinopterygian provisionally referred to *Moythomasia* (Figure 7.3P) is dated as Givetian (Gagnier *et al.*, 1989) and all the material of the rhenanid placoderm *Bolivosteus* (Figure 7.3J) are from a particular bed of the Chacoma locality, which is dated as Eifelian (P. Racheboeuf, pers. comm., 1991). Finally, isolated actinopterygian scales have been found in the Famennian Iquiri Formation of southeastern Bolivia (Figure 7.3O).

Figure 7.3 - cont.

Figure 7.3. Silurian and Devonian vertebrates of South America. A, acanthodian assemblage, Early Lochkovian Manacapuru Formation, Sucuriju Creek, Amazon Basin, Para, Brazil; spines and teeth in a borehole core (A1), isolated scale (A2), and tooth-whorl (A3); B, actinopterygian scales, uppermost Devonian or lowermost Carboniferous, Jandiatuba Basin, Upper Amazon Basin, northwestern Brazil. C–P, Silurian and Devonian vertebrates from Bolivia; C–H, specimens from the Lower Devonian (Lochkovian-Pragian) of Seripona, Chuquisaca; C, *Turinia gondwana* Turner, scales; D, *Sinacanthus boliviensis* Gagnier, fin spine; E, *Bolivosteus sagitalis* Gagnier, fin spine; F, *Antarctilamna seriponensis* Gagnier, fin spine; G, *Climatius enodicosta* Gagnier, fin spine; H, *Gyracanthus seriponensis* Gagnier, fin spine; I, *Antarctilamna* sp., Middle Devonian, Belén Formation, La Paz area, fin spine in anterior (I1) and lateral (I2) view; J, *Bolivosteus chacomensis* Goujet, Janvier and Suarez-Riglos, Eifelian, Belén Formation, Chacoma (La Paz), reconstruction of the neurocranium, palatoquadrate and Meckelian cartilage in ventral view; K, Ischnacanthiforme gen. et sp. indet., uppermost Silurian, Tarabuco Formation, Sucre area, fragment of jaw bone; L, *Climatius* sp., uppermost Silurian, Tarabuco Formation, Sucre area, fin spine and scapulocoracoid; M, *Pucapampella rodrigae* Janvier and Suarez-Riglos, Eifelian, Belén Formation, La Paz area, synarcual in ventral view; N, *Zamponiopteron triangularis* Janvier and Suarez-Riglos, Belén and Sicasica Formations, La Paz area, paired fin element in dorsal (N1), anterior (N2) and medial (N3) view, and in horizontal section (N4); O, Actinopterygian scale, Famennian Iquiri Formation, Sucre area; P, ?*Moythomasia* sp., Givetian, top of the Belén Formation, La Paz area, cleithrum; Q, chondrichthyan fin spine, Middle Devonian, Port Philomel beds, West Falkland Islands. Scales in mm. (A, from Janvier and Melo, 1987; B, from Janvier and Melo, 1987; C–H, from Gagnier *et al.*, 1988; I–O, from Janvier and Suarez-Riglos, 1986; P, from Gagnier *et al.*, 1989; Q specimen British Museum, Natural History, P.29309).

Beside these sparse findings, there is, in the lowermost part of the Santa Rosa Formation, a rich bone-bed which crops out in several localities, namely Seripona and La Higuera (Gagnier, 1987; Gagnier *et al.*, 1988, 1989), along the Rio Grande, and further south, Tabacal-Tomina (Vallegrande and Sucre areas, Central Bolivia, Figure 7.2 A3, 4). It has yielded the following vertebrates:

Thelodonti:
- *Turinia gondwana* Turner (in Gagnier *et al.*, 1988) (Figure 7.3C)

Acanthodii:
- *Climatius enodicosta* Gagnier (in Gagnier *et al.*, 1988) (Figure 7.3G)
- *Gyracanthus seriponensis* Gagnier (in Gagnier *et al.*, 1988) (Figure 7.3H)
- *Gyracanthus* sp.
- *Gomphonchus pluriformis* Gagnier (in Gagnier *et al.*, 1988)
- Ischnacanthida gen. et sp. indet.
- *Nodonchus rectus* Gagnier (in Gagnier *et al.*, 1988)
- *Onchus punctuatus* Gagnier (in Gagnier *et al.*, 1988)
- *Onchus sicaeformis* Gagnier (in Gagnier *et al.*, 1988)
- *Onchus* sp.

?Chondrichthyes:
- *Sinacanthus boliviensis* Gagnier (in Gagnier *et al.*, 1988) (Figure 7.3D)
- ?*Sinacanthus* sp.

Chondrichthyes:
- *Antarctilamna seriponensis* Gagnier (in Gagnier *et al.*, 1988) (Figure 7.3F)
- *Bolivacanthus sagitalis* Gagnier (in Gagnier *et al.*, 1988) (Figure 7.3E)

The age of this bone-bed is dated as Lochkovian–Pragian based on the chitinozoan assemblage (Gagnier *et al.*, 1989), yet the large thelodont *Turinia gondwana* rather compares to Middle Devonian forms from Antarctica and Australia (Turner in Gagnier *et al.*, 1988).

The biogeographical affinities of these Bolivian Siluro–Devonian vertebrates may be closer to Australia, Antarctica and possibly China, although many taxa, such as rhenanid placoderms or *Climatius*-like acanthodians may just have a worldwide distribution. *Zamponiopteron*, never recorded outside Bolivia, may well be an endemic form. In contrast, the *Sinacanthus* spines, possibly belonging to a chondrichthyan, are quite similar to those of the Silurian and Lower Devonian of China. The thelodont *Turinia gondwana* is suggestive of Middle Devonian thelodont scales from Victoria Land (Antarctica) and Australia. Finally, the xenacanthid chondrichthyan *Antarctilamna*, known from Antarctica, Australia and Iran, may well be another Gondwanan 'key taxon'. The rhenanid *Bolivosteus*, resembles more closely *Asterosteus*, from the Upper Devonian of North America than other forms of the group, yet the phylogeny of the latter is not worked out, and this resemblance may be due to symplesiomorphies.

2. Brazil

Katzer (1897) recorded *Machaeracanthus* spines and '*Pteraspis*' plates from the Maecuru Formation of the Middle Amazon Basin (Figure 7.2 A7) and, later, Kegel (1953), Santos (1961), and several other authors mentioned spines of *Ctenacanthus* (figured by Janvier and Melo, 1992), *Machaeracanthus* and *Devonchus* (sic) in the Givetian Pimenteira Formation of the Parnaiba Basin (Picos, Figure 7.2 A6). Leonardi (1983) recorded a possible tetrapod footprint referred to as *Notopus petri* in the Upper Devonian Ponta Grossa Formation of the Parana (Figure 7.2 A5). Janvier and Melo (1987) described some actinopterygian scales from borehole cores in the uppermost Devonian or lowermost Carboniferous (Jandiatuba Basin) of the Upper Amazon Basin, northwestern Brazil (Figure 7.2 A8). The most significant findings are the acanthodian remains from the Early Lochkovian Manacapuru Formation of the Trombeitas Group of Para, Amazon Basin (Figure 7.2 A7) (Janvier and Melo, 1988; note: these authors referred erroneously these remains to the underlying Pitinga Formation). These provide evidence of climatiids, ischnacanthids (Figure 7.3A) and possibly chondrichthyans. The spine assemblages from this locality are quite similar to that of Seripona in Bolivia (Figure 7.2 A3).

3. Venezuela

Indeterminate fish scales have been recorded by Théry (1982) from borehole cores in the Middle Devonian of the Maturin Basin of Venezuela (Figure 7.2 A9).

4. Falkland (Malvinas) Islands

Vertebrate remains have been recorded, but never figured, from the Devonian of the East Falkland Islands (Clarke, 1913; Greenway, 1972). Two samples from the top of the Middle Devonian Port Philomel Beds, West Falklands (Figure 7.2 B1), deposited in the collection of the Natural History Museum, London, bear respectively a fish spine preserved as a natural mould, and the impression of a tuberculated dermal plate. The spine (Figure 7.3Q) is devoid of ornamentation and possesses a large internal cavity, thereby recalling those from the Lower Devonian of the Amazon Basin, Brazil and Seripona, Bolivia. The dermal plate bears minute costulated tubercles which are suggestive of a placoderm, possibly a phlyctaeniid arthrodire.

AFRICA

Rare Silurian, but abundant Devonian vertebrate remains have been recorded from northern and southern Africa. Most of them occur in typically marine facies, in association with corals, goniatites, brachiopods, tribolites or conodonts.

1. North Africa

1.1 Morocco (including the former Spanish Sahara)

The three major Devonian outcrop areas of Morocco are the Tafilalt and Maider (Figure 7.2 D2), and the Dra Valley (Figure 7.2 D1). The former two belong to the westernmost part of the Béchar (Béni-Abbès) Basin (Figure 7.2 D3), and the latter to the easternmost part of the Tindouf Basin. The latter basin extends to the west into the former Spanish Sahara (Figure 7.2 D5) (now a disputed area), where Lehman (1977b) recorded some large Late Devonian ptyctodont tooth plates. In the Tafilalt and Maider, the Devonian is virtually complete from the Emsian to the Famennian, and vertebrate remains occur at random throughout the entire series. The Emsian of the Dra Valley (near Akka and Assa), Tafilalt and Maider has yielded the following vertebrates:

Acanthodii:
- *Machaeracanthus sulcatus* Lehman
- *Nostolepis* sp.
- *Haplacanthus* sp.
- Acanthodidae gen. et sp. indet.

Placodermi:
- *Atlantidosteus hollardi* Lelièvre (Figure 7.4D)
- *Antineosteus lehmani* Lelièvre (Figure 7.4A)
- ?Rhenanida gen. et sp. indet.

Both *Atlantidosteus* and *Antineosteus* are regarded as being closely related to the Australian genera *Goodradigbeeon* and *Taemasosteus*.

The Eifelian of the Dra Valley has yielded abundant onychodontiform remains (Lehman, 1976, 1977b; Aquesbi, 1988; Figure 7.4F), associated with *Machaeracanthus* cf. *peracutus*. In the Tafilalt-Maider Basin, Eifelian vertebrate remains occur also, in association with the tabulate *Favosites* and the goniatite *Anarcestes lateseptatus*:

Acanthodii:
- *Machaeracanthus* sp.

Placodermi:
- *Eastmanosteus* sp.
- *Hollardosteus marocanus* Lehman
- Coccosteidae gen. et sp. indet.

Hollardosteus is of uncertain affinities, yet it may be regarded as a large tuberculated brachythoracid, with a dunkleosteid-like inferognathal, bearing a well developed anterior tooth.

The Late Devonian outcrops of Morocco, in particular those of the Tafilalt and Maider, have yielded rich vertebrate faunas (Lehman, 1956,

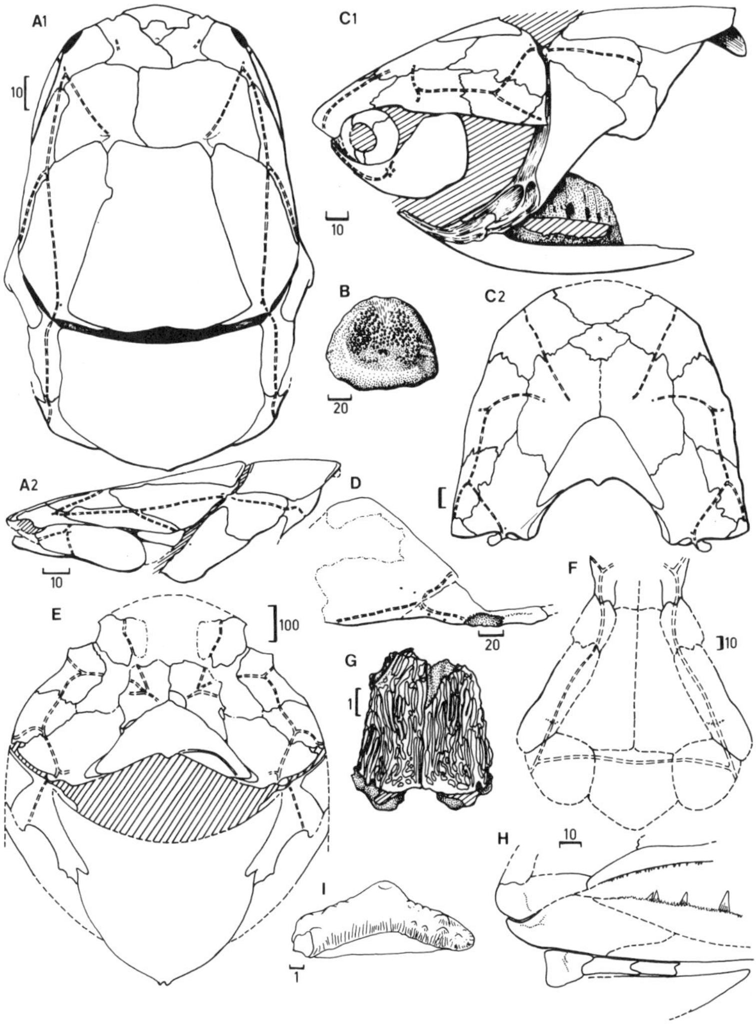

Figure 7.4. Devonian vertebrates of North Africa. All specimens from Morocco. A, B, *Antineosteus lehmani* Lelièvre; A, Emsian of Tafilalt and Dra Valley, reconstruction of the armour in dorsal (A1) and lateral (A2) view; B, Emsian of Beni-Abbès, Algeria, isolated parasphenoid. C, *Tafilalichthys lavocati* Lehman, Famennian of the Tafilalt, reconstruction of armour in lateral view (C1) and skull-roof in dorsal view (C2); D, *Atlantidosteus hollardi* Lelièvre, Emsian of the Dra Valley, suborbital plate in lateral view; E, *Titanichthys termieri* Lehman, Famennian of the Tafilalt, reconstruction of armour in dorsal view; F, Onychodontiforme gen. et sp. indet., Eifelian, Dra Valley, reconstruction of the skull-roof; G, Actinistia gen. et sp. indet., Famennian of the Tafilalt, orbito-ethmoidian part of braincase and skull-roof in dorsal view; H, Eusthenopteridae gen. et sp. indet., Famennian of the Tafilalt, reconstruction of cheek and mandible in lateral view; I, *Ctenacanthus devonicus*, Famennian of Tafilalt, isolated tooth. Scale in mm. (A–D, from Lelièvre, 1984b, 1988, 1991; E, from Lehman, 1956; F, from Aquesbi, 1988; G, H, from Lelièvre and Janvier, 1986, 1988; I, from Derycke, 1988).

1976, 1977b, 1978). The Late Devonian of the Dra Valley has hitherto yielded only a very few arthrodire and chondrichthyan remains.

The Frasnian of the Maider is relatively well exposed north-east of Fezzou, near the Jbel Issoumour, contrary to the Tafilalt where it is not so well exposed and until now devoid of vertebrate remains. Two vertebrate levels could be recognized: the lower one, containing the goniatites *Manticoceras* sp. and *Beloceras kayseri*, and the upper one, which consists of shales embedding concretions associated with the goniatite *Cheiloceras*. Both levels have yielded chondrichthyan remains (*Ctenacanthus* sp.) and various placoderms such as *Holonema* sp., *Aspidichthys* sp. and a Selenosteidae gen. et sp. indet.

The Famennian is well exposed in the Tafilalt, where vertebrate remains are relatively abundant but until now poorly diversified. They are associated with conodonts referred to the *Palmatolepis marginifera* zone, which extends from the Famennian 2 to 3. The vertebrate fauna recorded by Lehman (1956, 1976, 1977a, b, 1978a, b), Lelièvre (1991), Derycke (1988), Lelièvre and Janvier (1986, 1988) consists of:

Chondrichthyes:
- Phoebodontidae
  - *Thrinacodus ferox*
  - *Phoebodus politus* (= *P. gothicus*, Ginter, 1990)
  - *Phoebodus marocensis*, Derycke (in press)
- Protacrodontidae:
  - *Protacrodus* sp.
- Stethacanthidae:
  - *Stethacanthus thomasii*
  - *Stethacanthus* sp.
- Orodontidae:
  - *Orodus varicostatus*
  - *Hercynolepis*? sp.
  - *Orodus* sp.
- Ctenacanthidae:
  - *Ctenacanthus clarkii*
  - *C. venustus*
  - *C. devonicus* (Figure 7.4I)
- Placodermi:
  - *Dunkleosteus marsaisi* Lehman
  - *Dunkleosteus* sp.
  - *Tafilalichthys lavocati* Lehman (Figure 7.4C)
  - *Titanichthys termieri* Lehman (Figure 7.4E)
  - Selenosteidae gen. et sp. indet.
  - 'Mylostomid' gen. et sp. indet.
- Sarcopterygii:
  - Actinistia gen. et sp. indet. (Figure 7.4G)
  - Eusthenopteridae gen. et sp. indet. (Figure 7.4H)
  - Dipteridae gen. et sp. indet.

1.2 Algeria

Late Silurian (Pridoli, *Spathognathodus steinhornensis steinhornensis*-zone) microvertebrate remains have been described from the Oued Ali section of the Ougarta Chain in Algeria (Blieck, 1982, Blieck *et al.*, 1984) (Figure 7.2 D4). They consist of isolated acanthodian scales (*Nostolepis* sp., *Machaeracanthus* sp.), as well as possible fragments of heterostracan (?pteraspid) armour (one of the few heterostracan records from Gondwanan regions).

Lowermost Devonian (Lochkovian or Pragian) chondrichthyan scales (possibly *Machaeracanthus*) occur in the Idekel sandstone of southernmost Algeria, near the Niger border (Figure 7.2 D5).

Lelièvre (1988) recorded dermal plates of the primitive brachythoracid arthrodire *Antineosteus lehmani* (Figure 7.4B) associated with acanthodian scales referred to *Haplacanthus* from the Emsian (*Nowakia elegans* zone) of the Béni-Abbès area, in the Béchar Basin, to which also belong the Moroccan Tafilalt and Maider anticlines (Figure 7.2 D3).

1.3 Libya

Devonian vertebrate remains have been recorded from Libya by various authors. Lehman (1952b) mentioned the presence of 'arthrodire remains' from the Aouinet Ouenine area (Fezzan) (Figure 7.2 D6), and later referred to Famennian ptyctodonts and Siegenian petalichthyids and acanthodians from the same area. Freulon *et al.* (1953) recorded a *Ptyctodus* gnathal plate from the Lower Frasnian of the Hamada el Homra (along the Ghadamès–Sebha track, Figure 7.2 D6), associated with *Spirifer verneuili*. Burollet (1967) recorded plates of arthrodires and antiarchs (sic), associated with lingulids, in the Lower Devonian (Emsian) Emi-Magri Formation (= Ouan-Kasa Fm) of the Murzuk Basin (Figure 7.2 D7). Massa (1988) mentioned the presence of fish remains in various localities of the Devonian of Libya: the Ghadamès Basin (Aouinet Ouenine, Fezzan, Chatti) (Figure 7.2 D6), Murzuk Basin (Figure 7.2 D7) and Tihemboka (Figure 7.2 D8). These are distributed as follows:

Lower Devonian:

'Abundant placoderm remains' in the type section of the Emsian Ouan Kasa Formation of the Murzuk Basin (Figure 7.2 D7)

Middle Devonian:

'A placoderm bone-bed', with associated lycophyte stems in the Eifelian Aouinet-Ouenine I Formation of the West flange of the Murzuk Basin (Figure 7.2 D7).

Upper Devonian:

'Fish remains' associated with tentaculites and ostracods at the Frasnian/Famennian boundary of the Aouinet Ouenine Formation III–IV, Ghadamès Basin (Figure 7.2 D6).

'Numerous placoderm remains associated with lycophyte stems' in the Famennian Trarut Formation of the Chatti (Figure 7.2 D6).

'Placoderm remains' in the Frasnian–Famennian Ashkidah Formation of the Chatti.

'Vertebrate remains', including large cochliodontid bradyodonts, associated with *Cyrtospirifer verneuili*, in the ?Frasnian (more probably Famennian) Talagrouna Formation of Tihemboka (Figure 7.2 D8).

2. South Africa

The Middle Devonian Bokkeveld Group of Barrydale (Figure 7.2 C1), Cape Province, Republic of South Africa has yielded a vertebrate fauna containing the following taxa (Chaloner *et al.*, 1980):

Placodermi:
- *Barrydalaspis theroni* Chaloner *et al.* (Figure 7.5 A1, A2)
- Groenlandaspididae gen. et sp. indet.
- Arthrodira gen. et sp. indet.

Chondrichthyes:
- ?Chimaeroidea gen. et sp. indet. (egg capsule, may be belonging to a placoderm according to D. Goujet, pers. comm.).

Acanthodii:
- *Gyracanthides* sp. (Figure 7.5B)

The occurrence of *Gyracanthides* spines may be a typical Gondwanan 'signature', as it occurs largely in the Devonian of South America, Antarctica and Iran. The occurrence of groenlandaspids (*Barrydalaspis* and a

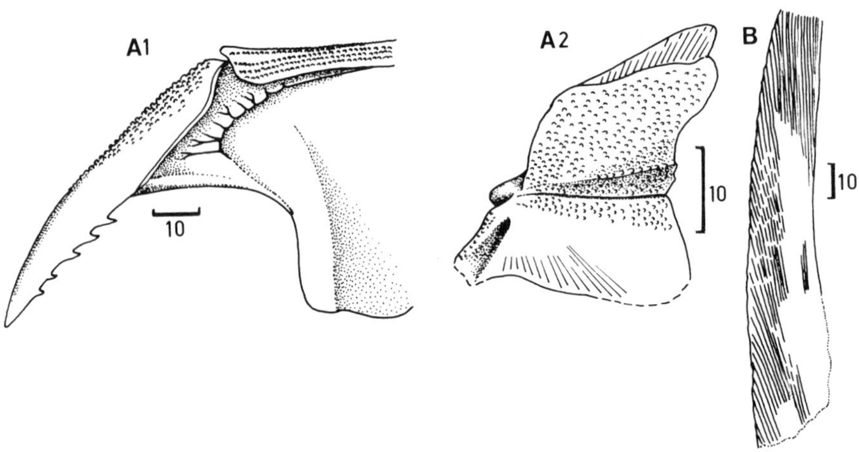

Figure 7.5. Devonian vertebrates of South Africa, Middle or Upper Devonian Bokkeveld Series, Barrydale. A, *Barrydalaspis theroni* Chaloner, Forey, Gardiner, Hill and Young; A1, lateral part of the thoracic armour in dorsal view; A2, anterolateral plate; B, *Gyracanthides* sp., fin spine. Scale in mm. (from Chaloner *et al.*, 1980).

second form) confirms the widespread distribution of this group (D. Goujet, pers. comm., 1991). Other localities within the Upper Witteberg Group, above the Bokkeveld Group, at the southern edge of the Karoo Basin, have yielded a rather diversified fauna of palaeoniscoid actinopterygians and acanthodians (*Mentzichthys*, *Australichthys*, *Æstuarichthys*, *Willomorichthys*, *Sundayichthys*, *Dwykia*, *Androichthys*, *Soetendalichthys*, *Acanthodes*; Jubb and Gardiner, 1975). On the basis of the familial components, Gardiner (1969) has compared the actinopterygian assemblage to the Lower Carboniferous one of Scotland. However, the Witteberg Group age has been recently reviewed as Givetian–Upper Devonian according to its spore assemblage (Stapleton, 1977; and on the basis of plants, M. Streel, pers. comm.) and its invertebrates (Theron and Loock, 1988; Hiller and Theron, 1988). So, the fish-bearing nodular levels of the Upper Witteberg Group are more probably Upper Devonian in age. They are characterized by an endemic actinopterygian fauna.

Recent discoveries by Eric Anderson (JLB Smith Institute) of late Devonian fishes from the Witteberg Group near Grahamstown include remains of *Bothriolepis* sp., phlyctaeniid arthrodires, a possible ptyctodontid or chondrichthyan, a dipnoan and sarcopterygian scales (J. Long, pers. comm.).

ARMORICA

The first record of vertebrate remains in the Siluro–Devonian of the Armorican component of West Gondwana was *Machaeracanthus* spines from the Middle Devonian Angers Limestone (Rouault, 1858). Since these early discoveries, only scarce specimens have occurred.

1. Portugal

A relatively rich acanthodian fauna was described by Priem (1910) from the Upper Silurian (Ludlow) of the neighbourhood of the Sao Felix chapel, Laundos area, north of Porto (Figure 7.2 E1). It contains mostly ischnacanthids and perhaps climatiids, associated with marine invertebrates [trilobite fragments, orthoceratids, molluscs (*Murchisonia*, *Modiolopsis*) and brachiopods (*Retzia*, *Obolus* and discinids)]. Priem referred these fish remains to the following taxa: *Onchus striatus* Ag. (Figure 7.6L), '*Byssacanthus*' (*Onchus*) aff. *arcuatus* Ag., *Plectrodus mirabilis* Ag. (Figure 7.6J), *Campylodus? delgadoi* (Figure 7.6K), *Ctenodus* sp.

Noteworthy is the presence of relatively large ornamented dermal bones, which may not all belong to acanthodians. In particular, some large but badly preserved rounded scales are suggestive of actinopterygian or sarcopterygian scales. This locality may thus be younger than the Late Silurian.

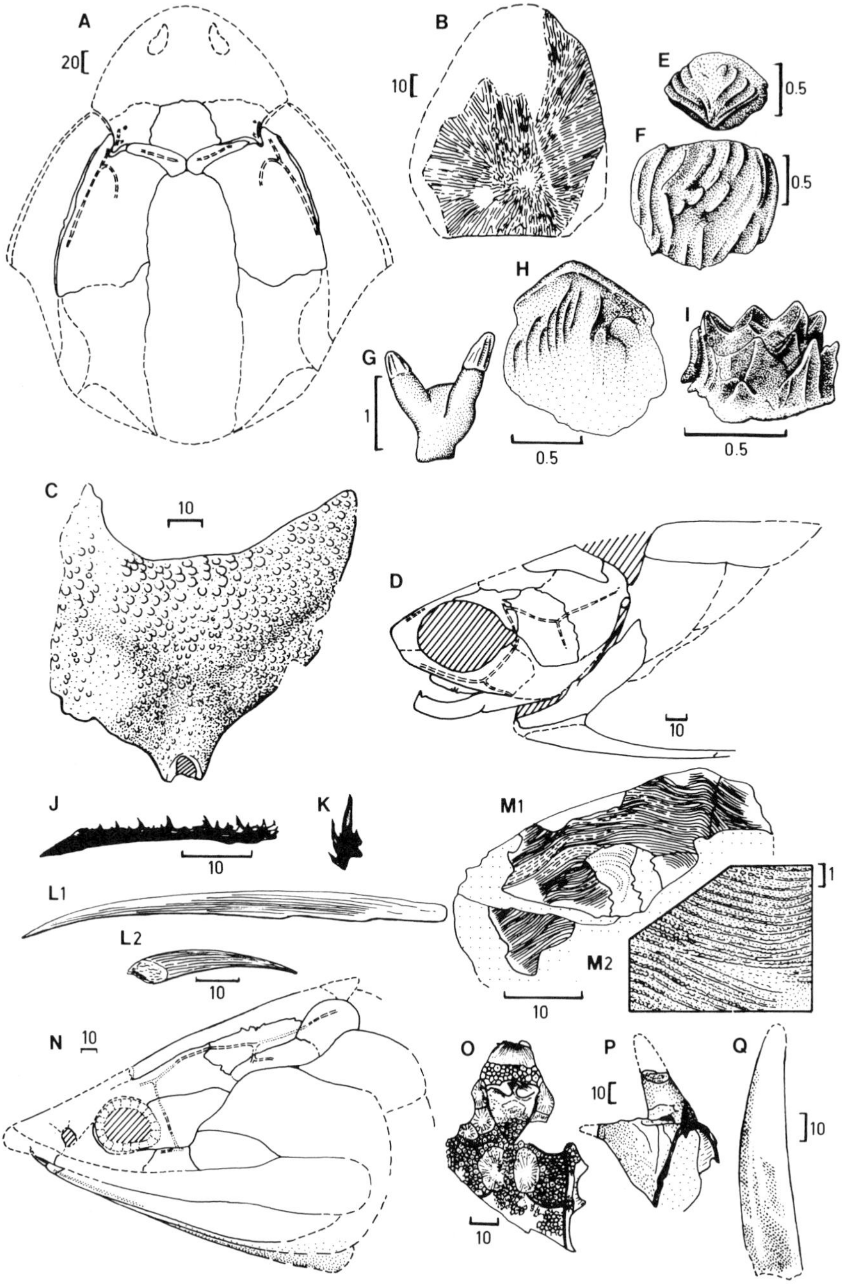
A
20
B
10
E
0.5
F
0.5
G
1
H
0.5
I
0.5
C
10
D
10
J
10
K
M1
1
L1
L2
10
M2
10
N
10
O
10
P
10
Q
10

2. Spain

Schultze (1973a) described a remarkably preserved skull of a large onychodontiform, *Grossius aragonensis* Schultze (Figure 7.6N), from the Middle Devonian Moyuela Formation along the Loscos–Monforte highway in Aragon (Figure 7.2 E3).

Thelodont scales (*Turinia pagei*) and *Machaeracanthus* spines have been recorded from the Pragian to Emsian Luesma and Noguerras Formations of the Celtiberian chain (Carls, 1977; Goujet 1976) (Figure 7.2 E2). Lehman (in Arbizu *et al.*, 1974) figured two plates from the Emsian of Oviedo (Figure 7.2E2), one could be referred to a petalichthyid, the other one to the problematic placoderm *Eczematolepis*. Blieck (1982) described some heterostracan fragments (Figure 7.6M) from the Early 'Gedinnian' Carazo Formation of the Palencia Province (Figure 7.2 E2), and he compared their ornamentation to that of *Allocryptaspis* and various protaspids, yet with reservations. Mader (1986) described numerous isolated teeth and scales of onychodontiforms (*Onychodus*), acanthodians (*Gomphonchus*, *Nostolepis*, *Gyracanthus*, ?*Cheiracanthoides*, ?*Euthacanthus*, *Machaeracanthus*) and chondrichthyans [*Pruemolepis*, *Leonodus* (Figure 7.6G), *Iberolepis* (Figure 7.6F), *Lunalepis* (Figure 7.6H), *Arauzia* (Figure 7.6I), *Ohiolepis*, *Cladolepis*, *Hercynolepis*] from the several Lower to Middle Devonian

---

Figure 7.6 - cont.

Figure 7.6. Silurian and Devonian vertebrates of Armorica and Bohemia. A, *Holonema radiatum* Obruchev, Middle Devonian of Oviedo, Spain, reconstruction of armour; B, *Holonema* cf. *H. radiatum* Obruchev, Upper Devonian of Brest, France, posterior ventrolateral plate; C, *Breizosteus armoricensis* Goujet, Pragian of Brest, France, median dorsal plate; D, *Melanosteus occitanus* Lelièvre and Goujet, Frasnian of Montagne Noire, France, reconstruction of armour in lateral view; E, *Machaeracanthus* sp., Emsian of St Céneré, France, isolated scale; F, *Iberolepis aragonensis* Mader, Lochkovian, Celtiberian chain, Spain, isolated scale (holotype); G, *Leonodus carlsi* Mader, Lochkovian–Pragian, Celtiberian chain, Spain, isolated tooth; H, *Lunalepis leonensis* Mader, Pragian, Celtiberian chain, Spain, isolated scale (holotype); I, *Arauzia federicoi* Mader, Lochkovian, Palencia, Spain, isolated scale (holotype). J–L, acanthodian remains from the uppermost Silurian of Laundos, Portugal; J, K, jaw-bone ('*Plectrodus mirabilis*' Ag.) and tooth-whorl (*Campylodus? delgadoi* Priem) of Ischnacanthiformes; L, fin spines referred to *'Onchus' tenuistriatus* Ag.; M, Heterostraci gen. et sp. indet., Lochkovian, Carazo Formation, Palencia, Spain, fragment of a ventral disc (M1) and detail of ornamentation (M2); N, *Grossius aragonensis* Schultze, Middle Devonian Moyuela Formation, Aragon, Spain, reconstruction of head in lateral view; O, *Radotina kosorensis* Gross, Lochkovian–Pragian of Prag, Czechoslovakia, skull-roof; P, *Grazosteus hoernesi* Gross, Middle Devonian of Graz, Austria, median dorsal plate in lateral view; Q, *Schizosteus perneri* Tarlo, Emsian, Trebotov limestone, Prague, Czechoslovakia, isolated branchial plate. Scale in mm. (A, from Lelièvre and Goujet, 1990; B, from Lelièvre *et al.*, 1983; C, from Goujet, 1980; D, from Lelièvre *et al.*, 1987; E, from Goujet, 1980; F–I, from Mader, 1986; J, L, from Priem, 1910; M, from Blieck, 1982; N, from Schultze, 1974; O, P, from Denison, 1979; Q, from Tarlo, 1965).

localities near León (Figure 7.2 E2), Palencia (Figure 7.2 E2) and Zaragoza (Figure 7.2 E3). Finally, an articulated specimen of *Holonema radiatum* partly preserved (Figure 7.6A) has been recently described by Lelièvre *et al.* (1990) from the Eifelian of Oviedo (Figure 7.2 E2); it was associated with holonematid scales previously known as *Artenolepis golshanii*, and with acanthodian scales of acanthodid type.

3. France

The Devonian of the Montagne Noire, Southern France (Figure 7.2 E7), has yielded some Pragian ischnacanthid acanthodian remains, and a Frasnian articulated selenosteid arthrodire, *Melanosteus occitanus* Lelièvre and Goujet (in Lelièvre *et al.*, 1987; Figure 7.6D). In addition, Lehman (1952b) described as *Thoralodus cabrieri* a large bradyodont (?psammodontid) tooth plate from the Famennian of Cabrières, in the same area.

In the Armorican Massif, vertebrate remains occur mainly in the Devonian of Brittany, with some more easterly finds in Anjou and Mayenne. Goujet (in Babin *et al.*, 1976) and Lelièvre *et al.* (1983) described respectively a large actinopterygian tail (?*Cheirolepis*) and plates of *Holonema* cf. *H. radiatum* (Figure 7.6B), associated with chondrichthyan and acanthodian scales from the Upper Devonian of the Brest area (western Brittany, Figure 7.2 E6). From the Pragian of the same area, Goujet (1980) also described a peculiar 'palaeacanthaspid' median dorsal plate referred to *Breizosteus armoricensis* Goujet (Figure 7.6C). Numerous acanthodian (*Machaeracanthus* sp., Figure 7.6E; *Nostolepis* sp.), chondrichthyan and placoderm remains (radotinid) have been recorded by Goujet (1976, 1980, 1981) from the Emsian of St-Céneré (Mayenne) (Figure 7.2 E4), a fauna found elsewhere in the Emsian of Anjou (Erblay, La Grange) (Figure 7.2 E5).

BOHEMIA AND THE ALPS

The Bohemian Massif is now regarded as a separate terrane, possibly closely attached to Armorica in Devonian times, but separated from Baltica. How far this block extends to the east and south is not clear, but it is likely that the Devonian outcrops of the Graz area, Austria, are to be referred to the same unit. In Czechoslovakia, the Prag area (Figure 7.2 E8) has long yielded vertebrate remains which were first referred to the Silurian (Gross, 1951) and are now regarded as Lochkovian to Pragian in age (see review in Chlupac, 1982 and Chlupac *et al.*, 1985). These are mostly acanthodian spines (*Machaeracanthus*, '*Onchus*') and the 'acanthothoracid' placoderms *Radotina kosoriensis* (Figure 7.6O) and *Kosoraspis peckai*. The Emsian Trebotov Limestone and the Eifelian Chotek Limestone have yielded psammosteid remains (*Schizosteus perneri*, Tarlo, 1965; Chlupac, 1982; Figure 7.6Q), a group which is widely distributed on Baltica. The Middle Devonian (Eifelian) has yielded some arthrodire remains (*Holonema*) and the antiarch *Gerdalepis*, the latter being known also in the Rhineland.

The Middle Devonian 'Lower Quartzite' of the Graz area, Austria (Figure 7.2 E9) has yielded a median dorsal plate of the 'acanthothoracid' placoderm *Grazosteus hoernesi* (Gross, 1958; Figure 7.6P) and the upper Emsian contains remains of acanthodians (*Cheiracanthus*, *Diplacanthus*), chondrichthyans (*Ohioaspis*) and sarcopterygians (*Onychodus*) (Poltnig, 1984).

In Italy, Pillola (pers. comm., 1988) found 'placoderm' remains in the Devonian of the east coast of Sardinia (Figure 7.2 E10).

THE MIDDLE EAST

1. Turkey

Only a few undetermined acanthodian scales have been recorded from the Early Devonian of the Hazro syncline, north-east of Diyarbakir, Turkey (Blieck *et al.*, 1984; Figure 7.2 F4). In contrast, rich vertebrate faunas have been recorded in the Late Devonian of two Turkish localities: the Armutgözlek Tepe (Figure 7.2 F2), near Kemer, Antalya Province, and Cukurca (Figure 7.2 F5), Hakkari Province.

The red sandstone of the Armutgözlek Tepe belongs to the Pamucak-Sapan Dere Unit of the Upper Antalya Nappe (Janvier and Marcoux, 1977; Janvier, 1980, 1983). It contains typically Late Devonian (probably Frasnian) vertebrates with:

Placodermi:
- *Groenlandaspis seni* Janvier and Ritchie (Figure 7.7A)
- *Holonema* cf. *H. radiatum* Obruchev (Figure 7.7D)
- *Bothriolepis* cf. *B. canadensis* (Whiteaves) (Figure 7.7C)
- Ptyctodontida gen. et sp. indet.
- Phyllolepida gen. et sp. indet. (Figure 7.7B)

Chondrichthyes:
- undetermined chondrichthyan tooth

Acanthodii:
- Ischnacanthiforme aff. *Persacanthus kermanensis* Janvier

Sarcopterygii:
- Megalichthyinae gen. et sp. indet. (Figure 7.7E)
- Dipteridae gen. et sp. indet. (Figure 7.7F)
- *Oervigia* sp. (Figure 7.7G)

The most remarkable features of this fauna are the presence of a phyllolepid placoderm and a primitive megalichthyine osteolepid. Both taxa in Euramerica would suggest a Famennian age, which contrasts with the presence of a *Holonema* species quite similar to the Eifelian or Frasnian *H. radiatum*, and unknown in the Famennian, as well as *Bothriolepis* cf. *B. canadensis*. However, phyllolepids are known to occur as early as the Late Givetian in Gondwanan areas (Australia; Long, 1984; Young, 1988a). Finally, the occurrence of *Holonema* in such a typical Old Red Sandstone

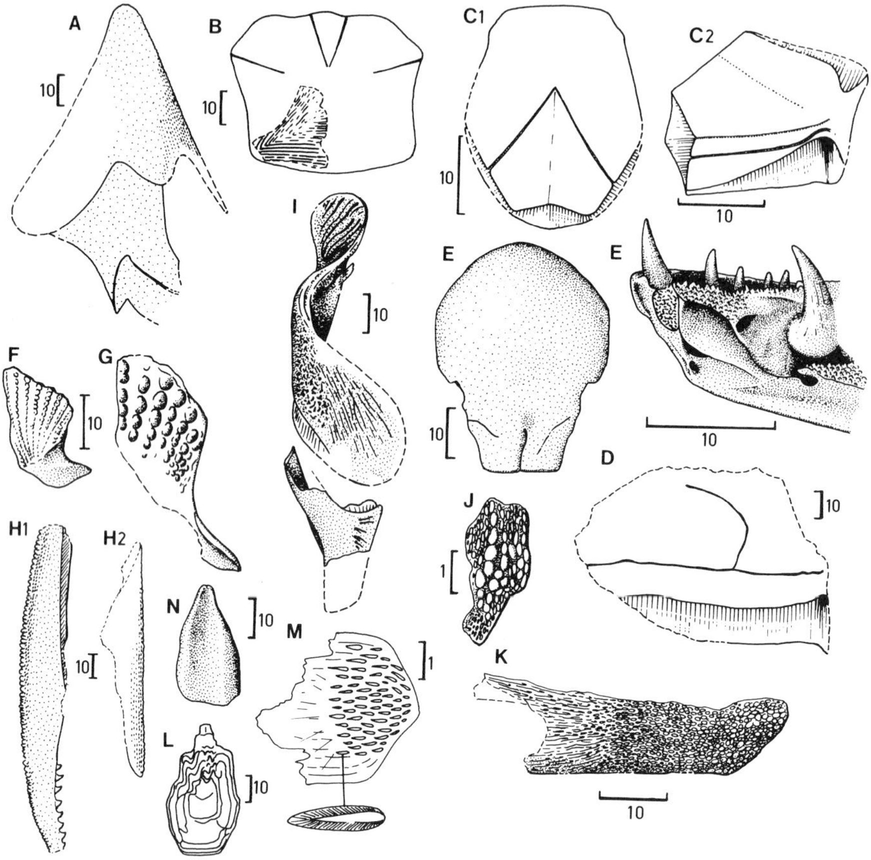

Figure 7.7. Devonian vertebrates from Turkey. A–G, specimens from the Frasnian of Armutgözlek Tepe, Kemer, Antalya. A, *Groenlandaspis seni* Janvier and Ritchie, median dorsal and posterior dorsolateral plates; B, Phyllolepida gen. et sp. indet., fragment of nuchal plate; C, *Bothriolepis* cf. *B. canadensis* Whiteaves, median dorsal (C1) and mixilateral (C2) plates; D, *Holonema* cf. *H. radiatum* Obruchev, anterior dorsolateral plate; E, Megalichthyinae gen. et sp. indet., orbito-ethmoidian part of braincase in dorsal view; F, Dipteridae gen. et sp. indet., upper tooth plate; G, *Oervigia* sp., upper tooth plate. H–N, specimens from the uppermost Famennian Köprülü Formation of Cukurca, Hakkari Province. H, *Groenlandaspis* sp., spinal (H1) and fragmentary median dorsal (H2) plates; I, *Strepsodus* sp., cleithrum and clavicle; J, Platysomidae gen. et sp. indet., pterygoid; K, *Howqualepis* sp., dentary; L, cf. *Andrejevichthys*, 'B' bone with Westoll-lines; M, Actinistia gen. et sp. indet., isolated scale; N, bradyodont tooth plate. Scale in mm. (A, from Janvier and Ritchie, 1977; B–G, from Janvier, 1983; H–N, from Janvier *et al.*, 1984).

facies is quite unusual, although it has been recorded in Scotland, and proves that the locality corresponds to a near-shore marine environment as, for instance, occurs also in the inter-reef facies of Gogo *Bothriolepis* and *Holonema*.

Not far from this locality, a small outcrop of autochthonous Upper Devonian (probably Frasnian) limestone at Armagan Tepe (Figure 7.2 F3), east of Bademli, between the Kupe Dag and the Seyran Dag, has yielded a typically marine fish fauna (unpublished) with ctenacanthid shark remains, *Onychodus* teeth and scales, and numerous *Moythomasia*-like actinopterygian scales.

The vertebrate remains of the Devonian of the Marmara Sea region (Figure 7.2 F1) have never been investigated in detail, with the exception of a short record of a rhenanid plate fragment from the Eifelian Gebze Formation on the coast of the Marmara Sea (Schmidt, 1963). The presence of abundant isolated remains (scales, plate fragments) in this area has nevertheless been briefly mentioned by geologists in the Pragian Kurtdojnurus Formation at Pendik Burnü, as well as in the Emsian Gebre Formation and in the Eifelian to Givetian Denizli Formation near Büyük Ada (Haas, pers. comm.).

In southeastern Turkey, close to the Iraqi border, the thin bedded black limestone of the Köprülü Formation is widely exposed around the town of Cukurca, Hakkari Province (Figure 7.2 F5). It is dated of the late Famennian (Tn 1a) on the basis of a rich ostracod fauna (Janvier *et al.*, 1984) and contains very abundant vertebrate remains with:

Placodermi:
- *Groenlandaspis* sp. (Figure 7.7H)

Acanthodii:
- *Acanthodes* sp.

Chondrichthyes:
- *Ctenacanthus* cf. *C. crenulatus* McCoy

Actinopterygii:
- *Canobius* sp.
- *Howqualepis* sp. (Figure 7.7K)
- Platysomidae gen. et sp. indet. (Figure 7.7J)
- ?Amblypteridae gen. et sp. indet.

Sarcopterygii:
- *Strepsodus* sp. (Figure 7.7I)
- Actinistia gen. et sp. indet. (isolated scale; Figure 7.7M)
- ?Onychodontiformes gen. et sp. indet.
- *Chirodipterus* sp.
- Dipnoi gen. et sp. indet. cf. *Andrejevichthys* (Figure 7.7L)

In the upper part of the Köprülü Formation, the bedded limestone passes to massive reef limestone ('Belek' facies) which is poor in vertebrate remains, but has yielded a typical *Cochliodus*-like bradyodont tooth (Figure 7.7N).

This fauna is unique in showing an association of an arthrodire

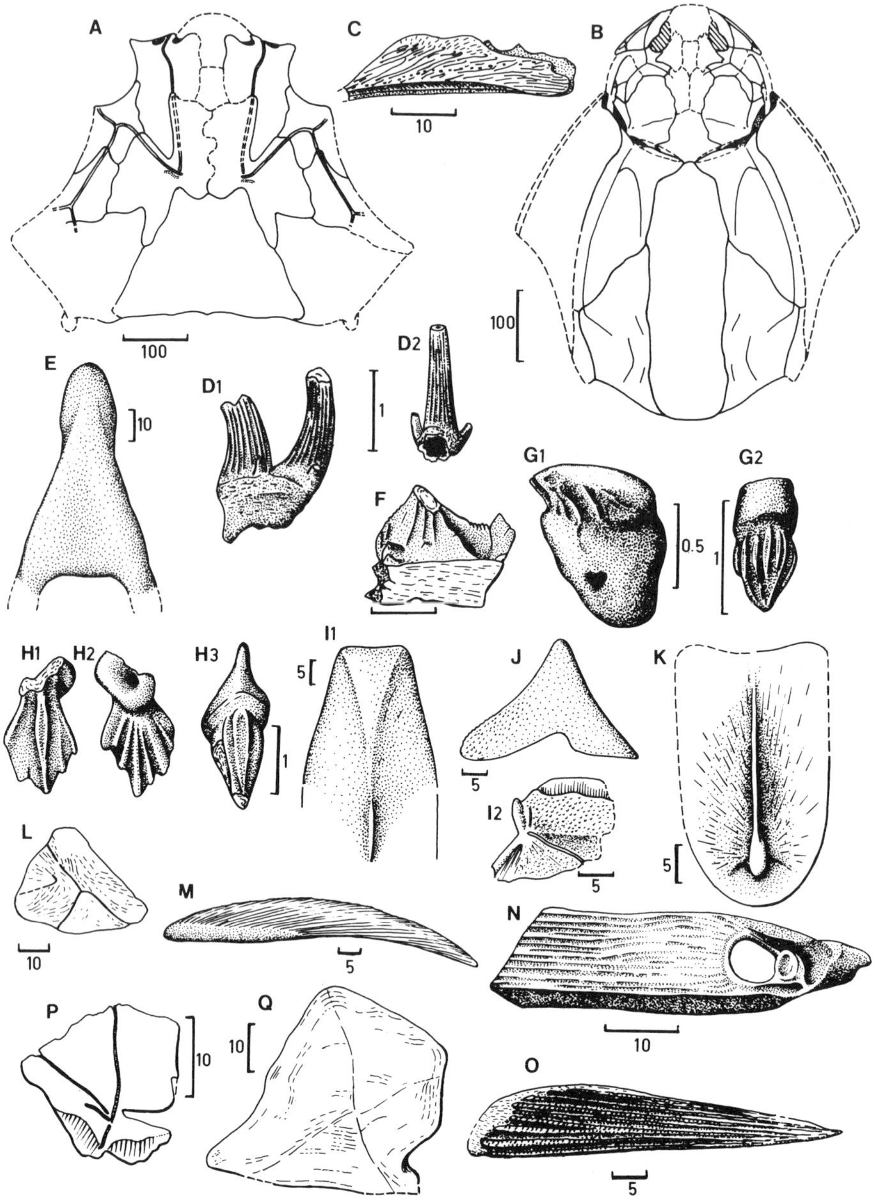

Figure 7.8. Devonian vertebrates from Iran and the Arabian Peninsula.

(*Groenlandaspis*) and cosmine-covered lungfishes, both characteristic of the Devonian, and *Strepsodus*, *Canobius*, platysomid and amblypterid actinopterygians, which are quite typical of Carboniferous faunas. Clearly, this fauna is representative of the transition between the latest Devonian and the earliest Carboniferous. The associated ostracod fauna suggests a progressive increase of the marine influence (from the underlying Yiginli red sandstone up to the 'Belek' facies of the Köprülü Formation) and the ostracod taxa are characteristic of the 'palaeo-tethysian' realm (Lethiers *in* Janvier *et al.*, 1984).

2. Nakhitchevan A.R. (Azerbaidjan)

Krupina (1979) described a lungfish, *Ganorhynchus caucasius*, from the Famennian of Nakhitchevan (Figure 7.2 F21), southern Caucasus, an area situated close to the Turkish border and which may pertain to either the Turkish or the Iranian Cimmerian terrane. This species looks quite similar to a large *Chirodipterus*-like form from the uppermost Famennian of the Hakkari Province of Turkey (see above).

Figure 7.8 - cont.

A–G, specimens from the Givetian and Frasnian of the Kerman region, Iran. A, *Golshanichthys asiatica* Lelièvre, Janvier and Goujet, Frasnian, Hootk, skull-roof; B, *Holonema* cf. *H. radiatum* Obruchev, Frasnian, Hootk, reconstruction of armour in dorsal view; C, *Diplocercides* sp., Frasnian, Chanaruh, incomplete lower jaw; D, *Strunius rolandi* (Gross), Frasnian, Tangil-e-Ab-Garm, Ravar, parasymphysial teeth in lateral (D1) and posterior (D2) view; E, *Iranorhynchus seyedemamii* Janvier and Martin, Frasnian, Chanaruh, lower jaw (holotype) in ventral view; F, *Persacanthus kermanensis* Janvier, Frasnian, Chanaruh, fragment of jaw-bone (holotype); G, *Turinia hutkensis*, Givetian or Frasnian, Hootk, isolated scales (G1, holotype). H–O, specimens from the Emsian or Eifelian of Khush-Yeilagh, Mazenderan, Iran. H, thelodont scales (from the marine layers overlying the bone-bed which yields the following specimens), *Skamolepis* sp. (H1, H2) and *Turinia* cf. *T. hutkensis* (H3); I, Phlyctaenioidei gen. et sp. indet., median dorsal plate (I1) and anterior dorsolateral plate (I2); J, Groenlandaspididae gen. et sp. indet., median dorsal plate; K, Coccosteidae gen. et sp. indet., median dorsal plate; L, *Holonema* sp. suborbital plate; M, *Gyracanthides* sp., fin spine; N, *Hyrcanaspis bliecki* Janvier and Pan, anterior ventrolateral plate; O, *Antarctilamna* sp., fin spine; P, Buchanosteidae gen. et sp. indet., Lower Emsian Jawf Formation, Saudi Arabia, central plate; Q, *Qataraspis deprofundis* White, Middle Devonian of Qatar (in borehole core), anterior lateral plate. Scale in mm. (A, B, from Lelièvre *et al.*, 1981; C, from Janvier, 1977; D, from Janvier and Martin, 1979; E, from Janvier and Martin, 1978; F, from Janvier 1974; G, from Blieck and Goujet, 1977; H, from Turner and Janvier, 1979; I–O, from Blieck *et al.*, 1980; P, from Lelièvre, 1989; Q, from White, 1969).

3. Iraq

Seilacher (1963) recorded a lungfish ('*Dipterus* sp.') tooth plate from the Late Devonian Pirispiki red beds of northern Iraq (Figure 7.2 F6), which are most probably to be correlated with the Yiginli Formation of Hakkari Province in Turkey. The latter underlies the vertebrate-bearing Köprülü Formation.

4. Arabian peninsula

Vertebrate remains have been recorded from the lower Emsian Jawf Formation (Figure 7.2 F19) of Saudi Arabia (Bahafzallah *et al.*, 1981; Lelièvre, 1989; Boucot *et al.*, 1989) and are referred to the following taxa:

Placodermi:
- ?*Stensioella heintzi* Broili (isolated scales)
- Buchanosteidae gen. et sp. indet. (Figure 7.8P)

Acanthodii:
- *Cheiracanthoides* cf. *C. comptus* Wells
- '*Acanthodes*' sp.
- *Nostolepis* cf. *N. costata* Goujet

Recently Forey and Young (1991) have reported on the discovery of fossil fish from the Lower Devonian Jawf and Wadjid Formations, the latter being situated in the southeastern part of the peninsula. These remains are referred to placoderms, acanthodians and chondrichthyans, which show affinities with taxa from both northern and southern hemispheres. Turner (1989) described microvertebrates from the Jawf Formation which included acanthodian and placoderm scales.

In addition, a thoracic plate of a phlyctaenioid arthrodire, *Qataraspis deprofundis* White (1969; Figure 7.8Q) has been found in a borehole core in Qatar (Figure 7.2 F20).

5. Iran

The first record of Devonian vertebrates from Iran was made by Rieben (1935) in the Upper Devonian of Zonuz (Figure 7.2 F7), Azerbaidjan, in the north-west of Iran. Schultze (1973b) re-examined Rieben's specimens and referred them to a *Holoptychius* sp. scale, a dendrodont tooth resembling those of *Laccognathus*, a bothriolepid plate, and various undetermined arthrodiran plates.

Huckriede *et al.* (1962) recorded some placoderm and actinopterygian remains (determined by W. Gross) from the Givetian and Frasnian of the Kerman area (Figure 7.2 F9, 11, 12), in eastern Central Iran, and Schultze (1973b) described arthrodiran remains from the Frasnian (*Eastmanosteus* sp., *Holonema* cf. *H. rugosum*, *Aspidichthys* sp.) and possibly also the

Famennian of Niaz, near Tabas (Figure 7.2 F8), and Hootk, near Kerman (Figure 7.2 F11). Janvier (1974, 1977a, 1979) and later Janvier and Martin (1978, 1979), Lelièvre *et al.* (1981) and Blieck and Goujet (1977) gave a thorough description of the Middle and Late Devonian faunas of the Kerman area, in particular the localities of Ab-e-Morad (Figure 7.2 F9), Chanaruh (= Bidu 1; Figure 7.2 F10), Hootk (Figure 7.2 F11) and Tangil-e-Ab-Garm gorge (Figure 7.2 F12), in the Kuh-e-band-e-Anâr, near Ravar. According to the conodont data provided by Huckriede *et al.* (1962), the age of these localities range from the late Givetian to the early Frasnian. The faunal list for these localities is as follows (the taxa occur in all four localities, unless indicated otherwise):

Thelodonti:
- *Turinia hutkensis* Blieck and Goujet (Hootk, Figure 7.8G)

Acanthodii:
- *Persacanthus kermanensis* Janvier (Figure 7.8F)
- ?Climatiida gen. et sp. indet.
- ?Acanthodida gen. et sp. indet.

Placodermi:
- Ptyctodontida aff. '*Ptyctodus*' sp. and '*Rhynchodus*' sp.
- ?Petalichthyida gen. et sp. indet. (uncertain, may rather be a spinal plate of *Groenlandaspis*)
- ?*Byssacanthus* sp. (Tangil-e-Ab-Garm)
- *Asterolepis* sp. (Chanaruh)
- *Holonema* cf. *H. radiatum* Obruchev [Figure 7.8B; the isolated scales described by Janvier (1974) as *Artenolepis golshanii* have turned out to be scales of *Holonema radiatum* (Lelièvre *et al.*, 1990)]
- ?*Aspidichthys* sp.
- *Plourdosteus* sp.
- *Golshanichthys asiatica* Lelièvre *et al.* (Figure 7.8A)
- *Eastmanosteus* sp.
- Dinichthyidae gen. et sp. indet.

Chondrichthyes:
- *Ctenacanthus* sp.
- *Protacrodus* sp.

Actinopterygii:
- *Moythomasia* sp.

Sarcopterygii:
- *Onychodus firouzi* Janvier and Martin
- *Strunius rolandi* (Gross; Figure 7.8D; Tangil-e-Ab-Garm)
- *Diplocercides* sp. (Figure 7.8C)
- *Iranorhynchus seyedemamii* Janvier and Martin (Figure 7.8E, Chanaruh)
- *Rhinodipterus* sp.
- ?*Dipterus* sp.
- ?*Chirodipterus* sp.
- ?Rhynchodipteridae gen. et sp. indet.
- Osteolepididae gen. et sp. indet.

This vertebrate fauna is typically marine, associated with brachiopods, corals and conodonts, with local exceptions, such as the fish remains in the intercalated red sandstone beds in Chanaruh, which, however, contain lingulids. Its noteworthy components are *Holonema radiatum*, a very abundant species which is also known from the Armorica and Baltica, *Golshanichthys asiatica*, a primitive dunkleosteid, *Iranorhynchus seyedemamii*, a long-snouted lungfish not found elsewhere (possibly a much derived relative of *Rhinodipterus*), and the late thelodont, *Turinia hutkensis*, which compares to Middle and Late Devonian thelodonts of Australia, Antarctica and South America. In summary, the Kerman fauna shares a large number of taxa with the Russian Platform (*H. radiatum*, *Plourdosteus*, *S. rolandi*, most lungfishes), but the large-scaled thelodont *T. hutkensis* is clearly of Gondwanan affinities, since no thelodont has ever been found in the Middle and Late Devonian of Euramerica.

In contrast to the Kerman localities, the Khush-Yeilagh locality, situated at the Khush-Yeilagh Pass of the Eastern Alborz Range, between Shahpasand and Chahrud (Figure 7.2 F14) consists of a centimetre thick bone-bed with a massive arkosic sandstone layer of the Khush-Yeilagh Formation. This bone-bed is extremely rich and extends over a distance of several kilometres. Its age is regarded as Emsian on the basis of conodonts (Hamdi and Janvier, 1981) to Eifelian (Blieck *et al.*, 1980; Goujet and Janvier, 1984) on the basis of fishes. The fauna has been the subject of a preliminary survey which yielded the following faunal list (from Blieck *et al.*, 1980, completed):

Acanthodii:
- *Gyracanthides* sp. (Figure 7.8M)
- '*Onchus*' sp.
- Ischnacanthiformes gen. et sp. indet.

Placodermi:
- Phlyctaenioidei gen. et sp. indet. (Figure 7.8 I)
- Groenlandaspididae gen. et sp. indet. (Figure 7.8J)
- *Holonema* sp. (Figure 7.8L)
- Coccosteidae gen. et sp. indet. (Figure 7.8K)
- Ptyctodontida gen. et sp. indet.
- *Hyrcanaspis bliecki* Janvier and Pan (Figure 7.8N)
- ?*Gerdalepis* sp.

Chondrichthyes:
- *Antarctilamna* sp. (Figure 7.8O)

Sarcopterygii:
- *Onychodus* sp.
- *Holoptychius* sp.
- Osteolepididae gen. et sp. indet. (cf. *Gyroptychius*?)

Noteworthy is the apparent absence of actinopterygians in this fauna. The small primitive euantiarch *Hyrcanaspis bliecki* is suggestive of Chinese Middle Devonian euantiarchs, in particular the primitive bothriolepid *Dianolepis*. The primitive *Holonema* sp. is quite similar to an undescribed form from the Eifelian Narva Beds of Estonia. In contrast, the arthrodire

Groenlandaspis is a widespread taxon. The rest of the fauna deserves a thorough study.

This bone-bed is overlain by a succession of marls and calcareous sandstone layers, one of which has yielded scales referred to *Skamolepis* sp. (Figure 7.8 H1–2) and *Turinia* cf. *hutkensis* (Figure 7.8 H3) (Turner and Janvier, 1979), in association with remains of actinopterygians (?*Orvikuina*), onychodontiforms, ptyctodonts, and possibly the placoderm *Buchanosteus* (Goujet and Janvier, 1984). The question of the age of this fauna, and consequently of the underlying bone-bed, is raised by the nature of the conodont assemblage found in association with it. Hamdi (in Hamdi and Janvier, 1981) recorded *Ozarkodina denkmani, Polygnathus linguiformis foveolatus, P. l. linguiformis, P.* cf. *P. dobrogensis, Icriodus fusiformis, Neoprionodus* cf. *bicurvatus* and *Hindeodella priscilla* and, therefore, referred this fauna to the Emsian. If this age is correct, then the underlying bone-bed would contain the earliest known representatives of many osteichthyan (osteolepids, *Onychodus*) and placoderm (coccosteids, *Holonema*) groups.

A few vertebrate remains have also been found in other minor localities of Central Iran, such as the Frasnian of Angireh, near Yazd (Figure 7.2 F13), the Frasnian and Famennian of Tabas (Figure 7.2 F8) (*Dunkleosteus, Holonema, Aspidichthys*, Schultze 1973b), the Frasnian of Zonuz, Azerbaidjan (Figure 7.2 F7, *Holoptychius, ?Laccognathus*, Schultze 1973b), and the late Famennian of Djeirud (Figure 7.2 F15, *Protacrodus*, platysomid actinopterygians).

6. Afghanistan

A few Upper Devonian vertebrate (Frasnian–Famennian) remains have been discovered during the Afghan–French mapping program in the years 1960–1976, in the Central Range, on either slope of the Tokya-Dewal anticline, south-west of Kabul (Figure 7.2 F17, 18). In addition, some 'fish teeth' are recorded from the Famennian of Robat-e-Pai, near Herat, in northwestern Afghanistan (Figure 7.2 F16). The material from the Central Range consists of remains of chondrichthyans, placoderms (*Holonema* sp.), lungfishes (*Rhinodipterus*?), actinopterygians (?*Moythomasia*), onychodontiformes (*Onychodus*?), quite similar to those from the Upper Devonian of the Kerman region of Iran (Janvier, 1973, 1977a; Blieck *et al.*, 1982).

CONCLUSIONS

From this review, it is clear that most Devonian vertebrate occurrences in western Gondwana are Emsian to Famennian in age, that is, a time span which is known to display less endemic groups than the Late Silurian and Early Devonian times. In addition, Silurian vertebrates are extremely rare in these areas, with a few records from Bolivia and Algeria. Late Devonian (Frasnian and Famennian) vertebrate faunas are reputedly quite homogeneous all over the world, in particular when marine fish faunas are considered. Typically widespread Late Devonian taxa are the placoderm genera *Holonema* and *Bothriolepis*, as well as the dipterid lungfishes. Therefore, any biogeographical analysis based on such taxa requires a good phylogenetic analysis at what is considered the species level. Unfortunately, the conditions of such an analysis, which requires well-preserved material, are rarely met with. To date, only one group, the antiarchs, has been the subject of a global biogeographical analysis (Young, 1988b). Yet many species represented by isolated plates had to be left aside.

The facies in which vertebrate remains are preserved, which reflect particular environments, may also be a cause of taxic segregation. Large arthrodires and lungfishes, for example, are far more abundant in typically marine limestone facies than in marginal sandstones where, in contrast, antiarchs may be more abundant. But this does not play a major role in the localities considered in this review.

The contents of these west Gondwanan localities may reveal three major ensembles:

1. The South American one, where chondrichthyans and acanthodians are extremely abundant but placoderms extremely rare (in fact, the only undoubted South American placoderm known to date is the rhenanid *Bolivosteus*).
2. The North African and Armorican one, characterized by the predominance of large arthrodires quite similar to those of eastern North America, and by the scarcity of antiarchs and lungfishes.
3. The Middle East ones, which display a great taxic diversity and a marked faunal resemblance to both the Russian Platform and eastern Gondwanan regions (Australia) and South China.

How to interpret these differences? In the case of the South American faunas, one may possibly invoke the presumed high latitude of Bolivia in Devonian times, a colder climate being perhaps less favourable to large arthrodires than to sharks and acanthodians. But the most striking feature of these faunas is the presence of some forms found elsewhere in Australia and Antarctica (*Turinia gondwana*) and possibly in China (*Sinacanthus*). This might be interpreted as a track following the eastern coast of the Gondwana, in the light of current palaeogeographical reconstructions (Figure 7.1), or as a transpacific track in the light of a non-plate tectonics model.

The case of the North African–Armorican faunas is more clear, since current palaeogeographical models put northwestern Africa close to northeastern North America (Figure 7.1), both areas being separated by a supposedly narrow ocean. The large Famennian dunkleosteids and titanichthyids occur abundantly in Morocco, the New York State and Ohio, but this Late Devonian community must not mask the close relationships between the Emsian Moroccan genera *Atlantidosteus* and *Antineosteus* on the one hand, and the Australian genera *Goodradigbeeon* and *Taemasosteus* on the other; this group of primitive brachythoracids has never been recorded elsewhere and the pattern of its distribution would be better interpreted in the light of a non-plate tectonics model, bringing

Australia close to West Gondwana. A new undescribed species of *Atlantidosteus* has recently been discovered in the Lower Devonian of Queensland, Australia (G. Young, pers. comm.).

Few conclusions can be drawn from the Armorican faunas, since articulated material remains relatively rare. The onychodontid *Grossius* resembles an unnamed form from the Eifelian of Morocco (Aquesbi, 1988), and the Frasnian selenosteid arthrodire *Melanosteus* looks quite similar to some forms of Wildungen in Germany. The occurrence of a psammosteid in Bohemia is an additional indication of a close proximity between the Armorican–Bohemian terranes and Baltica in Middle to Late Devonian times.

Among the Middle East vertebrate faunas, those from the Givetian–Frasnian of the Kerman area share several taxa with Baltica (Russian Platform) only, or even with the entire Laurussia area. These are for example the arthrodires *Plourdosteus* or *Holonema radiatum*, the antiarch *Byssacanthus*, the onychodontiform *Strunius rolandi* and the lungfish *Rhinodipterus*. The Iranian Emsian or Eifelian Khush Yeilagh fauna also contains some Baltican elements (a primitive *Holonema* species, *Gyroptychius*, and possibly *Holoptychius*), but associated with a small primitive euantiarch, *Hyrcanaspis*, which resembles in many respects the Chinese genus *Dianolepis*. This is also the conclusion arrived at by Forey and Young (1991) on the Jawf fauna.

In Turkey, both the supposedly Frasnian Armutgözlek Tepe (Antalya) and the Late Famennian Cukurca (Hakkari) faunas consist mostly of Baltican or worldwide elements, but they share also some forms with east Gondwanan (Australian) faunas. The former locality contains a pre-Famennian phyllolepid placoderm and a megalichthyiine osteolepid, both unknown respectively before the late Famennian and Carboniferous in Euramerica. The latter has yielded the actinopterygian *Howqualepis*, known elsewhere from Victoria, Australia (Long, 1988). In contrast, the most abundant lungfish of the Cukurca fauna looks closely similar to *Andrejevichthys* (Krupina, 1987), from the late Famennian of Toula, on the Russian Platform.

The Middle East Devonian vertebrate faunas seem thus to appear as a mixture of Baltican, east Gondwanan, and perhaps south Chinese components, with a large number of widespread taxa.

Finally, it is worthy noting that all the Middle Devonian west Gondwanan faunas yield abundant spines of the acanthodian *Gyracanthides* and the chondrichthyan *Antarctilamna*, both being also very abundant in east Gondwanan localities, but rare or absent in more northerly localities. These two taxa may possibly have been bound to the margins of the Gondwana in these times.

REFERENCES

Aquesbi, N., 1988, Étude d'un Onychodontiforme (Osteichthyes, Sarcopterygii) du Dévonien moyen (Eifélien) du Maroc, *Bulletin du Muséum national d'Histoire naturelle*, Paris, **1988**: 181–96.

Arbizu, M., Garcia-Alcalde, J.L., Soto, F. and Truyols, J., 1974, Nota sobre la fauna ictiologica del devonico cantabrico, *Breviora Geologica Asturica*, **18**: 28–32.

Babin, C., Goujet, D., Lardeux, H., Lejal-Nicol, A., Lethiers, F., Morzadec, P., Plusquellec, Y. and Weyant, M., 1976, La Formation des Schistes de Porsguen (Dévonien supérieur de la rade de Brest, Massif Armoricain). Lithologie, flore, faune, *Annales de la Société géologique du Nord*, **96**: 333–46.

Bahafzallah, A., Jux, U. and Omara, S., 1981, Stratigraphy and facies of the Jawf Formation, Saudi Arabia, *Neues Jahrbuch für Geologie und Paläontologie*, Monatshefte, **1**: 1–18.

Blieck, A., 1982, Les grandes lignes de la biogéographie des Hétérostracés du Silurien supérieur-Dévonien inférieur dans le domaine Nord-Atlantique, *Palaeogeography, Palaeoclimatology, Palaeoecology*, **38**: 283–316.

Blieck, A., Golshani, F., Goujet, D., Hamdi, B., Janvier, P., Mark-Kurik, E. and Martin, M., 1980. A new vertebrate locality in the Eifelian of the Khush-Yeilagh Formation, Eastern Alborz, Iran, *Palaeovertebrata*, **9**: 133–54.

Blieck, A. and Goujet, D., 1977, A propos de nouveau matériel de Thélodontes (Vertébrés, Agnathes) d'Iran et de Thaïlande: aperçu sur la répartition géographique et stratigraphique des Agnathes des 'régions gondwaniennes' au Paléozoïque moyen, *Annales de la Société géologique du Nord*, **97**: 363–72.

Blieck, A., Goujet, D., Janvier, P. and Lelièvre, H., 1984, Microrestes de vertébrés du Siluro-Dévonien d'Algérie, de Turquie et de Thaïlande, *Géobios*, **17**: 851–6.

Blieck, A., Janvier, P., Lelièvre, H., Mistiaen, B. and Montenat, C., 1982, Vertébrés du Dévonien supérieur d'Afghanistan, *Bulletin du Muséum national d'Histoire naturelle*, Paris, **4**: 3–19.

Boucot, A.J., McClure, H.A., Alvarez, F., Ross, J.R.P., Taylor, D.W., Struve, W., Savage, N.N. and Turner, S, 1989, New Devonian fossils from Saudi Arabia and their biogeographical affinities, *Senckenbergiana Lethaea*, **69**: 535–97.

Burollet, P.F., 1967, Sédimentologie du Dévonien inférieur en Libye. In *Colloque sur le Dévonien inférieur et ses limites* (Rennes, 1964), *Mémoires du Bureau de Recherches géologiques et minières*, **33**: 205–13.

Carls, P., 1977, Spain. In *The Silurian–Devonian Boundary*, IUGS, Ser.A, **5**: 143–58.

Chaloner, W.G., Forey, P.L., Gardiner, B.G., Hill, A.J. and Young, V.T., 1980, Devonian fish and plants from the Bokkeveld Series of South Africa, *Annals of the South African Museum*, **81**: 127–57.

Chlupac, I., 1982, Preliminary submission for Lower–Middle Devonian boundary stratotype in the Barrandian area, *Courier Forschungsinstitut Senckenberg*, **55**: 85–96.

Chlupac, I., Lukes, P., Paris, F. and Schönlaub, H.P., 1985, The Lochkovian–Pragian Boundary in the Lower Devonian of the Barrandian area (Czechoslovakia), *Jahrbuch für Geologie. B.-A.*, **128**: 9–41.

Clarke, J.M., 1913, Fosseis Devonianos do Parana, *Monografia do Serviço de geologia mineral do Brasil*, Rio de Janeiro, **1**: 1–353.

Derycke, C., 1988, *Sélaciens du Dévonien du Tafilalt (Maroc)*. Unpubl. Mémoire Univ. Paris VII, 41pp, 10pl.

Derycke, C., in press, Microrestes de Sélaciens et autres Vertébrés du Dévonien supérieur du Maroc, *Bulletin du Muséum National d'Histoire Naturelle*, T.14, Section C (1).

Forey, P.J. and Young, S., 1991, Devonian fish fauna of Saudi Arabia: the North and South of it. In Vezina, D. and Arsenault, M. (eds), *Abstracts, 7th International Symposium, Studies of Early Vertebrates*, Bibliotheque Nationale du Canada, p. 19.

Frenguelli, J., 1951, Floras devonicas de la Precordillera de San Juan, *Revista de l'Asociacion geologica Argentina*, **4**: 83–93.

Freulon, J.M., Lefranc, J.P. and Lehman, J.P., 1953, Sur la présence de Ptyctodontes dans le Dévonien supérieur d'Aouinet Ouenine (Fezzan), *Bulletin de la Société géologique de France*, **3**: 457–58.

Gagnier, P.Y., 1987, Datos preliminares sobre la ictiofauna de Seripona, formacion Catavi, Paleozoico de Bolivia. In *IV Congreso latinamericano de Paleontologia*, Santa-Cruz, **2**: 695–712.

Gagnier, P.Y., Paris, F., Racheboeuf, P., Janvier, P. and Suarez-Riglos, M., 1989, Les Vertébrés dévoniens de Bolivie: données biostratigraphiques et anatomiques complémentaires, *Bulletin de l'Institut français d'Etudes Andines*, Lima, **18**: 75–93.

Gagnier, P.Y., Turner, S., Friman, L., Suarez-Riglos, M. and Janvier, P., 1988, The Devonian vertebrate and mollusc fauna from Seripona (Chuquisaca, Bolivia), *Neues Jahrbuch für Geologie und Paläontologie, Abhandlungen*, **176**: 269–97.

Gardiner, B.G., 1969, New palaeoniscoid fish from the Witteberg Series of South Africa, *Zoological Journal of the Linnean Society*, **48**: 423–5.

Ginter, M., 1990, Late Famennian shark teeth from the Holy Cross Mountains, Central Poland, *Acta Geologica Polonica*, **40**: 69–81.

Golshani, F., Janvier, P., Brice, D., Corsin, P. and Lapparent, A.F. de, 1972, Découverte d'une faune de poissons et de restes de végétaux dans le Dévonien supérieur de Bidu, en Iran Central, *Compte-Rendus de l'Académie des Sciences*, Paris, **275**: 2103–6.

Goujet, D., 1976, Les Poissons. In *Les schistes et calcaires Dévoniens de Saint-Céneré (Massif Armoricain, France). Mémoires de la Société géologique et minéralogique de Bretagne*, **19**: 313–23.

Goujet, D., 1980, Les Poissons. In *Les schistes et calcaires de l'Armorique (Dévonien inférieur, Massif Armoricain). Mémoires de la Société géologique et minéralogique de Bretagne*, **23**: 309–13.

Goujet, D., 1981, Les Poissons. In *La Tranchée de la Lézais, Emsien supérieur du Massif Armoricain. Mémoires de la Société géologique et minéralogique de Bretagne*, **24**: 305–8.

Goujet, D. and Janvier, P., 1984, Un nouvel horizon à Vertébrés dans la coupe-type de la Formation de Khush-Yeilagh (Dévonien moyen), Alborz oriental, Mazenderan, Iran, *Géobios*, **17**: 245–50.

Goujet, D., Janvier, P. and Suarez-Riglos, M., 1984, Devonian vertebrates of South America, *Nature*, **312**: 311.

Goujet, D., Janvier, P. and Suarez-Riglos, M., 1985, Un nouveau Rhénanide (Vertebrata, Placodermi) de la Formation de Belén (Dévonien moyen), Bolivie, *Annales de Paléontologie*, **71**: 35–53.

Greenway, M.E., 1972, *The geology of the Falkland Islands*, British Antarctic Survey, Scientific Report n° 76.

Gross, W., 1951, Die paläontologische und stratigraphische Bedeutung der Wirbeltierfaunen des Old Reds und der marinen altpaläozoischen Schichten, *Abhandlungen der deutschen Akademie der Wissenschaften zu Berlin, matematische und naturwissenschaftliche Klasse*, Jahrgang **1949**: 1–130.

Gross, W., 1958, Über den von R. Hoernes 1891 entdeckten Arthrodiren-Rest aus dem Grazer Paläozoikum, *Jahrbuch der geologischen Bundesanstalt Wien*, **101**: 139–146.

Hamdi, B. and Janvier, P., 1981, Some conodonts and fish remains from the Lower Devonian (lower part of the Khoshyeylaq Formation), North East Shahrud,

Geological Survey of Iran Reports, **49**: 195–212.

Hiller, N. and Theron, J.N., 1988, Benthic communities in the South African Devonian. In N.J. McMillan, A.F. Embry and D.J. Glass (eds), *Devonian of the World, Canadian Society of Petroleum Geologists, Memoir* **14, III**: 229–42.

Huckriede, R., Kürsten, M. and Venzlaff, H., 1962, Zur Geologie des Gebietes zwischen Kerman und Sargand (Iran), *Beihefte des Geologischen Jahrbuch*, **51**: 1–197.

Janvier, P., 1974, Preliminary report on Late Devonian fishes from Central Iran, *Geological Survey of Iran Reports*, **31**: 5–48.

Janvier, P., 1976, Description de restes d'Elasmobranches (Pisces) du Dévonien moyen de Bolivie, *Palaeovertebrata*, **7**: 126–32.

Janvier, P., 1977a, Les Vertébrés Dévoniens de l'Iran central et de l'Afghanistan, *Mémoires de la Société géologique du France*, **8**: 277–89.

Janvier, P., 1977b, Vertébrés dévoniens de deux nouveaux gisements du Moyen-Orient. Le problème des relations intercontinentales au Paléozoïque moyen, vu à la lumière de la paléobiogéographie des Rhipidistiens ostéolépiformes et des premiers Tétrapodes, *Annales de la Société géologique du Nord*, **97**: 373–82.

Janvier, P., 1978a, Los peces devonicos de Bolivia, *Programa Internacional de Correlacion geologica* (progr. 42,44,120): 1–13, Academia Nacional de Ciencias, La Paz.

Janvier, P., 1978b, The Upper Devonian of the Middle East, with special reference to the Antiarchi of the Antalya 'Old Red Sandstone'. In E. Izdar and E. Nakoman (eds), *Sixth Colloquium on Geology of the Aegean Region, Piri Reis International Contributions Series*, **2**: 331–40.

Janvier, P., 1979, Les Vertébrés dévoniens de l'Iran Central. III: Antiarches, *Géobios*, **12**: 605–8.

Janvier, P., 1980, Osteolepid remains of the Middle East, with particular reference to the endoskeletal shoulder girdle. In A.L. Panchen (ed.), *The Terrestrial Environment and the Origin of Land Vertebrates*, Academic Press, London: 223–54.

Janvier, P., 1983, Les Vertébrés dévoniens de la Nappe Supérieure d'Antalya (Taurus lycien occidental, Turquie), *Géologie Méditerranéenne*, **10**: 1–13.

Janvier, P., 1987, Les Vertébrés siluriens et dévoniens de Bolivie: remarques particulières sur les Chondrichthyens. In *IV Congreso Latinamericano de Paleontologia*, Santa-Cruz, **1**: 159–78.

Janvier, P. and Dingerkus, G., 1991, Le synarcual de *Pucapampella* Janvier et Suarez-Riglos: une preuve de l'existence d'Holocéphales (Vertebrata, Chondrichthyes) dès le Dévonien moyen, *Compte-Rendus de l'Académie des Sciences*, Paris, **312**: 549–52.

Janvier, P., Lethiers, F., Monod, O. and Balkas, Ö., 1984, Discovery of a vertebrate fauna at the Devonian–Carboniferous boundary in SE Turkey, *Journal of Petroleum Geology*, **7**: 147–68.

Janvier, P. and Marcoux, J., 1977, Les grès rouges de l'Armutgözlek Tepe: leur faune de poissons (Antiarches, Arthrodires et Crossoptérygiens) d'âge dévonien supérieur (Nappes d'Antalya, Taurides occidentales, Turquie), *Géologie Méditerranéenne*, **4**: 183–88.

Janvier, P. and Martin, M., 1978, Les Vertébrés dévoniens de l'Iran Central. I: Dipneustes, *Géobios*, **11**: 819–33.

Janvier, P. and Martin, M., 1979, Les Vertébrés dévoniens de l'Iran Central. II: Coelacanthiformes, Struniiformes, Ostéolépiformes, *Géobios*, **12**: 497–511.

Janvier, P. and Melo, J.H.G. de, 1987, Late Devonian actinopterygian scales from the Upper Amazon Basin, Northwestern Brazil, *Anais da Academia brasileira de Ciencias*, **53**: 213–18.

Janvier, P. and Melo, J.H.G. de, 1988, Acanthodian fish remains from the Upper Silurian or Lower Devonian of the Amazon Basin, Brazil, *Palaeontology*, **31**: 771–7.

Janvier, P. and Melo, J.H.G. de, 1992, New acanthodian and chondrichthyan remains from the Lower and Middle Devonian of Brazil, *Neues Jahrbuch für Geologie und Paläontologie, Monatshefte*, **4**: 193–206.

Janvier, P. and Pan, J., 1982, *Hyrcanaspis bliecki* n.g. n.sp., a new primitive euantiarch (Antiarcha, Placodermi) from the Middle Devonian of northeastern Iran, with a discussion on antiarch phylogeny, *Neues Jahrbuch für Geologie und Paläontologie, Abhandlungen*, **164**: 364-92.

Janvier, P. and Ritchie, A., 1977, Le genre *Groenlandaspis* Heintz (Pisces, Placodermi, Arthrodira) dans le Dévonien d'Asie, *Compte-Rendus de l'Académie des Sciences*, Paris, **284**: 1385–8.

Janvier, P. and Suarez-Riglos, M., 1986, The Silurian and Devonian vertebrates of Bolivia, *Bulletin de l'Institut Français d'Etudes Andines*, Lima, **15**: 73–114.

Jubb, R.A. and Gardiner, B.G., 1975, A preliminary catalogue of identifiable fossil fish material from southern Africa, *Annals of the South African Museum*, **67**: 381–440.

Katzer, F., 1897, Das Amazonas-Devon und seine Beziehung zu den anderen Devongebieten der Erde, *Sitzungsberichte der Königlich böhmischen Gesellschaft der Wissenschaften (Mathematische-naturwissenschaftliche Klasse)*, **46**: 1–50.

Kegel, W., 1953, Contribuçao para o estudio do Devoniano de bacia do Parnaiba, *Boletim da Divisião de Geologia e Mineralogia*, **141**: 1–48.

Krupina, N.I., 1979, [A new dipnoan species from the Famennian of Caucasus]. *Paleontologischeskii Zhurnal*, **2**: 145–7, (in Russian).

Krupina, N.I., 1987, [A new dipnoan fish from the Upper Devonian of the Tula region]. *Paleontologischeskii Zhurnal*, **3**: 40–7, (in Russian).

Lehman, J.P., 1952a, Les Arthrodires du Maroc méridionnal, *Compte-Rendus de l'Académie des Sciences*, Paris, **234**: 2631–3.

Lehman, J.P., 1952b, Notes paléoichthyologiques, *Annales de Paléontologie*, **38**: 59–67.

Lehman, J.P., 1956, Les Arthrodires du Dévonien supérieur du Tafilalt (Sud Marocain), *Notes et Mémoires du Service géologique du Maroc*, **129**: 1–70.

Lehman, J.P., 1963, À propos de quelques Arthrodires et ichthyodorulites sahariens. In: Mélanges Ichthyologiques (Th. Monod ed.), *Mémoires de l'Institut Français d'Afrique noire*, **68**: 193–200.

Lehman, J.P., 1976, Nouveaux poissons fossiles du Dévonien du Maroc, *Annales de Paléontologie, Vertébrés*, **62**: 1–34.

Lehman, J.P., 1977a, Sur la présence d'un Ostéolépiforme dans le Dévonien supérieur du Tafilalt, *Compte-Rendus de l'Académie des Sciences*, Paris, **285D**: 151–3.

Lehman, J.P., 1977b, Nouveaux Arthrodires du Tafilalet et de ses environs, *Annales de Paléontologie, Vertébrés*, **63**: 105–32.

Lehman, J.P., 1978, A propos de deux poissons de Famennien du Tafilalt, *Annales de Paléontologie, Vertébrés*, **64**: 143–52.

Lelièvre, H., 1984a, *Atlanditosteus hollardi* n.g., n.sp., nouveau Brachythoraci (Vertébrés, Placodermes) du Dévonien inférieur du Maroc présaharien, *Bulletin du Muséum national d'Histoire naturelle*, Paris, **6**: 197–208.

Lelièvre, H., 1984b, *Antineosteus lehmani* n.g., n.sp., nouveau Brachythoraci du Dévonien du Maroc présaharien. Remarques sur la Paléobiogéographie des Homostéides de l'Emsien, *Annales de Paléontologie, Vertébrés*, **70**: 115–58.

Lelièvre, H., 1988, Nouveau matériel d'*Antineosteus lehmani* Lelièvre, 1984 (Placoderme, Brachythoraci) et d'Acanthodiens du Dévonien inférieur (Emsien)

d'Algérie, *Bulletin du Muséum national d'Histoire naturelle*, Paris, **10**: 287–302.
Lelièvre, H., 1989, Les Vertébrés de la Formation de Jawf du Dévonien inférieur d'Arabie Saoudite, *Bulletin du Muséum national d'Histoire naturelle*, Paris, **1989**: 123–30.
Lelièvre, H., 1991, New information on the structure and the systematic position of *Tafilalichthys lavocati* (Placoderm, Arthrodire) from the Late Devonian of Tafilalt, Morocco. In M.M. Chang, Y.H. Liu and G.R. Zhang (eds), *Early Vertebrates and Related Problems of Evolutionary Biology*, pp. 121–30. Science Press, Beijing.
Lelièvre, H., Feist, R., Goujet, D. and Blieck, A., 1987, Les Vertébrés Dévoniens de la Montagne Noire (Sud de la France) et leur apport à la phylogénie des Pachyostéomorphes (Placodermes, Arthrodires), *Palaeovertebrata*, **17**: 1–26.
Lelièvre, H. and Goujet, D., 1990, Un nouveau spécimen d'*Holonema radiatum* (Placodermi, Arthrodira) du Dévonien moyen de la région d'Oviedo, Espagne, *Bulletin du Muséum national d'Histoire naturelle*, Paris, **1990**: 53–83.
Lelièvre, H., Goujet, D. and Morzadec, P., 1983, Les poissons du Frasnien (Dévonien supérieur) de Traonliors en Plougastel-Daoulas (Bretagne occidentale), *Bulletin de la Société géologique et minéralogique de Bretagne*, **15**: 75–83.
Lelièvre, H., Janvier, P. and Goujet, D., 1981, Les Vertébrés dévoniens de l'Iran Central. 4: Arthrodires et Ptyctodontes, *Géobios*, 677–709.
Lelièvre, H. and Janvier, P., 1986, L'Eusthénopteridé (Osteichthyes, Sarcopterygii) du Famennian (Dévonien supérieur) du Tafilalt (Maroc): nouvelle description, *Bulletin du Muséum national d'Histoire naturelle*, Paris, **1986**: 351–65.
Lelièvre, H. and Janvier, P., 1988, Un Actinistien (Sarcopterygii, Vertebrata) dans le Dévonien supérieur du Maroc, *Compte-Rendus de l'Académie des Sciences*, Paris, **307**: 1425–30.
Leonardi, G., 1983, *Notopus petri*, n.g., n.sp., une empreinte d'Amphibien du Dévonien du Parana (Brésil), *Géobios*, **16**: 233–9.
Long, J.A., 1984, New phyllolepids from Victoria and the relationships of the group, *Proceedings of the Linnean Society of New South Wales*, **107**: 263–308.
Long, J.A., 1988, New palaeoniscoid fishes from the Late Devonian and early Carboniferous of Victoria, *Memoirs of the Association of Australasian Paleontologists*, **7**: 1–64.
Mader, H., 1986, Schuppen und Zähne von Acanthodiern und Elasmobranchiern aus dem Unter-Devon Spaniens (Pisces), *Göttinger Arbeiten in Geologie und Paläontologie*, **28**: 1–58.
Massa, D., 1988, *Paléozoïque de Libye occidentale: Stratigraphie et paléogéographie*. Unpubl. Thesis, Univ. Nice, 514pp.
McKerrow, W.S. and Scotese, C.R. (eds), 1990, *Palaeozoic Palaeogeography and Biogeography*, *Geological Society of London, Memoir* **12**.
Poltnig, W., 1984, Fischreste aus dem Unterdevon von Graz (Steiermark), *Mitteilungen der naturwissenschaftliche Verreinigung Steiermark*, **114**: 107–31.
Priem, F., 1910, Sur les poissons et autres fossiles du Silurien supérieur du Portugal, *Comunicações do Serviço Geologico de Portugal*, **8**: 1–11.
Racheboeuf, P., 1990, Paléobiogéographie de la marge nord-gondwanienne au Dévonien inférieur et moyen: nouvelles données déduites de l'étude des Brachiopodes Chonétacés, *Compte-Rendus de l'Académie des Sciences*, Paris, **310**: 1481–6.
Rieben, H., 1935, Contribution à la géologie de l'Azerbaidjan persan, *Bulletin de la Société Neuchâteloise de Sciences naturelles*, **59**: 19–144.
Rouault, M., 1858, Note sur les Vertébrés fossiles des terrains sédimentaires de l'Ouest de la France, *Compte-Rendus de l'Académie des Sciences*, Paris, **47**: 99–103.
Santos, R.S., 1961, Peixes fosseis do Devoniano Inferior de Picos, Estado do Piaui,

Anais da Academia brasileira de Ciências, **33**: 32.
Schmidt, W., 1963, Ein Rhenanide (Arthrodira) aus dem Mittel-Devon der Turkei, *Neues Jahrbuch für Geologie und Paläontologie, Monatshefte*, **118**: 217–30.
Schultze, H.P., 1973a, Crossopterygier mit heterozerker Schwanzflosse aus dem Oberdevon Kanadas, nebst einer Beschreibung von Onychodontida-Resten aus dem Mitteldevon Spaniens und aus dem Karbon der USA, *Palaeontographica A*, **143**: 188–208.
Schultze, H.P., 1973b, Large Upper Devonian Arthrodires from Iran, *Fieldiana: Geology*, **23**: 53–78.
Seilacher, A., 1963, Kaledonische Unterbau der Irakiden, *Neues Jahrbuch für Geologie und Paläontologie, Abhandlungen*, **10**: 527–42.
Sengör, A.M.C., 1984, The Cimmeride orogenic system and the tectonics of Eurasia. *Geological Society of America, Special Papers*, **195**.
Stapleton, R.T., 1977, Carboniferous unconformity in Southern Africa, *Nature*, **268**: 222–3.
Talent, J.A., Rajendra, K.G., Arvind, K.J. and Pickett, J.W., 1988, Silurian and Devonian of India, Nepal and Bhutan: Biostratigraphic and palaeobiogeographic anomalies, *Courier Forschungsinstitut Senckenberg*, **106**: 1–57.
Tarlo, L., 1965, Psammosteiformes (Agnatha) a review with descriptions of new material from the Lower Devonian of Poland. II. Systematic part, *Palaeontologia polonica*, **15**: 1–64.
Theron, J.N. and Loock, J.C., 1988, Devonian deltas of the Cape Supergroup, South Africa. In N.J. McMillan, A.F. Embry and D.J. Glass (eds), *Devonian of the World, Canadian Society of Petroleum Geologists, Memoir* **14, III**: 729–40.
Théry, J.M., 1982, *Constitution du Nord-Ouest du Continent sud-américain avant les tectoniques andines*. Unpubl. Thesis, University of Bordeaux III, 3 vol.
Turner, S., 1989, Microvertebrates. In Boucot, A.J., McClure, H.A., Alvarez, F., Ross, J.R.P., Taylor, D.W., Struve, W., Savage, N.N. and Turner, S. *New Devonian Fossils from Saudi Arabia and their Biogeographical Affinities, Senckenbergiana Lethaea*, **69**: 535–97.
Turner, S. and Janvier, P., 1979, Middle Devonian Thelodonti (Agnatha) from the Khush-Yeilagh Formation, North-East Iran, *Géobios*, **12**: 889–92.
White, E.I., 1969, The deepest vertebrate fossil and other arctolepid fishes, *Biological Journal of the Linnean Society*, **1**: 293–310.
Young, G.C., 1988a, New occurrences of phyllolepid placoderms from the Devonian of Central Australia, *Bureau of Mineral Resources Journal of Australian Geology and Geophysics*, **10**: 363–76.
Young, G.C., 1988b, Antiarchs (placoderm fishes) from the Devonian Aztec Siltstone, Southern Victoria Land, Antarctica, *Palaeontographica A*, **202**: 1–125.

Chapter 8

PALAEOZOIC MICROVERTEBRATE BIOSTRATIGRAPHY OF EASTERN GONDWANA*

Susan Turner

INTRODUCTION

Ordovician to Permian vertebrate microfossils (microvertebrates or ichthyoliths) from Australia include scales, teeth, spines and other bony and cartilaginous parts from agnathans, chondrichthyans, placoderms and teleostomes. Many of these are biostratigraphically useful and allow correlation of Palaeozoic rocks both within Australia and with those elsewhere.

Young (1981) defined the Eastern Gondwanan Province to include Australia, New Zealand and Antarctica. Subsequent work has shown the possibility of the close proximity or attachment of parts of the Shan-Thai terrane (Burrett *et al.*, 1990). Japan and New Guinea might also have been part of Eastern Gondwana (Young, 1990a). Late Devonian and younger microvertebrate faunas appear cosmopolitan but in earlier times, although there are some cosmopolitan forms, many endemic species occur in Eastern Gondwana. Early Devonian microvertebrate assemblages from New Zealand, Victoria, southern New South Wales and Tasmania differ from those elsewhere in Australia. This disparity might indicate a deeper water facies or reflect climatic differences. Variability in microvertebrate assemblages, certainly in the Devonian, sometimes reflect the major transgression–regression events which controlled sedimentation in Australia (e.g. Talent, 1989).

HISTORY

Woodward (1921) first identified vertebrate microremains from Eastern Gondwana, scales he regarded as belonging to an elasmobranch or

* A contribution to IGCP Project 328.

ostracoderm. These scales are now assigned to turiniid thelodonts (Turner and Young, 1992). Woodward (1906) also described scales and squamation of acanthodians from the Lower Carboniferous sediments near Mansfield, Victoria.

Pioneer work on Palaeozoic microvertebrates in Australia was carried out by Thomas (1957, 1959) who first recognised vertebrates, particularly cladodont and bradyodont shark teeth in the Early Carboniferous of the Canning Basin. Talent identified elasmobranch denticles from the Silverband Formation of the Grampians, western Victoria (Talent and Spencer-Jones, 1963), and reassessed the dating of the Silverband Formation from Lower Carboniferous to Middle–Upper Devonian. Subsequent work pushed the age further down to earliest Devonian or even late Silurian by re-identification of the scales as those of a turiniid agnathan (Turner, 1986).

Material from the Palaeozoic of New Zealand was first described when Macadie (1985) recognised acanthodian, placoderm and other remains from the Lower Devonian Lankey Limestone of Reefton in northern South Island. Based on the presence of buchanosteid plates, scales of the acanthodian *Nostolepis costata*, possible *Machaeracanthus* and shark scales, this fauna seems close to the late Pragian–Emsian assemblages of New South Wales and Tasmania (see below).

Another possible record for Eastern Gondwana is the presence of Lower Devonian(?) scale of either a nikoliviid thelodont or shark from Irian Jaya (Young and Nicoll, 1979). In 1992, J. Vergoossen (Groningen University) reported microvertebrates, acanthodian, turiniid thelodont and possible shark scales from a Late Ludlovian conodont sample (Van den Boogaard collection) from New Guinea.

In the mid 1960s, Peter Jones recognised thelodont scales in his shot point samples from the Toko Syncline of the Georgina Basin, western Queensland, and thus rocks previously thought to be Ordovician in age were placed in the Early Devonian (Jones, in Johnstone *et al.*, 1967; Turner *et al.*, 1981). In 1969 microvertebrates from Wilson's Cliffs No. 1 borehole of the southeastern Canning Basin, Western Australia, were sent by a French petroleum geologist to Professor Walter Gross, one of the century's leading palaeoichthyologists. He formally described the first agnathan remains from the southern hemisphere and named the thelodont, *Turinia australiensis*, which occurred with undetermined acanthodian scales (Gross, 1971a). In so doing, sediments were redated from Permian to Early Devonian. This trend has continued with the discovery of Devonian thelodont scales in presumed Ordovician rocks elsewhere in central Australia leading to a reassessment of dating (Young *et al.*, 1987). These examples illustrate one important use of microvertebrate assemblages—that of dating a rock sample (even a small one from a borehole) in the absence of other diagnostic fossils. This method was used when conodonts were undiagnostic to date the Tumblong Oolite of New South Wales (Pickett *et al.*, 1985).

Since 1980 there has been a systematic effort to recover microvertebrates from Palaeozoic rocks in Australia and search for them elsewhere in neighbouring Gondwana countries (e.g. Long, 1990; Turner, 1982a, b, 1991b; Wang and Turner, 1985). This effort is now part of a truly international research programme which began in 1987 and is coming to fruition under the banner of IGCP project 328 on Palaeozoic microvertebrate biochronology and global marine/non-marine correlation, from 1991 to 1995. Summaries of work in progress can be found in the group newsletter, *Ichthyolith Issues* (Turner, 1988, 1989, 1990b–d, 1991b, c, 1992).

BIOSTRATIGRAPHICAL ALIGNMENTS

Webby and Nicoll (1989), Strusz (1989), Young (1989), Jones (1989), and Archbold and Dickins (1989) have provided detailed summaries and timecharts for the Ordovician to Permian of Australia. These cover the basic information on invertebrate, plant and conodont faunas which are used to provide an international framework for dating Australian Palaeozoic sequences and assessing the microvertebrate faunas. Often the vertebrate data is the only usable biostratigraphical information, especially in central Australia. Mawson *et al.* (1985), Jell and Talent (1989), Talent and Yolkin (1987) and Talent (1990) have provided a general framework for the Silurian and Devonian within which we can assess the microvertebrate data.

Some sequences bearing microvertebrates are extremely well-dated using invertebrate faunas and conodonts. Examples include the Wee-Jasper–Taemas sequence in New South Wales and Buchan–Bindi succession of similar age with their abundant and diverse faunas of Emsian (*dehiscens* to *serotinus* Zones; Mawson, 1987a, b; Young, 1989). Conodonts from limestones from the late Lower Devonian to Late Devonian in New South Wales have been studied by Pickett (1978), Garratt and Wright (1988) and Wright *et al.* (1990). Other sequences in New South Wales such as those from the Darling Basin are more tentatively dated using invertebrates and conodonts (e.g. Pickett, 1983, 1989, 1992; Pickett and Ingpen, 1990; Pickett and McClatchie, 1991; Sherwin, 1992).

For the north Queensland Palaeozoic (Silurian, Devonian to Early Carboniferous) succession the biostratigraphy for the Broken River embayment and the Star and Burdekin Basins is underpinned by the work of Wyatt and Jell (e.g. 1980), Withnall *et al.* (1988), Telford (1975), Pickett (1981), Mawson *et al.* (1988) and Mawson and Talent (1989). There are rich coral and brachiopod faunas and conodonts which have been collected from detailed measured sections. This work by J.S. Jell, J. Talent and R. Mawson, and subsequent field collecting by the author, has produced a wealth of microvertebrate material (e.g. Turner, 1991a) and some macrovertebrate faunas (Young, 1990a), many of which await formal study.

For detailed stratigraphy of central Australia, Korsch and Kennard (1991) summarised the position for the Amadeus Basin, Draper (1976) and Turner *et al.* (1981) for the Georgina Basin.

Measured sections have now been collected for both conodonts and microvertebrates across the Devonian–Carboniferous boundary in Western Australia (J. Talent and R. Mawson, pers. comm.). These should underpin work on the rich macro- and microvertebrate faunas from the Canning and Bonaparte Basins gleaned from collections based on earlier work (e.g. Turner, 1982b, 1983; Druce and Radke, 1979; borehole information from petroleum exploration, Geological Survey of Western Australia archives).

Use of biostratigraphical schemes from the northern hemisphere can assist in understanding the east Gondwana occurrences (e.g. Valiukevicius, 1985, on acanthodians; Turner, 1973, on thelodonts; Blieck *et al.*, 1988). However, care needs to be taken because some taxa are endemic to Eastern Gondwana and others appear in that province earlier or later than elsewhere.

Problems can arise when microvertebrate faunas occur in presumed non-marine sediments where only relative dating is possible by reference to other vertebrate faunas in sediments above or below (e.g. in central Australia—Young, 1991a; Long *et al.*, 1988).

BIOSTRATIGRAPHY OF PALAEOZOIC MICROVERTEBRATE FAUNAS

Ordovician

Ritchie and Gilbert-Tomlinson (1977) described the first complete Ordovician vertebrates from Australasia (also Ritchie, 1985). They come from the Stairway Sandstone (Middle Ordovician: late Arenigian to early Llanvirnian) at Mt Watt and in the Charlotte Range of the Amadeus Basin, central Australia. The remains belong to primitive heterostracan-like fishes, *Arandaspis* and *Porophoraspis*. Other scale forms await description.

Arandaspis prionotolepis is represented by a complete cephalothorax with bone-like plates covered with a scale-like ornament of presumed dentine tubercles. There are diamond-shaped trunk scales arranged in regular rows and scales associated probably with the branchial region which have an anterior, narrow, smooth overlap zone, not unlike those of some anaspids. *Porophoraspis crenulata* is known from one small plate bearing numerous small, low, rounded tubercles with regular rows of pores opening onto the surface (Figure 8.1).

Disarticulated remains now are known from several other horizons and localities. Young (1991) described four assemblages: in the Carmichael Sandstone and Stokes Siltstone (younger than the Stairway Sandstone), and lower Horn Valley Siltstone and the top of the Pacoota Sandstone (Bendigonian: Assemblage 1 of Young, 1991a), the latter representing the oldest known verified vertebrates. There are scales with tesserate ornament similar to that of the mid-Ordovician heterostracan-like fish, *Saccabambaspis*, from Bolivia (Gagnier *et al.*, 1986). Other scales and bone fragments of possible early gnathostomes are recorded (Young, 1991a).

The known occurrences of Ordovician microfossils are apparently restricted to brackish/marginal marine shallow-water deposits in central Australia (Young, 1991a). The complete specimens from Australia, and North and South America have been found in the *Cruziana* facies.

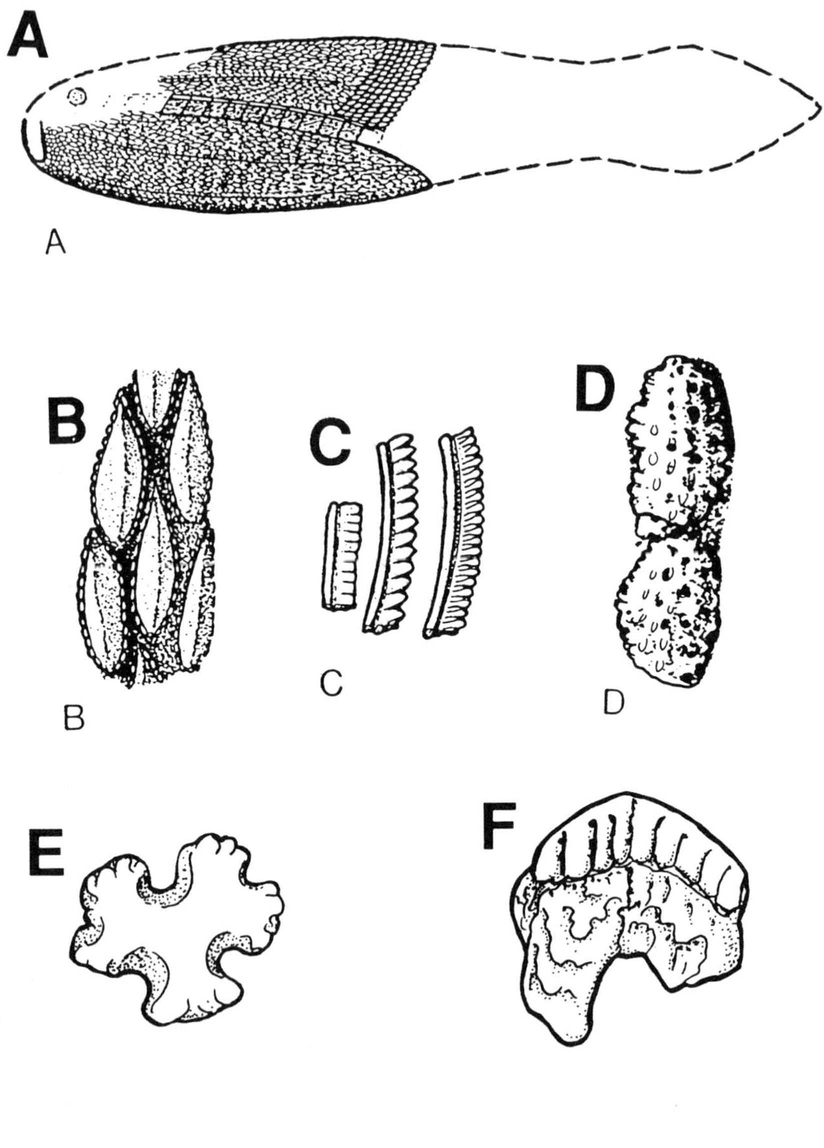

Figure 8.1. Ordovician microvertebrates from Eastern Gondwana. A–C, *Arandaspis prionotolepis*. A, restoration; B, dermal ornament; C, possible body scales; D, *Porophoraspis*, dermal tubercles. A–D from Ritchie and Gilbert-Tomlinson, 1977, not to scale. E, *Saccabambaspis*-like tubercles; F, primitive gnathostome? scale (E–F from G.C. Young photographs).

Silurian

Silurian vertebrates have been known in Eastern Gondwana for over a decade, nearly all from microvertebrate material. Acanthodian fishes dominate the Silurian microvertebrate assemblages. Scales, spines, jaws and tooth whorls appear in the Late Silurian of eastern Australia. Indeterminate bones probably also belong to acanthodians. Some plates and scales might belong to placoderms or sharks but not enough histological and comparative work has yet been done to determine this fact. Silurian placoderm macroremains are still relatively rare and thus comparative material is sparse. Reassessment of conodonts from some sites (Pickett, 1992) substantiates the appearance of thelodonts in the Late Silurian of Eastern Gondwana.

Microvertebrate assemblages

Silurian microvertebrates in Australia are known from New South Wales and Queensland (Turner and Pickett, 1982; Simpson, 1983), Australian Capital Territory (Henderson, 1982; Turner, 1991a) and Victoria (Strusz and Garratt, 1992; A. Simpson, pers. comm.). However, systematic searching of Silurian limestones for microvertebrates has only been underway for about a decade and any site where conodonts have been recorded, such as boreholes in the Canning Basin (e.g. Foreman and Wales, 1981), has potential for microvertebrates.

?Wenlockian
Turner and Pickett (1982) noted bone fragments which are probably acanthodian in Limestone D of the Mirrabooka Formation of the Cheesemans Creek near Orange, New South Wales. On the basis of graptolites the Mirrabooka Formation extends from late Wenlock to late Ludlow so the limestone sample could be late Wenlockian in age.

Ludlovian. Microvertebrate assemblage—siluricus–ploeckensis *zone*
The earliest definite record in mid-Ludlovian limestones was discovered during conodont sampling. Simpson, working on Silurian conodonts from the Broken River embayment of north Queensland (Simpson 1983), discovered a rich microvertebrate assemblage from this conodont zone in the Graveyard Creek Formation at Bullock Creek in the Jack Limestone Member. The fauna is predominantly acanthodian with scales, spines, jaws and teeth. Many scales are similar to those of the common Silurian–Early Devonian genus, *Nostolepis*, closely resembling *Nostolepis striata* (Figure 8.2A–D), as well as spines of climatiids (Denison, 1979). Ischnacanthids are represented by scales, including ones resembling those of *Gomphonchus*, and small jaws with ankylosed teeth (Denison, 1979). Closer inspection of the Queensland material suggests that some scales will have to be referred to new taxa (Long, pers. comm.). *Machaeracanthus* scales resembling those of *Machaeracanthus stonehousensis* (Legault, 1968) have been identified;

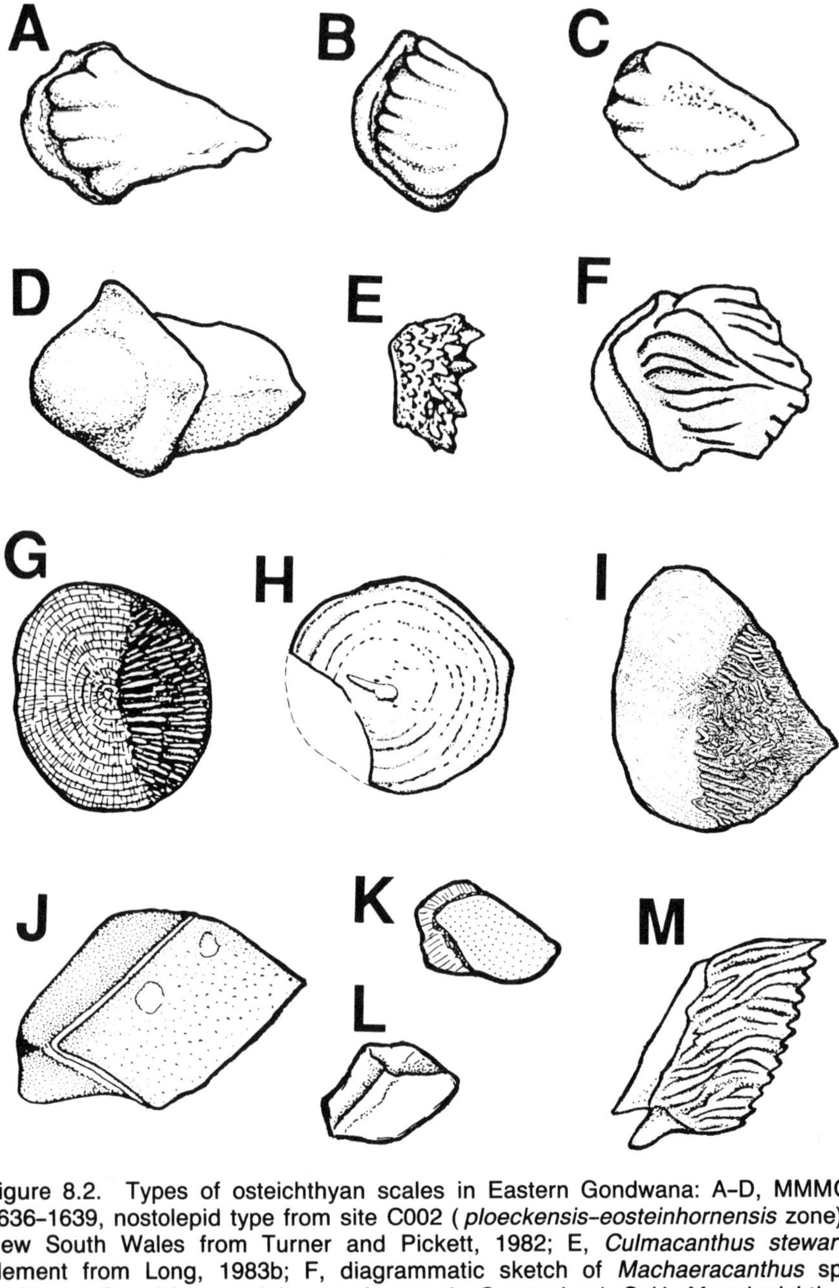

Figure 8.2. Types of osteichthyan scales in Eastern Gondwana: A–D, MMMC 1636–1639, nostolepid type from site C002 (*ploeckensis-eosteinhornensis* zone), New South Wales from Turner and Pickett, 1982; E, *Culmacanthus stewarti* element from Long, 1983b; F, diagrammatic sketch of *Machaeracanthus* sp. scale type found in *eosteinhornensis* zone in Queensland; G–H, *Marsdenichthys longioccipitus* cycloid scales dorsal (G, I) and ventral (H) views after Long, 1985; J, K–L, typical osteolepid-type rhombic cosmine scales; M, typical palaeoniscoid 'peg and socket' scale, *Howqualepis rostridens*, Frasnian, Mt Howitt, Victoria, typical flank scale (after Long, 1988a). Not to scale.

this scale form (Figure 8.2F) was first recorded from the *eosteinhornensis* zone of Arisaig, eastern Canada, and then in the early Lochkovian of Spain by Mader (1986). Thus the Eastern Gondwana record may be the oldest for the genus. Some of the bony platelets might belong to early placoderms, which have been found in the Silurian of China, and some are possible shark scales.

Microvertebrates also occur in the mid-Ludlovian Stony Creek Formation of eastern Victoria (A. Simpson, pers. comm.).

Crispa *zone*

Turner and Pickett (1982) referred some acanthodian scales from this site along with those from a limestone in the 'Bogan Gate Sandstone Group' now Cookeys Plains Formation (Pickett, 1992) to *Nostolepis striata* or cf. *N. striata* (Figure 8.2). The latter is now regarded by many as being a form taxon.

Pridoli

In the Pridoli the macro- and microvertebrate faunas include the appearance of an assemblage with turiniid thelodonts, acanthodians, placoderms (arthrodires with buchanosteid-type body scales), possible palaeoniscoid and onychodontid crossopterygians—a fauna which continues into the Early Devonian, possibly into the Emsian. This represents the first appearance for many of these fishes in Eastern Gondwana.

Microvertebrate assemblage—eosteinhornensis *zone*

The Silverband fauna of the Grampians (Turner, 1986), on evidence from the intrusion of Rocklands Rhyolite into what is thought to be equivalent sediments in the Black Range nearby, should be no younger than 400–406 million years (Young, 1989; C. Powell, pers. comm.). Turner (1986) reassessed the fish fauna as earliest Devonian principally on the presence of scales of a presumed primitive turiniid thelodont, *Turinia fuscina* (Figure 8.3) associated with acanthodian scales and spines of nostolepid and poracanthodid types (typical of Upper Silurian in Europe) and the similarity of acanthodian spines to those of *Sinacanthus* known in Silurian to Lower Devonian of China (Li, 1980). This fauna could be of mid to late Pridoli in age.

An assemblage of turiniid thelodont scales from the Kempfield 1 borehole in the Canning Basin, Western Australia (?Carribuddy Formation, locality shown in Foreman and Wales, 1981), might be Pridoli in age (Figure 8.3). Scales of *Turinia polita* Karatajute-Talimaa 1978, which they resemble, are typical of Gedinnian (basal Lochkovian) in Europe.

Limestones from the Mineral Hill Volcanics at Mineral Hill (C620, 622) and from Fifield in the Meloola Volcanics (C005), New South Wales, (probable *eosteinhornensis* but could be as old as *crispa* zone with elements of *Ozarkodina excavata*) (Pickett, 1989; Pickett and Ingpen, 1990) have yielded small assemblages including possible *Turinia* cf. *australiensis* (with relatively simpler scales), placoderm remains, onychodont teeth, and a possible *Machaeracanthus* scale.

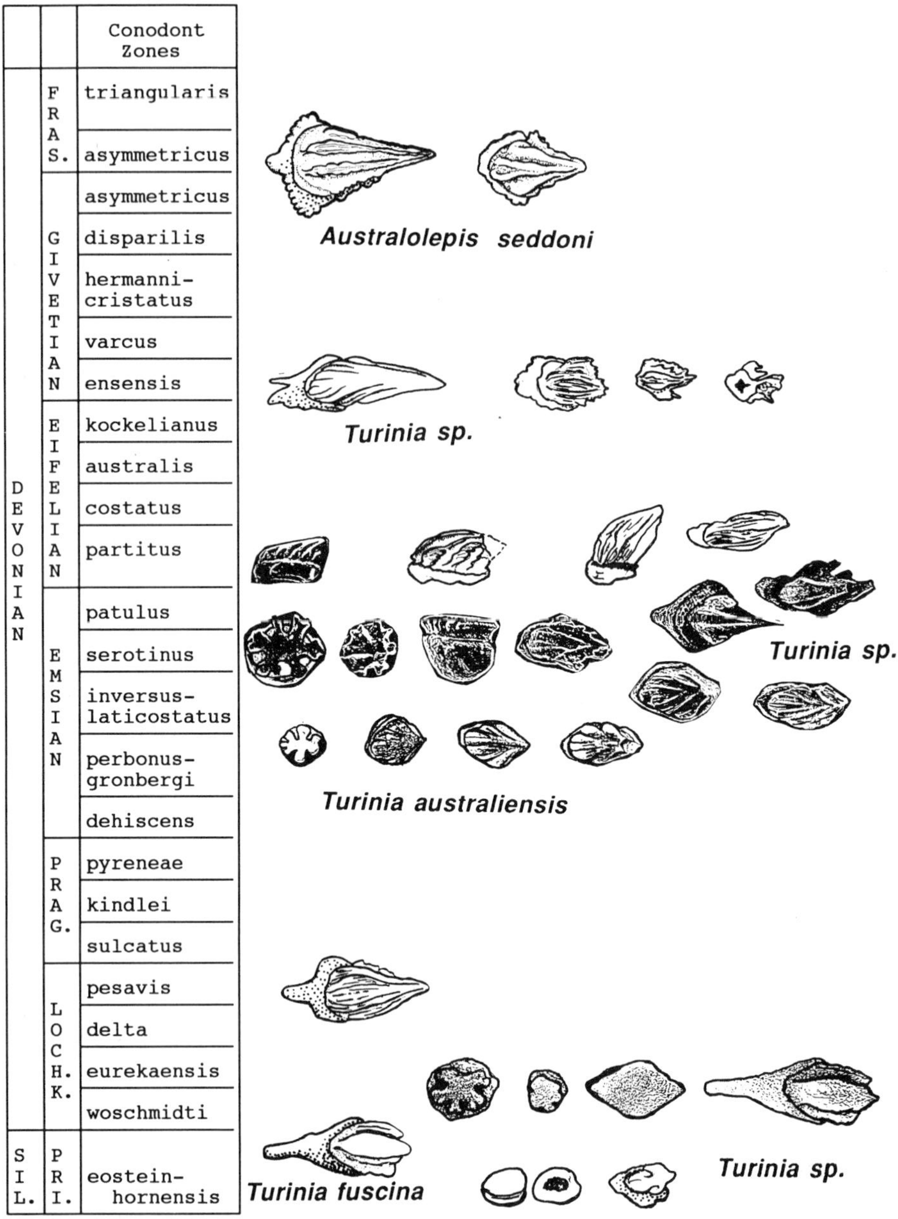

Figure 8.3. Thelodont scale ranges in Eastern Gondwana.

Devonian

Thelodonts

New information regarding the dating of certain limestone samples from New South Wales (see above) as well as data from other sources confirms that turiniid thelodonts appear in the late Silurian in Eastern Gondwana. Turiniid scales are then common in Devonian nearshore limestones in eastern Australia. The scales are generally well-preserved and can form distinct bone-beds, for instance in the Mulga Downs Formation, Darling Basin; in the Cravens Peak Beds, Georgina Basin (Turner *et al.*, 1981), and in the Parke Siltstone, Cleland Hills, western Amadeus Basin. Most Early Devonian scales belong to the *Turinia australiensis* morphotype. The scales of the type material from the southeastern Canning Basin (Gross, 1971a) were extremely rolled, making specific characters uncertain, except for the presence of minor parallel ribbing on the angled ribs of the crown in body scales. In New South Wales, the *T. australiensis* morphotype comes from sediments within the Bogan Gate and Molong Highs in a range of Pridoli and Emsian sediments, with a major concentration of rich microvertebrate faunas in the late Lochkovian–early Pragian of the Mineral Hill, Murda Syncline and Trundle regions.

Endemic thelodonts as well as the *T. australiensis* morphotype occur in the Lochkovian of the Broken River Embayment, north Queensland. Other endemic thelodonts occur in western Victoria—the supposed primitive turiniid, *Turinia fuscina* (Turner, 1986)—and in a possible Late Silurian borehole in the Canning Basin (see above). Simple thelodont scales from the Tumblong Oolite found in a thrust flake in the Tumut region (Pickett *et al.*, 1985; Figure 8.3) resemble those of *Turinia polita* Karatajute-Talimaa 1978, which are typical of basal Lochkovian in Europe.

Late Early or early Middle Devonian turiniids have recently been recovered from the Amadeus Basin from shallow-water, probably non-marine or brackish deposits, and from down a deep borehole in South Australia (Young *et al.*, 1987; Long *et al.*, 1988; Young and Turner, 1992). The scales are closely related to some from the Georgina Basin (Turner *et al.*, 1981). Middle Devonian turiniid and modified turiniid scales resembling those of nikoliviids are known from the Hatchery Creek Conglomerate Group of New South Wales (Young and Gorter, 1981) and the Broken River Formation of Queensland. The turiniid scales from these sites closely resemble those from west Yunnan (China), Antarctica and Bolivia (Wang *et al.*, 1986; Gagnier *et al.*, 1988; Turner and Young, 1992). The youngest thelodonts known in Australia, and possibly the world, are the turiniids called *Australolepis seddoni* from the early Frasnian Gneudna Formation of the Carnarvon Basin, Western Australia (Turner and Dring, 1981). The scales of this species exhibit characteristics such as lightening of scales and bifurcated distal crown ribs. The Gneudna Formation is overlain conformably by the Munabia Sandstone which has recently yielded a *Bothriolepis* fauna though to be of Frasnian age (Long, 1991a).

Whereas turiniids disappear from Laurentia (North America/Greenland/western Europe) near the beginning of the Middle Devonian, they flourished in near-shore Gondwanan environments until the early Late Devonian. As well as countries mentioned above, turiniid scales occur in the Early?–Late Devonian of China (west Yunnan, Wang *et al.*, 1986), Iran (Turner and Janvier, 1979) and Thailand (Blieck *et al.*, 1984). *Australolepis* has now been found in the Givetian (possibly late) of the Holy Cross Mountains in Poland (M. Ginter, pers. comm.). Turiniids can be expected in similar environments throughout Gondwana.

Nikoliviid-like thelodont scales may occur in Eastern Gondwana (Turner *et al.*, 1981) but some are definitely modified turiniid scales and some might belong to sharks. True nikoliviid scales, similar to those of Arctic Canada and western Europe, have recently been identified in Early Devonian Windmill Limestone (*delta* zone) of Nevada, associated with placoderm body scales resembling those from the Early Devonian of eastern Australia (Turner and Murphy, 1988). These are the first definite Devonian thelodonts from the USA, and they are closely related to those from Arctic Canada and Europe, and distinct from Gondwanan forms.

Australian turiniid stratigraphic occurrences are summarised in Figure 8.3.

Chondrichthyans

The first shark radiations took place in the Early Silurian to Early Devonian. These early sharks (i.e. pre-Middle Devonian) are known exclusively from microvertebrate remains. *Elegestolepis*, *Polymerolepis* and other as yet undescribed genera have been recovered from the Lower Silurian of Siberia, and *Mongolepis* from northern Mongolia (Karatajute-Talimaa *et al.*, 1990). Vieth (1980), Wang (1984), and Mader (1986) have found scales and possible teeth in the Early Devonian of Arctic Canada, China and Spain. Shark scales also occur in the Lower Silurian of eastern Canada (pers. obs.).

In Australia there are several simple scales in the Lower and Middle Devonian which cannot be referred to thelodonts, acanthodians or placoderms. Some of these neoselachian-like scales come from the Trundle Beds (late Lochkovian?) of western New South Wales (Turner, 1991a, Figure 8.4K). Interestingly, they lack the four neck canals which Reif (1978) defined as characteristic for the true placoid shark scale. Other possible shark scales occur in the Buchan Group of Victoria and in the Point Hibbs Limestone of Tasmania (called 'turiniids' in error, Turner, 1986, and Young, 1989).

The Cravens Peak Beds of the Georgina Basin (possible late Early Devonian) have yielded scales similar to those called *Gualepis* by Wang (1984) (Turner, 1991a, Figure 8.4G, H). This scale type is also seen in the Lankey Limestone of New Zealand (Macadie, 1985). Associated with the Cravens Peak scales are teeth and prismatic calcified cartilage from a shark which might have been a primitive hexacanthid (Turner, 1985). The teeth, *Mcmurdodus whitei* Turner and Young 1987, are similar to those of *M. featherensis* from the Aztec Siltstone of Antarctica (White, 1968; Young 1982). A tooth referred by Young (1982, CPC21229) to *Mcmurdodus*? is now known to belong to a new shark type from the Aztec Siltstone, new specimens of which were collected by J. Long over the 1991/1992 field season.

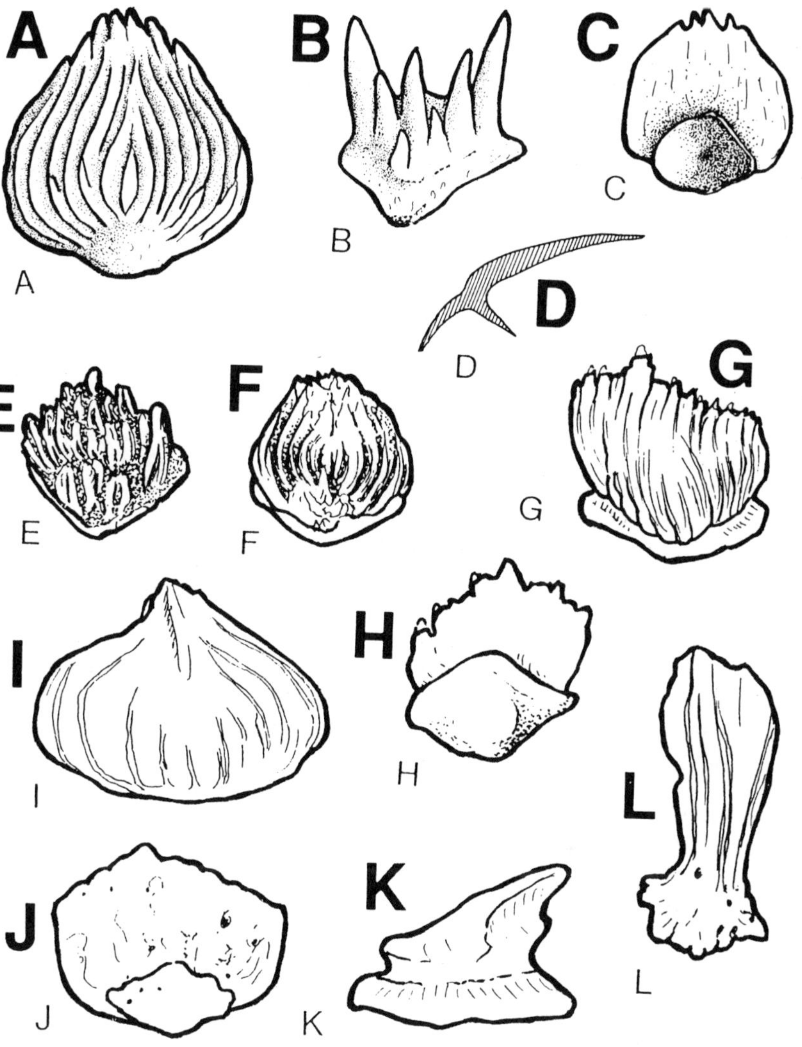

Figure 8.4. Shark scale types from Eastern Gondwana: A–D, *Antarctilamna prisca* Young 1982; E, *Ohiolepis* after Wells, 1944; F, *Cladolepis* after Wells, 1944; G–H, *Gualepis*-like scales; I–J, shark gen. et sp. indet.; K, primitive neoselachian? gen. et sp. indet.; L, *Skamolepis*-like scale. Not to scale, all around 1 mm in length.

One apparently cosmopolitan shark scale morphotype in the Early to early Middle Devonian is *Ohiolepis*, found originally in the USA (Wells, 1944; Turner, 1991a, Figure 5E), China (Wang, 1984) and Europe (e.g. Friman, 1983). It occurs also in some Lower Devonian limestones in Australia (e.g. Giffin, 1980). Scales from the same late Emsian Receptaculites Limestone Member of the Murrumbidgee Group, Taemas, New South Wales, were referred to '*Skamolepis*' by Giffin (1980). She compared them to those of supposed thelodont, *Skamolepis fragilis* Karatajute-Talimaa 1978, from the late Emsian to early Eifelian of Latvia and Spitsbergen. This taxon is now regarded as a chondrichthyan (Turner, 1991a). Such 'quasi thelodont' scales have also been found in the late Emsian Jesse Limestone (*perbonus–laticostatus* zone) and Mt Frome Limestone (*serotinus–patulus* zone) of New South Wales (Turner, 1982a, Figure 3F; Turner, 1991a, Figure 5L). They differ from turiniid scales in having a slender, high crown set on a thin diamond-shaped base.

Young (1982) described one of the earliest and apparently most primitive xenacanth sharks, *Antarctilamna prisca*, from the Middle Devonian Aztec Siltstone of Antarctica and Bunga Beds of New South Wales. Similar spines occur in the Khush-Yeilagh Beds of Iran (Blieck *et al.*, 1980) and Emsian/Eifelian Santa Rose Formation of Bolivia (Gagnier *et al.*, 1988, 1989). Various primitive xenacanthid teeth occur in the latest Devonian and Early Carboniferous of Australia (e.g. Turner, 1982c, Figure 8.4E).

Middle and Late Devonian shark scales and teeth are known from the limestones of the Broken River district of Queensland, from New South Wales and Western Australia. One of the earliest phoebodont teeth appears in a section (Storm Dam 146 of Mawson and Talent, 1989) in the Papilio Formation (Givetian) of Broken River, north Queensland. Ctenacanth and cladodont forms along with *Phoebodus sp.* (non *P. australiensis* Long 1990) and other phoebodonts and protacrodonts dominate Late Devonian (Frasnian and Famennian) assemblages (Turner, 1982b). Examples include limestones in the Mostyn Vale Formation of New South Wales, the Fairfield Group of Western Australia, the Teddy Mount and Myrtlevale Formations of north Queensland. These Late Devonian assemblages are remarkably similar worldwide; Gross (1973) and Ginter (1990) described examples from Europe and the USA, and a similar assemblage occurs in China (Wang and Turner, 1985), Thailand (Long, 1990), and Melville Island, Arctic Canada.

Another possible phoebodont is the Late Devonian (latest Famennian) to Early Carboniferous *Thrinacodus ferox* (Turner, 1982b, 1983). These teeth occur in Australia, Thailand, China, Europe and the USA. They can sometimes be mistaken for conodonts (see Wang and Turner, 1985). The type species, *Thrinacodus nanus* St John and Worthen 1875, is based on relatively large teeth (5–6 mm long) from the Late Devonian (*praesulcata* zone)–Lower Carboniferous(?) Kinderhook Formation of Iowa. *Thrinacodus ferox* is now known only from Late Devonian (*marginifera* zone) to earliest Carboniferous (early Visean) as microvertebrates of around 1–2 mm long, from Algeria, Austria, Australia, Belgium, China, Great Britain, Ireland, Poland, Portugal, Spain, and Thailand. Some may prove to be conspecific with the type species when more is known of its distribution in America.

Phoebodont teeth, known mostly as microvertebrates, are proving to be of great use in Late Devonian biostratigraphy. They may prove to be useful zone fossils from the Givetian through Famennian as they become better known. *Phoebodus s.s.* was first described from the Middle Devonian (now known to be Givetian) of the USA. Phoebodonts were present in Eastern Gondwana in the late Givetian. The relationship of younger Carboniferous sharks bearing true phoebodont teeth to the type genus is still uncertain.

Typical Palaeozoic shark teeth of 'cladodont' type are common in Late Devonian and Carboniferous deposits in Australia. Recent work (e.g. Hansen, 1988) is allowing a finer definition of these teeth into cladoselachian, stethacanth, ctenacanth and denaeaid forms. Associated with them are stethacanthid scales, pharyngobranchial scales of *Stemmatodus* type (St John and Worthen, 1875); modified head scales of *Lambdodus* type (St John and Worthen, 1875; Zangerl, 1981); and probably cladoselachian and ctenacanth scales.

Teeth of the enigmatic chondrichthyan, *Ageleodus pectinatus*, common as a macrovertebrate in the Late Carboniferous Coal Measures of Europe and North America, appears first in the Eastern Gondwana Province in the latest Famennian or earliest Tournaisian upper Telemon Formation of the Narrien Range, central Queensland (Turner and Hansen, pers. obs.). They are present also in the early Visean Raymond Formation bonebeds of the same region (Turner, 1982c). The series of bonebeds here were formed apparently under brackish conditions in an estuary or tidal flats. *Ageleodus* in Eastern Gondwana is known only as microvertebrates and the teeth possess fewer cusplets than in the type species.

Certain 'bradyodont' teeth occur as microvertebrates in Australia. Teeth resembling those of *Caseodus* occur in the Lower Carboniferous Namoi Formation of New South Wales (M. Leu collection), being the first record of this group of sharks in Eastern Gondwana. *Caseodus* is known from the Middle to Late Carboniferous in North America and possible advanced forms in the Permian of the USA and Greenland (Zangerl, 1981). A tooth from the Late Devonian or basal Carboniferous bonebed in the Telemon Formation of central Queensland resembles one of *Chiastodus* from the mid-Carboniferous of the Moscow region of Russia. This form might represent primitive caseodonts according to Zangerl (1981). Long (pers. comm.) has identified another so-called bradyodont tooth form, *Venustodus*, from Late Devonian limestones of Western Australia; the type species comes from the Lower Carboniferous of the USA (St John and Worthen, 1875).

Placoderms

Little is known about placoderm microfossils. In the Silurian and Early Devonian most placoderms appear to have had a primitive body covering of small bony tesserae and scales. By Late Devonian times most placoderms have severely reduced or lost their body scales. Jaws (gnathals) with teeth can be found as microfossils and are also potentially diagnostic, and examples are being found in the Early and Late Devonian of the Broken River area of north Queensland.

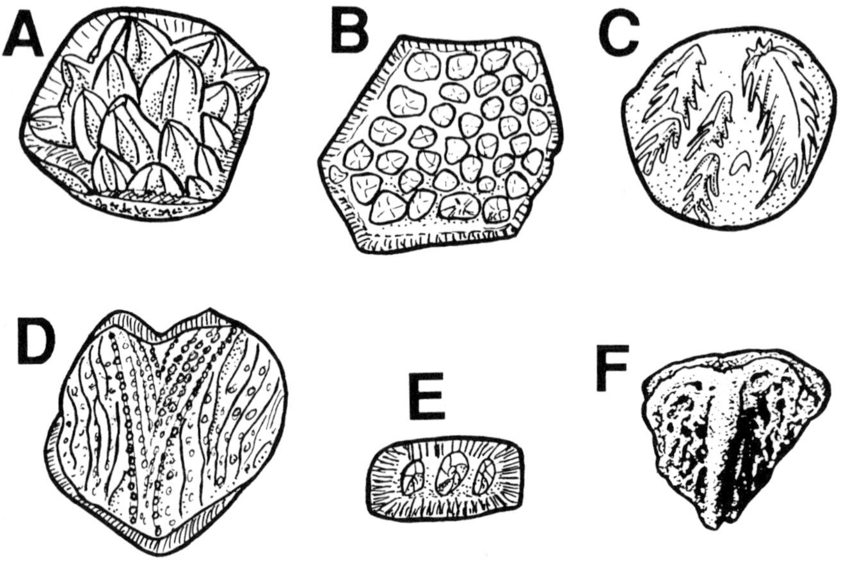

Figure 8.5. Placoderm body scale types. A, buchanosteid type. B, *Goodradigbeeon australianum* after White, 1978. C, Pragian arthrodire gen. et sp. indet. D, *Wuttagoonaspis*. E, cf. radotinid scale. F, asterolepidoid antiarch scale, after Young, 1984.

Body scales, sometimes referred to '*Ohioaspis*' (Denison, 1978; Giffin, 1980), are possibly some of the commonest remains in Early Devonian (Pragian–Emsian) marine limestones in Victoria, New South Wales, Queensland and Tasmania. By comparison with articulated specimens, many of these scales can be referred to *Buchanosteus* and related arthrodires such as *Goodradigbeeon* (White, 1978; Turner, 1991a; Figure 8.5A, B). Body scales of the Early Devonian acanthothoracid, *Murrindalaspis*, have been described by Long and Young (1988, Figure 9). These and other types of placoderm scales are also found in Early Devonian limestones in southern China, the western USA, and Europe (e.g. Goujet, 1976; Vieth-Schreiner, 1983; Poltnig, 1984).

In the Early Devonian Martins Well Limestone of Queensland, and in some Pragian localities in New South Wales, there are possible radotinid scales (Figure 8.5E) and rhenanid or radotinid tesserae (Turner, 1991a, pl. 4D). Other identifiable placoderm scales include lunaspid, wuttagoonaspid (Figure 8.5D) and groenlandaspid types. Long (1983a) and Long and Werdelin (1986) illustrated scales on an articulated bothriolepid tail from the Late Devonian of Victoria showing simple bony knubs with larger modified fulcrul scales.

Acanthodians

Acanthodian scales, tooth whorls, jawbones and spines have been found in nearshore Late Silurian to Carboniferous sediments (Long, 1991b; Turner, 1991a, Table 3). They are predominant in the limestones of the Buchan and Taemas districts and in the Early to Late Devonian marine sediments of eastern Australia. Complete acanthodians are known from Late Devonian continental sediments in Victoria (Long, 1983b, 1986a).

Woodward (1906) illustrated the first details of acanthodian scales, those of *Acanthodes australis*, *Eupleurogmus cresswelli*, and *Gyracanthides murrayi*, as he perceived them from the complete specimens from the Lower Carboniferous Mansfield fauna. The first record of isolated acanthodian scales from acid residues in Australia was that made by Philip (1965). He identified '*Nostolepis*' scales in the Coopers Creek Limestone of the Tyers district of Victoria, a scale type now known to be common in the Late Silurian to Lower Devonian. The Martins Well Limestone of the Broken River district of Queensland and several horizons in the Early Devonian (Pragian) of New South Wales (e.g. the Jerula Formation, Pickett and McClatchie, 1991) have a wide variety of scales and acanthodian platelets similar to those which Gross (1971b) and Goujet (1976) described from nostolepids in Europe and by Vieth (1980) from Arctic Canada. *Nostolepis costata* (Goujet, 1976) is common in the Pragian to early Emsian limestones of eastern Australia. Nostolepid scales are present in the Lower Devonian Reefton Beds (Macadie, 1985) and the Baton River Beds of New Zealand.

Ischnacanthid and climatiid acanthodian scales, spines and dentigerous jawbones have been reported from many localities (e.g. Young and Gorter, 1981; Long and Turner, 1984; Turner, 1991a). Some, such as *Rockycampacanthus* and *Taemascanthus*, are proving endemic to Australia (Long, 1986b). These have distinctive jawbones and teeth which might well turn up in microfossil assemblages. Some, however, such as *Cheiracanthoides comptus*, seem ubiquitous in Emsian–early Eifelian sediments worldwide (e.g. Wells, 1944; Vieth-Schreiner, 1983; Poltnig, 1984). Scales of *Machaeracanthus*, long known from the Middle to Late Devonian in Europe and America (e.g. Goujet, 1976; Zidek, 1981; Mader, 1986), occur in Early Devonian and in Silurian sites in New South Wales, Queensland, and Victoria. Philip (1965) figured one such from the Coopers Creek Beds (Turner, 1991a).

Long (1983b) described complete specimens of the diplacanthoid *Culmacanthus stewarti* from the Frasnian of Mt Howitt, Victoria, which have deep bases, a low neck and a thin crown with 6–7 thin ribs. Another form from Mt Howitt, which Long (1986a) called *Howittacanthus kentoni*, has *Acanthodes*-type scales with smooth crowns. Thus true acanthodiform fishes appear earlier in East Gondwana and their scales are present in Late Devonian assemblages. These would be indistinguishable from *Acanthodes s.s.* This incoming of this genus cannot, therefore, be a reliable indicator for the base of the Carboniferous as suggested by Wang and Turner (1985).

Gross (1971a) described acanthodian scales from the Wilson's Cliffs borehole horizon which could be Pragian–early Emsian. He could not assign the *Gomphonchus* and *Nostolepis*-like scales to a genus. The scales are of *Acanthodes* histological type and Gross found similarities in them to certain Middle Devonian scales (*Diplacanthus*, *Haplacanthus*, *Rhadinacanthus*). Long *et al.* (1988) figured a ?*Gomphonchus*-like scale from the Munyurai-1 borehole horizon which might be a worn *Cheiracanthoides comptus*, thus supporting the late Lower Devonian–basal Middle Devonian age suggested for that site.

Osteichthyans
Scales (Figure 8.2), fin ray supports (lepidotrichia, e.g. Long, 1988a, Figure 31, 32B), ribs, teeth and even otoliths of crossopterygians, dipnoans, and actinopterygians are found as microfossils in Devonian rocks (e.g. Giffin, 1980; Young and Gorter, 1981; Long, 1988a; Long *et al.*, 1988). Some scales, such as *Ligulalepis toombsi* Schultze 1968, are distinctive and seem restricted to Early Devonian limestones. Esin's (1990) preliminary studies on Late Devonian palaeoniscoid scales show clearly that these scales are recognisable to generic level, and that forms such as *Moythomasia* and *Mimia* are widespread in the Late Devonian. Crossopterygian scales (e.g. Young and Gorter, 1981; Long *et al.*, 1988; Figure 8.2) are still poorly known and much work has yet to be done on the potential use of these microremains in biochronology. One possible otolith has been found in a Late Devonian or Lower Carboniferous sample from the Canning Basin (RM21 Oscar Range, Geological Survey of Western Australia collection). Otoliths have also been described from natural impressions in Devonian palaeoniscoids and phyllolepid placoderms from Victoria (Long, 1984, 1988a).

Microvertebrate assemblages: Early Devonian

Lochkovian
Microvertebrate remains are found in limestones in the *woschmidti*, *eurekaensis*, possible *delta*, and *pesavis* zones, and are particularly rich in the latter and succeeding basal to mid-Pragian (*sulcatus* zone).

Samples include from the 'Derriwong Beds' (unnamed formation) (samples C227 and C228) and the Jack Limestone in the Broken River embayment of Queensland (*woschmidti*); the Amphitheatre Formation from the Wrightville district of New South Wales (Cobar Supergroup, C568-*eurekaensis*); the Windellama Limestone (*eurekaensis–delta*); the Basal Garra Limestone Formation, Wellington, NSW; and a whole suite of localities in the Nymagee, Mineral Hill, Murda, Trundle and Murga districts of NSW (*pesavis*).

The Lochkovian fauna includes the incoming of palaeoniscoid, *Ligulalepis toombsi*; *Turinia australiensis*-type scales, nostolepid scales including *N. costata*, and a possible new genus, onychodontid teeth, buchanosteid scales, ischnacanthids, possible shark scales.

Talent and Mawson (pers. comm.) extracted a microvertebrate assemblage from the Windellama Limestone (*eurekaensis*) which included buchanosteid scales, acanthodian and turiniid scales which are more akin to those from younger Emsian–Eifelian deposits. This material is being studied by Dr T. Märss of Tallinn.

The Martins Well Transgression begins at the top of the *pesavis* zone in north Queensland (Talent, 1990). In the Martins Well Limestone (*pesavis-sulcatus* zones: Withnall *et al.*, 1988; Mawson *et al.*, 1988) there are two rich microvertebrate horizons which have yielded a fauna including an endemic turiniid species (Turner, 1991a, Table 2), buchanosteid scales and a possible rhenanid (Turner, 1991, p. 450, pl. 4D), nostolepid acanthodians and onychodontid teeth.

Microfaunas within the Jerula Formation (e.g. C008, C600, C651, C654–655 of Pickett and McClatchie, 1991); Trundle Beds from the Narromine region (C287, C595–598, C605–607), the Dulladerry Rhyolite from west of the Murga–Parkes road (C609), and in the Mountain Dam Limestone near Tallebung (C641) include rich fauna of rare thelodonts, (*Turinia* cf. *australiensis*), cladodont? (*Ohiolepis*), a new shark, a nostolepid, (*Nostolepis costata*), and gomphonchid acanthodians, smooth-crowned scales of '*Acanthoides dublinensis*' Stauffer. Also included are new taxa, one of which appears to be a 'proto' *Cheiracanthoides*. The faunas also include *Machaeracanthus*, dominant placoderms scales including a radotinid or rhenanid, an asterolepidoid antiarch (cf. Young, 1984), *Lunaspis*, buchanosteids and other placoderms (one with 'Christmas tree' ornament Figure 8.5C), *Wuttagoonaspis*-like scales (C651), *Ligulalepis*, porolepiform scales (C657) and onychodonts. *Ohiolepis* was described by Wells (1944) as a chondrichthyan; however, the scales closely resemble those of primitive placoderms.

Samples from the *pesavis* zone can be rich in thelodont scales of *Turinia australiensis* type. Some show a tendency to a 'seddoni' type with more fragile indented crowns typical of the younger *Australolepis seddoni*. There are also simpler turiniid scales in some samples, in others acanthodian and placoderm scales predominate and no thelodont scales have been found. These differences, like the absence of polygnathids (Pickett, 1989) might reflect facies or ecological diversity.

Pragian

Microvertebrate assemblages in the *sulcatus* zone resemble those of the late Lochkovian *pesavis* zone. Samples from the Taylors Hill Formation of the Queens Pinch area of the Mudgee district of New South Wales (Tony Wright coll.) and the Garra Limestone (Unit II, Wellington golf course section—C137) include climatiid and machaerachanthid spines, and mostly placoderm remains including buchanosteid scales and a possible inferognathal.

The Coopers Creek Formation (Philip, 1965; Turner, 1991a) contains a rich microvertebrate assemblage which includes acanthodian, *Nostolepis costata* and *Machaeracanthus* scales, onychodontid teeth, and placoderm scales—some buchanosteid.

The Point Hibbs Limestone assemblage of Tasmania includes shark remains (not thelodonts, as stated by Turner, 1986, and in Young, 1989), placoderms, acanthodians and onychodonts comparable with those from the

Buchan Group and the Garra, and younger Taemas and Wee Jasper limestone formations.

The *kindlei–pirenae* zone was apparently a regressional and erosional interlude in Australia followed by a transgression in the *dehiscens* zone (Talent and Yolkin, 1987). Microvertebrate remains from the Jesse Limestone of Limekilns, New South Wales, might be as old as *pirenae* zone (Mawson *et al.*, 1988) but are probably younger (see below).

Emsian

Young (1989) placed the Tumblong oolite fauna (Pickett *et al.*, 1985) here because of *Ohiolepis* which also occurs at Taemas. However, it can occur in older sediments, such as the Cuifengshan Formation of south China (Wang, 1984) and in Australia (see above). The presence of simple turiniid scales of *T. fuscina* and *T. polita* types suggests that this fauna should be older (possibly Lochkovian).

The Mt Knowles limestone member of the Boogledie Formation, Dubbo region, New South Wales, has a microfauna of acanthodians, placoderms and onychodontids associated with conodonts of the *perbonus* zone (Pickett, 1992).

Young (1989) summarised the Wee Jasper–Taemas/Buchan–Bindi (*perbonus–serotinus* zones) fish fauna. The diverse microassemblage includes *Cheiracanthoides*, *Machaeracanthus*, placoderm scales *Buchanosteus*, *Murrindalaspis*, *Goodradigbeeon*, *Brindabellaspis*, crossopterygian teeth, bones and lepidotrichia, including onychodontids, palaeoniscoid *Ligulalepis toombsi*, *Ohiolepis*, and the '*Skamolepis*' form (Turner, 1991a). The Jesse Limestone from Limekilns, near Bathurst (*perbonus–laticostatus*), contains a similar microfauna (Garratt and Wright, 1988).

Microvertebrates from the Mt Frome Limestone (*serotinus–patulus* zone Dalejian and possibly across into early Eifelian–Pickett and Wright collections, Pickett, 1978; Garratt and Wright, 1988) include rare thelodont scales (*Turinia* sp. cf. the turiniid from Hatchery Creek Conglomerate of Young and Gorter, 1981), acanthodian scales—*Cheiracanthoides*, a diplacanthoid?, placoderm remains—some buchanosteid, *Ohiolepis*, '*Skamolepis*', onychodontid and palaeoniscoid remains.

Middle Devonian

Long *et al.* (1988) referred the *Turinia australiensis/Wuttagoonaspis* microvertebrate fauna from the Mulga Downs Group and Cravens Peak Beds as well as that from South Australian borehole at Munyurai 1 to the Lower–Middle Devonian boundary level. However, there appear to be two turiniid assemblages (Young, 1989), with the *T. australiensis* morphotype restricted to the Emsian (i.e. including some of the Mulga Downs Group and some of Cravens Peak Beds occurrences). The Munyurai-1, Toomba Range samples, Ross River, and Mt Winter turiniid faunas are younger, probably in the basal Eifelian (Figure 8.3). These turiniids are more similar to those from the Hatchery Creek Conglomerate, the Eifelian–Givetian *T. pagoda* assemblage of west Yunnan and the turiniid scales from the Middle Devonian Broken River Group and Mt Frome Limestone (end Lower Devonian–Zlichkovian or basal Eifelian).

Eifelian, partitus *zone (upper* patulus *zone)*
Some Cravens Peak Beds samples could be here with *Turinia* sp. (not *T. australiensis*), *Mcmurdodus whitei*, shark scales, *Machaeracanthus*, asterolepidoid plates and scales, and *Ligulalepis?*.

Microvertebrate assemblages are common in Eifelian and Givetian limestones of the Broken River Group of northern Queensland (Mawson *et al.*, 1988; Turner, 1991a). One important locality at Fish Hill near Broken River has yielded both macro- and microvertebrate remains in possible fore-reef facies. These microremains include species typical of late Emsian–Eifelian worldwide, such as *Cheiracanthoides comptus* scales, onychodontid teeth and palaeoniscoid remains. New shark scale types occur in these beds (e.g. Turner, 1991a).

Givetian
Young (1989) considered that the Hatchery Creek Conglomerate of New South Wales might be as young as Givetian, updating the late Eifelian age given earlier by Young and Gorter (1981). Here there is a unique association in Australia of turiniids, thelodonts and bothriolepids (*Monarolepis*), asterolepids (but also at Cravens Peak Beds?) and a phlyctaenid arthrodire, *Denisonosteus*. The *Turinia* species here is closely related to the Antarctic and Bolivian forms and possibly to forms from China and Thailand and to the older basal Eifelian? turiniids from central Australia. Young and Gorter (1981) pointed out the close similarity to *T. hutkensis* (now thought to be Givetian–basal Frasnian) of Iran.

Measured sections through the Broken River Formation have yielded many microvertebrates. One sample in the Papilio Formation within the *varcus* zone (Mawson and Talent, 1989) shows the presence of a phoebodont tooth at the same level as this group first appears in the USA and Europe.

Upper Devonian

Frasnian
There are few Frasnian microvertebrate sites as yet, surprising in view of the wealth of macrovertebrate fossils from sites such as the early Frasnian Gogo Formation in the Canning Basin. Long (1988b) reports that body scales of the arthrodire *Holonema* are found as microremains from the Gogo Formation (formerly named as *Artenolepis* scales; see Chapter 7).

Turner and Dring (1981) used invertebrate evidence for the basal Frasnian age of the microvertebrate assemblage from the Gneudna Formation of the Carnarvon Basin, Western Australia. This fauna includes *Australolepis seddoni* and was, with *T. hutkensis*, one of the youngest thelodonts. Young

(1989) reiterated a Givetian–Lower Frasnian? age for the Gneudna. *A. seddoni* shares characteristics with the older Eifelian turiniids and with *T. pagoda* of China. It appears to be paedomorphic in retaining the slight base and wide pulp cavity. A recent finding of *Australolepis*-like scales from the Givetian of the Holy Cross Mountains, Poland, points to a resolution of the age of some of these thelodonts.

Gigas *zone*
A limestone sample from the Mostyn Vale Formation (Wright *et al.*, 1990) of 'The Bulga' homestead, near Keepit, New South Wales, has a possible placoderm bone, several phoebodont or protacrodont teeth, a large tooth of cf. *Pristicladodus springeri* St John and Worthen 1875 (which may be a new phoebodontid), *Cladolepis* or *Maplemillia*-type scales, and an *Acanthodes*-type scale, as well as crossopterygian teeth. Of possible ecological significance is the apparent resorption or erosion of the bases of most of the shark teeth.

Limestone samples from the Baldwin Formation from near Gunnedal in the Tamworth Trough and from the Merrimbula Group of Ettrema Creek, New South Wales, are also apparently of this age (Pickett, 1972). Scanty microvertebrate remains include onychodontid teeth and indeterminate placoderm(?) bones.

Famennian, triangularis zone
The Coffee Hill Member has a cladodont and numerous dipterid lungfish teeth and palaeoniscoid scales in association with brachiopods and conodonts of this zone (Turner, 1991a; Kemp, 1991).

Marginifera *zone*
Long (1987a) retrieved a microvertebrate fauna associated with toII β conodonts from the Virgin Hills Formation of the Canning Basin. The fauna contains *Acanthodes*-type scales, a crossopterygian tooth and palaeoniscoid scales, and is associated with the skull of a dinichthyid placoderm (Long, 1987b). Pickett (1981) determined the conodonts from the mid Myrtlevale Formation of the Burdekin-Star Basin or shelf, north Queensland as late toII (C411 from Dotswood, C413 from Paynes Lagoon) to a minimum age of toIII (C410 from Dotswood) within the *marginifera* zone. Turner (1982b, Figure 8.7A–B, 7) identified phoebodont teeth which she compared with *Phoebodus politus* as figured by Gross (1973), protacrodont teeth ('*Protacrodus* sp. C') and scales of *Protacrodus wellsi* type, and a '*Cladodus*' tooth. She also mentioned complex shark scales of ctenacanth or cladoselachian type. New information of phoebodont teeth from Poland and Russia by Ginter and Ivanov suggests that these phoebodonts be placed into a new species (previously referred to as *P. australiensis* by Long, 1990).

Expansa *zone*
A thin limestone member of the Teddy Mount Formation of the Broken River embayment, northern Queensland, contains a rich fish microfauna with rare conodonts, ostracodes and phyllocarids. Fish include sharks—

Thrinacodus ferox, helodont teeth, stethacanthid teeth and scales, protacrodont teeth and scales, rare ctenacanth and neoselachian scales, possible placoderm platelets, acanthodians—large scales of *Holmesella* type, crossopterygian teeth, palaeoniscoid teeth, bones, and scales mainly of *Moythomasia* type predominate (Turner, in Lang, 1986, p. 55).

Microvertebrate assemblages from the Gumhole Formation of Western Australia contain predominantly shark and palaeoniscoid teeth and scales (J. Long, pers. comm.). The Formation ranges to the top of the Devonian.

Praesulcata *zone*

A sample (UQL4069) from the Bundock Creek region (now designated Teddy Mount Formation, Lang, 1986, p. 133) contained *Stethacanthus thomasi* (Figure 8.6C) and xenacanthid teeth (Figure 8.7E, possibly ?*Orthacanthus*, G. Johnston, pers. comm.). Conodonts from this site suggest latest Devonian to earliest Carboniferous (Mawson, in Lang, 1986).

Turner (1982b, Figures 2, 3, 8A–H) described a small microvertebrate assemblage from sample UQL4489 of the Slaty Creek region of Broken River, north Queensland. It included *Thrinacodus ferox* (Figure 8.7F), *Phoebodus* sp., and protacrodont species 'A' and 'B' (Figure 8.6A, B). Lang (1986) noted that this limestone is now part of the Teddy Mount Formation.

A small sample from the latest Devonian–basal Carboniferous Cann River Beds of Victoria has *Acanthodes australis* scales and scales of an unknown shark.

One possible Latest Devonian bonebed sample in the Narrien Range near Clermont, central Queensland, has a microfauna essentially the same as younger bonebeds from the region (see below, Turner and Long, 1987). This constitutes the oldest record for the shark *Ageleodus*.

Carboniferous

By the Early Carboniferous agnathans and placoderms had disappeared, and the Tournaisian microfossil assemblages usually are comprised of acanthodian, palaeoniscoid, crossopterygian and shark remains. Shark scales and teeth seem particularly useful for biostratigraphy in Carboniferous rocks; they are now being studied by micropalaeontologists, especially in the USA, and some oil companies are making use of them for work in Alaska (T. Morgan, pers. comm., 1983). There is a very complete succession of Carboniferous to Permian shark assemblages in North America, which can provide a standard. Hansen has been studying shark scales (Hansen, 1988), including those first described by Gunnell (1933) in his conodont studies.

Fish were among the first Carboniferous fossils found in Australia. An 'ichthyodorulite' described by Morris (in Strzelecki, 1845) from the Lower Carboniferous Booral Formation was compared with ones described by Prestwich from the Coal Measures of England. The spine belonged either to a xenacanth or a hybodont shark. This specimen has not been traced.

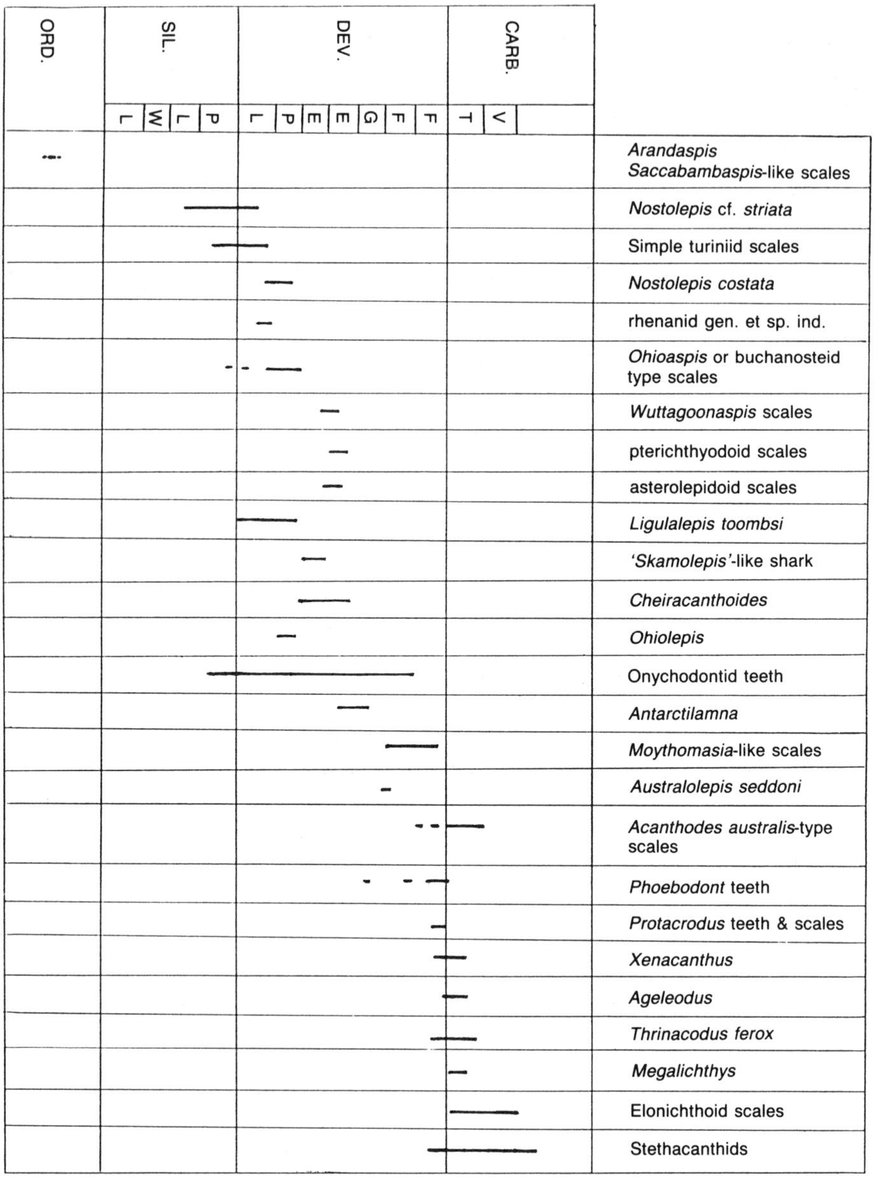

Figure 8.6. Useful age indicators in the Palaeozoic of Eastern Gondwana.

In Australia continental Early Carboniferous vertebrate faunas are known from Mansfield, Victoria (Mansfield Group) and the Narrien Range (Telemon and Raymond Fms) in Queensland (Woodward, 1906; Turner and Long, 1987; Long, 1988a, 1991b; Turner and Hansen, pers. obs.). These faunas include crossopterygians *Megalichthys* and *Barameda*, lungfish, acanthodians (*Acanthodes australis*, *Gyracanthides*), the palaeoniscoids *Mansfieldiscus* and *Novogonatodus*, and sharks. Long (1988a) illustrated the palaeoniscoid scale types. The sharks are all known as microfossils, including teeth of xenacanthids and *Ageleodus*. The microfauna tends to contain the same elements as the macrofauna, with the addition of shark remains.

New finds near Rockhampton are providing information on marine chimaeroid and cladodont sharks in the Tournaisian and Visean of Queensland (Turner, 1990a). The latter include helodont, the first Australian petalodont, psammodont and cochliodont teeth, and *Stethacanthus*. In limestones of the Namoi (*typicus* zone), Bingleburra (*anchoralis* zone) and underlying Dangarfield Formations of New South Wales, shark faunas exhibit a close similarity to those worldwide (e.g. St John and Worthen, 1875; Mader and Schultze, 1987). The age of these formations was recently reassigned to late Tournaisian. These assemblages include teeth of *Thrinacodus ferox*, stethacanthids, ctenacanths, and helodonts as well as orodonts and caseodonts. The New South Wales assemblages, as well as ones from late in the Early Carboniferous of Queensland, are nearly all microfossils.

Good Early Carboniferous shark faunas are also present in the Tournaisian aged Fairfield Group and Visean Utting Calcarenite of Western Australia. These assemblages include macro- and microfossils of *Thrinacodus ferox*, stethacanthids, helodonts and ctenacanths (Turner, 1982b, 1983). One nodule from the Utting Calcarenite in the Bonaparte Gulf, Western Australia (CP coll.), contains a stethacanthid exhibiting part of the jaw cartilages, a tooth battery, and external and internal scales of the jaws. Using this rare example, stethacanthid scales can be positively identified by association with the teeth. Rediscovery of the site of the nodules should yield further articulated specimens. *Ctenacanthus* spines, cf. *Orodus* and cf. *Mesmiodmus* remains have also been found in the Utting Calcarenite (Gilbert-Tomlinson, in Veevers and Roberts, 1968; J. Long, pers. comm.). These Early Carboniferous fish faunas contain elements in common with the classic North American and European assemblages, such as those from the midwestern USA (St John and Worthen, 1875).

Samples from the Late Visean Baywulla Formation from near Rockhampton, central eastern Queensland (G. Webb collection, University of Queensland, Geology Department) contain an interesting new xenacanthid tooth, associated with a small smooth-cusped cladodont tooth and a neoselachian(?) scale, and numerous palaeoniscoid teeth and other remains (GWK–9); all palaeoniscoid teeth (GW–55); and a hybodont scale with palaeoniscoid teeth and ostracodes (GWK–3).

The youngest Carboniferous vertebrate-bearing bed comes from the Barambah Limestone from near Murgon, southern Queensland. This bed contains microremains (scales and teeth) from denaeid and possible neoselachian sharks as well as palaeoniscoid teeth. The shark scales resemble those described by Gunnell (1933) from Late Carboniferous of the USA. The Barambah Limestone is thought to be of Namurian age based on conodonts and associated invertebrates (Palmieri, 1969). After this there are no known microvertebrate producing sites in Australia until the Early Permian—a time of probable glacial advance in Eastern Gondwana.

Permian

Permian vertebrate microremains are as yet rare, but Early Permian sites in Queensland have yielded palaeoniscoid (Turner, 1982c), and rare possible shark remains. The marine Permian sequences in Western Australia, which yielded *Helicoprion*, helodont teeth and *Crassidonta* (Teichert, 1940, 1943), and Queensland, are being investigated for microremains. The interesting freshwater Permian fish assemblage from Blackwater, Queensland, also has potential for microfossils. Michael Leu (Macquarie University) collected new articulated sharks from this site describing *Surcaudalus rostratus* (Leu, 1990), which has well-preserved scales, spines and 'phoebodontiform' teeth. Microvertebrates have recently been identified from the Artinskian Fossil Cliff Member, Holmswood Shale, of the Irwin River, Carnarvon Basin, Western Australia (D. Ferdinando, pers. comm.). The fauna includes palaeoniscoid remains with shark scales and denticles, and a tooth of ?*Campodus*.

SUMMARY

The study of vertebrate microfossils is an important source of biostratigraphical information not always available from invertebrates, conodonts, palynology, or even vertebrate macrofossils. Already the economic usefulness has been recognised by the application of vertebrate zonations in basin analysis in Siberia, the USA and Australia. In the Palaeozoic of Eastern Gondwana the analysis of the rich and numerous faunas is still in its infancy. Microvertebrates assist in providing data about the full geographical and temporal range of taxa and thus contribute to our knowledge of the early evolution of the vertebrates. They are necessary to the understanding of basin analysis and to the correlation of Palaeozoic rocks in Australasia because in continental rocks they are often the only age indicators preserved. Figure 8.6 summarises useful indicator fossils. Many Palaeozoic fish, especially chondrichthyans and some agnathans, and nearly all pre-Devonian fish, are known exclusively from microremains worldwide. The next decade should see greater application of microvertebrate remains in marine/non-marine correlations.

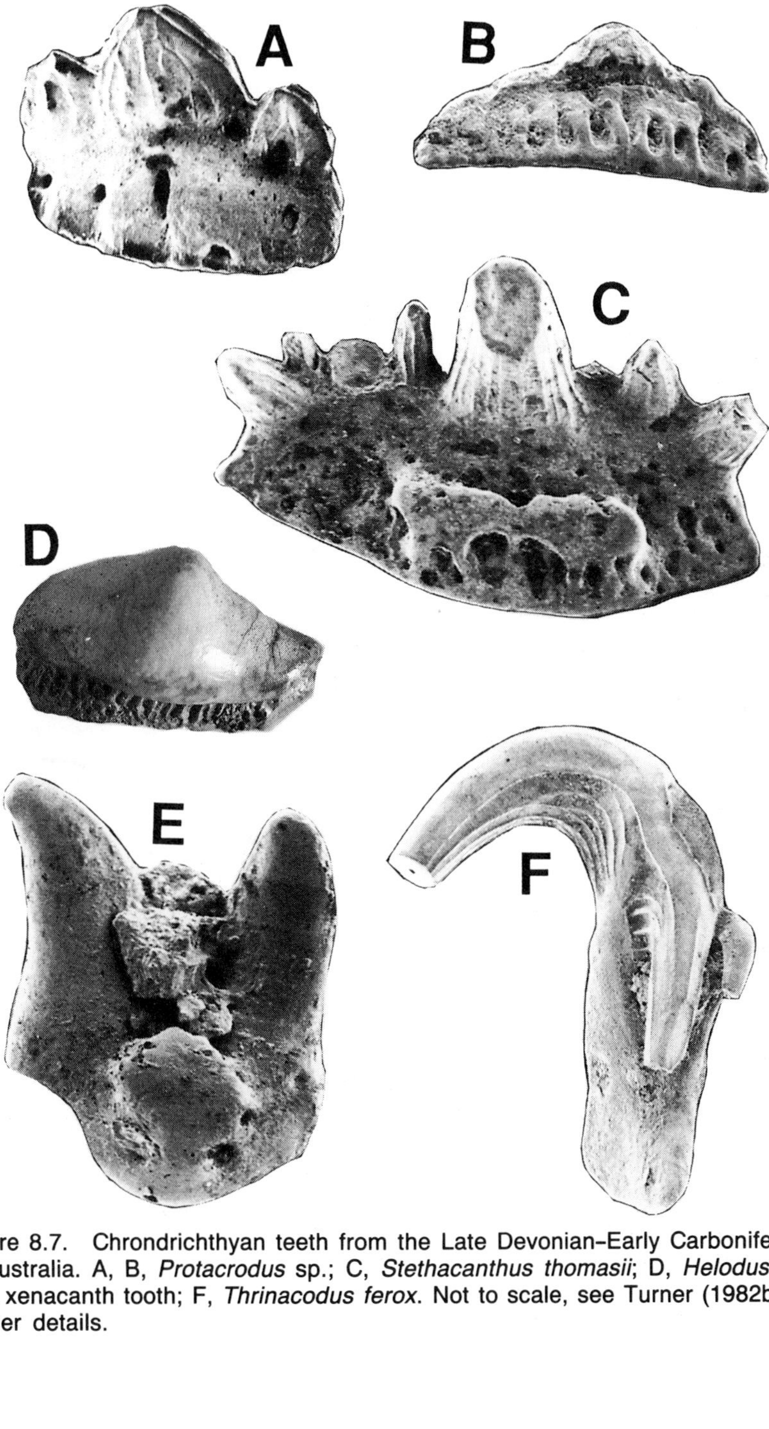

Figure 8.7. Chrondrichthyan teeth from the Late Devonian–Early Carboniferous of Australia. A, B, *Protacrodus* sp.; C, *Stethacanthus thomasii*; D, *Helodus* sp.; E, a xenacanth tooth; F, *Thrinacodus ferox*. Not to scale, see Turner (1982b) for further details.

ACKNOWLEDGEMENTS

I wish to thank the Australian Research Council and former Australian Research Grants Committee for grants (currently A39132723) towards this project over the past decade. This work would not be possible without the help of many colleagues who freely give time, specimens and information and to whom I give my grateful thanks. Many thanks to John Long who attacked the manuscript with gusto. 'C' numbered samples come from the collection of John Pickett (Geological Survey of New South Wales).

REFERENCES

Archbold, N. and Dickins, J.M., 1989, Australia Phanerozoic Timescales: Permian. Biostratigraphic Chart and Explanatory Notes, *Bureau of Mineral Resources Record,* **1989/36**.

Blieck, A., Golshani, F., Goujet, D., Hamdi, B., Janvier, P., Mark-Kurik, E. and Martin, M., 1980. A new vertebrate locality in the Eifelian of the Khush-Yeilagh Formation, Eastern Alborz, Iran, *Palaeovertebrata*, Montpelier, **9**: 133–54.

Blieck, A., Goujet, D., Janvier, P. and Lelièvre, H., 1984, Microrestes de vertebres du Siluro–Devonien d'Algerie, de Turquie et de Thailande, *Géobios*, **17**: 851–6.

Blieck, A., Mark-Kurik, E. and Märss, T., 1988, Biostratigraphical correlations between Siluro–Devonian invertebrate-dominated sequences: the East Baltic example. In N.J. McMillan, A.F. Embry and D.J. Glass (eds), *Devonian of the World*, Canadian Society of Petroleum Geologists, Calgary: 579–587.

Burrett, C., Long, J. and Stait, B., 1990, Early–Middle Palaeozoic biogeography of Asian terranes derived from Gondwana. In W.S. McKerrow and C.R. Scotese (eds), *Palaeozoic Palaeogeography and Biogeography, Geological Society Memoir*, **12**: 163–74.

Denison, R.H., 1978, *Placodermi*. In *Handbook of Paleoichthyology. Pt 2*, H.-P. Schultze (ed.), Gustav Fischer Verlag, Stuttgart, New York, 128pp.

Denison, R.H., 1979, *Acanthodii*. In *Handbook of Paleoichthyology. Pt 5*, H.-P. Schultze (ed.), Gustav Fischer Verlag, Stuttgart, New York, 62pp.

Draper, J.J., 1976, The Devonian rocks of the Toko Syncline, western Queensland, *Bureau of Mineral Resources Geology and Geophysics Record*, **1976/29**: 1–16.

Druce, E.C. and Radke, B.M., 1979, The geology of the Fairfield Group, Canning Basin, Western Australia, *Bureau of Mineral Resources Geology and Geophysics Bulletin*, **200**: 1–62.

Esin, D., 1990, Species of Devonian palaeoniscoid fishes of the world, *Ichthyolith Issues*, **3**: 14.

Foreman, D.J. and Wales, D.W., 1981, Geological evolution of the Canning Basin, Western Australia, *Bureau of Mineral Resources Bulletin*, **210**: 1–91.

Friman, L., 1983, *Ohiolepis*-Schuppen aus dem unteren Mitteldevon der Eifel (Rheinisches Schiefergebirge), *Neues Jahrbuch für Geologie und Palaontologie, Monatshefte*, **1983**: 228–36.

Gagnier, P.-Y., Blieck, A. and Rodrigo, G.S., 1986, First Ordovician vertebrate from South America, *Géobios*, **19**: 629–34.

Gagnier, P.-Y., Turner, S., Friman, L., Suarez-Riglos, M. and Janvier, P., 1988, The Devonian vertebrate and mollusc fauna from Seripona (Dept. of Chuquisaca, Bolivia), *Neues Jahrbuch für Geologie und Palaontologie, Abhandlungen*, **176**: 269–97.

Gagnier, P.-Y., Paris, F., Racheboeuf, P., Janvier, P. and Suarez-Riglos, M., 1989, Les Vertébrés de Bolivie: Donnes biostratigraphiques et anatomiques complémentaires, *Bulletin de l'Institute Français Etudes Andines 1989*, **18**: 75–93.

Garratt, M.J. and Wright, A.J., 1988, Late Silurian to Early Devonian biostratigraphy of southeastern Australia. In N.J. McMillan, A.F. Embry and D.J. Glass (eds), *Devonian of the World*, Canadian Society of Petroleum Geologists, Calgary: 647–62.

Giffin, E., 1980, Devonian vertebrates from Australia, *Postilla*, **80**: 1–15.

Ginter, M., 1990, Late Famennian shark teeth from the Holy Cross Mountains, Central Poland, *Acta Geologica Polonica*, **40**: 69–81.

Goujet, D., 1976, Les Poissons. In *Les Schistes et Calcaires Devoniens de Saint-Cenere (Massif Armoricain, France), sedimentologie, paleontologie, stratigraphie*, Babin, C., Bigey, F., Deunff, J., Drot, J., Goujet, D., Lardeaux, H., Le Menn, J., L'Hotellier, J., Melou, M., Moreau-Benoit, A., Morzadec, P., Paris, F., Plusqeullec, Y., Poncet, J, Racheboeuf, P., Taugoudeau, P. and Weyant, M. (eds), *Memoires de la Sociéte du géologie et mineralogie, Bretagne*, **19**: 313–23.

Gross, W., 1971a, Unterdevonische Thelodontier- und Acanthodier-Schuppen aus Westaustralien, *Paläontologisches Zeitschrifter*, **45**: 97–106.

Gross, W., 1971b, Downtonische und Dittonische Acanthodier-Reste des Ostseegebietes, *Palaeontographica*, **136A**: 1–82.

Gross, W., 1973, Kleinschuppen, Flossenstacheln und Zahne aus europäischen und nordamerikanischen Bonebeds des Devons, *Palaeontographica*, **142A**: 51–155.

Gunnell, F.H., 1933, Conodonts and fish remains from the Cherokee, Kansas City, and Wabaunsee Groups of Missouri and Kansas, *Journal of Paleontology*, **7**: 261–97.

Hansen, M., 1988, Microscopic chondrichthyan remains from Pennsylvianian marine rocks of Ohio and adjacent regions, *Ichthyolith Issues*, **1**: 6–7.

Henderson, G.A.M., 1982, Geology of Canberra, Queanbeyan, and environs. Notes to accompany the 1980 1:50000 geological map. Australian Government Printing Service: 47pp.

Jell, J.S. and Talent, J.A., 1989, The Silurian of Australia: the most instructive sections. In C.H. Holland and M.G. Bassett (eds), *A global standard for the Silurian System*, National Museum of Wales, Geological Series, **9**: 183–200.

Johnstone, M.H., Jones, P.J., Koop, W.J., Roberts, J., Gilbert-Tomlinson, J., Veevers, J.J. and Wells, A.T., 1967, The Devonian of Western and Central Australia, *International Symposium on the Devonian System*, Calgary: 599–612.

Jones, P.J., 1989, Australian Phanerozoic Timescales: Carboniferous. Biostratigraphic Chart and Explanatory Notes, *Bureau of Mineral Resources Record*, **1989/35**.

Karatajute-Talimaa, V.N., 1978, *Telodonti Silura i Devona S.S.S.R. i Shpitsbergena*, Mosklas, Vilnius, 334pp.

Karatajute-Talimaa, V.N., Novitskaya, L.I., Rozman, Kh.S. and Sodov, Z., 1990, *Mongolepis*—a new Lower Silurian genus of elasmobranchs from Mongolia, *Palaeontological Journal*, 1990, **1**: 37–48.

Kemp, A., 1991, Late Devonian lungfish microfossils from the Coffee Hill Member, Catombal Group, near Orange in New South Wales, Australia, *Ichthyolith Issues*, **8**: 15.

Korsch, R.J. and Kennard, J.M. (eds), 1991, Geological and geophysical studies in the Amadeus Basin central Australia, *Bureau of Mineral Resources Bulletin*, **236**: 594pp.

Lang, S.C., 1986, Devonian–Carboniferous stratigraphy of the southeastern Bundock Basin, Broken River area, north Queensland, *Geological Survey of*

Queensland, Record Series, Record 1986/5, vol. 1, 213pp, vol. 2, 61pp + maps.

Legault, J.A., 1968, Conodonts and fish remains from the Stonehouse Formation, Arisaig, Nova Scotia, *Geological Survey of Canada Bulletin*, **165**: 1–23.

Leu, M.R., 1990, A late Permian freshwater shark from eastern Australia, *Palaeontology*, **32**: 265–86.

Li Zuo-Cong, 1980, Age of the *Sinacanthus*-bearing beds in Hubei, *Dicengxue Zazhi* (*Journal of Stratigraphy*), **4**: 221–35.

Long, J.A., 1983a, New bothriolepid fishes from the Late Devonian of Victoria, Australia, *Palaeontology*, **26**: 295–320.

Long, J.A., 1983b, A new diplacanthoid acanthodian from the Late Devonian of Victoria, *Memoirs of the Association of Australasian Palaeontologists*, **1**: 51–65.

Long, J.A., 1984, New phyllolepids from Victoria and the relationships of the group, *Proceedings of the Linnean Society of New South Wales*, **107**: 263–308.

Long, J.A., 1985, The structure and relationships of a new osteolepiform fish from the Late Devonian of Victoria, Australia, *Alcheringa*, **9**: 1–22.

Long, J.A., 1986a, A new Late Devonian acanthodian fish from Mt Howitt, Victoria, Australia, with remarks on acanthodian biogeography, *Proceedings of the Royal Society of Victoria*, **98**: 1–17.

Long, J.A., 1986b, New ischnacanthid acanthodians from the Early Devonian of Australia, with a discussion of acanthodian interrelationships, *Zoological Journal of the Linnean Society*, **87**: 321–39.

Long, J.A., 1987a, Upper Devonian conodonts associated with a large placoderm fish skull from the Canning Basin, Western Australia, *Records of the Western Australian Museum*, **13**: 510–13.

Long, J.A., 1987b, A new dinichthyid fish (Placodermi, Arthrodira) from the Upper Devonian of Western Australia, with a discussion of dinichthyid interrelationships, *Records of the Western Australian Museum*, **13**: 515–40.

Long, J.A., 1988a, New palaeoniscoid fishes from the Late Devonian and Early Carboniferous of Victoria, *Memoirs of the Association of Australasian Palaeontologists*, **7**: 1–64.

Long, J.A., 1988b, Late Devonian fishes from the Gogo Formation, Western Australia, *National Geographic Research*, **4**: 436–50.

Long, J.A., 1990, Late Devonian chondrichthyans and other microvertebrate remains from northern Thailand, *Journal of Vertebrate Paleontology*, **10**: 59–71.

Long, J.A., 1991a, Devonian fish remains from the Munabia Sandstone, Carnarvon Basin, Western Australia, *Records of the Western Australian Museum*, **15**: 503–15.

Long, J.A., 1991b, The long history of fossil fish. In P. Vickers-Rich, J.N. Monaghan, R.F. Baird and T.H. Rich (eds), *Vertebrate Palaeontology in Australasia*, Pioneer Design Studios with Monash University Publications Committee, Melbourne: 336–428.

Long, J.A. and Turner, S., 1984, A checklist and bibliography of Australian fossil fishes. In M. Archer and G. Clayton (eds), *Vertebrate Zoogeography and Evolution in Australasia*, Hesperian Press, Carlisle, 235–54.

Long, J.A., Turner, S. and Young, G.C., 1988, A Devonian fish fauna from subsurface sediments in the eastern Officer Basin, South Australia, *Alcheringa*, **12**: 61–78.

Long, J.A. and Werdelin, L., 1986, A new species of *Bothriolepis* (Placodermi, Antiarchi) from Tatong, Victoria, with descriptions of others from the State, *Alcheringa*, **10**: 355–99.

Long, J.A. and Young, G.C., 1988, Acanthothoracid remains from the Early Devonian of New South Wales, including a complete sclerotic capsule and pelvic girdle, *Memoirs of the Association of Australasian Palaeontologists*, **7**: 65–80.

Macadie, C.I., 1985, Devonian fossil fishes of Reefton, New Zealand. In *Hornibrook Symposium, 1985, extended abstracts. Geological Survey of New Zealand*, **9**: 73–5.
Mader, H., 1986, Schuppen and Zahne von Acanthodiern und Elasmobranchiern aus dem Unter-Devon Spaniens (Pisces), *Göttinger Arbeiten zur Geologie und Paläontologie*, **28**: 1–59.
Mader, H. and Schultze, H.-P., 1987, Elasmobranchier-reste aus dem Unterkarbon des Rheinisches Schieferbirges und des Harzes (W-Deutchsland), *Neues Jahrbuch für Paläontologie und Geologie, Abhandlang*, **175**: 317–46.
Mawson, R., 1987a, Early Devonian conodont faunas from Buchan and Bindi, Victoria, Australia, *Palaeontology*, **30**: 251–97.
Mawson, R., 1987b, Documentation of conodont assemblages across the Early Devonian–Middle Devonian Boundary, Broken River Formation, north Queensland, Australia, *Courier Forschungsinstitut Senckenberg*, **92**: 251–73.
Mawson, R., Jell, J.S. and Talent, J.A., 1985, Stage boundaries within the Devonian: implications for application to Australian sequences, *Courier Forschungsinstitut Senckenberg*, **75**: 1–16.
Mawson, R., Talent, J.A., Bear, V.C., Benson, D.S., Brock, G.A., Farrell, J.R., Hyland, K.A., Pyemont, B.D., Sloan, T.R., Sorentino, L., Stewart, M.I., Trotter, J.A., Wilson, G.A. and Simpson, A.G., 1988, Conodont data in relation to resolution of stage and zonal boundaries for the Devonian of Australia. In N.J. McMillan, A.F. Embry and D.J. Glass (eds), *Devonian of the World*, Canadian Society of Petroleum Geologists, Calgary: 485–527.
Mawson, R. and Talent, J.A., 1989, Late Emsian–Givetian stratigraphy and conodont biofacies—carbonate slope and offshore shoal to sheltered lagoon and nearshore carbonate ramp—Broken River, north Queensland, Australia, *Courier Forschungsinstitut Senckenberg*, **117**: 205–59.
Palmieri, V., 1969, Upper Carboniferous conodonts from limestone near Murgon, south east Queensland, *Geological Survey of Queensland Papers in Palaeontology*, **17**: 1–13.
Philip, G.M., 1965, Lower Devonian conodonts from the Tyers area, Gippsland, Victoria, *Proceedings of the Royal Society of Victoria*, **75**: 95–117.
Pickett, J.W., 1972, Late Devonian (Frasnian) Conodonts from Ettrema, New South Wales, *Journal and Proceedings, Royal Society of New South Wales*, **105**: 31–7.
Pickett, J.W., 1978, Conodont faunas from the Mount Frome Limestone (Emsian/Eifelian), New South Wales. In D.J. Belford and V. Scheibnerova (compilers), *The Crespin Volume: Essays in honour of Irene Crespin. Bureau of Mineral Resources Geology and Geophysics, Bulletin*, **192**: 97–107.
Pickett, J.W., 1981, Late Devonian and Early Carboniferous conodonts from the Myrtlevale Beds and Hardwick Formation, Star Basin, *Queensland Government Mining Journal*, **82**: 24–7.
Pickett, J.W., 1983, Conodont assemblages from the Mountain Dam Limestone, *Geological Survey of New South Wales, Report No.* 1983/156.
Pickett, J.W., 1989, Review of fossil localities in the Meloola-Moorefield area, north of Condobolin, *Geological Survey of New South Wales, Report No.* 1989/358.
Pickett, J.W., 1992, Review of selected Silurian and Devonian conodont assemblages from the Mineral Hill-Trundle area, *Geological Survey of New South Wales, Report No.* 1992/024.
Pickett, J.W. and Ingpen, I.A., 1990, Ordovician and Silurian strata south of Trundle, New South Wales, *Geological Survey of New South Wales, Quarterly Notes*, **78**: 1–14.

Pickett, J.W. and McClatchie, L., 1991, Age and relations of stratigraphic units in the Murda Syncline area, *Geological Survey of New South Wales, Quarterly Notes*, **85**: 9–32.

Pickett, J.W., Turner, S. and Myers, B., 1985, The age of marine sediments from Tumblong, south west of Gundegai, *Geological Survey of New South Wales, Quarterly Notes*, **58**: cover photo + 1, 12–15.

Poltnig, W., 1984, Fischreste aus dem Unterdevon von Graz (Steiermark), *Mitteilungen der naturwissenschaftliche Verreinigung Steiermark*, **114**: 107–31.

Reif, W.-E., 1978, Types of morphogenesis of the dermal skeleton in fossil sharks, *Paläontologische Zeitscrifter*, **52**: 110–28.

Ritchie, A., 1985, *Arandaspis prionotolepis*. The Southern four-eyed fish. In P. Rich and G. van Tets (eds), *Kadimakara*, Pioneer Design Studios, Lilydale: 95–106.

Ritchie, A. and Gilbert-Tomlinson, J., 1977, First Ordovician vertebrates from the southern hemisphere, *Alcheringa*, **1**: 351–68.

Schultze, H.-P., 1968, Palaeoniscoidea-Schuppen aus dem Australiens und Kanadas und aus dem mitteldevon Spitzbergens, *Bulletin of the British Museum of Natural History (Geology)*, **16**: 343–68.

Sherwin, L., 1992, Siluro–Devonian biostratigraphy of central New South Wales, *New South Wales Geological Survey, Quarterly Notes*, **86**: 1–12.

Simpson, A., 1983, Silurian to basal Devonian conodont biostratigraphy of the Broken River area, north Queensland. Honours thesis, Macquarie Univ. (unpubl.).

St John, O.H. and Worthen, A.H., 1875, Description of fossil fishes, *Geological Survey of Illinois 1875*, **6**: 245–488.

Strusz, D., 1989, Australian Phanerozoic Timescales: Silurian. Biostratigraphic Chart and Explanatory Notes, *Bureau of Mineral Resources Record*, 1989/33.

Strusz, D. and Garratt, M., 1992, Australian brachiopod communities. In A. Boucot (ed.), *Project Ecostratigraphy*, Final Report Volume. Cambridge University Press.

Strzelecki, P.E. de, 1845, *Physical description of New South Wales and Van Diemen's Land: accompanied by a geological map, sections and diagrams and figures of organic remains*, Longman, Brown, Green and Longmans, London, 462pp.

Talent, J., 1989, Transgression–Regression Pattern for the Silurian and Devonian of Australia. In R.W. Le Maitre (ed.), *Pathways in Geology: essays in honour of Edwin Sherbon Hills*, distributed by Blackwells, Melbourne: 201–19.

Talent, J.A. and Spencer-Jones, D., 1963, The Devono–Carboniferous fauna of the Silverband Formation, Victoria, *Proceedings of the Royal Society of Victoria*, **76**: 1–11.

Talent, J.A. and Yolkin, E.A., 1987, Transgression–regression patterns for the Devonian of Australia and south west Siberia. *Courier Forschungsinstitut Senckenberg*, **92**: 235–49.

Teichert, C., 1940, *Helicoprion* in the Permian of Western Australia, *Journal of Paleontology*, **14**: 140–9.

Teichert, C., 1943, Bradyodont sharks in the Permian of Western Australia, *American Journal of Science*, **241**: 543–52.

Telford, P., 1975, Lower and Middle Devonian conodonts from the Broken River Embayment, north Queensland, Australia, *Special Papers in Palaeontology*, **15**: 100pp.

Thomas, G., 1957, Lower Carboniferous deposits in the Fitzroy Basin, Western Australia, *The Australian Journal of Science*, **19**: 160–1.

Thomas, G., 1959, The Lower Carboniferous Laurel Formation of the Fitzroy Basin, *Bureau of Mineral Resources Report*, **38**: 31–6.

Turner, S., 1973, Siluro–Devonian thelodonts from the Welsh Borderland, *Journal of the Geological Society of London*, **129**: 557–84.
Turner, S., 1982a, Thelodonts and correlation. In P.V. Rich and E.M. Thompson (eds), *The Fossil Vertebrate Record in Australasia*, Monash Offset Printing Unit, Clayton: 128–33.
Turner, S., 1982b, Middle Palaeozoic elasmobranchs remains from Australia, *Journal of Vertebrate Paleontology*, **2**: 117–31.
Turner, S., 1982c, A catalogue of fossil fish in Queensland, *Memoirs of the Queensland Museum*, **20**: 599–611.
Turner, S., 1983, Taxonomic note on 'Harpago', *Journal of Vertebrate Paleontology*, **3**: 38.
Turner, S., 1985, Remarks on the early history of chondrichthyans, thelodonts and some 'higher elasmobranchs'. In *Hornibrook Symposium, 1985, extended abstracts. Geological Survey of New Zealand Record*, **9**: 93–5.
Turner, S., 1986, Vertebrate fauna of the Silverband Formation, Grampians, western Victoria, *Proceedings of the Royal Society of Victoria*, **98**: 53–62.
Turner, S. (ed.), 1988, *Ichthyolith Issues No. 1.*
Turner, S. (ed.), 1989, *Ichthyolith Issues No. 2.*
Turner, S., 1990a, Lower Carboniferous shark remains from the Rockhampton district, Queensland. In S. Turner, R.A. Thulborn and R. Molnar (eds), *De Vis Symposium Volume, Memoirs of the Queensland Museum*, **28**: 65–73.
Turner, S. (ed.), 1990b, *Ichthyolith Issues No. 3.*
Turner, S. (ed.), 1990c, *Ichthyolith Issues No. 4.*
Turner, S. (ed.), 1990d, *Ichthyolith Issues No. 5.*
Turner, S., 1991a, Palaeozoic Vertebrate Microfossils in Australasia. In P. Vickers-Rich, J.N. Monaghan, R.F. Baird and T.H. Rich (eds), *Vertebrate Palaeontology in Australasia*, Pioneer Design Studios with Monash University Publications Committee, Melbourne: 429–64.
Turner, S. (ed.), 1991b, *Ichthyolith Issues No. 6.*
Turner, S. (ed.), 1991c, *Ichthyolith Issues No. 7.*
Turner, S. (ed.), 1992, *Ichthyolith Issues No. 8.*
Turner, S. and Dring, R.S., 1981, Late Devonian thelodonts (Agnatha) from the Gneudna Formation, Carnarvon Basin, Western Australia, *Alcheringa*, **5**: 39–48.
Turner, S., Jones, P.J. and Draper, J.J., 1981, Early Devonian thelodonts (Agnatha) from the Toko Syncline, western Queensland and a review of other Australian discoveries, *Bureau of Mineral Resources, Journal of Australian Geology and Geophysics,* **61**: 51–69.
Turner, S. and Janvier, P., 1979, Middle Devonian Thelodonti (Agnatha) from the Khush-Yeilagh Formation, North-east Iran, *Géobios*, **12**: 889–92.
Turner, S. and Long, J.A., 1987, Lower Carboniferous palaeoniscoids (Pisces: Actinopterygii) from Queensland, *Memoirs of the Queensland Museum*, **25**: 193–200.
Turner, S. and Murphy, M.A., 1988, Early Devonian vertebrate microfossils from the Simpsons Park Range, Eureka County, Nevada, *Journal of Paleontology*, **62**: 959–64.
Turner, S. and Pickett, J., 1982, Silurian vertebrates in Australia, *Search*, **13**: 11–12.
Turner, S. and Young, G.C., 1987, Shark teeth from the Early–Middle Devonian Cravens Peak Beds, *Alcheringa*, **11**: 233–44.
Turner, S. and Young, G.C., 1992, Thelodont scales from the Middle–Late Devonian Aztec Siltstone, southern Victoria Land, Antarctica, *Antarctic Science*, **4**: 89–109.

Valiukevicius, J.J., 1985, *Acanthodians from the Narva Regional Stage of the Main Devonian Field*, Mosklas, Vilnius, 143pp. [In Russian with English summary.]

Veevers, J.J. and Roberts, J., 1968, Upper Palaeozoic rocks, Bonaparte Gulf Basin of northwestern Australia, *BMR Geology and Geophysics, Bulletin*, **97**: 1–155.

Vieth, J., 1980, Thelodontier-, Acanthodier- und Elasmobranchier-Schuppen aus dem Unter-Devon der Kanadischen Arktis (Agnatha, Pisces), *Göttinger Arbeiten Geologie und Paläontologie*, **23**: 1–69.

Vieth-Schreiner, J., 1983, Fisch-Schuppen und -Zahne aus der Eifeler Kalkmulden-Zone (Emsium, Eifelium), *Senckenbergiana Lethaea*, **64**: 129–77.

Wang, N.-Z., 1984, Thelodont, acanthodian and chondrichthyan fossils from the Lower Devonian of southwest China, *Proceedings of the Linnean Society of New South Wales*, **107**: 419–41.

Wang, S.-T. and Turner, S., 1985, Vertebrate microfossils of the Devonian–Carboniferous boundary, Muhua section, Guizhou Province, *Vertebrata Palasiatica*, **23**: 223–34.

Wang, S.-T., Dong, Z.-Z. and Turner, S., 1986, Discovery of Middle Devonian Turiniidae (Thelodonti:Agnatha) from west Yunnan, China, *Alcheringa*, **10**: 315–25.

Webby, B.D. and Nicoll, R.S., 1989, Australian Phanerozoic Timescales: Ordovician. Biostratigraphic Chart and Explanatory Notes, *Bureau of Mineral Resources Record*, 1989/32.

Wells, J.W., 1944, Fish remains from the Middle Devonian Bone Beds of the Cincinnati Arch region, *Palaeontographica Americana*, **3**: 99–160.

White, E.I., 1968, Devonian fishes of the Mawson-Mulock area, Victoria land, Antarctica, *Transantarctic Expedition 1955–58 Scientific Reports No. 16*, Transantarctic Expedition Committee, London: 1–26.

White, E.I., 1978, The larger arthrodiran fishes from the area of the Burrinjuck Dam, NSW., *Transactions of the Zoological Society of London*, **34**: 149–262.

Withnall, I.W., Lang, S.C., Jell, J.S., McLennan, T.P.T., Talent, J.A., Mawson, J.A., Fleming, P.J.G., Law, S.R., Macansh, J.D., Savory, P., Kay, J.R. and Draper, J.J., 1988, Stratigraphy, sedimentology, biostratigraphy and tectonics of the Ordovician to Carboniferous, Broken River Province, north Queensland, *Australasian Sedimentologists Group Field Guide Series*, 5, Geological Society of Australia, Sydney.

Woodward, A.S., 1906, On a Carboniferous fish fauna from the Mansfield district, Victoria, *Memoirs of the National Museum of Victoria*, **1**: 1–32.

Woodward, A.S., 1921, Fish-remains from the Upper Old Red Sandstone of Granite Harbour, Antarctica. British Antarctic ('Terra Nova') Expedition, 1910. Natural History Report, *British Museum of Natural History (Geology)*, **1**: 51–62.

Wright, A.J., Pickett, J.W., Sewell, D., Roberts, J. and Jenkins, T.B.H., 1990, Corals and conodonts from the Late Devonian Mostyn Vale Formation, Keepit, New South Wales, *Memoirs of the Association of Australasian Palaeontologists*, **10**: 211–54.

Wyatt, D.H. and Jell, J.S., 1980, Devonian and Carboniferous stratigraphy of the northern Tasman Orogenic Zone in the Townsville hinterland, north Queensland. In R.A. Henderson and P.J. Stephenson (eds), *The Geology and Geophysics of Northeastern Australia*, Geological Society of Australia, Queensland Division, Brisbane: 201–28.

Young, G.C., 1981, Biogeography of Devonian vertebrates, *Alcheringa*, **5**: 225–43.

Young, G.C., 1982, Devonian sharks from south-eastern Australia and Antarctica, *Palaeontology*, **25**: 817–43.

Young, G.C., 1984, An asterolepidoid antiarch (placoderm fish) from the Early Devonian of the Georgina Basin, central Australia, *Alcheringa*, **8**: 65–80.

Young, G.C., 1989, Australian Phanerozoic Timescales 4: Devonian. Biostratigraphic Chart and Explanatory Notes, *Bureau of Mineral Resources Record*, 1989/34, 17pp + table.

Young, G.C., 1990, New antiarchs (Devonian placoderm fishes) from Queensland, with comments on placoderm phylogeny and biogeography. In S. Turner, R.A. Thulborn and R Molnar (eds), *De Vis Symposium Volume, Memoirs of the Queensland Museum*, **28**: 35–50.

Young, G.C., 1991, Palaeozoic vertebrates. In Shergold, J.H. (Coordinator) Late Proterozoic and early Palaeozoic palaeontology and biostratigraphy of the Amadeus Basin, *Geological and geophysical studies in the Amadeus Basin central Australia, Bureau of Mineral Resources Bulletin*, **236**: 97–111.

Young, G.C. and Gorter, J.D., 1981, A new fish fauna of Middle Devonian age from the Taemas/Wee Jasper region of New South Wales, *Bureau of Mineral Resources Bulletin*, **209**: 83–147.

Young, G.C. and Nicoll, R.S., 1979, Report on samples from Irian Jaya submitted for micropalaeontological examination, *Bureau of Mineral Resources Geology Report* 79/028, File No. 79/347: 1–5.

Young, G.C. and Turner, S., 1992, IGCP 328: New evidence for correlation of Palaeozoic fish between central Australia and the Canning and Georgina Basins. In 11th AGC, Ballarat, January 18–25, *Geological Society of Australia, Abstracts*, **32**: 177.

Young, G.C., Turner, S., Owen, M., Nicoll, R.S., Laurie, J.R. and Gorter, J.D., 1987, A new Devonian fish fauna, and revision of post-Ordovician stratigraphy in the Ross River Syncline, Amadeus Basin, central Australia, *Bureau of Mineral Resources Journal of Australian Geology and Geophysics*, **10**: 23–42.

Zangerl, R., 1981, Paleozoic Chondrichthyes. In *Handbook of Paleoichthyology. Pt 3A*, H.-P. Schultze (ed.), Gustav Fischer Verlag, Stuttgart, New York, 114pp.

Zidek, J., 1981, *Machaeracanthus* Newberry (Acanthodii: Ischnacanthiformes)—morphology and systematic position, *Neues Jahrbuch für Paläontologie und Geologie, Abhandlang*, **H12**: 742–8.

Chapter 9

MIDDLE PALAEOZOIC MACROVERTEBRATE BIOSTRATIGRAPHY OF EASTERN GONDWANA

G.C. Young

INTRODUCTION

East Gondwana included Australia, New Zealand, and Antarctica in the Palaeozoic, but only Australia and South Victoria Land, Antarctica, are sufficiently well studied to provide biostratigraphically useful data. Devonian vertebrate faunas are known from many localities in Australia (Figure 9.1), from sequences which are predominantly marine in the west (Carnarvon, Canning and Bonaparte Basins), predominantly non-marine in the centre (Amadeus, Georgina, and Officer Basins), and both non-marine and marine in eastern Australia.

Eastern Australia (the Tasman Foldbelt) formed part of the Pacific margin of Gondwana during the Palaeozoic. It was inundated to a greater or lesser extent by various marine transgressions responding to the global eustatic cycle, and the extensive orogenic activity and volcanism resulted in thick Devonian sequences of redbeds intercalated with marine and volcanic horizons. This diversity of lithological types and depositional regimes makes available a variety of data sets which can be integrated to provide age control for the vertebrate faunal succession. For this reason the better studied faunas of southeastern Australia and Antarctica are used to provide a biostratigraphic framework for interpreting other sequences.

The occurrence of Middle Palaeozoic vertebrate fossils in the region of East Gondwana has been known for almost 150 years. Palaeozoic vertebrates were discovered in eastern Australia in the early days of exploration, but only in recent decades have detailed systematic studies provided a taxonomic database sufficient to support useful biozonation. The first

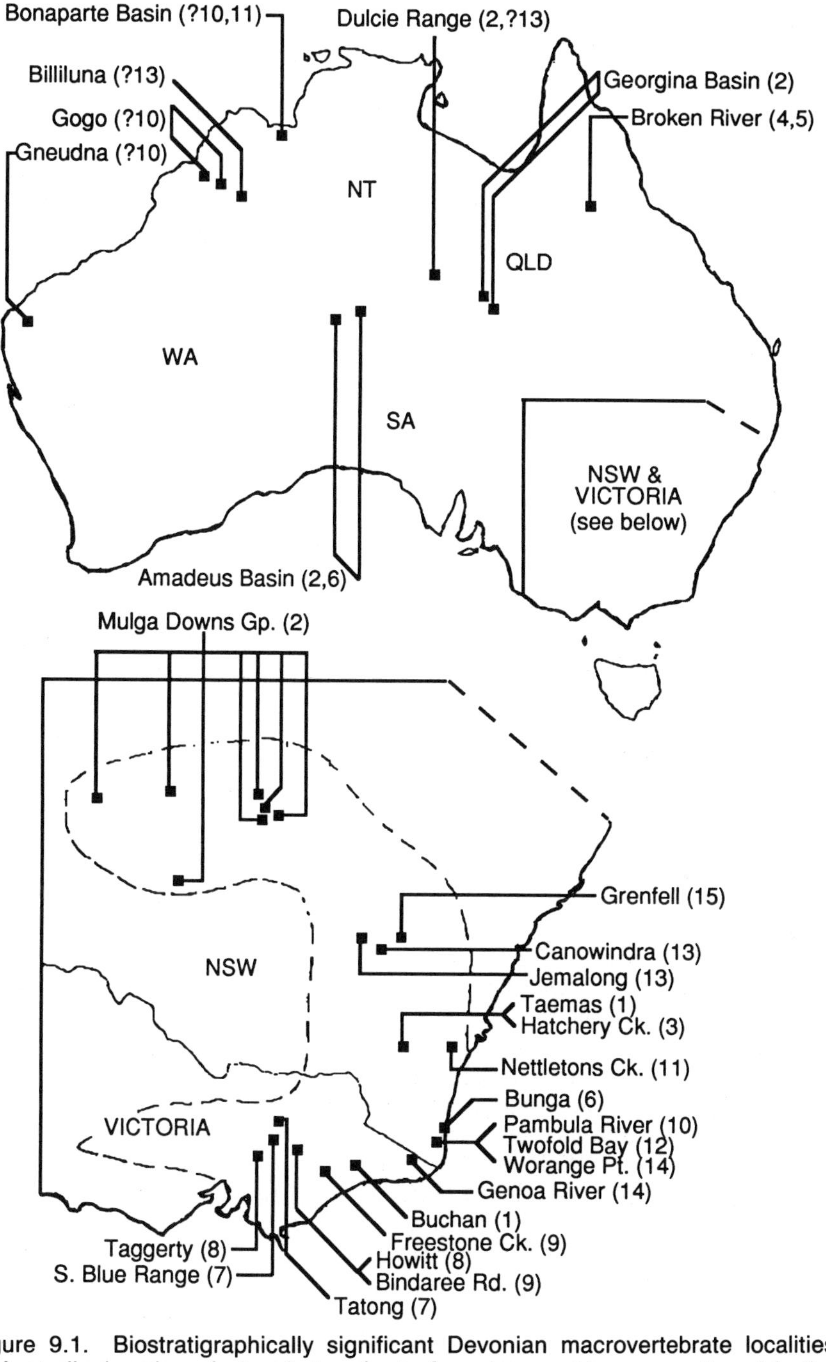

Figure 9.1. Biostratigraphically significant Devonian macrovertebrate localities of Australia (numbers in brackets refer to faunal assemblages mentioned in the text). Adapted from Long (1991b).

Carboniferous fossils collected by the explorer Strzelecki in 1842, described by Morris (in Strzelecki, 1845), included fish remains (an 'ichthyodorulite'). Dana (1848) recorded Permian fish remains from New South Wales, and McCoy (1876) described a placoderm plate from Buchan in Victoria from an assemblage now known to be of late Early Devonian age. Etheridge (1890) recorded fish remains from the Early Carboniferous of Queensland, and Woodward (1902, 1906) reported on a diverse Early Carboniferous fish fauna from Mansfield in Victoria, and later (Ferguson, 1917) another supposed Carboniferous occurrence in the Grampians of western Victoria (now known to be of Devonian age). Hills (1929, 1931) was the first to describe a fish fauna of Late Devonian age. A fish fauna of comparable age from South Victoria Land, Antarctica, discovered by the British Antarctic 'Terra Nova' expedition of 1910–13, was described by Woodward (1921). Permian fish remains were first reported from Western Australia by Teichert (1940, 1943), and subsequently (Teichert, 1949) he recorded Devonian fish remains from northwestern Australia (Canning Basin).

Systematic contributions on these faunas, leading up to the review of Hills (1958), concentrated on the marine Early Devonian Taemas–Buchan fauna (e.g. Chapman, 1916; Hills, 1936a, 1941; Woodward, 1941; White, 1952), and the Late Devonian faunas of the 'Old Red Sandstone' type 'Lambie' facies in New South Wales (Hills, 1932, 1936b) and central Australia (Hills, 1959). Hills (1958) summarised the older Taemas–Buchan fauna as cosmopolitan at family level, and the Late Devonian fishes of eastern Australia as 'entirely cosmopolitan at the generic level'. He noted, however, that the association of three Late Devonian placoderm genera (*Bothriolepis*, *Phyllolepis* and *Remigolepis*), indicated that they did not have the same relative temporal significance in Australia as had been established for these taxa in Europe.

The prevailing opinion at that time was that new discoveries in areas like Australia tended to confirm the results of studies in the Northern Hemisphere. This opinion was based on very superficial taxonomic studies, and led to some notable mistakes in attempting to use fossil fish to date the non-marine rocks of East Gondwana, as had been successfully carried out in Europe for the Old Red Sandstone. An example is the fish fauna from the Mulga Downs Group of western New South Wales, reported by Rade (1964) to contain a typical 'Old Red Sandstone' Late Devonian fish fauna including the placoderm genera *Phyllolepis*, *Holonema* and *Groenlandaspis*. Accordingly the Mulga Downs Group was included as a Late Devonian–Carboniferous sequence in Packham's (1969) comprehensive review of New South Wales geology. However, the first investigation by a trained vertebrate palaeontologist (Ritchie, 1969) revealed a fauna consisting almost entirely of endemic genera, which by broad comparison with Northern Hemisphere faunas suggested instead an Early or Middle Devonian age (Ritchie, 1973).

Over the last twenty years the need to provide a more reliable taxonomic database which can be applied to the solution of geological problems has been recognised (Young, 1987b). Detailed systematic analysis of some groups has enabled fossil vertebrates to be utilised in age control of sedimentary sequences, biogeographic input to reconstructions of palaeogeography, and palaeoenvironmental analysis. The first of these is the topic of this chapter. It cannot be over-emphasised that a workable biostratigraphy for East Gondwana using Middle Palaeozoic macrovertebrates is now emerging only because of the groundwork established by detailed taxonomic studies. These are still in their early stages, and the improvements in our local taxonomic data must be seen in the context of great expansion of the global taxonomic database for Palaeozoic vertebrates—as important for biogeographic studies (Young, 1990b) as for biostratigraphy.

TAXONOMIC DATA BASE

Eight major groups of vertebrates (four subgroups of agnathans or jawless fishes, and four of gnathostomes or jawed fishes) had already appeared by Early Devonian time. Approximate numbers of named Palaeozoic genera were summarised by Young (1990b) as follows (data from Blieck, 1984; Campbell and Barwick, 1987; Carroll, 1988; Denison, 1978, 1979; Janvier, 1985b; Young, unpubl.): osteostracans, 35; galeaspids, 30; anaspids, 10; thelodontids, 16; heterostracans, 110; acanthodians, 60; placoderms, 250; osteichthyans, 85; chondrichthyans, 22; tetrapods, 5. The major radiations of the last three groups occurred in post-Devonian times. As noted by Young (1990b), the taxonomic descriptions of faunas from extra-European areas over the last 25 years has led to a major increase in the database of described vertebrate fossils. It is still a very incomplete sample of the fossil record, but available data give a global coverage for the first time, and preliminary global correlations are now possible (e.g. Mark-Kurik, 1991). Many earlier studies concentrated on the 'Old Red Continent' of Europe and North America in a search for widespread taxa of biostratigraphic utility, but the modern approach acknowledges the endemic component in many faunas, and uses biogeographic data integrated with biostratigraphic analysis to explain the appearance and disappearance of taxa in particular successions.

BIOSTRATIGRAPHIC CONTROLS

Vertebrate macrofossils tend to be much rarer than invertebrates, but once discovered their scarcity is balanced by their high information content when they are studied in detail. Marine vertebrate faunas may be accurately dated on associated invertebrates, including conodonts (Figure 9.2), as is the case with the Early Devonian Taemas/Buchan fauna, the Middle Devonian Broken River assemblages, and the Late Devonian Gogo fauna (see below). However, the main biostratigraphic significance of Middle Palaeozoic fish remains stems from the many occurrences in non-marine strata, where dating using marine fossils cannot be directly applied. Internal evidence from the vertebrate assemblages themselves will provide increased resolution

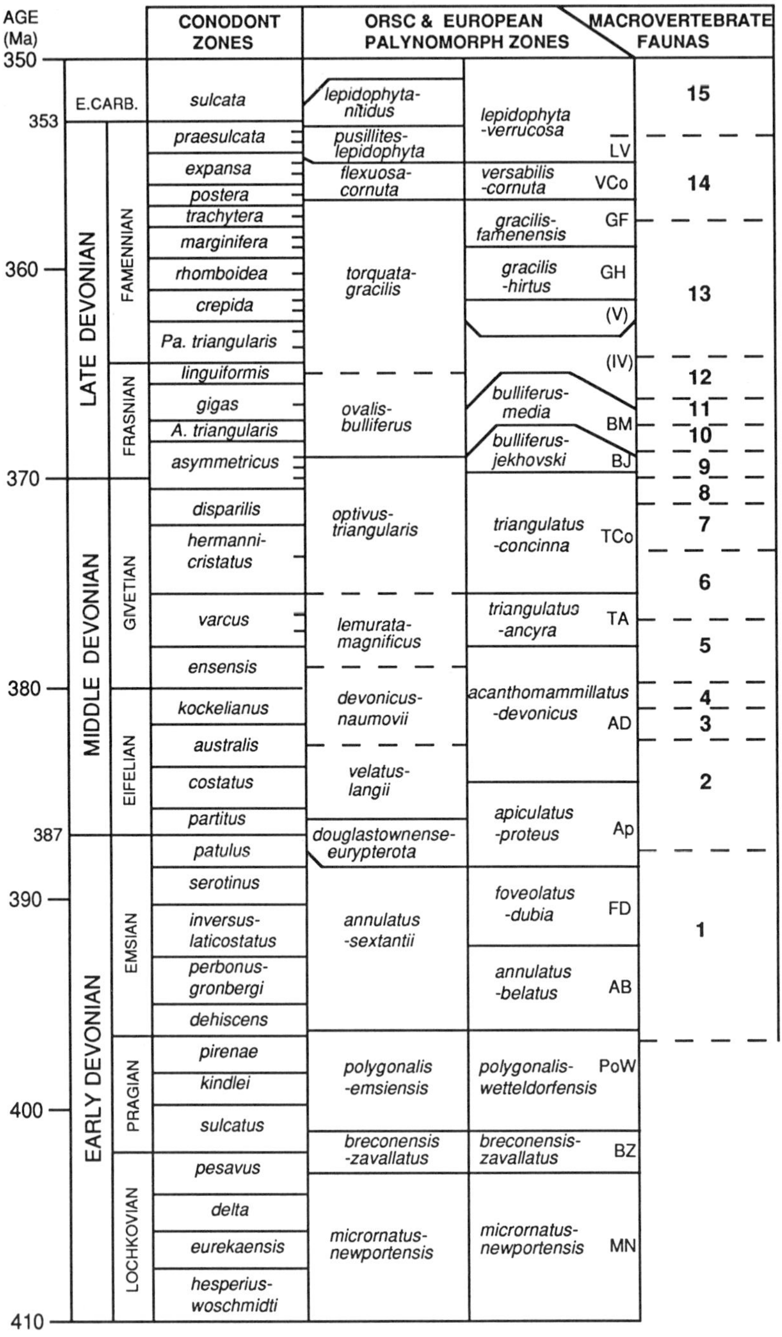
AGE (Ma)
CONODONT ZONES
ORSC & EUROPEAN PALYNOMORPH ZONES
MACROVERTEBRATE FAUNAS
350
353
360
370
380
387
390
400
410
E.CARB.
LATE DEVONIAN
MIDDLE DEVONIAN
EARLY DEVONIAN
FAMENNIAN
FRASNIAN
GIVETIAN
EIFELIAN
EMSIAN
PRAGIAN
LOCHKOVIAN
sulcata
praesulcata
expansa
postera
trachytera
marginifera
rhomboidea
crepida
Pa. triangularis
linguiformis
gigas
A. triangularis
asymmetricus
disparilis
hermanni-cristatus
varcus
ensensis
kockelianus
australis
costatus
partitus
patulus
serotinus
inversus-laticostatus
perbonus-gronbergi
dehiscens
pirenae
kindlei
sulcatus
pesavus
delta
eurekaensis
hesperius-woschmidti
lepidophyta-nitidus
pusillites-lepidophyta
flexuosa-cornuta
torquata-gracilis
ovalis-bulliferus
optivus-triangularis
lemurata-magnificus
devonicus-naumovii
velatus-langii
douglastownense-eurypterota
annulatus-sextantii
polygonalis-emsiensis
breconensis-zavallatus
micrornatus-newportensis
lepidophyta-verrucosa LV
versabilis-cornuta VCo
gracilis-famenensis GF
gracilis-hirtus GH
(V)
(IV)
bulliferus-media BM
bulliferus-jekhovski BJ
triangulatus-concinna TCo
triangulatus-ancyra TA
acanthomammillatus-devonicus AD
apiculatus-proteus Ap
foveolatus-dubia FD
annulatus-belatus AB
polygonalis-wetteldorfensis PoW
breconensis-zavallatus BZ
micrornatus-newportensis MN
15
14
13
12
11
10
9
8
7
6
5
4
3
2
1

in dating non-marine occurrences only as the interrelationships of taxa, and the ecological and biogeographic factors controlling their distribution in space, become better understood. To develop a system of faunal successions based on this internal evidence requires a framework of broad time control provided by several types of external evidence, such as palynology, magnetostratigraphy, geochronology and eustasy.

Palynomorphs associated with non-marine fossil fish assemblages, and marine interbeds containing datable invertebrates, can provide correlation with established palynozones or marine fossil zonations. However, the deeply oxidised surface exposures of Devonian redbeds across Australia rarely contain palynomorphs, which so far have been used primarily for subsurface age control. The occurrence of many fish faunas in redbeds means they are potentially datable by magnetostratigraphy, but to date there has been insufficient research in this field for reliable results. Marine horizons in predominantly non-marine sequences may represent global eustatic events, which can provide detailed correlations, even if the contained invertebrate assemblages are poorly preserved or not age-diagnostic. In addition, interbedded volcanics can be dated isotopically to provide a numerical calibration for vertebrate fossil zones, and both these methods have been applied in southeastern Australia and Antarctica.

VERTEBRATE FAUNAL SUCCESSION IN SOUTHEASTERN AUSTRALIA AND ANTARCTICA

For eastern Gondwana, a framework of broad time control for Devonian non-marine strata is provided by the non-marine fish faunas of post-Emsian age from southeastern Australia and Antarctica. Well-preserved fish faunas

---

Figure 9.2. Integration of conodont and palynomorph zonations for the Devonian (after Young, 1989a; Grey, 1991), shown against the macrovertebrate zones for East Gondwana put forward here. On the left is the international standard conodont zonation on which Devonian stage boundaries were defined by Ziegler and Klapper (1985), based on the work of Ziegler (1962, 1971), Ziegler and Sandberg (1984), and Sandberg and Dreesen (1984). A revised Late Devonian zonal scheme recently proposed by Ziegler and Sandberg (1990) is regarded as provisional and subject to testing in other sequences, but the Frasnian–Famennian boundary follows their usage. The middle column shows two Northern Hemisphere zonal schemes for Devonian palynomorphs derived from Richardson and McGregor (1986; left side), and Streel *et al.* (1987; right side). The macrovertebrate faunas in the right column are numbered in the text as follows: 1, Taemas-Buchan fauna; 2, *Wuttagoonaspis* fauna; 3, Hatchery Creek fauna; 4, *Wurungulepis-Atlantidosteus* fauna; 5, *Nawagiaspis* fauna; 6, Aztec fauna; 7, Tatong-South Blue Range fauna; 8, Taggerty-Howitt fauna; 9, Bindaree Road–Freestone Creek fauna; 10, Pambula River fauna; 11, Nettletons Creek fauna; 12, Twofold Bay fauna; 13, Jemalong-Canowindra fauna; 14, Worange Point fauna; 15 Grenfell fauna.

occur in eastern Victoria (e.g. Long, 1983a, b, 1984a, b, 1985a, 1986b, 1987c, 1988a, 1989a, b; Long and Werdelin, 1986), and on the far south coast of NSW (e.g. Fergusson *et al.*, 1979; Young, 1982, 1983, 1989c). The Victorian faunas are excellently preserved, and interbedded with isotopically dated volcanics (e.g. Richards and Singleton, 1981; Williams *et al.*, 1982). The NSW faunas are less well known, but the sequence includes a marine interval (the Bellbird Creek Formation), with fishes above and below, which permits correlation with marine zones. The Antarctic succession from the Aztec Siltstone (Young, 1988a, 1989b, c; Young *et al.*, 1992) bridges the gap between these faunas and older thelodont-bearing assemblages (e.g. the Hatchery Creek fauna of NSW; see Young and Gorter, 1981).

Isotopic dating of the fish faunas

Early isotopic work (Evernden and Richards, 1962; McDougall *et al.*, 1966) in eastern Victoria erroneously assumed the fish faunas approximated to the Devonian–Carboniferous boundary, whereas the palaeontological evidence suggested a somewhat older age within the Late Devonian (Young, 1974; Marsden, 1976; Boucot, 1975). Williams *et al.* (1982) reanalysed the dates for the Cerberean Cauldron using both K-Ar and Rb-Sr methods, and concluded an age of 367 ± 2 Ma for the Taggerty fish fauna (see below), based on very close determinations for the Snobs Creek rhyolite beneath the fauna (368.6 ± 2.3 Ma) and for the Cerberean volcanics above it (367.1 ± 2.3 Ma). The age of the fish fauna is now considered to be early Frasnian, based on a date of 370 Ma for the Givetian–Frasnian boundary (Young, 1989a).

Invertebrate and palynological evidence

In the Eden–Pambula sequence of southeastern NSW Young (in Fergusson *et al.*, 1979) identified four preliminary assemblages of fishes, three beneath and one above a marine horizon (the Bellbird Creek Formation). The lower fish faunas in the Pambula River area occur in sediments interbedded with rhyolites and basalts of the Boyd Volcanic Complex (Fergusson *et al.*, 1979). Above this, the Merrimbula Group rests with local angular unconformities. The base of the Merrimbula Group contains another fish assemblage in the Twofold Bay Formation, and this is overlain by the marine Bellbird Creek Formation containing an invertebrate fauna. The Worange Point Formation at the top of the sequence contains the youngest fish fauna.

Similar sequences containing marine sediments and fish faunas occur to the north in the Minuma and Budawang Ranges (the Long Swamp Creek and Deua Formations, and correlatives of the Merrimbula Group, respectively). Fish faunas are known from beneath these formations in the Eden and Budawang sequences, and from above them in the Eden and Minuma sequences. Evidence of the age of the marine incursions is provided by a range of invertebrates, including brachiopods, tentaculitids, and conodonts.

Talent (in Boucot, 1975, p. 63) cited the occurrence of the brachiopod *Cyphoterorhynchus* as suggesting a 'Frasnian, perhaps late Frasnian age' for the Bellbird Creek Formation in the Eden–Pambula district. A more recent faunal list for this formation in the Eden district (Fergusson *et al.*, 1979, Table 2) includes other brachiopods (?*Hypothyridina*, *Atrypa reticularis*), bryozoans, and bivalves (*Phthonia*, ?*Aviculopecten*). New fossil localities in the Braidwood district to the north were reported by Pickett (1981, 1983), to contain brachiopods (*Cyrtospirifer* sp., *Cyphoterorhynchus* sp., *Productella* sp.), bivalves (*Leptodesma* sp.), gastropods (a bellerophontid), crinoids, cephalopods (a michelinocerid), and bryozoans, suggesting a Frasnian age. In the Krawarree area south of Braidwood Johnston (1964, unpublished thesis) identified a similar fauna of brachiopods (including *Atrypa* sp., and cf. *Pleurodon*) and bivalves from the Deua Formation. The sequence here (Wyborn and Owen, 1986) shows two separate marine episodes (the Long Swamp Creek and Deua Formations), not manifested elsewhere, and suggesting a marginal position relative to the main area of marine incursion. These may represent two of the five sea level oscillations indicated for the Frasnian of the Canning Basin by Becker *et al.* (in press).

The marine sequence north of Braidwood is much thicker (up to 3.2 km according to Powell, 1983). The Nettletons Creek fish horizon northeast of Braidwood (see below), is overlain by some 700 metres of sediment in the Comerong Volcanics, followed conformably by marine sediments of the Merrimbula Group, again with invertebrates in its lower parts. A similar fauna including tentaculitids was reported by Wood (in McElroy and Rose, 1962; also Pickett, 1983). Ritchie (1984, p. 326) reported a measured section above his fish horizon, with marine invertebrates at the 920, 1390, and 1520 m levels. The higher two assemblages (including the brachiopods *Cyrtospirifer* cf. *subdisjunctus* and *Sinotectirostrum*) were said to show some similarity to marine faunas in central NSW of possible Famennian age. The presence of tentaculitids and atrypid brachiopods in these faunas indicate a Frasnian rather than Famennian age. At the northern end of the Budawang Synclinorium the Ettrema Limestone member has yielded the oldest reliable age for a Late Devonian marine transgression, a late Frasnian age based on conodonts of the upper *gigas* Zone (Pickett, 1972).

Late Devonian sea-level curve

There is some agreement from eustatic curves for the Devonian that a major global transgression occurred in the late Frasnian. The Devonian eustatic curve for Euramerica of Johnson *et al.* (1985; see also Sandberg *et al.*, 1988) shows an increase in sea level throughout the Frasnian to a maximum in the upper *gigas* conodont Zone (*rhenana* and *linguiformis* Zones in the new scheme of Ziegler and Sandberg, 1990). Evidence of this event is seen on the Pacific and northern margins of Gondwana from both eastern and western Australia (Talent and Yolkin, 1987; Talent, 1989). In the Canning Basin of Western Australia a maximum transgression in the late Frasnian (*asymmetricus*–upper *gigas* Zone) is documented at the end of the Pillara reef cycle (e.g. Playford, 1980, 15), followed by a major regression. Recent more detailed work in the Canning Basin by Becker *et al.* (in press) shows five fluctuations in sea level within an overall increase through Frasnian time, to a highstand in the upper *asymmetricus* to base of the lower *gigas* Zone, followed by rapid fluctuations in the upper *gigas* at the end of the Frasnian. In the Bonaparte Basin a limestone horizon within the mainly terrigenous Cockatoo Formation (the Westwood Member near the coast, the Hargreaves and Kunnunurra Members inland), is referable to the same conodont zones (Roberts, 1972).

In eastern Australia, limestones additional to the Ettrema Limestone mentioned above which have yielded Frasnian conodonts include that from the Dee Volcanics (Mount Morgan, Queensland; Druce, 1970), and the allochthonous blocks of the Mostyn Vale Formation (*gigas* Zone) in NSW (Wright *et al.*, 1990). In the Broken River area of Queensland a Frasnian marine horizon with the brachiopod *Cyrtospirifer* is now identified within the non-marine Bundock Creek Formation (Mawson *et al.*, 1985). Talent (1989) equated the Ettrema and Mount Morgan limestones with the upper *gigas* Zone sea level maximum, but the latter may be slightly older (*asymmetricus* Zone, early Frasnian; see Druce, 1974; Mawson *et al.*, 1985). Farther inland, the Late Devonian sediments of central and western NSW (Hervey Group) become largely non-marine. Connolly (1965) recognised three major lithological sequences in the Hervey Group—a lower Beargamil Subgroup, a middle Nangar Subgroup, and an upper Cookamidgera Subgroup. Lithological correlation with his eastern province (Lambie and Catombal Groups), in which the basal equivalents of the Beargamil Subgroup are marine, was assisted by the occurrence of placoderm plates from various localities and horizons (mainly within the sandstones of the Nangar Subgroup), referred to by Connolly as the 'bothriolepid fish assemblage'. Hills (1932, 1936b) had earlier identified the antiarchs *Bothriolepis* and *Remigolepis* in a fish fauna from Gingham Gap in the Hervey Syncline. Campbell and Bell (1977, 1982) have since described tetrapods and lungfishes from the Hervey Group near Forbes, but the widespread placoderm element in these faunas still awaits description. Nevertheless many localities have been visited and recollected, and there is no confirmed occurrence of fish remains older than the marine deposits forming the basal part of the sequence in the eastern province. All the evidence is consistent with this marine horizon being of late Frasnian age (see discussions in Campbell and Bell, 1977, p. 375; Mawson *et al.*, 1985, p. 11). Accepting the overall shape of the Late Devonian eustatic curve in the current literature, the age of basal transgressive sequences in different successions are likely to vary within the Frasnian according to geographic position and local tectonism. However, on present indications the top of any Frasnian marine interval should be better constrained within the *rhenana* and *linguiformis* Zones at the end of the Frasnian. Based on the invertebrate and conodont age control and eustatic events summarised above the non-marine fish faunas of southeastern Australia may, therefore, be considered in two groups: pre-late Frasnian, and Famennian. In the following section these are integrated with marine occurrences to develop a preliminary chronological sequence of vertebrate assemblages.

DEVONIAN MACROVERTEBRATE FAUNAL SUCCESSION

Presence–absence data of various key taxa can be used to develop a preliminary biostratigraphic scheme, based on the systematically best studied group (antiarch placoderms, Figures 9.3, 9.4). It has been assumed in the first instance that faunal differences amongst the various localities are biostratigraphically controlled, allowing faunal components to be

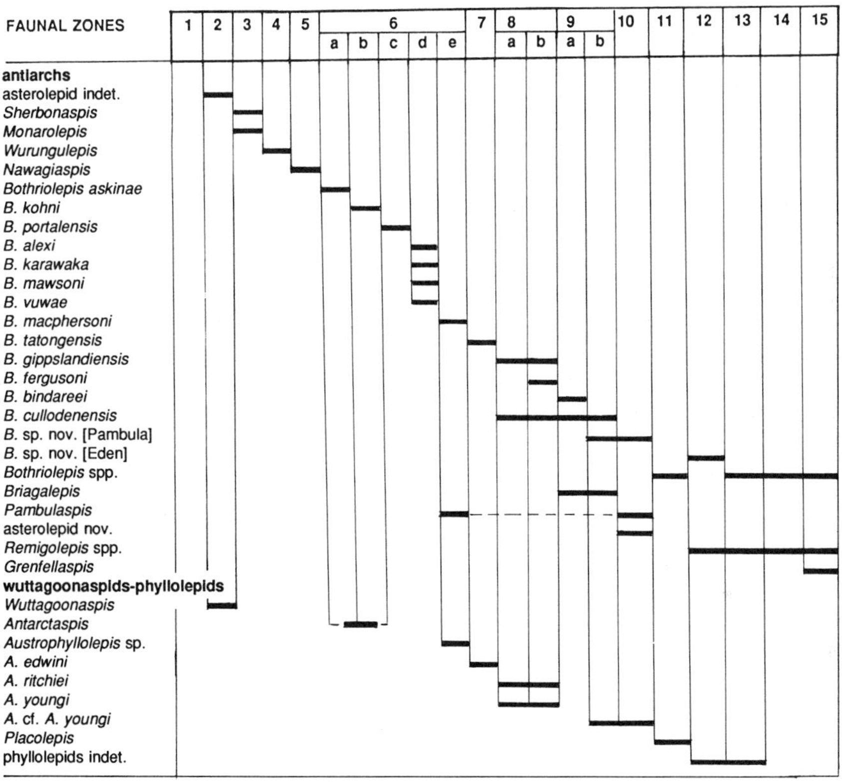

Figure 9.3. Stratigraphic distribution of key placoderm taxa (antiarchs, wuttagoonaspids, phyllolepids) against a preliminary biozonation of 15 macrovertebrate faunas, numbered as in the text: 1, Taemas-Buchan fauna; 2, *Wuttagoonaspis* fauna; 3, Hatchery Creek fauna; 4, *Wurungulepis-Atlantidosteus* fauna; 5, *Nawagiaspis* fauna; 6, Aztec fauna (6a, *askinae* zone; 6b, *kohni* zone; 6c, *portalensis* zone; 6d, *karawaka* zone; 6e, *Pambulaspis*/phyllolepid zone); 7, Tatong-South Blue Range fauna; 8, Taggerty-Howitt fauna (8a, Taggerty assemblage; 8b, Howitt assemblage); 9, Bindaree Road-Freestone Creek fauna (9a, Bindaree Road assemblage; 9b, Freestone Creek assemblage); 10, Pambula River fauna; 11, Nettletons Creek fauna; 12, Twofold Bay fauna; 13, Jemalong-Canowindra fauna; 14, Worange Point fauna; 15 Grenfell fauna.

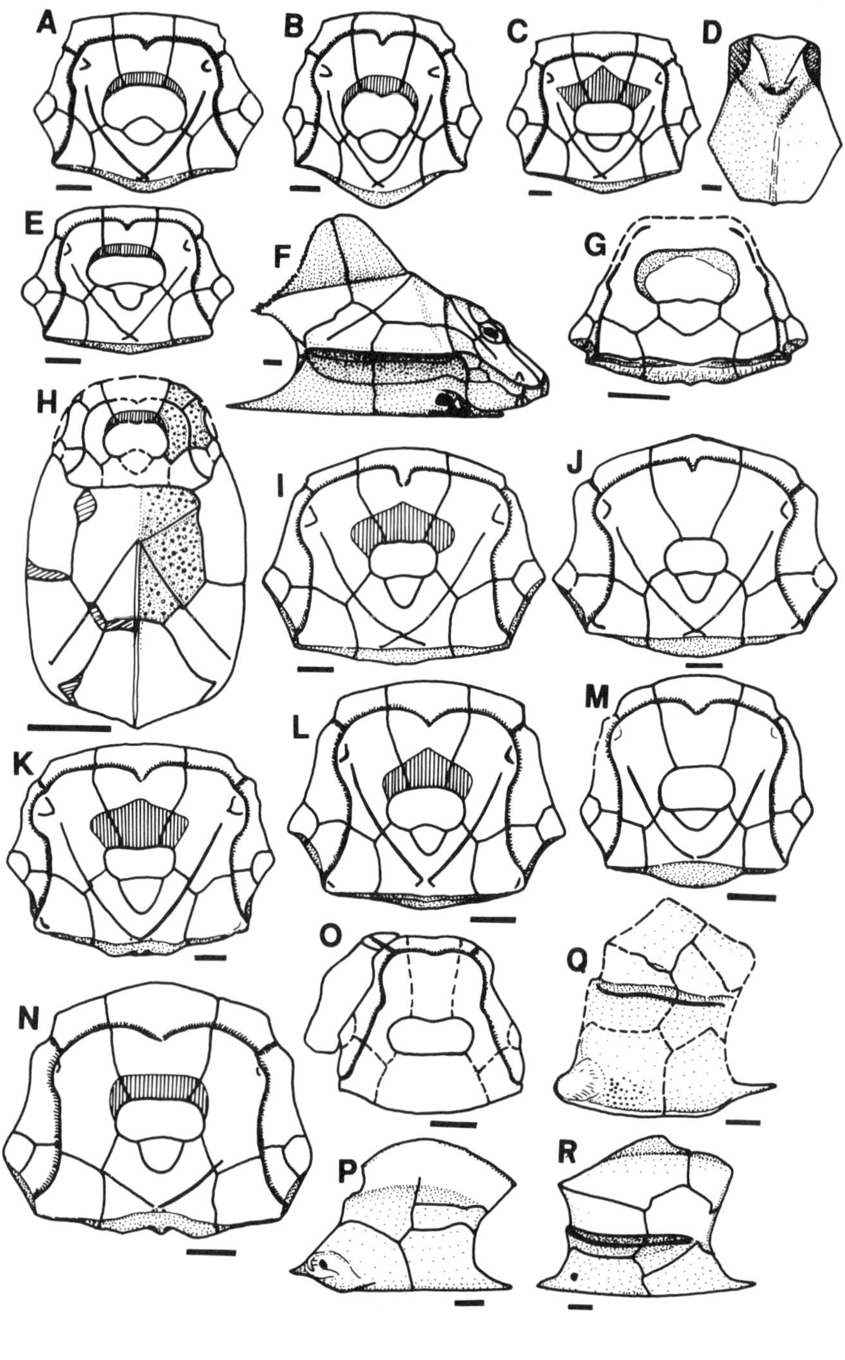
A
B
C
D
E
F
G
H
I
J
K
L
M
N
O
P
Q
R

organised in a sequence involving minimum overlap. Remaining inconsistencies may be attributed to palaeogeographic or palaeoecologic rather than biostratigraphic factors.

Evidence providing constraints on age are summarised below for each of 15 macrovertebrate assemblages making up a provisional succession from eastern Gondwana. Faunal lists are given in Appendix 1, and distribution of various taxa through the provisional biozonation are summarised in Figures 9.3, 9.5, 9.7 and 9.9. The chronological resolution of marine (M) and non-marine (NM) faunas of about the same age which are faunally dissimilar is problematic. There has been some discussion about whether classical localities from the Old Red Sandstone of Europe are non-marine, as traditionally assumed (e.g. Janvier, 1985a). The use of these terms here is based on the presence or absence of marine invertebrates (e.g. corals, brachiopods), with no more specific environmental connotation. However, prevailing local opinion regards most 'non-marine' Late Devonian fish assemblages as either fluviatile (NSW, central Australia) or lacustrine (eastern Victoria) deposits. Marine faunas for the Early and Middle Devonian are placed in a chronological sequence, but for the Late Devonian, diverse and well-studied faunas such as that from the Gogo Formation in the Canning Basin (Appendix 1), are difficult to accommodate within the essentially non-marine scheme worked out for southeastern Australia, on which Figures 9.3, 9.5, 9.7 and 9.9. are based. For a different reason (insufficient taxonomic study) the Middle–Late Devonian faunas of central Australia can only be broadly correlated with the southeast, as indicated below.

Occurrences represented by fragmentary or undescribed remains only are mentioned but not dealt with as assemblages, pending further taxonomic work on these faunas. Excluding Ordovician occurrences from central Australia, the oldest macrovertebrate remains are from the southeast, but they are poorly known and undescribed. The tail of an indeterminate acanthodian fish from the Yea district of Victoria is associated with the plant *Baragwanathia* in the uppermost beds of the Yea Formation, which is regarded as Ludlovian in age (e.g. Garratt and Wright, 1988). Of similar Late Silurian age, associated with volcanics dated isotopically as Ludlovian

---

Figure 9.4. Endemic East Gondwana antiarchs (headshields in dorsal view, mostly, or trunkshields in lateral view); bar scales equal one centimetre. A, *Bothriolepis gippslandiensis*; B, *B. cullodenensis*; C, *B. tatongensis*; D, *Monarolepis verrucosa* (anterior median dorsal plate, ventral view); E, *B. fergusoni*; F, *B. cullodenensis*; G, *Pambulaspis cobandrahensis*; H, *Briagalepis warreni*; I, *B. othriolepis karawaka*; J, *B. vuwae*; K, *B. macphersoni*; L, *B. portalensis*; M, *B. barretti*; N, *B. askinae*; O, P, *Nawagiaspis wadeae*; Q, *Wurungulepis denisoni*; R, *Sherbonaspis hillsi*. (A, B, C, E, F, H, after Long and Werdelin, 1986; D, R, after Young and Gorter, 1981; G, after Young, 1983; I–N, after Young, 1988a; O–Q, after Young, 1990b).

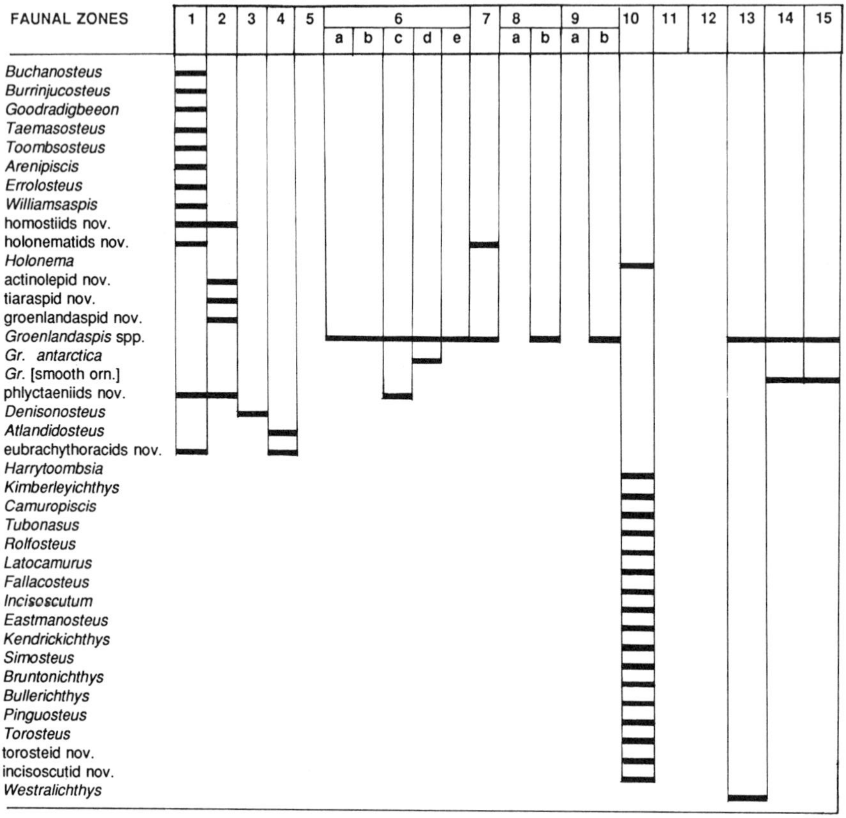

Figure 9.5. Stratigraphic distribution of various arthrodires against a preliminary biozonation of 15 macrovertebrate faunas, numbered as in the text: 1, Taemas-Buchan fauna; 2, *Wuttagoonaspis* fauna; 3, Hatchery Creek fauna; 4, *Wurungulepis-Atlantidosteus* fauna; 5, *Nawagiaspis* fauna; 6, Aztec fauna (6a, *askinae* zone; 6b, *kohni* zone; 6c, *portalensis* zone; 6d, *karawaka* zone; 6e, *Pambulaspis*/phyllolepid zone); 7, Tatong-South Blue Range fauna; 8, Taggerty-Howitt fauna (8a, Taggerty assemblage; 8b, Howitt assemblage); 9, Bindaree Road-Freestone Creek fauna (9a, Bindaree Road assemblage; 9b, Freestone Creek assemblage); 10, Pambula River fauna; 11, Nettletons Creek fauna; 12, Twofold Bay fauna; 13, Jemalong-Canowindra fauna; 14, Worange Point fauna; 15 Grenfell fauna.

(Wyborn *et al.*, 1982), are acanthodian spines from near Canberra (Tomlinson, 1968; Henderson, 1980). Acanthodian spines are also reported from the early Lochkovian Mount Ida Formation of central Victoria by Neil (1981). Of slightly younger (Pragian) age (based again on isotopic evidence) are possible tetrapod trackways from the Early Devonian Grampians

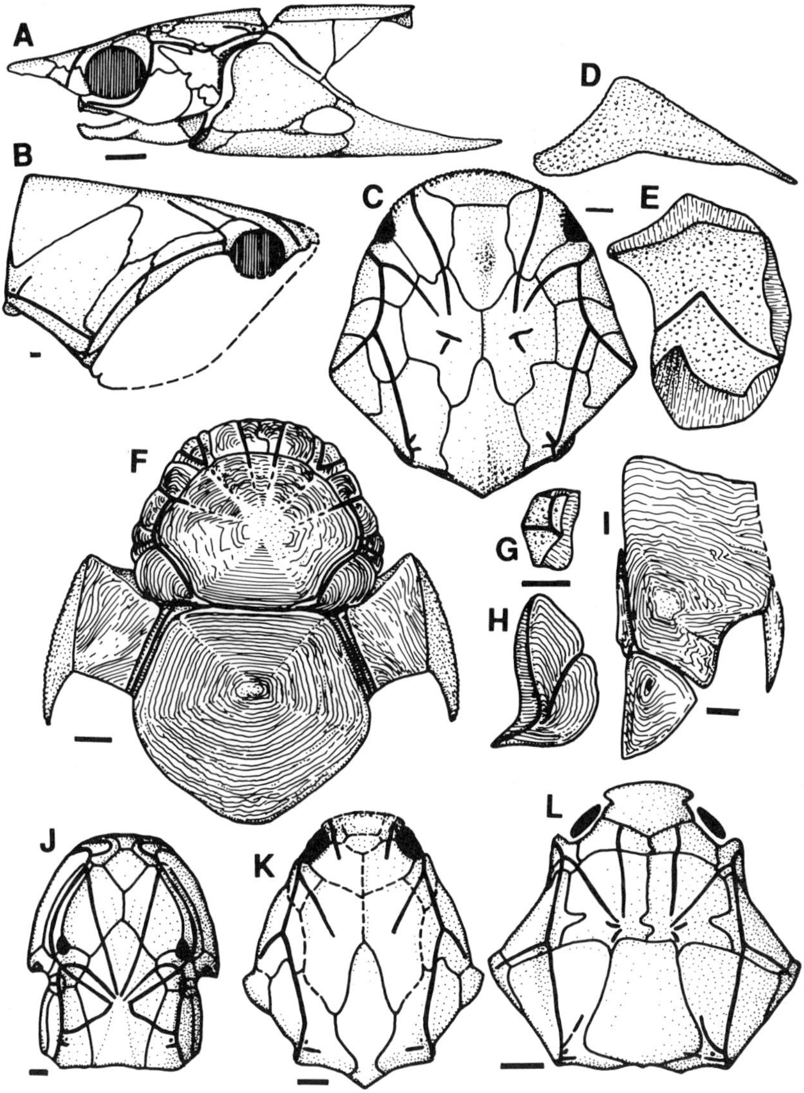

Figure 9.6. Some endemic East Gondwana arthrodires, all bar scales are 1 centimetre. A, *Fallacosteus turnerae*; B, *Westralichthys uwagadensis*; C–E, *Groenlandaspis antarcticus* (C, headshield; D, median dorsal plate; E, posterior dorsolateral plate); F, *Placolepis budawangensis*; G–I, *Austrophyllolepis* (G, *A. ritchiei*, marginal plate; H, *A. youngi*, paranuchal plate, I, *A. youngi*, left side of ventral armour); J, *Wuttagoonaspis fletcheri*; K, *Denisonosteus weejasperensis*; L, *Buchanosteus confertituberculatus*. (A, after Long, 1990b; B, after Long, 1987a; C–E, after Ritchie, 1975; F, after Ritchie, 1984; G–I, from Long, 1984; J, from Ritchie, 1973; K, from Young and Gorter, 1981; L, from Young, 1979).

Group of western Victoria (Warren *et al.*, 1986), which if confirmed would be the oldest tetrapod fossil evidence known to date. A lungfish jaw and placoderm plates have recently been discovered in the Bell Point Limestone of eastern Victoria (K.S.W. Campbell and J.A. Long, pers. comm.), which contains conodonts and microvertebrates of late Pragian age (see Chapter 8). Rare placoderm remains occur in the Garra Formation in New South Wales, of late Lochkovian–early Pragian age (*pesavus–sulcatus* Zones; Strusz, 1972; Mawson *et al.*, 1985). An unnamed acanthothoracid is the only form so far identified.

1. Taemas–Buchan fauna (M)

The Taemas–Buchan assemblage (Emsian) from marine limestone in the Burrinjuck Dam area (Taemas–Wee Jasper) of NSW, Buchan in Victoria, and some other small isolated outcrops (Tarago, Ravine in NSW), is the oldest significant Devonian macrovertebrate fauna from east Gondwana. Various major groups are represented, of which placoderms (White, 1978; Young, 1979, 1980, 1981, 1985b; Long and Young, 1988) are the most abundant and diverse (arthrodires, acanthothoracids, petalichthyids and ptyctodontids). The arthrodire *Buchanosteus* is shown in Figure 9.6L. Less common are lungfishes (Campbell and Barwick, 1982, 1983, 1984, 1985), acanthodians, onychodontids, and porolepiforms (see Young, 1985b; Long, 1984a, 1986a). The associated microvertebrate assemblage (see Turner, Chapter 8) includes *Cheiracanthoides*, *Ohiolepis*, *Ligulalepis*, '*Skamolepis*', and abundant buchanosteid, acanthothoracid and other placoderm scales (including *Ohioaspis*). Undescribed climatiid acanthodian spines, listed in Appendix 1, occur beneath the limestone sequence near Wee Jasper in the Kirawin Shale within the underlying volcanics.

Age constraints
Dehiscens to *serotinus* Zone conodonts are known from the limestones (Mawson *et al.*, 1985, 1988), placing this fauna in the Emsian.

2. Wuttagoonaspis *fauna (NM)*

The *Wuttagoonaspis* fauna of the Mulga Downs Group in western New South Wales contains an abundant and diverse placoderm fauna, largely undescribed except for the nominate form (Ritchie, 1973). Also present are crossopterygians, acanthodians, and the thelodont *Turinia*. This macrofauna was apparently widespread in Australia; it occurs in the Georgina (Turner *et al.*, 1981; Young, 1984) and Amadeus Basins (Young, 1985a, 1988b; Young *et al.*, 1987), and the associated microfauna containing the thelodont *Turinia* is recorded from the Officer (Long *et al.*, 1988) and Canning Basins (Gross, 1971; note that Turner, 1991, regards the latter as a slightly older assemblage). Taxa not yet identified in the Mulga Downs Group but associated with *Wuttagoonaspis* (Figure 9.6J) in the Cravens Peak Beds of the Georgina Basin include asterolepid antiarchs (Young, 1984), the shark *Mcmurdodus* (Turner and Young, 1987), and the new agnathans *Pituriaspis* and *Neeyambaspis* (Young, 1991b). Newly discovered faunas in the western Amadeus Basin associated with *Wuttagoonaspis* (e.g. Young, 1988b, Figure 13) include osteolepids, dipnoans (?dipterid) and groenlandaspids, associated with abundant thelodonts.

Age constraints

A maximum age for the Mulga Downs Group is provided by the underlying marine Cobar Supergroup, containing various conodont faunas, but apparently lacking *Polygnathus*, so the Cobar Supergroup is probably older (pre-Emsian) than any *Polygnathus* zone (see Pickett, 1980; Pickett and McClatchie, 1991; Sherwin, 1991). At its oldest, therefore, the *Wuttagoonaspis* fauna could be contemporaneous with the Taemas–Buchan fauna, but the presence or duration of a hiatus at the base of Mulga Downs Group is uncertain. The underlying Winduck Group is a shallowing-up marine sequence above which the fluviatile Mulga Downs Group shows bedding concordance, but with local angular unconformities, and there is no palaeontological control on the duration of a hiatus, if present.

A younger age limit is also problematic. In the type area there is little palaeontological control from higher in the sequence. Long *et al.* (1988) tentatively suggested an early Eifelian age for the *Wuttagoonaspis–Turinia* assemblage, but an age as young as Givetian was not excluded. An Eifelian age was also originally proposed for the Hatchery Creek fauna (see below), but apart from turiniid thelodonts and osteolepids this is a completely different vertebrate assemblage, which is assumed for the present to be of younger age, because it contains bothriolepid antiarchs (see below). However, the occurrence of the shark *Mcmurdodus* in the Cravens Peak fauna (Turner and Young, 1987) might be evidence for a younger age, since this form is otherwise only known from the Aztec Siltstone fauna of southern Victoria Land, Antarctica. The biostratigraphic significance of similarities between the Aztec fauna and that from the Hatchery Creek Conglomerate (Young and Gorter, 1981) is discussed below.

3. Hatchery Creek Fauna (NM)

This is a non-marine fauna (Young and Gorter, 1981) in which bothriolepid antiarchs (*Monarolepis*, Figure 9.4D) and turiniid thelodonts (see Figure 8.3) are associated with the asterolepid antiarch *Sherbonaspis* (Figure 9.4R), the phlyctaeniid arthrodire *Denisonosteus* (Figure 9.6K) and an osteolepid (*'Gyroptychius' australis*). The bothriolepid–turiniid association (Appendix 1) is otherwise only encountered in the Aztec fauna of Antarctica, and there are also similarities in the osteolepids from the two faunas (see below).

Age constraints

The Hatchery Creek fauna was tentatively assigned a late Eifelian age by Young and Gorter (1981), taking into consideration that 1800 metres of section separate the fish bed from the underlying Taemas Limestone. An older age limit of late Emsian, based on conodonts in the upper part of the underlying limestone sequence, assumes there is no hiatus between these formations, as did Talcnt (1989), who cquatcd thc changc from marinc to non-marine sedimentation with a regressive event in the late Emsian (*serotinus* Zone). However, in view of the coarseness of the basal units of the Hatchery Creek Formation (conglomerates and pebble sandstones) some tectonic influence to the change in sedimentation seems likely.

There is no good control on a younger age limit, the overall character of the fish fauna suggesting only a probable Middle Devonian age. The similarities to the Aztec fauna of Antarctica in the bothriolepid–turiniid association and the osteolepids suggests that they may be broadly contemporaneous, both faunas capturing the overlap in stratigraphic range of youngest freshwater turiniids with the earliest bothriolepid antiarchs.

4. Wurungulepis–Atlantidosteus *fauna (M)*

A fauna of probable Eifelian age occurs in the Broken River Formation of Queensland, but only the asterolepid antiarch *Wurungulepis* (Figure 9.4Q) has been described (Young, 1990a). Associated remains of large eubrachythoracid arthrodires include a specimen of *Atlantidosteus* (see Figure 7.4D) and other remains which are more advanced than most of the arthrodires from the Taemas–Buchan fauna (the only confirmed eubrachythoracid occurs high in the Taemas sequence at Wee Jasper). A large dipnoan is also present (S. Turner, pers. comm.). Systematic work on these remains should substantiate the existence of a discrete fauna as provisionally proposed here. Associated microvertebrates include turiniid thelodonts and cheiracanthid and machaeracanthid acanthodians (Turner, 1982, 1986, and pers. comm.).

Age constraints
The Broken River Formation has an age range of Lochkovian (*pesavis* conodont zone) to Givetian (Wyatt and Jell, 1980; Mawson *et al.*, 1985), and analysis of conodont faunas in the Middle Devonian part of the formation is still incomplete. The lowermost Eifelian *partitus* conodont zone has been identified not far from the type locality of *Wurungulepis*, and an Eifelian age is given to the surrounding vertebrate localities, including 'Fish Hill' (J.S. Jell, pers. comm.). These all come from the limestone lithofacies of the Burges Formation (Withnall *et al.*, 1988, p. 149). Conodont analysis of this section is hampered by the complex geology (Mawson *et al.*, 1985), but the Burges Formation ranges into the early Givetian (Withnall *et al.*, 1988, Figure 71).

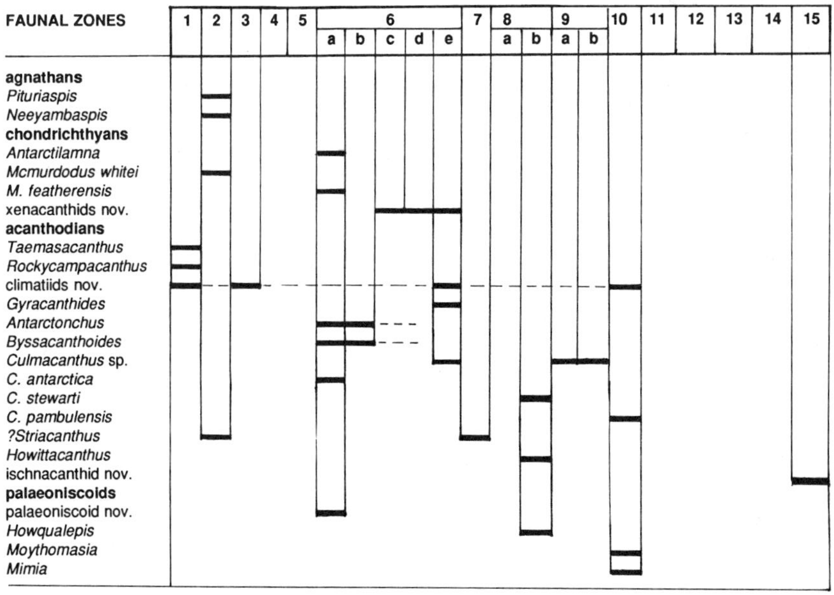

Figure 9.7. Stratigraphic distribution of various agnathan, chondrichthyan, acanthodian and actinopterygian taxa against a preliminary biozonation of 15 macrovertebrate faunas, numbered as in the text: 1, Taemas-Buchan fauna; 2, *Wuttagoonaspis* fauna; 3, Hatchery Creek fauna; 4, *Wurungulepis-Atlantidosteus* fauna; 5, *Nawagiaspis* fauna; 6, Aztec fauna (6a, *askinae* zone; 6b, *kohni* zone; 6c, *portalensis* zone; 6d, *karawaka* zone; 6e, *Pambulaspis*/phyllolepid zone); 7, Tatong-South Blue Range fauna; 8, Taggerty-Howitt fauna (8a, Taggerty assemblage; 8b, Howitt assemblage); 9, Bindaree Road-Freestone Creek fauna (9a, Bindaree Road assemblage; 9b, Freestone Creek assemblage); 10, Pambula River fauna; 11, Nettletons Creek fauna; 12, Twofold Bay fauna; 13, Jemalong-Canowindra fauna; 14, Worange Point fauna; 15 Grenfell fauna.

5. Nawagiaspis *fauna (M)*

A younger Givetian fauna in the Broken River Formation is so far represented only by the bothriolepidoid *Nawagiaspis* (Figure 9.4O–P; Young, 1990a).

Age constraints
This occurrence, in a limestone nodule from an area mapped as Papilio Mudstone (Withnall *et al.*, 1988, Figure 80), is probably Givetian in age (J.S. Jell, pers. comm.), but again detailed conodont determinations are not available.

6. Aztec fauna (NM)

The diverse fauna from the Aztec Siltstone of Southern Victoria Land, Antarctica (Appendix 1), was originally regarded as of Late Devonian age (Woodward, 1921; Ritchie, 1975), but more detailed study of the fauna now supports the view of White (1968) that the fauna is largely Middle Devonian (Young, 1989b, 1991a). Young (1988a) proposed a biostratigraphic scheme of six zones based on the association of different species of the antiarch *Bothriolepis* with other elements in the fish fauna. This scheme has been expanded with subsequent taxonomic studies on other groups: acanthodians (Young, 1989c), rhipidistians (Young *et al.*, 1992), and thelodonts (Turner and Young, 1992). The lower zones are characterised by an association of *Bothriolepis* with the thelodont agnathan *Turinia antarctica*, not recorded from any other sequence, whilst the upper zone contains phyllolepid placoderms, which could indicate a Late Devonian age, although their lower age limit (?Givetian) is uncertain (Young, 1989b). These associations indicate that the Aztec fauna represents a transitional succession apparently linking the youngest non-marine turiniid thelodonts with the first appearance of phyllolepid placoderms. In central Australia, the equivalent time interval (unfossiliferous) is probably represented in the Pertnjara Group of the Amadeus Basin. Turiniid thelodonts are associated with *Wuttagoonaspis* in the basal Deering Siltstone member of the Parke Siltstone, and a *Bothriolepis*–phyllolepid assemblage occurs just 100 m above in the overlying Harajica Sandstone member (Young, 1985a, 1988b). The spore *Geminospora lemurata* is associated in the later assemblage (?early Givetian; see below). The time equivalent of the basal Aztec *Turinia–Bothriolepis* association must lie within the unfossiliferous interval between these horizons. Faunal lists for these assemblages are given in Appendix 1, and faunal associations in five biostratigraphic zones in the Aztec succession are summarised below from oldest to youngest.

6a. askinae *zone*
This zone is characterised by the antiarch species *Bothriolepis askinae* (Figure 9.4N) associated with the thelodont *Turinia antarctica*, the sharks *Antarctilamna* (spine illustrated Figure 1.6G) and *Mcmurdodus*, and the acanthodian *Culmacanthus antarctica* (Figure 9.8B). Rhipidistian material from this zone (Young *et al.*, 1992) includes *Platyethmoidia antarcticus*, and indeterminate porolepiform and osteolepid remains. Apart from the nominate species and *Turinia*, the upper range of the other taxa is not reliably defined.

6b. kohni *zone*
This zone contains the more advanced species *Bothriolepis kohni*, again associated with the thelodont *Turinia antarctica*. The megalichthyinid rhipidistian *Mahalalepis* (Figure 9.8F) and various isolated osteolepid remains are also recorded.

6c. portalensis *zone*
Characterised by the species *Bothriolepis portalensis* (Figure 9.4L) in an

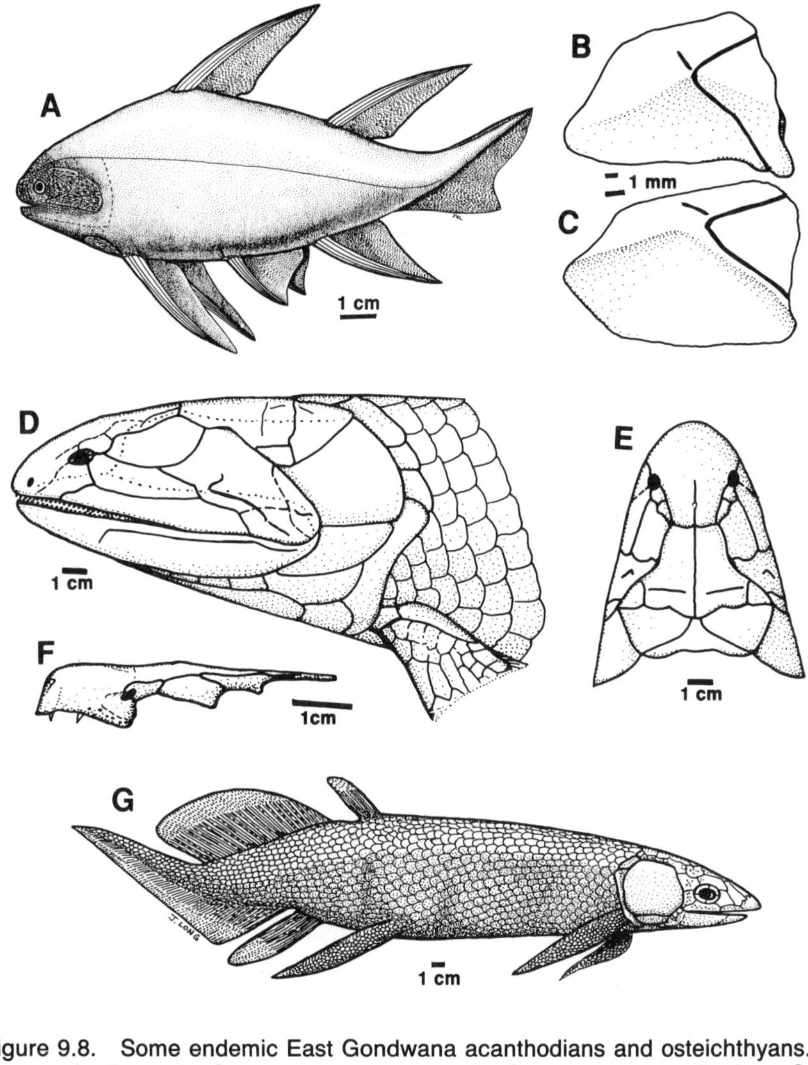

Figure 9.8. Some endemic East Gondwana acanthodians and osteichthyans. A–C, acanthodians, A, *Culmacanthus stewarti*; B, *C. antarctica* cheek plate; C, *C. pambulensis* cheek plate; D–F, rhipidistians, D, *Koharalepis jarviki*; E, *Canowindra grossi*; F, *Mahalalepis resima*; G, the dipnoan *Barwickia downunda*. (A, after Long, 1983b; B, C, from Young, 1989c; D, F, from Young *et al.*, 1992; E, from Long, 1985b; G, from a sketch by J. Long, original).

assemblage which also contains abundant xenacanth shark teeth, and phlyctaeniid arthrodires. Noteworthy is the absence of thelodont scales from this zone, and higher in the succession. Rhipidistian material from this zone includes the canowindrid *Koharalepis* (Figure 9.8D), and indeterminate eusthenopterid and osteolepid remains.

6d. karawaka *zone*

Characterised by the co-occurrence of four species of *Bothriolepis* (*B. karawaka*, Figure 9.4I; *B. vuwae*, Figure 9.4J; *B. alexi*, *B. mawsoni*), the last two poorly known, and several indeterminate species. *Groenlandaspis antarctica* (Figure 9.6C–E) is an associated species at the type locality, and *Bothriolepis barretti* (Figure 9.4M) is provisionally placed at this level, although it has not been found in association with the other species. The assemblage also includes gyracanthidid acanthodians, eusthenopterid and rhizodontid (?*Notorhizodon*) remains, as well as the osteolepids *Platyethmoidia* and *Vorobjevaia*. J. Long (pers. comm.) recently collected samples of a vertebrate bone-bed from the *B. barretti* horizon at Mt Crean that can be acid prepared, and has already yielded an assemblage of several different acanthodian scales, and dipnoan and shark teeth.

6e. Pambulaspis/*phyllolepid zone*

The holotypes of *Bothriolepis macphersoni* (Figure 9.4K), *Pambulaspis antarctica* and *Notorhizodon mackelveyi* occur in this zone. The four species of *Bothriolepis* from the previous zone are interpreted to overlap the basal part of this zone. The distinctive remains of phyllolepids were not recorded from the previous two zones at their type locality (Mount Ritchie, southern Warren Range), and a separate phyllolepid zone was proposed by Young (1988a) as the youngest in the Aztec sequence, by comparison with southeastern Australia, where phyllolepids are common (see below). However, more recent discoveries by J.A. Long (1991/92 season, pers. comm.) indicate rare phyllolepid remains lower in the Aztec sequence, and on the charts this zone is considered to overlap the range of *Pambulaspis* (Figure 9.4G) as is the case in southeastern Australia. Indeterminate species of *Bothriolepis*, *Gyracanthides* and *Groenlandaspis*, and indeterminate rhizodontid and osteolepid remains are associated with the phyllolepids from several localities (e.g. Alligator Peak, Boomerang Range). For the present the Harajica fauna from the Amadeus Basin, central Australia (Young, 1985a), which also contains a phyllolepid/*Bothriolepis* assemblage not far above the range of turiniid thelodonts, is provisionally regarded as equivalent to this zone, as proposed by Young (1988b, Tables 2,3). The Harajica fauna includes climatiid acanthodian spines, but similar spines occur in both older (Hatchery Creek) and younger (Pambula River) assemblages (see Young, 1985a, Figure 8I; 1989c, Figure 2D; Young and Gorter, 1981, pl. 9, Figure 8), so their apparent absence from the Aztec fauna is not considered to have biostratigraphic significance.

Age constraints

Spore evidence provides some external age control on the Aztec sequence. Kyle (1977) described a poorly preserved assemblage from the Terra Cotta Sandstone (basal Taylor Group) which suggested a possible Early Devonian (Emsian) age, so this does not constrain the age of the Aztec fauna within the Middle Devonian. Helby and McElroy (1969) proposed a Late Devonian (Frasnian) age for the Aztec based on an abundance of the spore *Geminospora lemurata* in a palynoflora from Aztec Mountain. However, the biostratigraphical significance of this taxon depends on its assigned range in local sequences—mid-Givetian to early Famennian (Playford, 1983), or Eifelian to mid-Frasnian (Balme, 1988). Grey (1991) has followed Young's (1989a) application of the schemes of Richardson and McGregor (1986) and Streel *et al.* (1987) in which *G. lemurata* indicates an early Givetian age, equivalent to the *ensensis* or *varcus* conodont Zones (see also Young, 1989b, 1991a).

Young (1988a) discussed broad correlations between the Aztec succession and other sequences linked to the standard conodont zonation using bothriolepid antiarchs. The association of bothriolepids with turiniid thelodonts in the *askinae* and *kohni* zones is otherwise only known from the Hatchery Creek fauna of southeastern Australia, which must be post-Emsian (see above). In China the oldest *Bothriolepis* occurrences are dated as late Eifelian (see Chapter 10), and the Chinese species *B. shaokuanensis*, which resembles *B. portalensis*, approximates to the early Givetian *ensensis* conodont Zone (overlain by *varcus–disparilis* Zone conodonts according to Wang and Ziegler, 1983).

A younger age limit for the Aztec in the early Frasnian may be indicated by the phyllolepid placoderms which are dated radiometrically in southeastern Australia (see below), and links with southeastern Australia are strengthened with the recent recognition of the form *Austrophyllolepis* in Antarctica (J. Long, pers. comm.), but an older age limit (?Givetian) for Australian phyllolepids is not yet established. The Bunga Beds fauna from southeastern NSW has the early xenacanth shark *Antarctilamna* in common with the lower Aztec (Young, 1982), and may be of similar age. Associated acanthodians and crossopterygians from the Bunga Beds are undescribed. Thus specific constraints for the age limits of the Aztec sequence remain unresolved.

7. Tatong–South Blue Range fauna (NM)

The Tatong fish fauna is interpreted as the oldest vertebrate zone in the faunal succession of eastern Victoria proposed by Long and Werdelin (1986). The type locality for *Bothriolepis tatongensis* (Figure 9.4C) is near the top of the basal conglomerate succession of the Hollands Creek Rhyodacite, about 5 km southwest of Tatong, Victoria. The fauna of the South Blue Range near Mansfield (Hills, 1936b) is provisionally correlated with this zone because of the similar ornament and large size of the *Bothriolepis* remains (faunas f1 and f2 of Long, 1988a, Figure 2). It is possible that the Tatong fauna is slightly older, but further described taxa are needed to distinguish it clearly from the South Blue Range fauna. The phyllolepid from the latter was described by Long (1989b) as a new species *Austrophyllolepis edwini*. Associated are groenlandaspid, rhipidistian and acanthodian remains (Appendix 1).

Age constraints
As noted above, an older age limit for the appearance of phyllolepid placoderms in southeastern Australia is not yet established. Richards and Singleton (1981) dated post-Emsian intrusives which are older than the volcanics and fish beds of eastern Victoria at 381 ± 7 Ma, which could be anywhere in the Middle Devonian using the calibration of Young (1989a). However new isotopic dates for granites in southern NSW indicate that these dates may be too young (see below).

8. Taggerty–Howitt fauna (NM)

This is the zone characterised by the distinctive cristate antiarch *Bothriolepis gippslandiensis* (Figure 9.4A) in the analysis of Long (1983a). *Bothriolepis gippslandiensis* was described by Hills (1931), but Stensiö (1969) referred it to a new genus (*Hillsaspis*). The basis for this subsequently proved to be erroneous (Young and Gorter, 1981). More complete descriptions as a species of *Bothriolepis* (Long, 1983a; Long and Werdelin, 1986) are based on articulated specimens from the Mount Howitt locality in central Victoria (Avon River Group, 'lower mudstone' of Marsden, 1976), which was designated fauna f3 by Long (1988a, Figure 2). The type locality is in the Blue Range Formation, near Taggerty, Victoria. The fauna (Appendix 1) includes additional species of *Bothriolepis* (e.g. *B. cullodenensis*, Figure 9.4B; *B. fergusoni*, Figure 9.4E) and *Austrophyllolepis* (Figure 9.6G–I), rhipidistians, a palaeoniscoid, dipnoans and acanthodians (Long, 1983a, b, 1985a, 1986b, 1987c, 1988a, 1989b, 1992). In Figures 9.3, 9.5, 9.7 and 9.9. the fauna is divided into slightly older (8a, Taggerty) and younger (8b, Howitt) assemblages.

Age constraints
The Taggerty fish site is constrained by isotopically dated rhyolites giving an age of 367 ± 2 Ma (Williams *et al.*, 1982), regarded as early Frasnian in age (see Young, 1989a). This must be regarded as a minimum age (I. Williams, pers. comm.).

9. Bindaree Road–Freestone Creek fauna (NM)

The *cullodenensis* zone, based on the analysis of Long (1983b), is designated for the slightly younger Victorian faunas from Bindaree Road and Freestone Creek (lower and upper horizons within the Wellington Rhyolite). The species *Bothriolepis cullodenensis* (Figure 9.4B, F) is associated with *B. gippslandiensis* (Figure 9.4A) at Taggerty and the main fish locality at Mt Howitt, but the latter species is absent from the upper conglomerate unit in the Howitt section, and at the other localities. *Bothriolepis bindareei* is the sole endemic element of the upper conglomerate fauna. The Freestone Creek assemblages were suggested to be slightly younger by Long (1988a, faunas f5, f6), and are separated (9b)

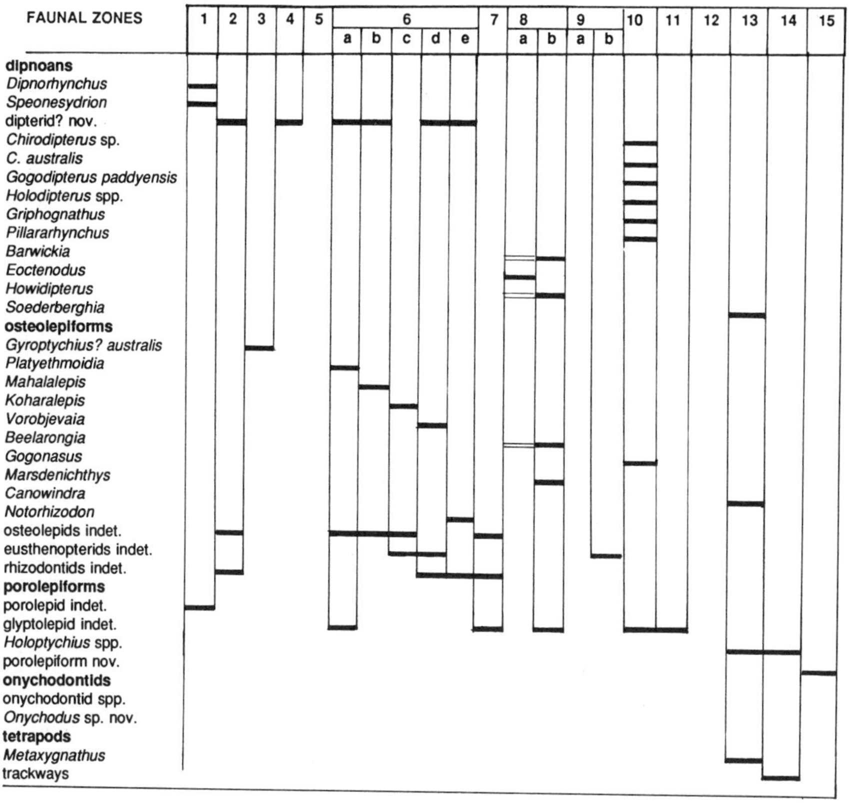

Figure 9.9. Stratigraphic distribution of various osteichthyans (dipnoans, rhipidistians, onychodontids) and tetrapods against a preliminary biozonation of 15 macrovertebrate faunas, numbered as in the text: 1, Taemas-Buchan fauna; 2, *Wuttagoonaspis* fauna; 3, Hatchery Creek fauna; 4, *Wurungulepis-Atlantidosteus* fauna; 5, *Nawagiaspis* fauna; 6, Aztec fauna (6a, *askinae* zone; 6b, *kohni* zone; 6c, *portalensis* zone; 6d, *karawaka* zone; 6e, *Pambulaspis*/phyllolepid zone); 7, Tatong-South Blue Range fauna; 8, Taggerty-Howitt fauna (8a, Taggerty assemblage; 8b, Howitt assemblage); 9, Bindaree Road-Freestone Creek fauna (9a, Bindaree Road assemblage; 9b, Freestone Creek assemblage); 10, Pambula River fauna; 11, Nettletons Creek fauna; 12, Twofold Bay fauna; 13, Jemalong-Canowindra fauna; 14, Worange Point fauna; 15 Grenfell fauna.

from the Bindaree Road assemblage (9a) in the figures. They may be more readily differentiated with further taxonomic study. The faunal lists include *Culmacanthus* sp., *Howqualepis* sp., *Groenlandaspis* spp., *Briagalepis warreni* (Long *et al.*, 1990; Figure 9.4H), *Austrophyllolepis* cf. *A. youngi* (an AVL plate provisionally referred to this species; Long, 1989), all occurring also in the previous fauna.

Age constraints

An older age limit for the Bindaree Road assemblage is based on correlation of the underlying main Howitt locality with the isotopically dated Taggerty fauna (see above). A younger age limit within the Frasnian may be inferred from a broad correlation between Freestone Creek and the diverse Pambula River fauna from facies 2 of the Boyd Volcanic Complex in southeastern NSW (Young, in Fergusson *et al.*, 1979; Young, 1983, 1989c), as discussed below. The Pambula River fauna is overlain by marine beds of the Bellbird Creek Formation (late Frasnian; see above).

10. Pambula River fauna (NM)

This diverse fauna (Appendix 1) is largely undescribed, but a provisional correlation with the previous two Victorian faunas is indicated by various common elements, including the acanthodian *Culmacanthus* (Figure 9.8C, closely related species; Young, 1989c), an undescribed *Bothriolepis* species with a large lateral pit closely similar to *B. cullodenensis*, other similar remains of *Bothriolepis*, osteolepid and glyptolepid rhipidistians (*Marsdenichthys* from Victoria; Long, 1985a), and phyllolepids (*Austrophyllolepis* in Victoria, the Pambula form apparently similar to the Freestone Creek material). However, there are differences with the better known main Howitt fauna which may indicate a slightly younger age, and/or differences in depositional environment, including the following: the apparent absence at Pambula River of dipnoans, which are diverse in the Howitt fauna; the presence of two asterolepid antiarchs (*Pambulaspis cobandrahensis*, Figure 9.4G, and a new genus, Young and Zhang, in press), and of climatiid acanthodian spines at Pambula River; the absence at Pambula River of crested bothriolepids, and *Groenlandaspis*; and the different species of *Bothriolepis* and *Culmacanthus*. Other elements reported in the Howitt fauna are acanthodiform acanthodians and palaeoniscoid osteichthyans, but these may be present but unrecognised in the Pambula River fauna, which comprises generally disarticulated remains. In Figure 9.2 this fauna is placed chronologically above the Freestone Creek fauna, with some overlapping taxa. It is considered to be younger than the fauna from the Bunga Beds (mentioned above under the Aztec fauna), which was also assigned to a sedimentary facies ('Flyschoid Facies 2') within the Boyd Volcanic Complex (Fergusson *et al.*, 1979). The Bunga Beds fauna is of limited diversity and lacks placoderms, so it is difficult to place within the scheme developed here.

The very significant Gogo fauna from the Canning Basin, WA (Appendix 1), which is associated with *asymmetricus* Zone conodonts (Druce, 1976), is provisionally referred to this zone, but as noted above, even diverse and well-studied faunas such as this are difficult to place within the essentially non-marine scheme worked out for southeastern Australia. Endemic higher taxonomic groups occurring at Gogo include the incisoscutid and camuropiscid arthrodires (Figure 9.6A; Long, 1988b), and several new species of placoderms and dipnoans await description at present (J. Long, pers. comm.). In the Carnarvon Basin the Munabia Sandstone has yielded a fragmentary fauna of three taxa (Appendix 1), which on the evidence of *Holonema* may be assumed equivalent in age to the Gogo fauna (Long, 1991a). A diverse macrovertebrate fauna from the underlying Gneudna Formation contains *Groenlandaspis* sp., eubrachythoracid arthrodires, *Chirodipterus australis* and other chirodipterids (Long, 1987c), and osteolepid jaws similar to *Gogonasus* (Long, 1985c) from the contemporaneous Gogo fauna (J. Long, pers. comm.).

Age constraints
A maximum age for the Devonian vertebrate faunas of the NSW south coast is provided by isotopically dated intrusives of the Bega Batholith which are nonconformably overlain by the Boyd Volcanic Complex. A Rb/Sr age of 381 ± 7 Ma (anywhere in the Middle Devonian using the timescale of Young, 1989a) was given by Fergusson *et al.* (1979) for the Kameruka Granodiorite of the Bega Batholith beneath the volcanics. A younger K/Ar age (363 ± 12) for the Gabo Island Granite implied a similar age for the volcanics and their contained fish faunas, which is consistent with geochemical evidence that the intrusives and volcanics were comagmatic (D. Wyborn, pers. comm.). However, this must now be questioned, because new data of 395 ± 4 Ma (95 per cent confidence level) by Williams (in press) on the Gabo Island Granite ('Watergum Granite', No. 1 of B.W. Chappell) makes the fish faunas unacceptably older (Emsian on the timescale of Young, 1989a). The older Kameruka Granodiorite is still older on the new zircon dates (about 420 Ma, at the older end of the emplacement ages for the Bega Batholith, but not well constrained; I. Williams, pers. comm.). On these new data the intrusives provide no meaningful older age constraint for the fish sequence. A younger age constraint is provided by the overlying Frasnian marine transgression of the Merrimbula Group.

11. Nettletons Creek fauna (NM)

Only the phyllolepid *Placolepis budawangensis* Ritchie 1984 (Figure 9.6F) has been described from this fauna. This taxon was regarded as a primitive phyllolepid (e.g. Long, 1984b; Young, 1987b), but has since been discovered in probable Famennian deposits in central Australia (Young, 1988b, table 3), and may be more derived in skull morphology than other known phyllolepids. Associated *Bothriolepis* material is similar to but probably not conspecific with *Bothriolepis* from Pambula River and Freestone Creek. The Nettletons Creek fauna is provisionally regarded as slightly younger than those just mentioned (Figure 9.2).

Age constraints
Marine horizons 920 m above the fish occurrence presumably represent the late Frasnian marine transgression, and an isotopic date of 355 ± 13 Ma for the Donovan Basic Complex (Wyborn and Owen, 1986) provides a younger

age constraint. The Comerong Volcanics unconformably overlie Ordovician, so there is no meaningful older age constraint.

12. Twofold Bay fauna (NM)

Another species of *Bothriolepis* occurs in the Twofold Bay Formation near Eden on the NSW south coast. Phyllolepid and possible remigolepid remains are associated.

Age constraints

The Twofold Bay Formation is the basal formation of the Merribula Group which overlies the Boyd Volcanic Complex. The fish fauna comes from near the top of the formation, and beneath the marine Bellbird Creek Formation, which provides a younger age limit (late Frasnian).

13. Jemalong–Canowindra fauna (NM)

Typical of this fauna is a *Bothriolepis*–phyllolepid–*Remigolepis*–*Groenlandaspis* association of placoderms in non-marine rocks (e.g. Hervey Group, Khan Yunis Formation, in central and southeastern New South Wales, Trapyard Hill fauna in eastern Victoria). The genus *Remigolepis* is represented by many species from southeastern Australia, all of which await description. Holoptychiid scales are commonly associated with the placoderms, and dipnoans and a probable tetrapod also occur (Appendix 1). A similar placoderm association occurs in central Australia (upper Dulcie Sandstone in the Georgina Basin; Young, 1988b). In the Canning Basin antiarchs (indeterminate) extend into the latest Famennian (Yellow Drum Sandstone), and a fragmentary assemblage from the Knobby Sandstone near Billiluna includes *Bothriolepis* and an asterolepid, probably *Remigolepis*, presumed to be of Famennian age (Young, 1987a). An anterior median dorsal plate of *Remigolepis* is recorded from the middle Famennian Piker Hills Formation (J. Long, pers. comm.), and the dinichthyid arthrodire *Westralichthys* (Figure 9.6B) is probably of similar age, based on associated conodonts (Long, 1987a, b). Phyllolepid placoderms have yet to be recorded from the west, and the only *Groenlandaspis* comes from the marine Gneudna Formation, of Frasnian age (Long, 1991a; these marine occurrences of *Remigolepis* and *Groenlandaspis* are to be expected, given their cosmopolitan distribution). *Bothriolepis* occurs with *Remigolepis* at most localities in the southeast, often fragmentary or disarticulated, but even well preserved articulated material from central NSW (Canowindra, Jemalong Range; see Long, 1991b, pl. 3A) remains undescribed.

Age constraints

Connolly (1965) provided the only detailed consideration of the stratigraphic relationships of the Upper Devonian sediments in central NSW.

Since that time preliminary studies of most occurrences of 'fishplates' he mentioned have shown *Remigolepis* to be of common occurrence. As noted above there is no confirmed occurrence in beds older than the late Frasnian marine transgression, which is represented in the eastern part of the outcrop. In the absence of contrary evidence all these occurrences are regarded as of Famennian age, which is consistent with the Canning Basin marine occurrence of *Remigolepis*, and assessments of the age of this taxon overseas (e.g. East Greenland; Stensiö, 1931). A younger age limit of latest Famennian may be indicated by palynomorphs of the *Retispora lepidophyta* palynofloral zone from western New South Wales, but these are provisionally regarded as dating the next highest zone (see below).

14. Worange Point fauna (NM)

The Worange Point Formation contains the youngest fauna in the Eden–Pambula region, with a *Bothriolepis–Remigolepis–Groenlandaspis* association, but lacking phyllolepids. Large *Remigolepis* are typical, but *Bothriolepis* remains tend to be less common and often fragmentary. Holoptychiid scales and other crossopterygian remains are common at some locations. A species of the arthrodire *Groenlandaspis* is characterised by unornamented dermal bones, and similar remains occur in an undescribed fauna from the Genoa River Beds on the Victorian border (A. Ritchie, pers. comm.), which is associated with the tetrapod trackway reported by Warren and Wakefield (1972). The Genoa River remains include large *Remigolepis* bones and *Holoptychius*(?) scales and teeth, and is presumed to be another occurrence of the Worange Point fauna. Similar scales (*Holoptychius* sp.) occur in the Jemalong–Canowindra fauna in central NSW, but the arthrodire *Groenlandaspis* is represented by a different species in this fauna. *Groenlandaspis* remains with unornamented dermal bones also occur at Grenfell, but the presence of a sinolepid antiarch and different porolepiform crossopterygian indicates that this is a younger fauna (see below).

Age constraints
A maximum age is provided by the underlying marine Bellbird Creek Formation (late Frasnian). The younger age limit is unconstrained, and there is no evidence to preclude extension into the Tournaisian. A younger age limit for the *Bothriolepis–Remigolepis* association, extending at least into the late Famennian, is suggested by palynomorphs from Pondie Range No. 1 well in western New South Wales, which Evans (1968) referred to the *Retispora lepidophyta* palynofloral zone of Playford (1976, 1982). The same horizon contains very fragmentary fish remains, including crossopterygians and antiarchs, the latter suggestive of the *Bothriolepis–Remigolepis* assemblage as it occurs in the Worange Point Formation. The *Retispora lepidophyta* palynofloral zone crosses the Devonian–Carboniferous boundary (Young, 1989a). The apparent absence of phyllolepids is taken to indicate a younger age than the placoderm assemblages of the Hervey Group

discussed above. The fact that phyllolepid remains occur lower in the south coast sequence, but not in the Worange Point fauna (Young, 1983), suggests that this fauna, and the Grenfell fauna discussed next, represent younger horizons than the upper biostratigraphic range of this taxon. As discussed below, the different crossopterygian, and absence of *Grenfellaspis*, suggest that the Worange Point fauna may be slightly older than the Grenfell fauna.

15. Grenfell fauna (NM)

This fauna (Appendix 1) from the Hervey Group in the Grenfell area (the Hunter Siltstone of Connolly, 1965) contains a diverse assemblage characterised by the sinolepid antiarch *Grenfellaspis branagani* (Ritchie *et al.*, 1992), which is associated with the antiarchs *Bothriolepis* and *Remigolepis* (undescribed species), and two species of the arthrodire *Groenlandaspis*, one with unornamented dermal bones, which also occurs in the Worange Point Formation (A. Ritchie, pers. comm.). Associated are crossopterygians (a holoptychiid porolepiform), and acanthodians (a new ischnacanthid genus; Long, 1991b).

Age constraints

Age control for this fauna is based solely on the evidence of the fish faunas. The Grenfell fauna is inferred to be slightly younger than the Worange Point fauna just described. As in that case, the absence of phyllolepid placoderms, which occur widely in other Famennian fish faunas in NSW, is considered of biostratigraphic significance, and the same species of *Groenlandaspis* in the two assemblages indicate a closely similar age. The holoptychiid porolepiforms in the two faunas may be readily distinguished by their different scale morphology. In view of the latest Famennian age inferred above for the remigolepid–phyllolepid placoderm association, an earliest Carboniferous age for the Grenfell assemblage cannot be excluded. It is worth noting that a sinolepid antiarch assemblage (including *Bothriolepis* and *Remigolepis*) represents the youngest Famennian non-marine vertebrate assemblage known from China (Pan, 1981; Pan and Dineley, 1988), an age supported by good palynological evidence (Ritchie *et al.*, 1992). Janvier *et al.* (1984) regarded *Groenlandaspis* as also extending possibly into the earliest Carboniferous.

ACKNOWLEDGEMENTS

I thank Prof. K.S.W. Campbell, Drs J.A. Long, A. Ritchie, P. Jones and S. Turner for advice on unpublished material and other assistance, and Dr J.A. Long for editorial advice and supplying Figures 9.4, 9.6 and 9.8. Published with the permission of the Director, Australian Geological Survey Organisation.

REFERENCES

Balme, B.E., 1988, Miospores from Late Devonian (early Frasnian) strata, Carnarvon Basin, Western Australia, *Palaeontographica*, **B209**: 109–66.

Becker, R.T., House, M.R. and Kirchgasser, W.T., in press, Devonian goniatite biostratigraphy and timing of facies movements in the Frasnian of the Canning Basin, Western Australia. In Hailwood, E.A. and Kidd, P.R. (eds), *High Resolution of Marine Sequences. Geological Society of London, Special Publication* (in press).

Blieck, A., 1984, *Les Heterostraces Pteraspidiformes, agnathes du Silurien–Devonien du continent nord-atlantique et de blocs avoisinants: revision systematique, phylogenie, biostratigraphie*, Cahiers de Paleontologie (section vertebres), Editions du CNRS, Paris, 199pp.

Boucot, A.J., 1975, *Evolution and extinction rate controls*, Elsevier, Amsterdam, 427pp.

Campbell, K.S.W. and Barwick, R.E., 1982, The neurocranium of the primitive dipnoan *Dipnorhynchus sussmilchi* (Etheridge), *Journal of Vertebrate Paleontology*, **2**: 286–327.

Campbell, K.S.W. and Barwick, R.E., 1983, Early evolution of dipnoan dentitions and a new genus *Speonesydrion*, *Memoirs of the Australasian Association of Palaeontologists*, **1**: 17–49.

Campbell, K.S.W. and Barwick, R.E., 1984, *Speonesydrion*, an Early Devonian dipnoan with primitive toothplates, *Palaeoichthyologica*, **2**: 1–48.

Campbell, K.S.W. and Barwick, R.E., 1985, An advanced massive dipnorhynchid lungfish from the Early Devonian of New South Wales, Australia, *Records of the Australian Museum*, **37**: 301–16.

Campbell, K.S.W. and Barwick, R.E., 1987, Paleozoic lungfishes—a review, *Journal of Morphology, Supplement*, **1**: 93–131.

Campbell, K.S.W. and Barwick, R.E., 1990, Paleozoic dipnoan phylogeny: functional complexes and evolution without parsimony, *Paleobiology*, **16**: 143–69.

Campbell, K.S.W. and Barwick, R.E., 1991, Teeth and toothplates in primitive lungfish and a new species of *Holodipterus*. In M.M. Chang, Y.H. Liu and G.R. Zhang (eds), *Early Vertebrates and Related Problems of Evolutionary Biology*, Science Press, Beijing, pp. 429–40.

Campbell, K.S.W. and Bell, M.W., 1977, A primitive amphibian from the Late Devonian of New South Wales, *Alcheringa*, **1**: 369–81.

Campbell, K.S.W. and Bell, M.W., 1982, *Soederberghia* (Dipnoi) from the Late Devonian of New South Wales, *Alcheringa*, **6**: 143–9.

Carroll, R.L., 1988, *Vertebrate Paleontology and Evolution*, Freeman & Co., New York.

Chapman, F., 1916, On the generic position of '*Asterolepis ornata* var. *australis*' McCoy, with description of a new variety, *Proceedings of the Royal Society of Victoria*, **28**: 211–15.

Connolly, J., 1965, The stratigraphy of the Hervey Group in central New South Wales, *Journal and Proceedings, Royal Society of New South Wales*, **98**: 37–83.

Dana, J.D., 1848, Fossils of the exploring expedition under the command of Charles Wilkes, USN: a fossil fish from Australia, and a belemnite from Tierra del Fuego, *American Journal of Science*, **5**: 433–5.

Denison, R.H., 1978, *Placodermi. Handbook of Paleoichthyology, volume 2*, H.-P. Schultze (ed.), Gustav Fischer Verlag, Stuttgart, New York.

Denison, R.H., 1979, *Placodermi. Handbook of Paleoichthyology, volume 5*, H.-P. Schultze (ed.), Gustav Fischer Verlag, Stuttgart, New York.

Dennis, K.D. and Miles, R.S., 1979a, A second eubrachythoracid arthrodire from Gogo, Western Australia, *Zoological Journal of the Linnean Society*, **67**: 1–29.

Dennis, K.D. and Miles, R.S., 1979b, Eubrachythoracid arthrodires with tubular rostral plates from Gogo, Western Australia, *Zoological Journal of the Linnean Society*, **67**: 297–328.

Dennis, K.D. and Miles, R.S., 1980, New durophagous arthrodires from Gogo, Western Australia, *Zoological Journal of the Linnean Society*, **69**: 43–85.

Dennis, K.D. and Miles, R.S., 1981, A pachyosteomorph arthrodire from Gogo, Western Australia, *Zoological Journal of the Linnean Society*, **73**: 213–58.

Dennis, K.D. and Miles, R.S., 1982, A eubrachythoracid arthrodire with a snub-nose from Gogo, Western Australia, *Zoological Journal of the Linnean Society*, **75**: 153–66.

Dennis-Bryan, K., 1987, A new species of eastmanosteid arthrodire (Pisces: Placodermi) from Gogo, Western Australia, *Zoological Journal of the Linnean Society*, **90**: 1–64.

Dennis-Bryan, K. and Miles, R.S., 1983, Further eubrachythoracid arthrodires from Gogo, Western Australia, *Zoological Journal of the Linnean Society*, **77**: 145–73.

Druce, E.C., 1970, Frasnian conodonts from Mount Morgan, Queensland, *Bureau of Mineral Resources, Australia, Bulletin*, **108**: 75–84.

Druce, E.C., 1974, Australian Devonian and Carboniferous conodont faunas. In Bouckaert J. and Streel, M. (eds), *International Symposium on Belgian Micropalaeontological Limits from Emsian to Visean. Namur, 1974, Publication* **5**: 1–18.

Druce, E.C., 1976, Conodont biostratigraphy of the Upper Devonian reef complexes of the Canning Basin, Western Australia, *Bureau of Mineral Resources, Australia, Bulletin*, **158**.

Etheridge, R., 1890, Note on the occurrence of fish-remains in the rock of the Drummond Range, central Queensland, *Records of the Geological Survey of New South Wales*, **2**: 71–2.

Etheridge, R., 1906, The cranial buckler of a dipnoan fish, probably *Ganorhynchus*, from the Devonian beds of the Murrambidgee River, New South Wales, *Records of the Australian Museum*, **6**: 129–32.

Evans, P.R., 1968, Upper Devonian and Lower Carboniferous miospores from the Mulga Downs Beds, NSW, *Australian Journal of Science*, **31**: 45–6.

Evernden, J.F. and Richards, J.R., 1962, Potassium-argon ages in eastern Australia, *Journal of the Geological Society of Australia*, **9**: 1–49.

Ferguson, W.H., 1917, The discovery of fossils in the Grampians Sandstones with general notes on the formation, *Records of the Geological Survey of Victoria*, **4**: 5–9.

Fergusson, C.L., Cas, R.A., Collins, W.J., Craig, G.Y., Crook, K.A.W., Powell, C.McA., Scott, P.A. and Young, G.C., 1979, The Late Devonian Boyd Volcanic Complex, Eden, NSW, *Journal of the Geological Society of Australia*, **26**: 87–105.

Gardiner, B.G. and Bartram, A.W.H., 1977, The homologies of ventral cranial fissures in osteichthyans. In S.M. Andrews, R.S. Miles and A.D. Walker (eds), *Problems in Vertebrate Evolution, Linnean Society Symposium Series*, **4**: 227–45.

Gardiner, B.G. and Miles, R.S., 1990, A new genus of eubrachythoracid arthrodire from Gogo, Western Australia, *Zoological Journal of the Linnean Society*, **99**: 159–204.

Garratt, M.J. and Wright, A.J., 1988, Late Silurian to Early Devonian biostratigraphy of southeastern Australia. In N.J. McMillan, A.F. Embry and D.J. Glass (eds), *Devonian of the World, Canadian Society of Petroleum Geologists, Calgary*, **III**: 647–62.

Grey, K., 1991, A mid-Givetian miospore age for the onset of reef development on the Lennard Shelf, Canning Basin, Western Australia, *Review of Palaeobotany and Palynology*, **68**: 37–48.

Gross, W., 1971, Unter devonische Thelodontier- und Acanthodier-Schuppen aus Westaustralien, *Paläontologischen Zeitschrift*, **45**: 97–106.

Helby, R.J. and McElroy, C.T., 1969, Microfloras from the Devonian and Triassic of the Beacon Group, Antarctica, *New Zealand Journal of Geology and Geophysics*, **12**: 376–82.

Henderson, G.M., 1980, Geology of Canberra, Queanbeyan and Environs. Notes to accompany the 1980 1:50000 Geological Map. *Bureau of Mineral Resources, Geology and Geophysics*, Canberra.

Hills, E.S., 1929, The geology and palaeontography of the Cathedral Range and Blue Hills in north western Gippsland, *Proceedings of the Royal Society of Victoria*, **41**: 176–201.

Hills, E.S., 1931, The Upper Devonian fishes of Victoria, Australia, and their bearing on the stratigraphy of the state, *Geological Magazine*, **68**: 206–31.

Hills, E.S., 1932, Upper Devonian fishes from New South Wales, *Quarterly Journal of the Geological Society of London*, **88**: 850–58.

Hills, E.S., 1936a, On certain endocranial structures in *Coccosteus*, *Geological Magazine*, **73**: 213–26.

Hills, E.S., 1936b, Records and descriptions of some Australian Devonian fishes, *Proceedings of the Royal Society of Victoria*, **48**: 161–71.

Hills, E.S., 1941, The cranial roof of *Dipnorhynchus sussmilchi* (Eth. fil.), *Records of the Australian Museum*, **21**: 45–55.

Hills, E.S., 1958, A brief review of Australian fossil vertebrates. In *Studies on fossil vertebrates*, T.S. Westoll (ed.), Athlone Press, London: 86–107.

Hills, E.S., 1959, Record of *Bothriolepis* and *Phyllolepis* (Upper Devonian) from the Northern Territory of Australia, *Journal and Proceedings of the Royal Society of New South Wales*, **92**: 174–5.

Janvier, P., 1985a, Environmental framework of the diversification of the Osteostraci during the Silurian and Devonian, *Philosophical Transactions of the Royal Society of London*, **B309**: 259–72.

Janvier, P., 1985b, *Les Cephalaspides du Spitsberg*, Cahiers de Paleontologie, Section Vertebres, Paris: Editions du Centre National de la Recherche Scientifique, 244pp.

Janvier, P., Lethiers, F., Monod, O. and Balkas, O., 1984, Discovery of a vertebrate fauna at the Devonian–Carboniferous boundary in south-eastern Turkey (Hakkari Province), *Journal of Petroleum Geology*, **7**: 147–68.

Johnson, J.G., Klapper, G. and Sandberg, C.A., 1985, Devonian eustatic fluctuations in Euramerica, *Geological Society of America Bulletin*, **96**: 567–87.

Kyle, R., 1977, Devonian palynomorphs from the basal Beacon Supergroup of south Victoria Land, Antarctica (note), *New Zealand Journal of Geology and Geophysics*, **20**: 1147–50.

Long, J.A., 1983a, A new diplacanthoid acanthodian from the Late Devonian of Victoria, *Memoirs of the Association of Australasian Palaeontologists*, **1**: 51–65.

Long, J.A., 1983b, New bothriolepid fish from the Late Devonian of Victoria, Australia, *Palaeontology*, **26**: 295–320.

Long, J.A., 1984a, New placoderm fishes from the Early Devonian Buchan Group, eastern Victoria, *Proceedings of the Royal Society of Victoria*, **96**: 173–86.

Long, J.A., 1984b, New phyllolepids from Victoria and the relationships of the group, *Proceedings of the Linnean Society of New South Wales*, **107**: 263–308.

Long, J.A., 1985a, The structure and relationships of a new osteolepiform fish from the Late Devonian of Victoria, Australia, *Alcheringa*, **9**: 1–22.

Long, J.A., 1985b, New information on the head and shoulder girdle of *Canowindra grossi*, from the Late Devonian Mandagery Sandstone, New South Wales, *Records of the Australian Museum*, **37**: 91–9.

Long, J.A., 1985c, A new osteolepid fish from the Upper Devonian Gogo Formation, Western Australia, *Records of the Western Australian Museum*, **12**: 361–77.

Long, J.A., 1986a, New ischnacanthid acanthodians from the Early Devonian of Australia, with comments on acanthodian interrelationships, *Zoological Journal of the Linnean Society of London*, **87**: 321–39.

Long, J.A., 1986b, A new Late Devonian acanthodian fish from Mt Howitt, Victoria, Australia, with remarks on acanthodian biogeography, *Proceedings of the Royal Society of Victoria*, **98**: 1–17.

Long, J.A., 1987a, A new dinichthyid fish (Placodermi: Arthrodira) from the Upper Devonian of Western Australia, with a discussion of dinichthyid interrelationships, *Records of the Western Australian Museum*, **13**: 515–40.

Long, J.A., 1987b, Upper Devonian conodonts associated with a large placoderm fish skull from the Canning Basin, Western Australia, *Records of the Western Australian Museum*, **13**: 510–13.

Long, J.A., 1987c, A redescription of the lungfish *Eoctenodus* Hills 1929, with reassessment of other Australian records of the genus *Dipterus* Sedgewick and Murchison 1828, *Records of the Western Australian Museum*, **13**: 297–314.

Long, J.A., 1987d, An unusual osteolepiform fish from the Late Devonian of Victoria, Australia, *Palaeontology*, **30**: 839–52.

Long, J.A., 1988a, New palaeoniscoid fishes (Osteichthyes, Actinopterygii) from the Late Devonian and Early Carboniferous of Victoria, *Memoirs of the Association of Australasian Palaeontologists*, **7**: 1–64.

Long, J.A., 1988b, Late Devonian fishes from the Gogo Formation, Western Australia, *National Geographic Research*, **4**: 436–50.

Long, J.A., 1988c, New information on the arthrodire *Tubonasus* from Gogo, Western Australia, *Memoirs of the Association of Australasian Palaeontologists*, **7**: 81–5.

Long, J.A., 1988d, A new camuropiscid arthrodire (Pisces: Placodermi) from Gogo, Western Australia, *Zoological Journal of the Linnean Society*, **94**: 233–58.

Long, J.A., 1989a, A new rhizodontiform fish from the Early Carboniferous of Victoria, Australia, with remarks on the phylogenetic position of the group, *Journal of Vertebrate Paleontology*, **9**: 1–17.

Long, J.A., 1989b, Phyllolepid placoderm fishes from the Late Devonian of Eastern Victoria. In R.W. Le Maitre (ed.), *Pathways in Geology: Essays in Honour of Edwin Sherbon Hills*, Blackwells Scientific Publications, 52–8.

Long, J.A., 1990a, Late Devonian chondrichthyans and other microvertebrate remains from northern Thailand, *Journal of Vertebrate Paleontology*, **10**: 59–71.

Long, J.A., 1990b, Two new arthrodires (placoderm fishes) from the Upper Devonian Gogo Formation, Western Australia, *Memoirs of the Queensland Museum*, **28**: 51–63.

Long, J.A., 1991a, Devonian fish remains from the Munabia Sandstone, Carnarvon Basin, Western Australia, *Records of the Western Australian Museum*, **15**: 503–15.

Long, J.A., 1991b, The long history of fossil fish on the Australian continent. In P. Vickers-Rich, J.N. Monaghan, R.F. Baird and T.H. Rich (eds), *Vertebrate Palaeontology of Australasia*, Pioneer Design Studios with Monash University Publications Committee, Melbourne: 336–428.

Long, J.A., 1992a, Cranial anatomy of two new Late Devonian lungfishes, from Mt Howitt, Victoria, *Records of the Australian Museum*.

Long, J.A., 1992b, *Gogodipterus paddyensis* (Miles) gen. nov., a new chirodipterid lungfish from the Late Devonian Gogo Formation, Western Australia, *The Beagle*, **14**.

Long, J.A., Burrett, C.F., Ngan, P.K. and Janvier, P., 1990, A new bothriolepid antiarch (Pisces: Placodermi) from the Devonian of Do Son peninsula, northern

Vietnam, *Alcheringa*, **14**: 181–94.

Long, J.A., Turner, S. and Young, G.C., 1988, A Devonian fish fauna from subsurface sediments in the eastern Officer Basin, South Australia, *Alcheringa*, **12**: 61–78.

Long, J.A. and Werdelin, L., 1986, A new Late Devonian bothriolepid (Placodermi, Antiarcha) from Victoria, with descriptions of other species from the state, *Alcheringa*, **10**: 355–99.

Long, J.A. and Young, G.C., 1988, Acanthothoracid remains from the Early Devonian of New South Wales, including a complete sclerotic capsule and pelvic girdle, *Memoirs of the Australasian Association of Palaeontologists*, **7**: 65–80.

Mark-Kurik, E., 1991, Contribution to the correlation of the Emsian (Lower Devonian) on the basis of placoderm fishes, *Newsletters on Stratigraphy*, **25**: 11–23.

Marsden, M.A.H., 1976, Upper Devonian–Carboniferous. In Douglas, J.G. and Ferguson, J.A. (eds), *Geology of Victoria, Special Publication of the Geological Society of Australia*, **5**: 77–124.

Mawson, R., Jell, J.S. and Talent, J.A., 1985, Stage boundaries within the Devonian: implications for application to Australian sequences, *Courier Forschungsinstitut Senckenberg*, **75**: 1–16.

Mawson, R., Talent, J.A., Bear, V.C., Benson, D.S., Brock, G.A., Farell, J.R., Hyland, K.A., Pyemont, B.D., Sloan, T.R., Sorentino, L., Stewart, M.I., Trotter, J.A., Wilson, G.A. and Simpson, A.G., 1988, Conodont data in relation to resolution of stage and zonal boundaries for the Devonian of Australia. In N.J. McMillan, A.F. Embry and D.J. Glass (eds), *Devonian of the World, Canadian Society of Petroleum Geologists, Calgary*, **III**: 485–527.

McCoy, F., 1876, *Asterolepis ornata* (Eichwald) var. *australis* (McCoy), *Geological Survey of Victoria. Prodromus of the Palaeontology of Victoria*, **4**: 19–20.

McDougall, I., Compston, W. and Bofinger, V.M., 1966, Isotopic age determinations on Upper Devonian rocks from Victoria, Australia: a revised estimate of the age of the Devonian–Carboniferous boundary, *Bulletin of the Geological Society of America*, **77**: 1075–8.

McElroy, C.T. and Rose, G., 1962, Reconnaissance geological survey: Ulladulla 1-mile military sheet, and southern part of Tianjara 1-mile military sheet, *Bulletin of the Geological Survey of New South Wales*, 17–67.

Miles, R.S., 1971, The Holonematidae (placoderm fishes), a review based on new specimens of *Holonema* from the Upper Devonian of Western Australia, *Philosophical Transactions of the Royal Society*, **263**: 101–234.

Miles, R.S., 1977, Dipnoan (lungfish) skulls and the relationships of the group: a study based on new specimens from the Devonian of Western Australia, *Zoological Journal of the Linnean Society*, **61**: 1–328.

Miles, R.S. and Dennis, K., 1979, A primitive eubrachythoracid arthrodire from Gogo, Western Australia, *Zoological Journal of the Linnean Society*, **66**: 31–62.

Miles, R.S. and Young, G.C., 1977, Placoderm interrelationships reconsidered in the light of new ptyctodontids from Gogo, Western Australia. In S.M. Andrews, R.S. Miles and A.D. Walker (eds), *Problems in Vertebrate Evolution*, *Linnean Society Symposium Series*, **4**: 123–98.

Neil, J.V., 1981, A slab of Devonian fossils from Heathcote and a little palaeoecology, *Victorian Naturalist*, **98**: 204–9.

Ørvig, T., 1969, Vertebrates from the Wood Bay Group and the position of the Emsian–Eifelian boundary in the Devonian of Vestspitsbergen, *Lethaia*, **2**: 273–328.

Packham, G.H., 1969, *The geology of New South Wales*, *Journal of the Geological Society of Australia*, **16**: 1–654.

Pan, J., 1981, Devonian antiarch biostratigraphy of China, *Geological Magazine*, **118**: 69–75.

Pan, J. and Dineley, D.L., 1988, A review of early (Silurian and Devonian) vertebrate biogeography and biostratigraphy of China, *Proceedings of the Royal Society of London*, **B235**: 29–61.

Pickett, J., 1972, Late Devonian (Frasnian) conodonts from Ettrema, New South Wales, *Journal and Proceedings of the Royal Society of New South Wales*, **105**: 31–7.

Pickett, J., 1980, Conodont assemblages from the Cobar Supergroup (Early Devonian), New South Wales, *Alcheringa*, **4**: 67–88.

Pickett, J., 1981, Marine fossils from the Cobargo 1:25,000 Sheet area, *Geological Survey of New South Wales, Palaeontological Report* 81/12 (unpublished).

Pickett, J., 1983, Late Devonian marine fossils from the Merrimbula Group, *Geological Survey of New South Wales, Palaeontological Report* 83/7 (unpublished).

Pickett, J.W. and McClatchie, L., 1991, Age and relations of stratigraphic units in the Murda Syncline area, *New South Wales Geological Survey, Quarterly Notes*, **85**: 9–32.

Playford, G., 1976, Plant microfossils from the Upper Devonian and Lower Carboniferous of the Canning Basin, Western Australia, *Palaeontographica*, **158B**: 1–71.

Playford, G., 1982, A latest Devonian palynoflora from the Buttons Beds, Bonaparte Gulf Basin, Western Australia, *Bureau of Mineral Resources Journal of Australian Geology and Geophysics*, **7**: 149–57.

Playford, G., 1983, The Devonian miospore genus *Geminospora* Balme 1962: a reappraisal based on topotypic *G. lemurata* (type species), *Memoirs of the Association of Australasian Palaeontologists*, **1**: 311–25.

Playford, P.E., 1980, Devonian 'Great Barrier Reef' of Canning Basin, Western Australia, *American Association of Petroleum Geologists Bulletin*, **64**: 814–40.

Powell, C. McA., 1983, Geology of NSW south coast, *Geological Society of Australia, Specialist Group in Tectonics and Structural Geology*, Field Guide No. 1.

Rade, J., 1964, Upper Devonian fish from the Mount Jack area, New South Wales, Australia, *Journal of Paleontology*, **38**: 929–32.

Richards, J.R. and Singleton, O.P., 1981, Palaeozoic Victoria, Australia: igneous rocks, ages and their interpretation, *Journal of the Geological Society of Australia*, **28**: 395–421.

Richardson, J.B. and McGregor, D.C., 1986, Silurian and Devonian spore zones of the Old Red Sandstone Continent and adjacent regions, *Geological Survey of Canada, Bulletin*, **364**.

Ritchie, A., 1969, Ancient fish of Australia, *Australian Natural History*, **16**: 218–23.

Ritchie, A., 1973, *Wuttagoonaspis* gen. nov., an unusual arthrodire from the Devonian of Western New South Wales, Australia, *Palaeontographica*, **143A**: 58–72.

Ritchie, A., 1975, *Groenlandaspis* in Antarctica, Australia and Europe, *Nature*, **254**: 569–73.

Ritchie, A., 1984, A new placoderm, *Placolepis* gen. nov. (Phyllolepidae) from the Late Devonian of New South Wales, Australia, *Proceedings of the Linnean Society of New South Wales*, **107**: 321–53.

Ritchie, A., Wang, S.T., Young, G.C. and Zhang, G.R., 1992, The Sinolepidae, a family of antiarchs (placoderm fishes) from the Devonian of South China and eastern Australia, *Records of the Australian Museum*.

Roberts, J., 1972, Correlation of the Upper Devonian rocks of Australia, *Journal of the Geological Society of Australia*, **18**: 467–90.

Sandberg, C.A. and Dreesen, R., 1984, Late Devonian icriodontid biofacies models and alternate shallow-water conodont zonations, *Geological Society of America, Special Paper*, **196**: 143–78.

Sandberg, C.A., Ziegler, W., Dreesen, R. and Butler, J.L., 1988, Late Frasnian mass extinction: conodont stratigraphy, global changes, and possible causes, *Courier Forschungsinstitut Senckenberg*, **102**: 267–307.

Sherwin, L., 1991, Siluro-Devonian biostratigraphy of central New South Wales, *New South Wales Geological Survey, Quarterly Notes*, **86**: 1–12.

Stensiö, E.A., 1931, Upper Devonian vertebrates from East Greenland, collected by the Danish Greenland Expedition in 1929 and 1930, *Meddelelser om Gronland*, **86**: 1-212.

Stensiö, E.A., 1969, *Elasmobranchiomorphi Placodermata Arthrodires*. In J. Piveteau (ed.), *Traite de Paleontologie*, 4(2): 71–692, Masson, Paris.

Streel, M., Higgs, K., Loboziak, S., Riegel, W. and Steemans, P., 1987, Spore stratigraphy and correlation with faunas and floras in the type marine Devonian of the Ardenne–Rhenish regions, *Review of Palaeobotany and Palynology*, **50**: 211–29.

Strusz, D.L., 1972, Correlation of the Lower Devonian rocks of Australasia, *Journal of the Geological Society of Australia*, **18**: 427–55.

Strzelecki, P.E., De, 1845, Physical Description of New South Wales and Van Diemen's Land: accompanied by a geological map, sections and diagrams and figures of organic remains, *Longman, Brown, Green and Longmans*, London, 462pp.

Talent, J., 1989, Transgression–Regression Pattern for the Silurian and Devonian of Australia. In R.W. Le Maitre (ed.), *Pathways in Geology: essays in honour of Edwin Sherbon Hills*, distributed by Blackwells, Melbourne: 201–19.

Talent, J.A. and Yolkin, E.A., 1987, Transgression-regression patterns for the Devonian of Australia and southern West Siberia, *Courier Forschungsinstitut Senckenberg*, **92**: 235–49.

Teichert, C., 1940, *Helicoprion* in the Permian of Western Australia, *Journal of Paleontology*, **14**: 140–9.

Teichert, C., 1943, Bradyodont sharks from the Permian of Western Australia, *American Journal of Science*, **241**: 534–52.

Teichert, C., 1949, Observations on the stratigraphy and palaeontology of the Devonian, western portion of the Kimberley Division, Western Australia, *Bureau of Mineral Resources, Report 2*.

Thomson, K.S., 1973, Observations on a new rhipidistian fish from the Upper Devonian of Australia, *Palaeontographica*, **143A**: 209–20.

Tomlinson, J.G., 1968, A new record of *Bothriolepis* in the Northern Territory of Australia, *Bureau of Mineral Resources, Australia, Bulletin*, **80**: 191–224.

Turner, S., 1982, Thelodonts and correlation. In P.V. Rich and E.M. Thompson (eds), *The Fossil Vertebrate Record of Australia*, Monash University Offset Printing Unit, Melbourne: 128–32.

Turner, S., 1986, Vertebrate fauna of the Silverband Formation, Grampians, western Victoria, *Proceedings of the Royal Society of Victoria*, **98**: 53–62.

Turner, S., 1991a, Palaeozoic Vertebrate Microfossils in Australia. In Rich, P., Monaghan, J.M. *et al.* (eds), *Vertebrate Palaeontology of Australasia*, Pioneer Design Studios, Lilydale, Victoria: 429–64.

Turner, S., Jones, P.J. and Draper, J.J., 1981, Early Devonian thelodonts (Agnatha) from the Toko Syncline, western Queensland, and a review of other Australian discoveries, *Bureau of Mineral Resources Journal of Australian Geology and Geophysics*, **6**: 51–69.

Turner, S. and Young, G.C., 1987, Shark teeth from the Early–Middle Devonian Cravens Peak Beds, Georgina Basin, Queensland, *Alcheringa*, **11**: 233–44.

Turner, S. and Young, G.C., 1992, Thelodont scales from the Middle–Late Devonian Aztec Siltstone, southern Victoria Land, Antarctica, *Antarctic Science*, **4**: 89–105.

Wang, C.Y. and Ziegler, W., 1983, Devonian conodont biostratigraphy of Guangxi, South China, and the correlation with Europe, *Geologica et Palaeontologica*, **17**: 75–107.

Warren, A., Jupp, R. and Bolton, B., 1986, Earliest tetrapod trackway, *Alcheringa*, **10**: 183–6.

Warren, J.W. and Wakefield, N.A., 1972, Trackways of tetrapod vertebrates from the Upper Devonian of Victoria, Australia, *Nature*, **228**: 469–70.

White, E.I., 1952, Australian Arthrodires, *Bulletin of the British Museum of Natural History (Geology)*, **1**: 249–304.

White, E.I., 1968, Devonian fishes of the Mawson-Mulock area, Victoria land, Antarctica, *Trans-Antarctic Expedition 1955–1958. Scientific Reports, Geology*, **16**: 1–26.

White, E.I., 1978, The larger arthrodiran fishes from the area of the Burrinjuck Dam, NSW, *Transactions of the Zoological Society of London*, **34**: 149–262.

Williams, I.S., Tetley, N.S., Compston, W. and McDougall, I., 1982, A comparison of K-Ar and Rb-Sr ages of rapidly cooled igneous rocks: two points in the Palaeozoic time scale re-evaluated, *Journal of the Geological Society of London*, **139**: 557–68.

Williams, I.S., in press, Some observations on the use of zircon uranium-lead geochronology in the study of granitic rocks, *Transactions of the Royal Society of Edinburgh*.

Withnall, I.W., Lang, S.C., Jell, J.S., McLennan, T.P.T., Talent, J.A., Mawson, J.A., Fleming, P.J.G., Law, S.R., Macansh, J.D., Savory, P., Kay, J.R. and Draper, J.J., 1988, Stratigraphy, sedimentology, biostratigraphy and tectonics of the Ordovician to Carboniferous, Broken River Province, north Queensland, *Australasian Sedimentologists Group Field Guide Series*, 5, Geological Society of Australia, Sydney.

Woodward, A.S., 1902, Preliminary note on a Carboniferous fish fauna from Victoria, *Geological Magazine*, **9**: 471–3.

Woodward, A.S., 1906, On a Carboniferous fish fauna from the Mansfield district, Victoria, *Memoirs of the National Museum, Melbourne*, **1**: 1–32.

Woodward, A.S., 1921, Fish-remains from the Upper Old Red Sandstone of Granite Harbour, Antarctica, *British Antarctic ('Terra Nova') Expedition, 1910. Natural History Report (Geology)*, **1**: 51–62.

Woodward, A.S., 1941, The headshield of a new macropetalichthyid fish (*Notopetalichthys hillsi*, gen. et sp. nov.) from the Middle Devonian of Australia, *Annals and Magazine of Natural History, Series 2*, **8**: 91–6.

Wright, A.J., Pickett, J.W., Sewell, D., Roberts, J. and Jenkins, T.B.H., 1990, Corals and conodonts from the Late Devonian Mostyn Vale Formation, Keepit, New South Wales, *Memoirs of the Association of Australasian Palaeontologists*, **10**: 211–54.

Wyatt, D.H. and Jell, J.S., 1980, Devonian and Carboniferous stratigraphy of the northern Tasman orogenic zone in the Townsville hinterland, North Queensland. In R.A. Henderson and P.J. Stephenson (eds), *The Geology and Geophysics of northeastern Australia*, Geological Society of Australia, Queensland Division: 201–28.

Wyborn, D. and Owen, M., 1986, Araluen, New South Wales. 1:100000 map commentary, *Bureau of Mineral Resources, Geology and Geophysics*, Canberra.

Wyborn, D., Owen, M., Compston, W. and McDougall, I., 1982, The Laidlaw Volcanics: a Late Silurian point on the geological time scale, *Earth and Planetary Science Letters*, **59**: 90–100.

Young, G.C., 1974, Stratigraphic occurrence of some placoderm fishes in the Middle and Late Devonian, *Newsletters on Stratigraphy*, **3**: 243–61.

Young, G.C., 1978, A new Early Devonian petalichthyid fish from the Taemas/Wee Jasper region of New South Wales, *Alcheringa*, **2**: 103–16.

Young, G.C., 1979, New information on the structure and relationships of *Buchanosteus* (Placodermi, Euarthrodira) from the Early Devonian of New South Wales, *Zoological Journal of the Linnean Society*, **66**: 309–52.

Young, G.C., 1980, A new Early Devonian placoderm from New South Wales, Australia, with a discussion of placoderm phylogeny, *Palaeontographica*, **167A**: 10–76.

Young, G.C., 1981, New Early Devonian brachythoracids (placoderm fishes) from the Taemas–Wee Jasper region of New South Wales, *Alcheringa*, **5**: 247–71.

Young, G.C., 1982, Devonian sharks from southeastern Australia and Antarctica, *Palaeontology*, **25**: 817–43.

Young, G.C., 1983, A new antiarchan fish (Placodermi) from the Late Devonian of southeastern Australia, *Bureau of Mineral Resources Journal of Australian Geology and Geophysics,* **8**: 71–81.

Young, G.C., 1984, An asterolepidoid antiarch (placoderm fish) from the Early Devonian of the Georgina Basin, central Australia, *Alcheringa*, **8**: 65–80.

Young, G.C., 1985a, New discoveries of Devonian vertebrates from the Amadeus Basin, central Australia, *Bureau of Mineral Resources Journal of Australian Geology and Geophysics,* **9**: 239–254.

Young, G.C., 1985b, Further petalichthyid remains (placoderm fishes, Early Devonian) from the Taemas–Wee Jasper region of New South Wales, *Bureau of Mineral Resources Journal of Australian Geology and Geophysics*, **9**: 121–31.

Young, G.C., 1987a, Devonian fish remains from Billiluna, eastern Canning Basin, Western Australia, *Bureau of Mineral Resources Journal of Australian Geology and Geophysics,* **10**: 179–92.

Young, G.C., 1987b, Devonian palaeontological data and the Armorica problem, *Palaeogeography, Palaeoecology, Palaeoclimatology*, **60**: 283–304.

Young, G.C., 1988a, Antiarchs (placoderm fishes) from the Devonian Aztec Siltstone, southern Victoria Land, Antarctica, *Palaeontographica*, **202A**: 1–125.

Young, G.C., 1988b, New occurrences of phyllolepid placoderms from the Devonian of central Australia, *Bureau of Mineral Resources Journal of Australian Geology and Geophysics*, **10**: 363–76.

Young, G.C., 1989a, Australian Phanerozoic Timescales 4. Devonian. Biostratigraphic chart and explanatory notes, *Bureau of Mineral Resources, Australia, Record*, 1989/34.

Young, G.C., 1989b, The Aztec fish fauna of southern Victoria Land—evolutionary and biogeographic significance. In Crame, J.A. (ed.), *Origins and Evolution of the Antarctic Biota, Geological Survey of London Special Publication*, **47**: 43–62.

Young, G.C., 1989c, New occurrences of culmacanthid acanthodians (Pisces, Devonian) from Antarctica and southeastern Australia, *Proceedings of the Linnean Society of New South Wales*, **111**: 12–25.

Young, G.C., 1990a, Devonian vertebrate distribution patterns, and cladistic analysis of palaeogeographic hypotheses. In W.S. McKerrow and C.R. Scotese (eds), *Palaeozoic biogeography and palaeogeography*, *Geological Society of London Memoir*, **12**: 243–55.

Young, G.C., 1990b, New antiarchs (Devonian placoderm fishes) from Queensland, with comments on placoderm phylogeny and biogeography, *Memoirs of the Queensland Museum*, **28**: 35–50.

Young, G.C., 1991a, Fossil fishes from Antarctica. In R.J. Tingey (ed.), *The Geology of Antarctica*, Oxford University Press, pp. 538–67.

Young, G.C., 1991b, The first armoured agnathan vertebrates from the Devonian of Australia. In Chang, M.M., Liu, Y.H. and Zhang, G.R. (eds), *Early Vertebrates and Related Problems of Evolutionary Biology*, Science Press, Beijing, China: 67–85.

Young, G.C. and Gorter, J.D., 1981, A new fish fauna of Middle Devonian age from the Taemas/Wee Jasper region of New South Wales, *Bureau of Mineral Resources Geology and Geophysics, Bulletin*, **209**: 83–147.

Young, G.C., Long, J.A. and Ritchie, A.R., 1992, Crossopterygian fishes from the Devonian of Antarctica: systematics, relationships and biogeographic significance, *Records of the Australian Museum Supplement*, **14**: 1–77.

Young, G.C., Turner, S., Owen, M., Nicoll, R.S., Laurie, J. and Gorter, J., 1987, A new Devonian fish fauna, and revision of post-Ordovician stratigraphy in the Ross River Syncline, Amadeus Basin, central Australia, *Bureau of Mineral Resources Journal of Australasian Geology and Geophysics*, **10**: 233–42.

Young, G.C. and Zhang, G.R, in press, A new antiarch (placoderm fish; Devonian) from the south coast of New South Wales, *Proceedings of the Linnean Society of New South Wales*.

Ziegler, W., 1962, Taxionomie und Phylogenie Oberdevonischer Conodonten und ihre stratigraphische Bedeutung, *Notizblatt Hessisches Landesamt. Bodenforsch*, **38**: 1–166.

Ziegler, W., 1971, Conodont stratigraphy of the European Devonian. In Sweet, W.C. and Bergstrom, S.M. (eds), *Symposium of Conodont biostratigraphy, Memoir of the Geological Society of America*, **127**: 227–84.

Ziegler, W. and Klapper, G., 1985, Stages of the Devonian System, *Episodes*, **8**: 104–9.

Ziegler, W. and Sandberg, C.A., 1984, *Palmatolepis*-based revision of upper part of standard Late Devonian conodont zonation, *Geological Society of America, Special Paper*, **196**: 179–94.

Ziegler, W. and Sandberg, C.A., 1990, The Late Devonian standard conodont zonation, *Courier Forschungsinstitut Senckenberg*, **121**: 1–115.

APPENDIX 1: FAUNAL LISTS OF EASTERN GONDWANA DEVONIAN MACROVERTEBRATE ASSEMBLAGES

Compiled by G.C. Young, J.A. Long and S. Turner.

Microvertebrates [M] are dealt with by Turner (Chapter 8).

All cited references can be found at the ends of Chapters 8 and 9.

1. TAEMAS–BUCHAN FAUNA

Acanthodii

Cheiracanthoides comptus Giffin 1980 [M]

Taemasacanthus erroli Long 1986a

climatiid indet.

Placodermi

Buchanosteus confertituberculatus (Chapman 1916)

Burrinjucosteus asymmetricus White 1978

Goodradigbeeon australianum White 1978

Taemasosteus novaustrocambricus White 1952

Toombsosteus denisoni White 1978

Arenipiscis westolli Young 1981

Errolosteus goodradigbeensis Young 1981

Williamsaspis bedfordi White 1952

holonematid nov.

homosteid nov.

eubrachythoracid nov.

Weejasperaspis gavini White 1978
Brindabellaspis stensioi Young 1980
Murrindalaspis wallacei Long 1984
M. bairdi Long 1984
Notopetalichthys hillsi Woodward 1941
Wijdeaspis warrooensis Young 1978
Shearsbyaspis oepiki Young 1985b
Lunaspis sp. Young 1985b
ptyctodontid gen. nov.
'*Ohioaspis tumulosa*' Giffin 1980 [M]
Chondrichthyes
Ohiolepis Giffin 1980 [M]
'*Skamolepis*' Giffin 1980 [M]
Osteichthyes
Dipnorhynchus sussmilchi Etheridge 1906
D. kiandrensis Campbell and Barwick 1982
D. kurikae Campbell and Barwick 1985
Speonesydrion iani Campbell and Barwick 1983
Ligulalepis toombsi Schultze 1968 [M]
onychodontid indet. Ørvig 1969
porolepiform indet. Young 1985b

2. *WUTTAGOONASPIS* FAUNA
Agnatha
Turinia australiensis Turner *et al.* 1981 [M]
T. cf. *pagei* Turner *et al.* 1981 [M]
Gampsolepis? sp. Turner *et al.* 1981 [M]
Pituriaspis doylei Young 1991b
Neeyambaspis enigmatica Young 1991b
Acanthodii
Nostolepis Turner *et al.* 1981 [M]
Gomphonchus Turner *et al.* 1981 [M]
Machaeracanthus [M]
Placodermi
Wuttagoonaspis fletcheri Ritchie 1973
Wuttagoonaspis spp. Ritchie 1973
actinolepid nov. Ritchie 1973
tiaraspid nov. Ritchie 1973
groenlandaspid nov. Ritchie 1973
?homostiid nov.
pterichthyodid indet. Young 1984
Chondrichthyes
Mcmurdodus whitei Turner and Young 1987 [M]
cf. *Gualepis* Turner 1991 [M]
Osteichthyes
onychodontid indet.
rhizodontid indet.

3. HATCHERY CREEK FAUNA
Agnatha
Turinia cf. *hutkensis* Young and Gorter 1981 [M]
Placodermi
Monarolepis verrucosa (Young and Gorter 1981)
Sherbonaspis hillsi Young and Gorter 1981
Denisonosteus weejasperensis Young and Gorter 1981
Acanthodii
climatiid indet. Young and Gorter 1981
ischnacanthid indet. Young and Gorter 1981
Osteichthyes
Gyroptychius? australis Young and Gorter 1981

4. *WURUNGULEPIS–ATLANTIDOSTEUS* FAUNA
Acanthodii
?Cheiracanthoides [M]
?Machaeracanthus [M]
Placodermi
Wurungulepis denisoni Young 1990b
Atlantidosteus sp.
eubrachythoracid nov.
Osteichthyes
dipterid? indet.

5. NAWAGIASPIS FAUNA
Placodermi

Nawagiaspis wadeae Young 1990b

6. AZTEC FAUNA

Agnatha

Turinia antarctica Turner and Young 1992 [M]

Acanthodii

Cheiracanthoides comptus (J. Long, obs.)
Gyracanthides warreni White 1968
Antarctonchus glacialis White 1968
Byssacanthoides debenhami Woodward 1921
Culmacanthus antarctica Young 1989c

Placodermi

Bothriolepis antarctica Woodward 1921
Bothriolepis alexi Young 1988a
Bothriolepis askinae Young 1988a
Bothriolepis barretti Young 1988a
Bothriolepis karawaka Young 1988a
Bothriolepis kohni Young 1988a
Bothriolepis macphersoni Young 1988a
Bothriolepis mawsoni Young 1988a
Bothriolepis portalensis Young 1988a
Bothriolepis vuwae Young 1988a
Bothriolepis sp. indet. 1–13
Pambulaspis antarctica Young 1988a
Antarctolepis gunni White 1968
Groenlandaspis antarcticus Ritchie 1975
Groenlandaspis spp.
phlyctaeniid spp. nov.
Austrophyllolepis sp.
Antarctaspis mcmurdoensis White 1968

Chondrichthyes

Mcmurdodus featherensis White 1968
Antarctilamna prisca Young 1982
xenacanthid gen. nov. 1
xenacanthid gen. nov. 2

Osteichthyes

?ctenodontid Campbell and Barwick 1987
?Eoctenodus sp. (J. Long, obs.)
Gyroptychius? antarcticus (Woodward 1921)
Koharalepis jarviki Young *et al.* 1992
Mahalalepis resima Young *et al.* 1992
Platyethmoidia antarcticus Young *et al.* 1992
Vorobjevaia dolonodon Young *et al.* 1992
Notorhizodon mackelveyi Young *et al.* 1992
eusthenopterid indet.
rhizondontid indet.
Palaeoniscoid gen. nov.
?palaeoniscoid indet.

HARAJICA FAUNA

Placodermi

Bothriolepis spp.
phyllolepid

Osteichthyes

dipnoan indet.
osteolepid indet.

Acanthodii

Striacanthus sp.
climatiid indet.

7. TATONG–SOUTH BLUE RANGE FAUNA

TATONG

Placodermi

Bothriolepis tatongensis Long and Werdelin 1986

Osteichthyes

osteolepiform gen. indet.

SOUTH BLUE RANGE

Acanthodii

Striacanthus sp.

Placodermi

Bothriolepis cf. *B. tatongensis* Long and Werdelin 1986
Austrophyllolepis edwini Long 1989b
Groenlandaspis sp.

Osteichthyes
Osteolepiform gen. nov.

8. TAGGERTY–HOWITT FAUNA
TAGGERTY
Placodermi
Bothriolepis gippslandiensis Hills 1929
Austrophyllolepis cf. *A. ritchiei* Long 1984b
Osteichthyes
Eoctenodus microsoma Hills 1929

MT HOWITT
Acanthodii
Howittacanthus kentoni Long 1986a
Culmacanthus stewarti Long 1983b
Placodermi
Bothriolepis gippslandiensis Hills 1929
Bothriolepis cullodenensis Long 1983b
Bothriolepis fergusoni Long 1983b
Austrophyllolepis ritchiei Long 1984b
Austrophyllolepis youngi Long 1984b
Groenlandaspis sp. nov.
Osteichthyes
Marsdenichthys longioccipitus Long 1985a
Beelarongia patrichae Long 1987d
Howqualepis rostridens Long 1988a
Howidipterus donnae Long 1992a
Barwickia downunda Long 1992a
Porolepiformes cf. *?Glyptolepis*
Actinistian

9. BINDAREE ROAD–FREESTONE CREEK FAUNA
BINDAREE ROAD
Placodermi
Bothriolepis bindareei Long 1983b
Briagalepis warreni (Long 1983b)
phyllolepid indet.
Groenlandaspis sp.

FREESTONE CREEK
Acanthodii
Striacanthus sicaeformis Hills 1931
Culmacanthus cf. *C. stewarti* Long 1983a
Placodermi
Bothriolepis cullodenensis Long 1983b
Briagalepis warreni (Long 1983b)
Austrophyllolepis cf. *A. youngi* Long 1984b
Groenlandaspis spp.
Osteichthyes
Howqualepis cf. *H. rostridens* Long 1988a
Eusthenopterid cf. *Marsdenichthys* Long 1985a

10. PAMBULA RIVER FAUNA
PAMBULA RIVER
Placodermi
Bothriolepis sp.
Pambulaspis cobandrahensis Young 1983
new genus antiarch
cf. *Austrophyllolepis*
Osteichthyes
cf. *Marsdenichthys* Long 1985a
Acanthodii
Culmacanthus pambulensis Young 1989c
Striacanthus sp.
climatiid indet.

GOGO FORMATION, W.A.
Placodermi
Holonema westolli Miles 1971
Harrytoombsia elegans Miles and Dennis 1979
Kimberleyichthys whybrowi Dennis-Bryan and Miles 1983
K. bispicatus Dennis-Bryan and Miles 1983
Camuropiscis concinnus Dennis and Miles 1979a
C. laidlawi Dennis and Miles 1979a
Tubonasus lennardensis Dennis and

Miles 1979b, see also Long 1988c
Rolfosteus canningensis Dennis and Miles 1979b
Latocamurus coulthardi Long 1988d
Fallacosteus turnerae Long 1990b
Incisoscutum ritchiei Dennis and Miles 1981
new genus incisoscutid (J. Long)
Eastmanosteus calliaspis Dennis-Bryan 1987
Kendrickichthys cavernosus Dennis and Miles 1980
Simosteus tuberculatus Dennis and Miles 1980
Bruntonichthys multidens Dennis and Miles 1980
Bullerichthys fascidens Dennis and Miles 1980
Pinguosteus thulborni Long 1990b
Torosteus pulchellus Gardiner and Miles 1990
Torosteus tuberculatus Gardiner and Miles 1990
Torosteidae new genus (J. Long)
Bothriolepis sp.
Campbellodus decipiens Miles and Young 1977
Ctenurella gardineri Miles and Young 1977
Osteichthyes
Actinopterygii
Mimia toombsi Gardiner and Bartram 1977
Moythomasia durgaringa Gardiner and Bartram 1977
Crossopterygii
Onychodontiformes
Onychodus sp. nov. Andrews in prep.
Osteolepiformes
Gogonasus andrewsae Long 1985c
Dipnoi
Chirodipterus australis Miles 1977
Gogodipterus paddyensis (Miles) Long 1992b
Pillararhynchus longi Campbell and Barwick 1990
Holodipterus gogoensis Miles 1977
Holodipterus longi Campbell and Barwick 1991
Griphognathus whitei Miles 1977

GNUEDNA FM, W.A.
Agnatha
Australolepis seddoni Turner and Dring 1981
Placodermi
Groenlandaspis sp.
Eubrachythoraci indet.
Osteichthyes
Chirodipterus australis Miles 1977
chirodipterid indet.
Crossopterygii
cf. *Gogonasus*
Palaeoniscoidei [M]
onychodont indet.
Acanthodii indet.

MANUBIA SANDSTONE, W.A.
Placodermi
Bothriolepis sp.
Holonema sp.
Osteichthyes
Osteolepiformes indet.

11. NETTLETONS CREEK FAUNA
Acanthodii
'*Striacanthus*' sp.
Placodermi
Bothriolepis sp.
Placolepis budawangensis Ritchie 1984
Osteichthyes
glyptolepid

12. TWOFOLD BAY FAUNA
Placodermi
Bothriolepis sp.
?remigolepid
phyllolepid

13. JEMALONG–CANOWINDRA FAUNA

Placodermi
Bothriolepis sp.
Remigolepis sp.
Groenlandaspis sp.
Osteichthyes
Crossopterygii
Canowindra grossi Thompson 1973

14. WORANGE POINT FAUNA

Placodermi
Bothriolepis sp.
Remigolepis sp.
Groenlandaspis sp.
Osteichthyes
Holoptychius? sp.

15. GRENFELL FAUNA

Acanthodii
Ischnacanthid new genus
Placodermi
Grenfellaspis branagani Ritchie *et al.* 1992
Bothriolepis sp.
Remigolepis sp.
Groenlandaspis spp.
Osteichthyes
?Eusthenodon sp.
Ctenodus?
porolepiform gen. nov.

Chapter 10

VERTEBRATE BIOSTRATIGRAPHY OF THE MIDDLE PALAEOZOIC OF CHINA*

Wang, S.-T.

INTRODUCTION

In this chapter the vertebrate biostratigraphy of the Silurian of South China, the Devonian throughout China, and the Lower Carboniferous of Guizhou Province will be discussed. In recent years there has been much new work resulting in many new discoveries including both macrovertebrates and microvertebrates. Notable recent finds of macrovertebrates include the discovery of Silurian and Early Devonian antiarchs (e.g. *Silurolepis*, *Dayaoshania*), which increases the members of the endemic family Sinolepidae; primitive lungfish in the late Lower Devonian in South China (*Sorbitorhynchus*, in Campbell and Barwick, 1990), and the agnathan *Dayongaspis* has shown that the endemic group Eugaleaspida occurred in the Lower to Middle Silurian (Pan, 1985, 1986). Microvertebrate faunas have only recently been described in detail from China. Discoveries in the Silurian and Devonian of China such as *Nostolepis striata*, *Nostolepis* sp., *Gomphonchus* sp., *Ohiolepis* sp., *Ohioaspis* sp., *Ligulalepis yunnanensis*, *Thelodus sinensis*, *Turinia asiatica*, *T. pagoda*, *Gualepis*, *Changolepis*, *Naxilepis*, *Hanilepis*, *Phoebodus* and *Thrinacodus* (Wang, N., 1984; Wang and Dong, 1989, Wang *et al.*, 1986; Wang and Turner, 1985) extend the age range of some taxa and enable refinement of biostratigraphic correlations within China and to neighbouring terranes.

PALAEOENVIRONMENTAL SETTINGS

Silurian vertebrates of China occur mostly in marine sequences with few fishes occurring in non-marine sandstones (Pan and Dineley, 1988).

* A contribution to IGCP Project 328.

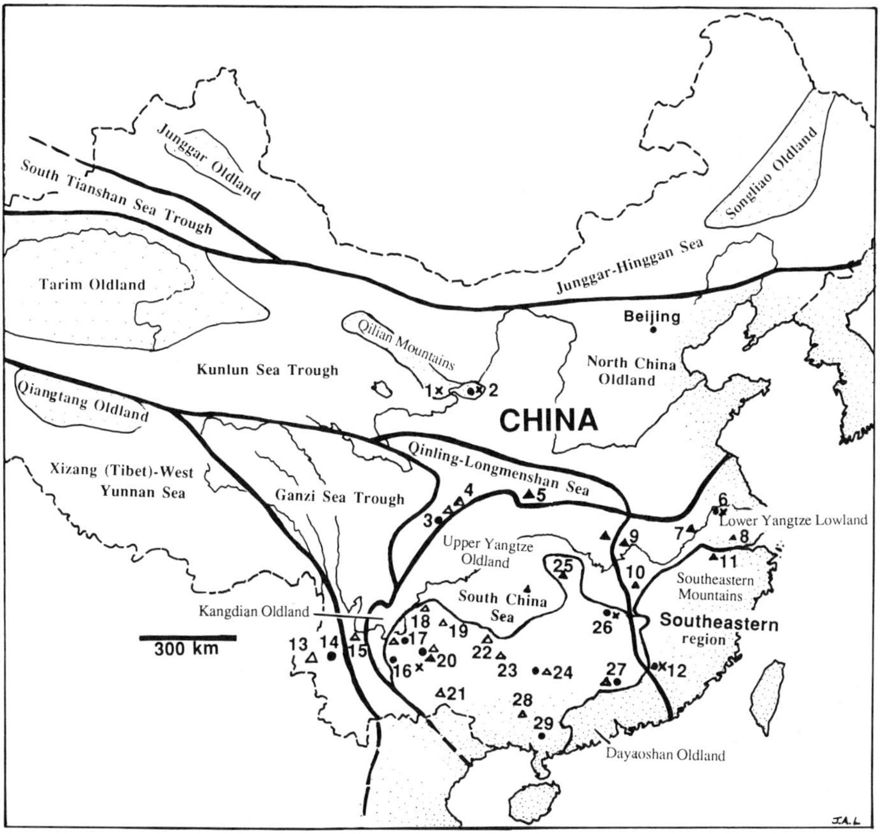

Figure 10.1. The Devonian paleogeography and distribution of Silurian and Devonian vertebrates in China. ▲ - Silurian, △ - Early Devonian, ● - Middle Devonian, X - Late Devonian. Localities: 1, Jingtai; 2, Zhongwei, Zhongning; 3, Jiangyou; 4, Yanmenba; 5, Ziyang; 6, Nanjing; 7, Chaoxian; 8, Changxing; 9, Wuhan; 10, Xiushui; 11, Ningguo; 12, Yudu; 13, Tengchong; 14, Shidian; 15, Lijiang; 16, Kunming; 17, Wuding; 18, Zhaotong; 19, Hezhang; 20, Qujing; 21, Guangnan; 22, Guiyang; 23, Duyun; 24, Xiangzhou; 25, Dayong; 26, Liling; 27, Pingle; 28, Liujing; 29, Bobai.

Figure 10.1 shows the major Siluro–Devonian vertebrate localities and their palaeogeographic settings. The Middle Silurian agnathans *Hangyangaspis* and *Latirostraspis* occur in the marine yellowish-green siltstone and mudstone associated with a marine invertebrate fauna containing trilobites and brachiopods, or are intercalated with marine lithologies. The Upper Silurian agnathans *Sinogaleaspis* and *Xiushuiaspis* occur in shallow marine facies in north Jiangxi, south Anhui and west Zhejiang (Figures 10.2, 10.3).

According to the palaeogeographic and tectonic framework of Pan and

Dineley (1988) there are no sediments containing fishes in the Lower–Middle Devonian in the Middle–Lower Yangtze valley (north Jiangxi, south Anhui and west Zhejiang) as there was uplift of mountain belts here after the Upper Silurian. In the Upper Devonian (Famennian) there are some special antiarch placoderms, *Sinolepis* and *Jiangxilepis* (Zhang and Liu, 1991), occurring in the deltaic facies of this region. Although the cosmopolitan genus *Asterolepis* was described from this area by Pan Jiang in 1958, recent new material from the Upper Devonian of south Jiangxi, consisting of many specimens, indicates that it constitutes a new genus and that *Asterolepis* is not present here. Zhang and Liu (1991) believe the new form, *Jianxilepis*, should be placed in a new family, Jiangxilepidae, and differs from *Asterolepis* in its longer pectoral fins, large armour plates of the pectoral fin and larger postpineal plate of the skull roof.

In Qujing of east Yunnan, Longmenshan district of Sichuan, Xiangzhou and Liujing area of Guangxi, the Lower Devonian vertebrate fauna, which includes *Yunnanolepis*, *Lianhuashanolepis*, *Orientolepis*, *Qujinolepis*, and *Liujiangolepis*, occur in the marine facies which were littoral environments having lingulid brachiopods occurring with the fishes. In the Wudang Formation of Guizhou Province the brachiopod *Acrospirifer* is associated with *Kuiechowlepis*, *Sinopetalichthys* and antiarchans. Similarly, in the upper part of the Danlin Formation of Guizhou Province the agnathan *Duyunolepis paoyangensis* is associated with brachiopods *Howellella* and *Acrospirifer*. However, in the Hexian Formation (upper part of Dayaoshan Group) of Guangxi there are no marine invertebrate fossils associated with the vertebrate faunas (that includes *Dayaoshania*, a new sinolepid), perhaps because the facies represents non-marine or lagoonal environments. The author has described the Lower Devonian palaeocommunities of South China (Wang, S., 1991) which comprise ten palaeocommunities from Lianhuashanian to Sipaian (Lochkovian–Emsian equivalents). Most of these represent littoral to shallow marine environments, with a few representing non-marine environments.

Most of the well-known Middle Devonian vertebrates (see Figure 10.4) were found in non-marine inner basin or deltaic facies, including the antiarchan placoderms *Wudinolepis*, *Xichonolepis*, *Dianolepis*, *Bothriolepis*, and *Hunanolepis* might have lived in the deltaic environment, whereas *Bothriolepis niushoushanensis* probably inhabited the inner part of the sedimentary basin (Pan *et al.*, 1987).

The Upper Devonian vertebrate faunas mostly come from the non-marine deposits, such as the *Remigolepis* and *Sinolepis* faunas in the Zhongning Formation of Ningxia (Pan *et al.*, 1987) and in the Wutung Formation of Nanjing (Liu and Pan, 1958). The antiarchs *Sinolepis* and *Jiangxilepis* occur in the Sanmentan Formation of South Jiangxi (Zhang and Liu, 1991) where a few brachiopods are associated with the fish; perhaps this is a littoral environment of deposition (Hou *et al.*, 1988). The Wutung Formation has no invertebrate microfossils, only some acritarchs collected by Gao Lianda, and it is suggested that this may represent a littoral environment further from the sea shore than that of the Sanmentan Formation.

The only Carboniferous vertebrate faunas from China are of Lower

Carboniferous (Tournasian) age from Guizhou Province (Mahua Section, Hou *et al.*, 1985; and Dapoushang Section, Ji *et al.*, 1989). These are limestones representing deep marine environments associated with small invertebrates such as ammonoids, brachiopods and trilobites.

BIOSTRATIGRAPHY

Silurian

Correlation of the Ludlovian deposits of the Lower Yangtze valley (Figure 10.2) are based on the presence of the agnathans *Hanyangaspis* (Figure 10.3G), *Latirostraspis*, and the acanthodians *Asiacanthus* and *Neoasiacanthus*. According to the opinions of Pan (1986) and Pan and Dineley (1988), *Latirostraspis chaohuensis* (Wang *et al.*, 1980) may be amended to *Hanyangaspis chaohuensis*, and the spines of *Neoasiacanthus*, similar in form to those of *Asiacanthus*, are thought to be the spinal plates of Arthrodira. The author disagrees with Pan's opinion for three reasons. First, the type specimens have been re-examined by the authors and they believe that *Neoasiacanthus* should be referred to the Acanthodii because there is no shoulder girdle characters indicating it is an arthrodire. Second, in the same bed there are no plates or fragments of arthrodires, and third, in the Lower Yangtze valley from Wuhan to West Zhejian no arthrodire fossils occur in Fentou Formation and other Silurian formations, so why should there be arthrodires only in the Chaoxian (N. Anhui) area? At present the author thinks that the genus *Latirostraspis* should be referred to the Hanyangaspidae, and that *Neoasiacanthus* is an acanthodian.

In the Wenlockian, correlations may be made using *Sinogaleaspis* (Figure 10.3E) and *Xiushuiaspis* (Pan and Wang, 1980, 1983) in N. Jiangxi, S. Anhui (Ningguo) and W. Zhejiang (Changxing). *Sinacanthus* may be used for biostratigraphic correlations in W. Hunan, S. Shanxi and the northern border of Tarim (collected by Fourth Petroleum Team of Guiyang).

In the Rongxi Formation, of Llandovery age, Dayong County, W. Hunan, the polybranchiaspid *Dayongaspis* was found and described by Pan and Zeng (1985). Galeaspidis were found in the Wujiahe Formation of Silurian age in Ziyang County, S. Shanxi Province (Pan and Dineley, 1988). It is unusual that no fossils have been found in E. Yunnan or Longmenshan (Sichuan) indicating the ancestors that may have given rise to the polybranchiaspid or galeaspid lineages. The evolution of these groups may be postulated as occurring in the Early–Middle Silurian of the Middle–Lower Yangtze area (W. Hunan, from Pan and Zeng, 1985; N. Jiangxi from Pan and Wang, 1980; W. Zhejiang from Pan, 1986, 1987) and north to Upper Yangtze Oldland (S. Shanxi from Pan and Dineley, 1988).

In South China the earliest antiarchan, *Silurolepis platydorsalis* (Zhang and Wang, 1992) has been found in the Upper part of the Guandi Formation, corresponding to the Upper Wenlockian to Lower Ludlovian in Qujing County, eastern Yunnan. This antiarch does not belong to the Sinolepidae, as it has a median ventral plate in the trunkshield. Wang, J.

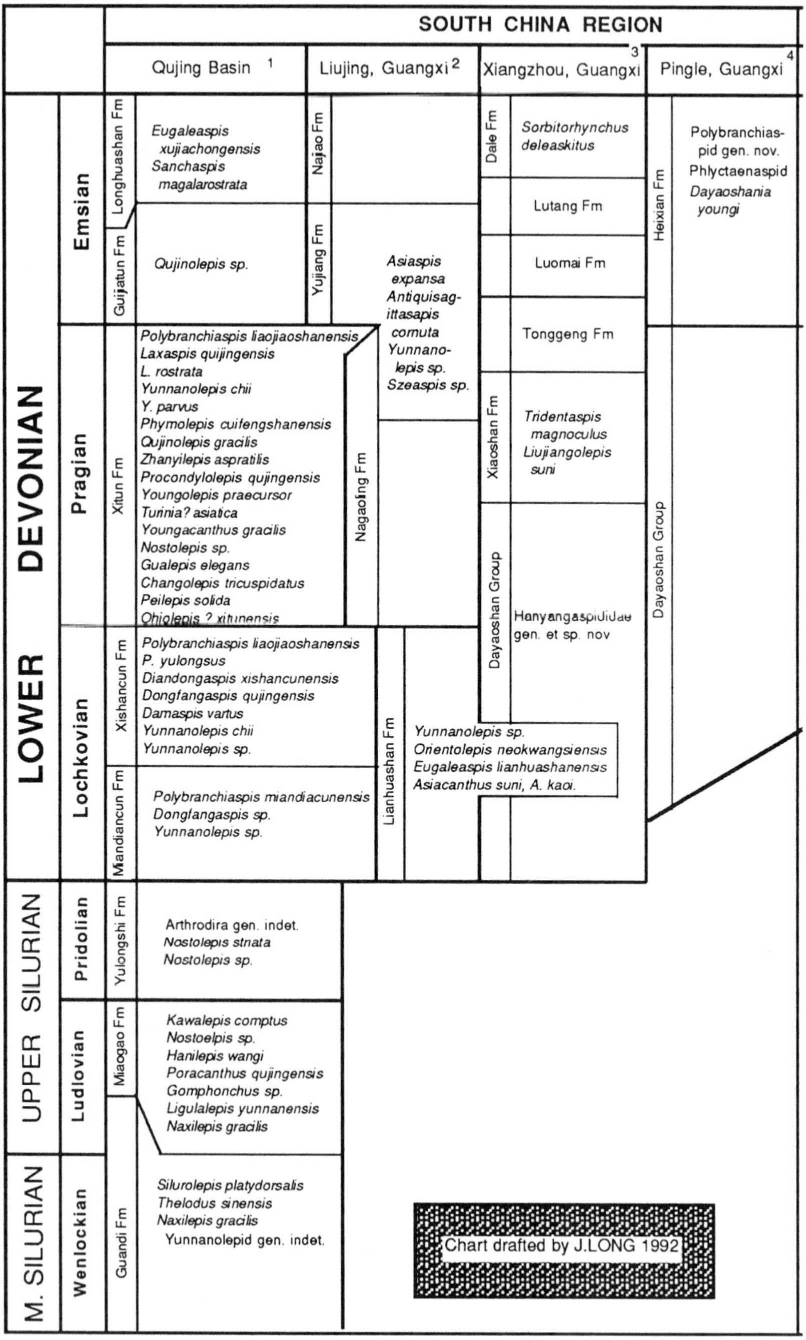

Figure 10.2. Chart showing stratigraphic sections for the Silurian and Early Devonian of China with principal biostratigraphic genera and species of vertebrates (macro and micro).

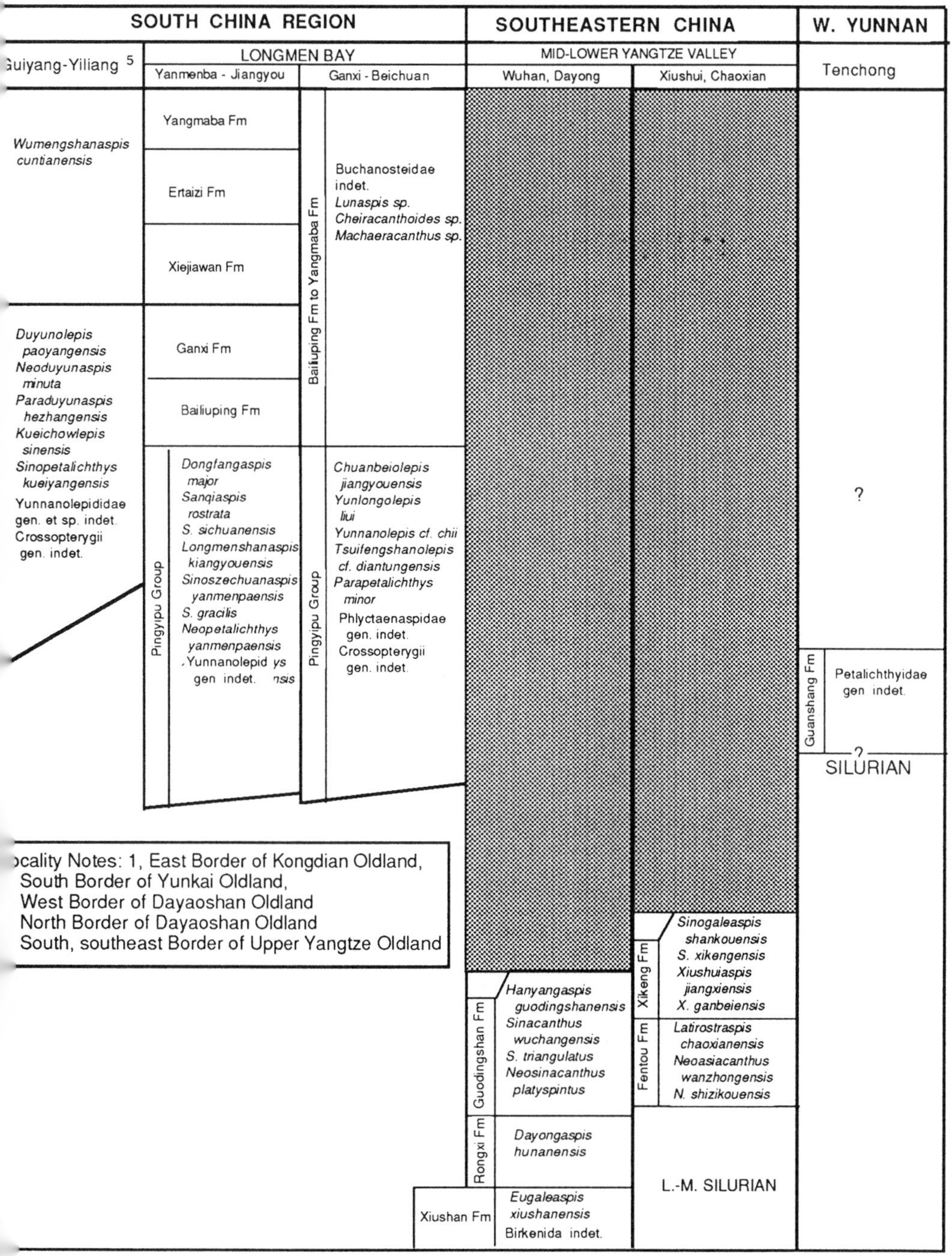

SOUTH CHINA REGION
SOUTHEASTERN CHINA
W. YUNNAN
;uiyang-Yiliang 5
LONGMEN BAY
Yanmenba - Jiangyou
Ganxi - Beichuan
MID-LOWER YANGTZE VALLEY
Wuhan, Dayong
Xiushui, Chaoxian
Tenchong
Yangmaba Fm
Ertaizi Fm
Xiejiawan Fm
Ganxi Fm
Bailiuping Fm
Bailiuping Fm to Yangmaba Fm
Wumengshanaspis cuntianensis
Buchanosteidae indet.
Lunaspis sp.
Cheiracanthoides sp.
Machaeracanthus sp.
Duyunolepis paoyangensis
Neoduyunaspis minuta
Paraduyunaspis hezhangensis
Kueichowlepis sinensis
Sinopetalichthys kueiyangensis
Yunnanolepididae gen. et sp. indet.
Crossopterygii gen. indet.
Pingyipu Group
Dongfangaspis major
Sanqiaspis rostrata
S. sichuanensis
Longmenshanaspis kiangyouensis
Sinoszechuanaspis yanmenpaensis
S. gracilis
Neopetalichthys yanmenpaensis
Yunnanolepid ys gen indet. nsis
Chuanbeiolepis jiangyouensis
Yunlongolepis liui
Yunnanolepis cf. chii
Tsuifengshanolepis cf. diantungensis
Parapetalichthys minor
Phlyctaenaspidae gen. indet.
Crossopterygii gen. indet.
?
Guanshang Fm
Petalichthyidae gen. indet.
?
SILURIAN
ocality Notes: 1, East Border of Kongdian Oldland,
South Border of Yunkai Oldland,
West Border of Dayaoshan Oldland
North Border of Dayaoshan Oldland
South, southeast Border of Upper Yangtze Oldland
Xikeng Fm
Sinogaleaspis shankouensis
S. xikengensis
Xiushuiaspis jiangxiensis
X. ganbeiensis
Fentou Fm
Latirostraspis chaoxianensis
Neoasiacanthus wanzhongensis
N. shizikouensis
Guodingshan Fm
Hanyangaspis guodingshanensis
Sinacanthus wuchangensis
S. triangulatus
Neosinacanthus platyspintus
Rongxi Fm
Dayongaspis hunanensis
Xiushan Fm
Eugaleaspis xiushanensis
Birkenida indet.
L.-M. SILURIAN

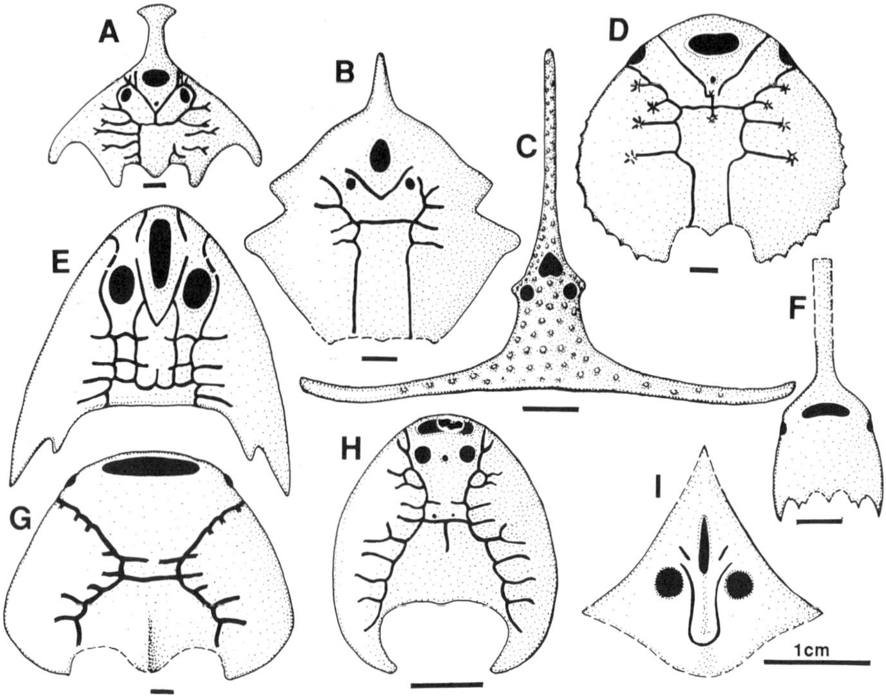

Figure 10.3. Endemic forms of galeaspid agnathans from the Siluro-Devonian South China terrane faunas. A, *Sanchaspis magalarostrata* (after Pan and Wang, 1981); B, *Nanpanaspis microculus*; C, *Lungmenshanaspis kiangyouensis* (after Pan *et al.*, 1975); D, *Cyclodiscaspis ctenus*; E, *Sinogaleaspis shankonensis*; F, *Sanqiaspis zhaotongensis* (B, D, F, after Liu, Y., 1975); G, *Hanyangaspis guodingshanensis* (after Pan, 1986); H, *Changxingaspis gui* (E, H, after Wang, N., 1991); I, *Tridenaspis magnoculus* (after Liu, S., 1986). All scales equal 1 cm.

(1991a) has also described fragmentary antiarch remains, named as *Shimenolepis graniferans* from the Early Silurian Xiushan Formation in Lixian County, Hunan.

The microvertebrates of Middle–Late Silurian of South China include some endemic acanthodians (*Hanilepis wangi*, Figure 10.5E), thelodontids (*Kawalepis comptus*, *Thelodus sinensis*, Figure 10.5B, C), actinopterygians (*Naxilepis gracilis*, Figure 10.5K) and some cosmopolitan genera (*Gomphonchus*, *Nostolepis*, *Ligulalepis yunnanensis*, *Poracanthodes qujingensis* and *Thelodus sinensis*; Wang and Dong, 1989; see Figure 10.5). Correlations may be made in South China using this fauna.

Early Devonian

The Early Devonian vertebrate faunas of China include many species of endemic Agnatha, Placodermi, Acanthodii and Osteichthyes. Some of these represent higher taxonomic groups endemic only to China, such as the galeaspid and polybranchiaspid agnathans (Figure 10.3), the yunnanolepidoid antiarchs (Figure 10.7D) and the *Youngolepis*-like crossopterygians. Microvertebrate faunas include both endemic forms (e.g. *Changolepis*, Figure 10.5I; *Youngacanthus*, Figure 1.7J; Wang, N., 1984) as well as genera occurring elsewhere (*Turinia*, Figure 10.5A; *Nostolepis*, Figure 10.5D). Sue Turner (pers. comm.) believes, however, that *Turinia asiatica* may turn out to be a new genus of thelodont once more material is described.

The only macrovertebrates known from the Early Devonian of China which occur elsewhere are the placoderms *Lunaspis* (Liu, 1981, also known from Germany, Gross, 1961) and Australia (Young, 1985) and *Buchanosteus* (= *Kueichowlepis sinensis* Pan *et al.*, 1975), also known from Australia (Young, 1979), with closely allied genera known from Armorica (Goujet, 1981) and the Middle East (Goujet and Janvier, 1984). Biostratigraphic correlations across China in the Early Devonian are based on the presence of mainly yunnanolepidoid antiarchs and galeaspid agnathans.

Yunnanolepids occur in East Yunnan (Liu, Y., 1965; Zhang, G., 1978, Longmenshan region (Hou *et al.*, 1988), Guangxi (Pan *et al.*, 1975; Pan and Wang, 1978), and southern Hunan (collections of the Geological Survey of Hunan), ranging from Upper Silurian (East Yunnan, Guandi Formation) to early Emsian. The genus *Yunnanolepis* (Figure 10.7D, see also Chapter 11) was first described from the Xishancun Formation (upper Gedinnian–lower Seigennian) of east Yunnan (Qujing Basin) by Liu (1963). It has now been found also in the Longmenshan region, Beichuan County, from the Pingyipu Group of Lower Devonian age (Wang, 1987), close to the Gedinnian/Seigennian boundary, in a fauna containing *Chuanbeiolepis jiangyouensis*, *Yunlongolepis liui*, *Yunnanolepis* cf. *chii* and *Tsuifengshanolepis* cf. *diantungensis*. It is also known from the Pragian Yujiang Formation of Liujing, Guangxi (Liu, 1974) with *Orientolepis neokwangsiensis*, and from the Pragian Xitun Formation fauna with the associated antiarchs *Phymolepis cuifengshanensis*, *Qujinolepis gracilis*, *Zhanjilepis aspratilis* (Zhang, 1978), and *Procondylolepis qujingensis* (Zhang, 1984). A few specimens of *Yunnanolepis* sp. have come from the Lochkovian Miandliancun Formation of Qujing, east Yunnan. Yunnanolepids also occur in northern Vietnam, the southernmost extension of the South China terrane (Thanh and Janvier, 1987, 1990).

Eugaleaspids of early Devonian age occur in eastern Yunnan (Liu, 1962). The genus *Eugaleaspis* was first found and described from the Xitun Formation (Liu, 1963), but has also been found in the Longhuashan region of eastern Yunnan (Xujiachong Formation, upper Pragian to Emsian; Liu, 1975), and in the Lianhuashan Formation of Liujing (Liu, 1986). *Yunnanogaleaspis* was found and described from the middle part of the

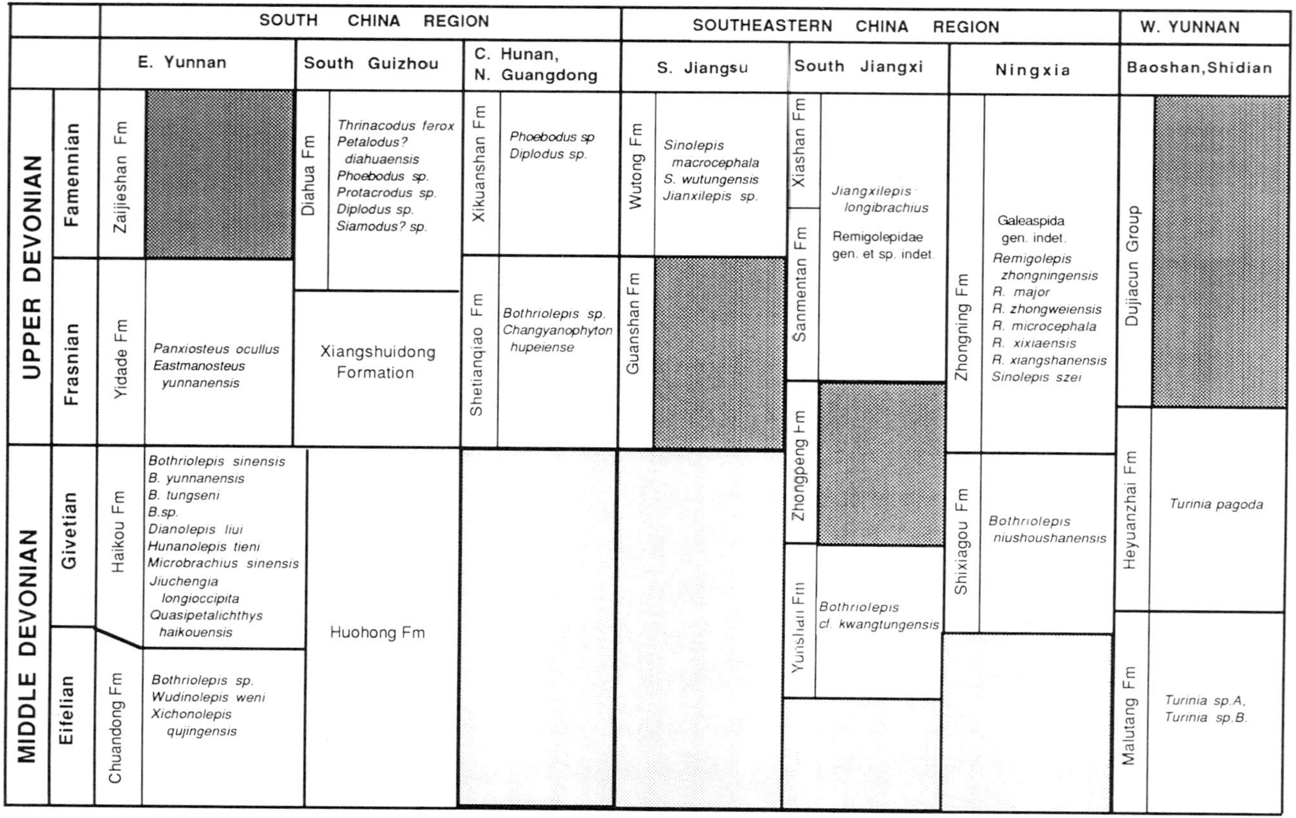

Figure 10.4. Chart showing stratigraphic sections for the Middle Devonian–Lower Carboniferous of China with principal biostratigraphic vertebrates.

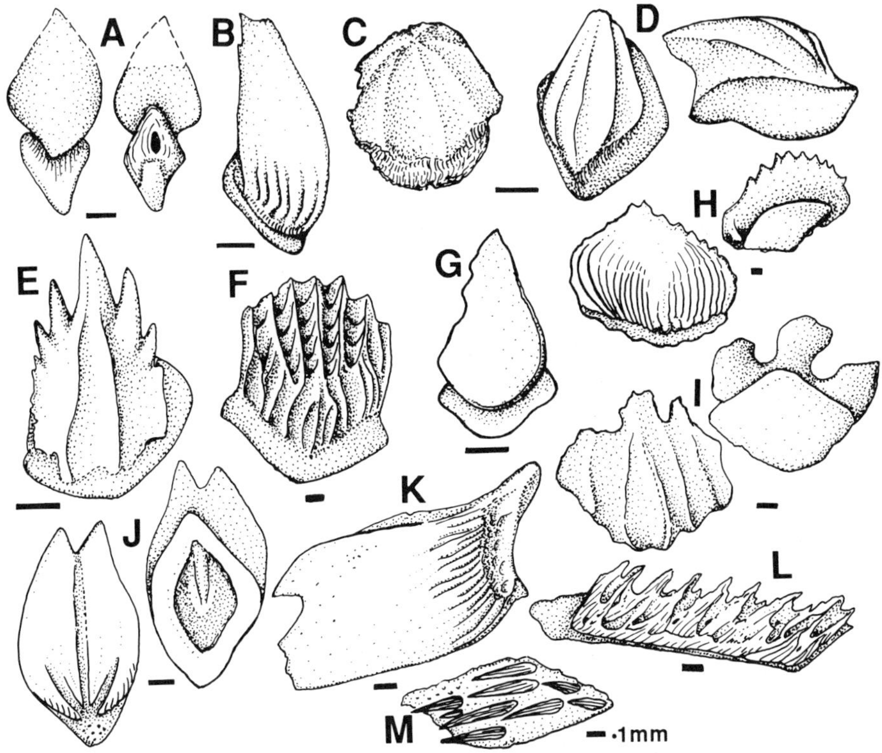

Figure 10.5. Late Silurian-Early Devonian microvertebrate scales useful in biostratigraphic correlations within China. A-C, Thelodonti. A, *Turinia asiatica*; B, *Kawalepis comptus*; C, *Thelodus sinensis*. D-G, Acanthodii. D, *Nostolepis* sp.; E, *Hanilepis wangi*; F, *Poracanthodes qujingensis*; G, ischnacanthid gen. indet. H-J, Chondrichthyes. H, *Gualepis elegans*; I, *Changolepis tricuspidatus*; J, *Peilepis solida*. K-L, Actinopterygii. K, *Naxilepis gracilis*; L, *Ligulalepis yunnanensis*. M, Placodermi? *Ohiolepis? xitunensis* (all from Wang, N., 1984, and Wang and Dong, 1989).

Xishancun Formation in eastern Yunnan by Pan and Wang (1980).

The genus *Polybranchiaspis* (for illustration see Chapter 11.5H) was first described from the Xishancun Formation of Lochkovian age in the Qujing Basin of east Yunnan, but also occurs in the Nagaoling Formation of Liujing in Guangxi Province (Liu, 1983a). Polybranchiaspids such as *Polybranchiaspis liaojiaoshanensis*, *Laxaspis qujingensis* and *L. rostrata* occur in the Xitun Formation, Qujing County, eastern Yunnan (Liu, 1965, 1973). *Polybranchiaspis liaojiaoshanensis* occurs in the Xishancun Formation northeastern Yunnan (Liu, 1975; Pan and Wang, 1975, 1978), Guangxi (Liu, S., 1983a, b, 1986). Polybranchiaspids are also found in Guizhou (Pan *et al.*,

1975; Pan and Wang, 1978) and the Longmenshan region of Sichuan (Pan and Wang, 1978) and southeastern Yunnan (Cao, 1979). *Polybranchiaspis* also occurs on the South China terrane in northern Vietnam, in the Sika Formation and Bac Bun Suite at Trang Xa and Dong Mo (Thanh and Janvier, 1987, 1990) and to the north in Ha Giang Province.

In the Lower Devonian of Guangxi there are two important antiarchans, *Liujiangolepis* and *Dayaoshania*. *Liujiangolepis* was described by Wang (1987) as a possible asterolepidid, although recent reinvestigations of the material have shown that the anterior ventrolateral plate is narrow, as is the anterior end of the posterior ventrolateral plate, and there is no median ventral plate (Zhang Guorui and Wang Shitao, pers. obs.). This indicates that, like *Xichonolepis*, *Dayaoshania* and *Sinolepis* (Figure 10.7A), *Liujiangolepis* is a member of the family Sinolepidae. Sinolepids are now thought to have evolved from ancestral stock like *Liujiangolepis*, an Early Devonian form from Xiangzhou. Figure 10.6 shows the age ranges of sinolepid antiarchs in China.

The crossopterygian *Youngolepis* was described by Zhang and Yu (1981) from the Xitun Formation of Qujing, eastern Yunnan, and the braincase and skull described in detail by Zhang (1982). Recently Chang has described the shoulder girdle and head exoskeleton in detail (Chang, 1991).

Two Early Devonian dipnoans are now known from South China, *Diabolepis speratus* (Chang and Yu, 1984) from the Xitun Formation of Qujing (Pragian), and *Sorbitorhynchus* (Wang *et al.*, 1990, in Campbell and Barwick, 1990), from the Emsian Dale Formation (Liuhui member) of the Ma'anshan section, Zhongping district, Xiangzhou County, Guangxi.

Although some debate has been going on as to whether *Diabolepis* is a true dipnoan or an ancestral 'protodipnoan' (e.g. Campbell and Barwick, 1987; Forey, 1987), *Sorbitorhynchus* is clearly a dipnoan with close affinities to *Pillararhynchus longi*, from the Upper Devonian Gogo Formation of Western Australia (Campbell and Barwick, 1990). Smith and Chang's (1990) study of the histology of *Diabolepis* toothplates indicates it has dentition that forms and grows in a manner unique to the Dipnoi, so it should now be regarded as a primitive dipnoan.

In the Upper part of the Lower Devonian from Bailiuping Formation to the Xiejiawan Formation in the Longmenshan Section, there are many microvertebrate remains, including acanthodians, chondrichthyans and placoderms. Some of these genera are used for biostratigraphic correlations, such as *Gomphonchus*, *Nostolepis*, *Cheiracanthoides* and *Machaeracanthus*. Endemic chondrichthyan microfossils occur in the Xitun Formation in Qujing, such as *Gualepis*, *Peilepis* and *Changolepis* (Figure 10.5H–J; Wang, N., 1984). *Nostolepis* scales also occur in the Xitun Formation as well as in the Ganxi Formation of Longmenshan. *Gualepis* scales are currently under study by the author, Sue Turner and Gavin Young.

The Baoshan area of West Yunnan is part of the Shan-Thai terrane, and not on the South China terrane. There are no microvertebrates described, only one macrofossil of a petalichthyid placoderm similar to *Wijdeaspis* from Europe, Spitsbergen and Australia. The specimen comes from the Lower Devonian Shizishan Formation in Tengchong County (Wang and Cao, 1988).

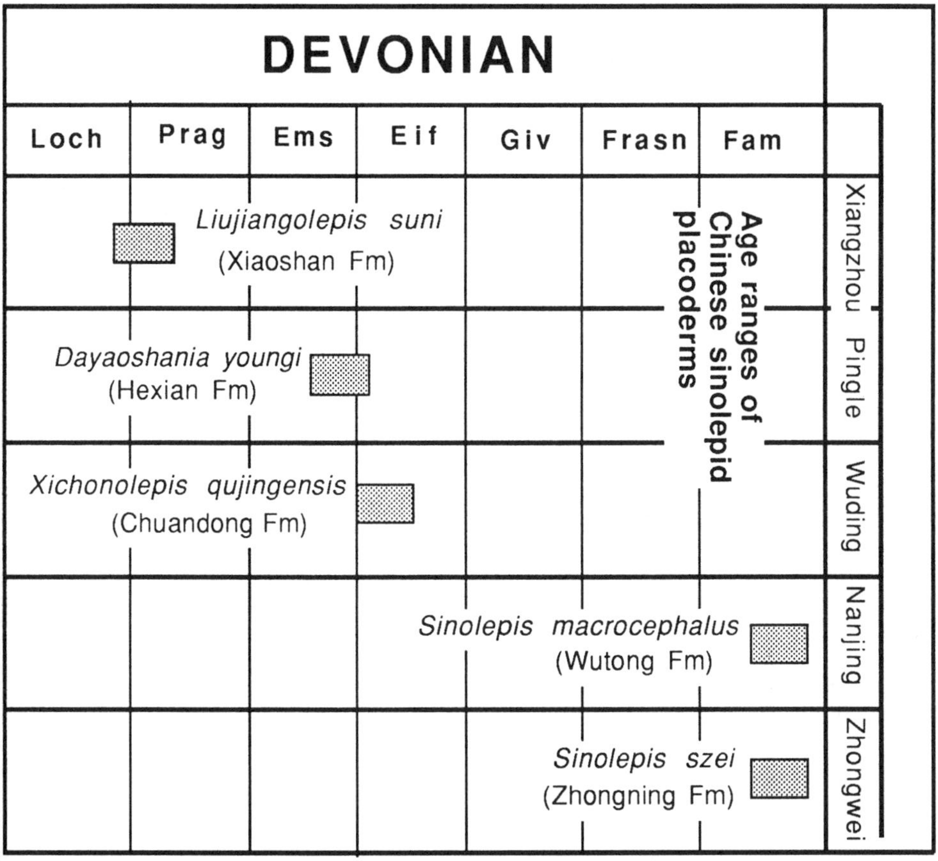

Figure 10.6. Age ranges of sinolepid placoderms in China.

The arthrodire *Kueichowlepis* occurs in the Lower Devonian Wudang Formation in Guiyang, and in the Xiaoshan Formation of Dale, Xiangzhou and Qinzhou Counties of Guangxi. It is a buchanosteid (Young, 1979), a group now known from Australia, Armorica (Goujet, 1981) and the Middle East (Goujet and Janvier, 1984). *Sinopetalichthys* occurs in the Lower Devonian Wudang Formation of Guiyang associated with *Kueicholepis*, antiarchans (yunnanolepids collected by Wang Shitao and Pan Jiang) and a crossopterygian. The petalichthyids *Xinanapetalichthys*, *Neopetalichthys* and *Diandongpetalichthys* are found in eastern Yunnan and the Longmenshan district of Sichuan Province. *Lunaspis* occurs in the Lower Devonian Cuifengshan Group of eastern Yunnan (Liu, 1981), with fragmentary remains of *Lunaspis*? sp. in the Ganxi and Xiejiawan Formations of Sichuan (collected by the author).

The diverse macrovertebrate faunas from the Gedinnian–Emsian Cuifengshan Group of Qujing have been divided into vertebrate assemblages separately by Wang, J. (1984), Liu, S. (1984) and Pan and Dineley (1988).

Wang, J. (1984) has erected three assemblages of Lower Devonian age from lower to upper as follows:

1. *Polybranchiaspis liaojiaoshanensis–Laxaspis qujingensis* assemblage.
2. *Yunnanolepis chii–Youngolepis praecursor* assemblage.
3. *Sanchaspis magalarostrata–Qingmenaspis microculus* assemblage.

Liu, S. (1984) has erected three vertebrate assemblages from lower to upper as follows:

1. *Polybranchiaspis–Laxaspis* assemblage.
2. *Yunnanolepis–Qujinolepis* assemblage.
3. *Sanchaspis–Asiaspis* assemblage.

These assemblages were erected based only on the genera of fishes, and essentially coincided with each other and between the assemblages of Wang Junqing.

Pan and Dineley (1988) also erected three Lower Devonian vertebrate assemblages, based on agnathan species only, from lower to upper as follows:

1. *Yunnanogaleaspis major–Dongfangaspis qujingensis* assemblage.
2. *Eugaleaspis changi–Nanpanaspis microculus* assemblage.
3. *Eugaleaspis xujiachongensis–Sanchaspis megalorostrata* assemblage.

In addition the author has erected ten palaeocommunities from the Lianhuashanian to the Sipaian (Lochkovian–Emsian) in the South China biogeographic province (Wang, S., 1991).

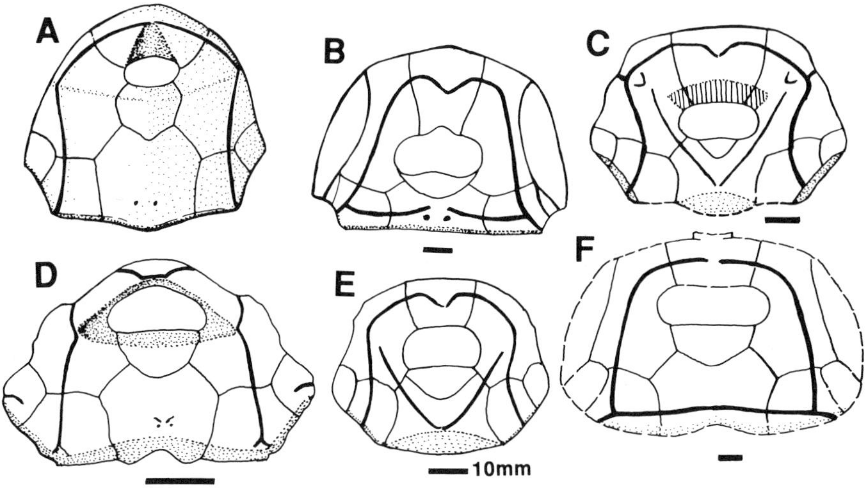

Figure 10.7. Antiarch headshields from the Devonian of China. A, *Sinolepis* (after Young, 1984); B, *Hunanolepis tieni* (after Wang, J., 1991b); C, *Bothriolepis niushoushanensis*; D, *Yunnanolepis parvus* (after Zhang, 1980); E, *Dianolepis* (after Long, 1983); F, *Remigolepis zhongningensis* (C, F, after Pan and Wang, 1980).

Middle Devonian

In the Middle Devonian *Bothriolepis* is the dominant antiarchan which is widely distributed in both South China and North China (Figure 10.4). The earliest appearance of *Bothriolepis* worldwide may be in the Chuandong Formation of eastern Yunnan (Eifelian, P'an, 1981). The first record of *Bothriolepis* from China was *B. sinensis*, described from central Hunan by Chi (1940). *Hunanolepis tieni* (Figure 10.7B) occurs with *B. sinensis* in central Hunan, in the Haikou Formation of eastern Yunnan (Pan *et al.*, 1978) and Bobai County of southeastern Guangxi (Pan and Dineley, 1988). *Hunanolepis* was recently redescribed by Wang, J. (1991b) as belonging to the Asterolepiformes. Other *Bothriolepis* species are known from the Dahepo Formation of Lochang County, northern Guangdong Province (*B. shaokuangensis*, Chang, 1963), and from the Shixiagou Formation of Zhongning County, Ningxia Province in North China (*B. niushoushanensis*, Figure 10.7C; Pan and Wang, 1980; Pan *et al.*, 1987). In the early Givetian there is *Bothriolepis tungseni* (Chang, 1965) and *B. yunnanensis* (Liu, 1963), and in the late Givetian there is *B. lochangensis* and *B. kwangtungensis* (upper part of Dahepo Formation) in Lochang County of northern Guangdong Province (P'an, 1964).

The bothriolepid *Dianolepis liui* (Figure 10.7E; Chang, 1965) occurs in the probable Eifelian–early Givetian. It was placed as the sister taxon to

Bothriolepis by Long (1983). The microbrachiid *Wudinolepis weni* (Chang, 1965) occurs in the lower part of the Haikou Formation of Wuding County, eastern Yunnan. It is similar to *Hohsienolepis hsientuensis* (Pan, 1978) from the Xindu Formation (late Eifelian–early Givetian) of Hexian County, northeastern Guangxi Province.

The sinolepid *Xichonolepis qujingensis* (Pan and Wang, 1978) is widely distributed in the lower part of the Haikou Formation (and its equivalent the Chuandong Formation, Yang *et al.*, 1981; P'an, 1981) of Qujing and Wuding localities. In Wuding Basin it occurs in calcareous mudstones or muddy limestones that may represent lagoonal facies, but in Qujing the environment it is found in was far from the seashore.

Microbrachius sinensis (Pan, 1984) occurs in the Qujing Formation (early Givetian) of Dongshan, Qujing County. Its discovery enables correlations between China and the Middle Old Red Sandstone of Europe.

There are several endemic genera of arthrodires known from the Middle Devonian of China. The Givetian Haikou Formation of eastern Yunnan has produced *Jiuchengia longiocciputa* (Wang and Wang, 1983), *Yinosteus major* (Wang and Wang, 1984), *Kungmingolepis lucaowanensis* (Liu and Wang, 1981) and *Yangaspis jinningensis* (Liu and Wang, 1981). One arthrodire is known from the Middle Devonian of China as well as in other countries—*Holonema*, known from Europe, USSR and Australia (Miles, 1971) is recorded from east Yunnan, with *Wudinolepis* (Wang, J., 1984) and Guangxi in China (Changcun Formation, late Eifelian–early Givetian; Wang, in press). The dinichthyid *Eastmanosteus yunnanensis* was described by Wang (1991c) from the upper Middle Devonian of Panxi (formerly described as a species of *Dunkleosteus*). It has close resemblances to *E. calliaspis* from the lower Frasnian of Western Australia.

The petalichthyid *Quasipetalichthys haikouensis* (Liu, 1973) occurs in the lower part of the Haikou Formation from the Middle Devonian of Haikou, Kunming, in association with *Bothriolepis sinensis* and *Hunanolepis*. *Quasipetalichthys* cf. *haikouensis* (Pan *et al.*, 1987) has also been reported from the middle part of the Shixiagou Formation, Middle Devonian of Zhongning County, Ningxia Province, in north China. This enables an age of Late Givetian to be placed on the Shixiagou Formation, based on the correlation with the South China *Bothriolepis–Hunanolepis* fauna. The petalichthyid *Eurycaraspis* was recently described from the Givetian Dolomite Member (Haikou Formation) of Longhua Hill, Zhanyi, Yunnan (Liu, Y., 1991). *Eucaryaspis* occurs with *Bothriolepis tungseni*, *Dianolepis liui* and *Xichonolepis*, above the Lower Sandstone containing *Bothriolepis tungseni* and *Xichonolepis*, and below the Upper Sandstone containing *Bothriolepis sinensis*, *B. yunnanensis*, *Hunanolepis* and *Quasipetalichthys*.

In Shidian County, western Yunnan (Shan-Thai terrane), microvertebrates have been described which include the thelodont *Turinia pagoda* from the lower part of the Upper Givetian Heyuanzhai Formation (Wang *et al.*, 1986), and *Turinia* sp. A and sp. B, from the Eifelian Malutang Formation, based on associated conodont assemblages. Recently *Turinia* cf. *T. pagoda* was described from Australia (Long *et al.*, 1988) and similar thelodontids are known from Thailand (Blieck *et al.*, 1984) and Iran (Turner and Janvier, 1979).

Late Devonian

In the Late Devonian the sinolepid antiarchs *Sinolepis wutungensis* and *S. macrocephala* occur in the upper part of the Wutung Group (Leigutai Formation, Famennian; Liu and Pan, 1958) and *S. szei* in the upper part of the Zhongning Formation of Ningxia, associated with several species of *Remigolepis* (*R. microcephala*, *R. major*, *R. zhongningensis*, Figure 10.7F; *R. zhongweiensis*, *R. xixiaensis* and *R. xiangshanensis*; Pan *et al.*, 1987). *Remigolepis* sp. is also recorded from the Xikuangshan Formation of Liling County, central Hunan Province (collected by Pan Jiang), and *Sinolepis* occurs in from the Sanmentan Formation of Yudu County, south Jiangxi Province, associated with *Jiangxilepis longibrachius* (Zhang and Liu, 1991).

Jiangxilepis longibrachius has a dorsal spine on the anterior median dorsal plate, long pectoral fins and the pectoral armour plates are larger than for *Asterolepis*. The '*Asterolepis*' described by Pan from the Wutung Group in Nanjing has longer pectoral fins, similar to those of *Jiangxilepis*, and in this paper the material is referred to *Jiangxilepis* sp. Hopefully, future discoveries, especially of trunk armours from the Wutung Group, may resolve this problem.

The only known agnathan fossil, a eugaleaspid, occurs in the Zhongning Formation of Ningxia, north China (Pan *et al.*, 1987), associated with *Remigolepis* and *Sinolepis*. The specimens consist of four incomplete cephalic shields showing branchial fossae and impressions of tubercles of the dermal ornament. They represent very large fishes with cephalic shields some 30 cm long by 25 cm wide, but are presently of indeterminate identification.

Microvertebrates in the Upper Devonian occur in the Daihua Formation (Muhua and Dapoushang Sections) on the Devonian/Carboniferous boundary in Changshun County, Guizhou Province. This fauna was first described by Wang and Turner (1985) from the upper Famennian of the Muhua Section, and Wang (1989) described further material from the Dapoushang Section. The fauna of the Daihua Formation (Muhua Section) includes the chondrichthyans *Phoebodus* sp., *Petalodus? daihuanensis*, *Protacrodus* sp., *Thrinacodus ferox*, *Diplodus* cf. *priscus*, ?*Siamodus* sp. (Long, 1990), *Ctenacanthus* sp. and *Fungulodus rotundus*. *Fungulodus* is an enigmatic form, whose Class, Order and Family is *incertae sedis*. In the Dapoushang Section *Phoebodus gothicus*, (not *P. 'politus'* as in Wang and Turner, 1985; Ginter, 1990) and *Petalodus? daihuaensis* are found only in the upper part of the Daihua Formation (Famennian). Other chondrichthyans, such as *Thrinacodus ferox*, *Protacrodus* sp., '*Diplodus*' sp., *Symmorium* sp. and *Stethacanthus* (*Cladodus*) sp. are distributed in the Famennian Diahua Formation to the Lower Carboniferous Wangyou Formation (Wang, 1989).

Microvertebrates also occur in the Upper Devonian Changlai Formation and Lower Carboniferous Dasaiba Formation from Shaoguan County, northern Guangdong Province, and in the Famennian Xikuangshan Formation of central Hunan (specimens in the Geological Survey of Hunan). These faunas contain scales and teeth of actinopterygians and

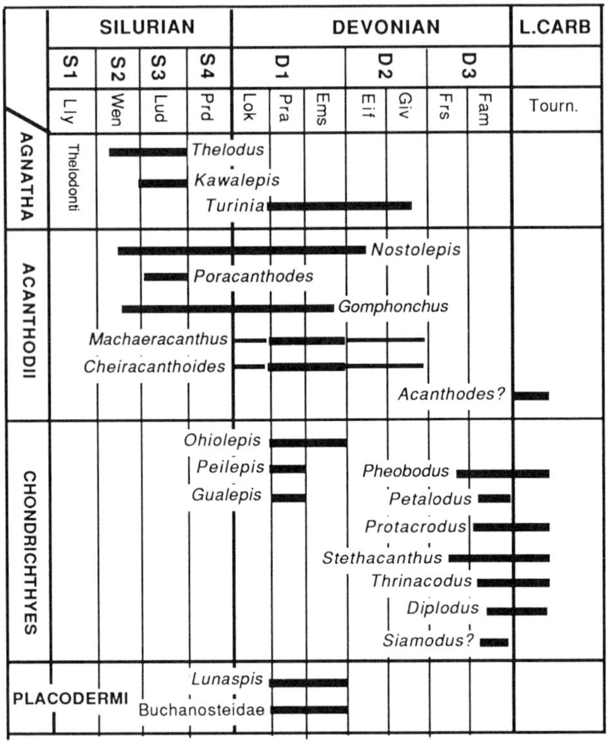

Figure 10.8. Age ranges of important biostratigraphic genera for the Middle Palaeozoic of China (macro and micro).

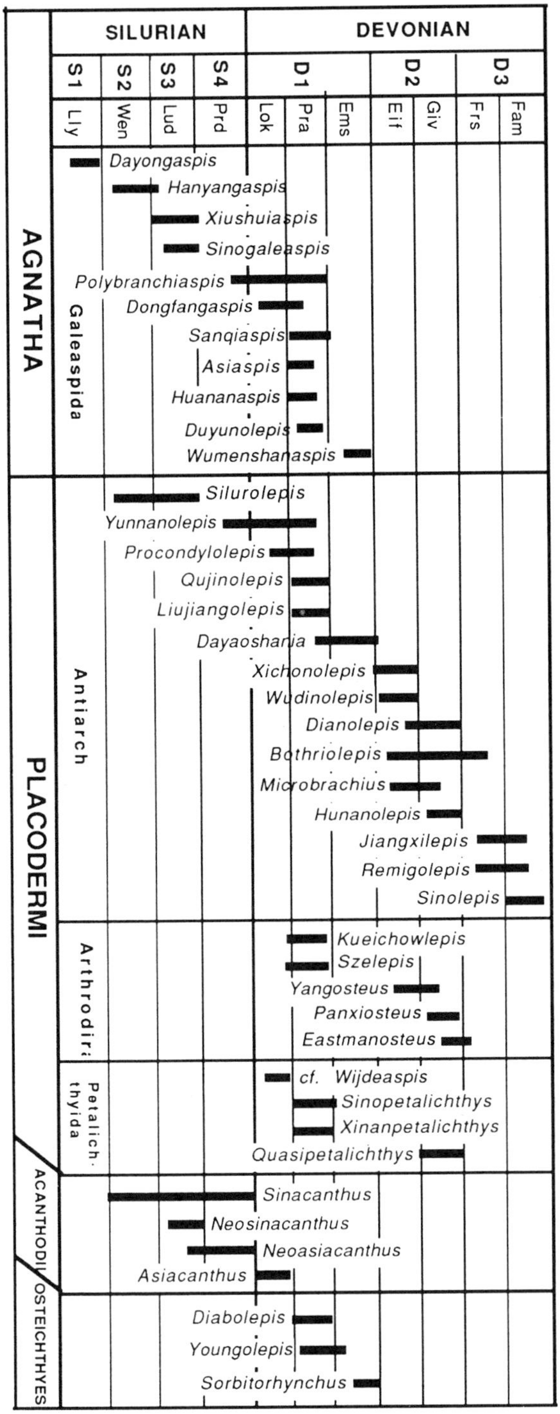
SILURIAN
DEVONIAN
S1
S2
S3
S4
D1
D2
D3
Lly
Wen
Lud
Prd
Lok
Pra
Ems
Eif
Giv
Frs
Fam
AGNATHA
Galeaspida
Dayongaspis
Hanyangaspis
Xiushuiaspis
Sinogaleaspis
Polybranchiaspis
Dongfangaspis
Sanqiaspis
Asiaspis
Huananaspis
Duyunolepis
Wumenshanaspis
PLACODERMI
Antiarch
Silurolepis
Yunnanolepis
Procondylolepis
Qujinolepis
Liujiangolepis
Dayaoshania
Xichonolepis
Wudinolepis
Dianolepis
Bothriolepis
Microbrachius
Hunanolepis
Jiangxilepis
Remigolepis
Sinolepis
Arthrodira
Kueichowlepis
Szelepis
Yangosteus
Panxiosteus
Eastmanosteus
Petalichthyida
cf. Wijdeaspis
Sinopetalichthys
Xinanpetalichthys
Quasipetalichthys
ACANTHODII
Sinacanthus
Neosinacanthus
Neoasiacanthus
Asiacanthus
OSTEICHTHYES
Diabolepis
Youngolepis
Sorbitorhynchus

crossopterygians, but no chondrichthyans, and may represent littoral or shallow marine facies.

Lower Carboniferous

Many macrovertebrates such as *Phoebodus* sp., *Thrinacodus ferox*, *Protacrodus* sp., *Diplodus* sp., *Symmorium* sp., *Stethacanthus* sp. occur in the Tournaisian Wangyou Formation, as discussed in the previous section. The presence of the acanthodian *Acanthodes guizhouensis*, from the Muhua Section (Gedongguan Beds, Wangyou Formation; also Dapoushang member, Wangyou Formation, Dapoushang Section) may be taken as representing the base of the Carboniferous (Hou and Wang, 1985). In Europe, however, *Acanthodes* may occur in the Upper Devonian of France, Spain and Belgium (A. Blieck, pers. comm.). Figure 10.8 summarises the known ranges of some biostratigraphically useful fossil vertebrates from the Silurian–Carboniferous of China.

DEVONIAN VERTEBRATE BIOGEOGRAPHY OF CHINA

South China has a diverse endemic Early Devonian vertebrate fauna including over 60 unique genera of eugaleaspids, yunnanolepids, polybranchiaspids, yunnanolepidoid antiarchs, which Young (1981) named the Yunnanolepid–polybranchiaspid Biogeographic Province, also called the Eugaleaspid–Polybranchiaspid–Yunnanolepid Realm by Pan and Dineley (1988). Few cosmopolitan forms are known, such as the petalichthyid *Lunaspis*, also known from Germany and Australia; the presence of Buchanosteidae (arthrodire *Kueichowlepis*). The crossopterygian *Youngolepis* has close affinities to *Powichthys* from Arctic Canada (Zhang, M. and Yu, 1981, Chang, 1982). Recently Young (1991) has described agnathans from central Australia (e.g. *Pituriaspis*) that have their closest affinity with the endemic Chinese forms. The dipnoan *Sorbitorhynchus*, from the Emsian of Guangxi, has closest affinity with *Pillararhynchus*, from the Gogo Formation of Australia (Campbell and Barwick, 1990). So, in the early part of the Devonian, although the South China Block was isolated by the Tethyan Sea it may not have been too wide for some forms to reach it from Australia and western Gondwana.

The Qujing Basin of eastern Yunnan was situated slightly north of the equator in the Early Devonian (Hou and Wang, 1985). Eastern Gondwana (Australia, Antarctica) may have been linked to South China by a series of smaller blocks, such as Thailand, Iran and Afghanistan. As indicated earlier, the antiarchs may have originated from the South China block in the Early–Middle Silurian (Pan and Dineley, 1988).

In the Middle–Late Devonian there are many endemic forms of placoderms, but some widespread genera are also known, such as the antiarchs *Bothriolepis*, *Remigolepis* and *Microbrachius*, and the arthrodires *Eastmanosteus* and *Holonema* (Wang, J., 1984, 1991c). Recently a sinolepid antiarch has been found in the Famennian Hunter Siltstone of New South Wales, Australia, the only record of the group outside China (Ritchie *et al.*, in press). These vertebrate assemblages fall into the Middle Devonian *Quasipetalichthys–Hunanolepis–Bothriolepis* Realm and the Upper Devonian *Sinolepis–Remigolepis* Realm (Pan and Dineley, 1988).

The author thinks that these realms may also include areas outside of China, for example Australia, Canada, Greenland and extend to most of the world. Therefore, I prefer to use biogeographic provinces, in the sense of Young (1981) rather than 'realms' within China. Besides, the divisions of vertebrate biogeography should be based on the Lower Devonian which shows the highest degree of endemism, and, therefore, I concur with Young (1981, also Chapter 12) on the nature and boundaries of vertebrate biogeographic provinces during the Devonian.

REFERENCES

Cited references used in text and charts as well as additional references documenting the Palaeozoic fish faunas of China.

Blieck, A., Goujet, D., Janvier, P. and Lelievre, H., 1984, Microrestes de vertebres du Siluro-Devonien d'Algerie, de Turquie et de Thailande, *Géobios*, **17**: 851–6.

Campbell, K.S.W. and Barwick, R.E., 1984, The choanae, maxillae, premaxillae and anterior palatal bones of early dipnoans, *Proceedings of the Linnean Society of New South Wales*, **107**: 147–70.

Campbell, K.S.W. and Barwick, R.E., 1987, Palaeozoic lungfishes—a review, *Journal of Morphology, Supplement*, **1**: 93–131.

Campbell, K.S.W. and Barwick, R.E., 1990, Paleozoic dipnoan phylogeny: functional complexes and evolution without parsimony, *Paleobiology*, **16**: 143–69.

Cao, R., 1979, A Lower Devonian Agnathan of South-eastern Yunnan, *Vertebrata Palasiatica*, **17**: 118–20.

Chang, K., 1963, A new species of *Bothriolepis* from Kwangtung, *Vertebrata Palasiatica*, **7**: 342–51.

Chang, K., 1965, New antiarchs from the Middle Devonian of Yunnan, *Vertebrata Palasiatica*, **9**: 1–14.

Chang, M., 1982, The braincase of *Youngolepis*, a Lower Devonian crossopterygian from Yunnan, south-western China, *Papers in the Department of Geology, University of Stockholm*, Sweden: 1–113.

Chang, M., 1991, Head exoskeleton and shoulder-girdle of *Youngolepis*. In M.M. Chang, Y.H. Liu and G.R. Zhang (eds), *Early Vertebrates and Related Problems of Evolutionary Biology*, Science Press, Beijing, pp. 355–78.

Chang, M. and Yu, X., 1984, Structure and phylogenetic significance of *Diabolichthys speratus* gen. et sp. nov., a new Dipnoi-like form from the Lower Devonian of eastern Yunnan, China, *Proceedings of the Linnean Society of New South Wales*, **107**: 171–84.

Chi, Y., 1940, On the discovery of *Bothriolepis* in the Devonian of central Hunan, *Bulletin of the Geological Society of China*, **20**: 57–62.

Fang, R. *et al.*, 1985, *The Middle Silurian and Early Devonian Stratigraphy and Palaeontology in Qujing District, Yunnan*, Institute of Geological Sciences, Yunnan, People's Publishing House, Yunnan.

Forey, P.L., 1987, Relationships of lungfishes, *Journal of Morphology, Supplement*, **1**: 75–91.
Ginter, M., 1990, Late Famennian shark teeth from the Holy Cross Mts, central Poland, *Acta Geologica Polonica*, **40**: 69–81.
Goujet, D., 1981, Les poissons. *Memoires de la Societé Geologique et Mineralogique, Bretagne*, **24**: 305–8.
Goujet, D. and Janvier, P., 1984, Un nouvel horizon a vertebres dans la coupe-type de la Formation de Khush-Yeilagh (Devonien moyen), Alborz oriental, Mazenderan, Iran, *Géobios*, **17**: 245–50.
Gross, W., 1961, *Lunaspis broilli* und *L. heroldi* aus dem Hunsruckshiefer (Unterdevon; Rheinland), *Notizblatt Hessisches Landesamt Bodenforsch, Weisbaden*, **89**: 17–43.
Hou, H. and Wang, S., 1985, Devonian palaeogeography of China, *Acta Palaeontologica Sinica*, **24**: 186–97.
Hou, H. and Wang, S., 1986, Palaeogeographic map of the late Early Devonian of South China. In Wang Hongzen (ed.), *Atlas of Paleogeography of China*, Cartographic Publishing House, Beijing: 60.
Hou, H., Wan, Z. and Xian, S. (eds), 1988, *The Devonian biostratigraphy and sedimentary facies of Longmenshan, Sichuan Province*, Geological Publishing House, Beijing: 1–487.
Ji, Q. *et al.* (eds), 1989, The Dapoushang Section—an excellent section for the Devonian–Carboniferous boundary stratotype in China, Science Press, Beijing, China: 1–148.
Liu, H.-T., 1955, *Kiangyouosteus*, a new arthrodiran fish from Szechuan, China, *Acta Palaeontologica Sinica*, **3**: 261–74.
Liu, S., 1973a, New materials of *Bothriolepis shaokuanensis* and the age of the fish bearing beds, *Vertebrata Palasiatica*, **11**: 36–42.
Liu, S., 1973b, Some new acanthodian fossil materials from Devonian of South China, *Vertebrata Palasiatica*, **11**: 144–7.
Liu, S., 1974, The significance of the discovery of yunnanolepidoid fauna from Guangxi, *Vertebrata Palasiatica*, **12**: 144–8.
Liu, S., 1981, Occurrence of *Lunaspis* in China, *Kexue Tongbao*, **26**: 829–830.
Liu, S., 1982a, Preliminary note on the Arthrodira from Guangxi, China, *Vertebrata Palasiatica*, **20**: 106–14.
Liu, S., 1982b, An arthrodire endocranium, *Vertebrata Palasiatica*, **20**: 271–5.
Liu, S., 1983a, Agnatha from Sichuan, China, *Vertebrata Palasiatica*, **21**: 97–102.
Liu, S., 1983b, Biogeography of Silurian and Devonian vertebrates in China, *Vertebrata Palasiatica*, **21**: 292–300.
Liu, S., 1984, The sequence of Early Devonian vertebrate assemblages in China, *Vertebrata Palasiatica*, **22**: 103–8.
Liu, S., 1986, Fossil Eugaleaspida from Guangxi, *Vertebrata Palasiatica*, **24**: 1–9.
Liu, S., 1991, Biogeography of early vertebrates in China. In M.M. Chang, Y.H. Liu and G.R. Zhang (eds), *Early Vertebrates and Related Problems of Evolutionary Biology*, Science Press, Beijing: 499–504.
Liu, T.S. and Pan, K., 1958, Devonian fishes from the Wutung series near Nanking, China, *Palaeontologica Sinica*, **15**: 1–41.
Liu, Y., 1962, A new species of *Bothriolepis* from Yunnan, *Vertebrata Palasiatica*, **6**: 80–5.
Liu, Y., 1963, On the Antiarchi from Chutsing, Yunnan, *Vertebrata Palasiatica*, **7**: 39–45.
Liu, Y., 1965, New Devonian agnathans of Yunnan, *Vertebrata Palasiatica*, **9**: 125–34.
Liu, Y., 1973, On the new forms of Polybranchiaspiformes and Petalichthyida from the Devonian of Southwest China, *Vertebrata Palasiatica*, **11**: 132–43.

Liu, Y., 1975, Lower Devonian agnathans of Yunnan and Sichuan, *Vertebrata Palasiatica*, **13**: 202–16.
Liu, Y., 1979, On the arctolepid arthrodirea from Lower Devonian of Yunnan, *Vertebrata Palasiatica*, **17**: 23–34.
Liu, Y., 1981, A nomenclatural note of *Szelepis* for *Szeaspis* Liu 1979, *Vertebrata Palasiatica*, **19**: 294.
Liu, Y., 1991, On a new petalichthyid, *Eurycaraspis incilis* gen. et sp. nov., (Placodermi, Pisces) from the Middle Devonian of Zhanyi, Yunnan. In M.M. Chang, Y.H. Liu and G.R. Zhang (eds), *Early Vertebrates and Related Problems of Evolutionary Biology*, Science Press, Beijing: 139–77.
Liu, Y. and Wang, J., 1981, On three new arthrodires from the Middle Devonian of Yunnan, *Vertebrata Palasiatica*, **19**: 295–304.
Liu, Y. and Wang, J., 1985, A galeaspid (Agnatha) *Antiquisagittaspis cornuta* gen. et sp. nov. from the Lower Devonian of Guangxi, China, *Vertebrata Palasiatica*, **23**: 257–54.
Long, J.A., 1983, New bothriolepid fishes from the Late Devonian of Victoria, Australia, *Palaeontology*, **26**: 295–320.
Long, J.A., 1990, Late Devonian chondrichthyans and other microvertebrate remains from northern Thailand, *Journal of Vertebrate Paleontology*, **10**: 59–71.
Long, J.A., Turner, S. and Young, G.C., 1988, A Devonian fish fauna from subsurface sediments in the eastern Officer Basin, South Australia, *Alcheringa*, **11**: 69–88.
Miles, R.S., 1971, The Holonematidae (placoderm fishes), a review based on new specimens from the Upper Devonian of Western Australia, *Philosophical Transactions of the Royal Society of London*, **263**: 101–234.
P'an, K., 1964, Some Devonian and Carboniferous fishes from South China, *Acta Palaeontologica Sinica*, **12**: 139–68.
P'an, K., 1981, Devonian antiarch biostratigraphy of China, *Geological Magazine*, **118**: 69–75.
Pan, J., 1984, The phylogenetic position of the Eugaleaspida in China, *Proceedings of the Linnean Society of New South Wales*, **107**: 309–19.
Pan, J., 1985, Dayongaspidae, a new family of Polybranchiaspiformes (Agnatha) from the Early Silurian of Hunan, China, *Vertebrata Palasiatica*, **23**: 207–13.
Pan, J., 1986a, A new species of *Microbrachius* from the Middle Devonian of Yunnan, *Vertebrata Palasiatica*, **22**: 8–13.
Pan, J., 1986b, New findings of Silurian vertebrates from China, *Professional Papers of the Department of Geology, Peking University* (Presented to Prof. Yoh Sen-Shing), 67–76.
Pan, J., 1987, Note on Silurian vertebrates of China, *Bulletin of the Institute of Geology, Chinese Academy of Sciences*, **15**.
Pan, J. and Dineley, D.L., 1988, A review of early (Silurian and Devonian) vertebrate biogeography and biostratigraphy of China, *Proceedings of the Royal Society of London*, **235(B)**: 29–61.
Pan, J., Hou, F., Cao, J., Gu, Qi., Liu, S.Y., Wang, J., Gao, L. and Liu, C., 1987, *Continental Devonian of Ningxia and its biotas*, Geological Publishing House, Beijing.
Pan, J., Wang, S. and Liu, Y., 1975, The Lower Devonian Agnatha and Pisces of South China, *Professional Papers of Stratigraphy and Palaeontology 1*, Geological Publishing House, Beijing: 135–69.
Pan, J. and Wang, S., 1978, Devonian Agnatha and Pisces of South China, Geological Publishing House, Beijing, 1974: 292–333.
Pan, J. and Wang, S., 1980, New findings of Galeaspiformes in South China, *Acta Palaeontologica Sinica*, **19**: 1–7.

Pan, J. and Wang, S., 1981, New discoveries of polybranchiaspids of Yunnan Province, *Vertebrata Palasiatica*, **19**: 113–21.
Pan, J. and Wang, S., 1983, Xiushuiaspidae, a new family of Polybranchiaspiforms from Xiushui of Jiangxi Province, *Acta Palaeontologica Sinica*, **22**: 505–9.
Pan, J., Wang, S., Gao, L. and Hou, J., 1978, *Devonian continental and sedimentary formations of South China*, Geological Publishing House, Beijing, 1974: 240–69.
Pan, J. and Zeng, X., 1985, Dayongaspidae, a new family of Polybranchiaspiformes (Agnatha) from early Silurian of Hunan, China, *Vertebrata Palasiatica*, **23**: 207–13.
Ritchie, A., Wang, S.T., Young, G.C. and Zhang, G.R., in press, The Sinolepidae, a family of antiarchs (placoderm fishes) from the Devonian of South China and eastern Australia, *Records of the Australian Museum* (ms 112pp, 36 text-Figs. 12., 5 tables).
Smith, M.M. and Chang, M., 1990, The dentition of *Diabolepis speratus* Chang and Yu, with further considerations of its relationships and the primitive dipnoan dentition, *Journal of Vertebrate Paleontology*, **10**: 420–433.
Thanh, T.-D. and Janvier, P., 1987, Les vertebres Devoniens du Vietnam, *Annals de Paleontologie*, **73**: 165–194.
Thanh, T.-D. and Janvier, P., 1990, Les vertebres du Devonien inferieur du Bac Bo oriental (provinces de Bac Thai et Lang Son, VietNam), *Bulletin du Museum d'Histoire Naturelle*, 4 ser, **12**: 143–223.
Turner, S. and Janvier, P., 1979, Middle Devonian Thelodonti (Agnatha) from the Khush-Yeilagh Formation, northeast Iran, *Géobios*, **12**: 889–92.
Turner, S. and Tarling, D.H., 1982, Thelodont and other agnathan distributions as a test of Lower palaeozoic continental reconstructions, *Palaeogeography, Palaeoclimatology and Palaeoecology*, **39**: 295–311.
Wang, C., Ruan, Y., Mu, D., Wang, Z., Rong, J., Yin, B., Kuang, G. and Su, Y., 1979, Subdivision and correlation of the Lower and Middle Devonian series in different facies of Guangxi, *Acta Palaeontologica Sinica*, **3**: 305–11.
Wang, J., 1981, A toothplate of Dipnoi from Qujing, Yunnan, *Vertebrata Palasiatica*, **19**: 198–9.
Wang, J., 1982, New materials of Dinichthyidae, *Vertebrata Palasiatica*, **20**: 181–6.
Wang, J., 1984, Geological and palaeogeographic distribution of Devonian fishes in China, *Vertebrata Palasiatica*, **22**: 219–29.
Wang, J., 1991a, The Antiarchi from the early Silurian of Hunan, *Vertebrata Palasiatica*, **29**: 240–4.
Wang, J., 1991b, New material of *Hunanolepis* from the Middle Devonian of Hunan. In M.M. Chang, Y.H. Liu and G.R. Zhang (eds), *Early Vertebrates and Related Problems of Evolutionary Biology*, Science Press, Beijing: 213–47.
Wang, J., 1991c, A fossil Arthrodira from Panxi, Yunnan, *Vertebrata Palasiatica*, **29**: 264–75.
Wang, J. and Wang, N., 1983, A new genus of Coccosteidae, *Vertebrata Palasiatica*, **21**: 1–8.
Wang, J. and Wang, N., 1984, New materials of Arthrodira from the Wuding Region, Yunnan, *Vertebrata Palasiatica*, **22**: 1–7.
Wang, N., 1984, Thelodont, acanthodian, and chondrichthyan fossils from the Lower Devonian of southwest China, *Proceedings of the Linnean Society of New South Wales*, **107**: 419–441.
Wang, N., 1991, Two new Silurian galeaspids (jawless craniates) from Zhejiang Province, China, with a discussion of galeaspid–gnathostome relationships. In M.M. Chang, Y.H. Liu and G.R. Zhang (eds), *Early Vertebrates and Related Problems of Evolutionary Biology*, Science Press, Beijing: 41–65.

Wang, N. and Dong, Z., 1989, Discovery of Late Silurian microfossils of Agnatha and fishes from Yunnan, China, *Acta Palaeontologica Sinica*, **8**: 192–206.

Wang, S., 1986, Palaeogeographic map of the Early Devonian of South China. In Wang Hongzen (ed.), *Atlas of Paleogeography of China*, Cartographic Publishing House, Beijing: 60.

Wang, S., 1987, A new antiarch from the Early Devonian of Guangxi, *Vertebrata Palasiatica*, **25**: 81–90.

Wang, S., 1989, Biostratigraphy of vertebrate microfossils. In Ji Qiang *et al.* (eds), *The Dapoushang Section—an excellent section for the Devonian-Carboniferous Boundary Stratotype in China*, Science Press, Beijing, China: 36–8, 103–7.

Wang, S., 1991, Lower Devonian vertebrate palaeocommunities from South China. In M.M. Chang, Y.H. Liu and G.R. Zhang (eds), *Early Vertebrates and Related Problems of Evolutionary Biology*, Science Press, Beijing: 487–97.

Wang, S. and Cao, R., 1988, Discovery of Macropetalichthyidae from the Lower Devonian in Western Yunnan, *Vertebrata Palasiatica*, **26**: 73–5.

Wang, S., Dong, Z. and Turner, S., 1986, Middle Devonian Turiniidae (Thelodont, Agnatha) from Western Yunnan, China, *Alcheringa*, **10**: 315–25.

Wang, S. and Hou, H., 1986, Palaeogeographic map of the Early Devonian of South China. In Wang Hongzen (ed.), *Atlas of Paleogeography of China*, Cartographic Publishing House, Beijing: 57–8.

Wang, S. and Lan, C., 1984, New discovery of polybranchiaspids from Yiliang County, Northeastern Yunnan, *Bulletin of the Geological Institute, Chinese Academy of Geological Sciences*, **9**: 113–23.

Wang, S. and Turner, S., 1985, Vertebrate microfossils of the Devonian-Carboniferous boundary, Muhua Section, Guizhou Province, *Vertebrata Palasiatica*, **23**: 223–34.

Wang, S., Xia, S., Chen, L. and Du, S., 1980, On the discovery of Silurian Agnatha and Pisces from Chaoxian County, Anhui Province and its stratigraphical significance, *Bulletin of the Geological Institute, Chinese Academy of Geological Sciences*, **2**: 101–12.

Yang, S., P'An, K. and Hou, H., 1981, The Devonian System in China, *Geological Magazine*, **118**: 113–24.

Young, G.C., 1979, New information on the structure and relationships of *Buchanosteus* (Placodermi: Euarthrodira) from the Early Devonian of New South Wales, *Zoological Journal of the Linnean Society*, **66**: 309–52.

Young, G.C., 1981, Biogeography of Devonian vertebrates, *Alcheringa*, **5**: 225–43.

Young, G.C., 1984, Comments on the phylogeny and biogeography of antiarchs (Devonian placoderm fishes), and the use of fossils in biogeography, *Proceedings of the Linnean Society of New South Wales*, **107**: 443–73.

Young, G.C., 1985, Further petalichthyid remains (placoderm fishes, Early Devonian) from the Taemas-Wee Jasper region, New South Wales, *Bureau of Mineral Resources Journal of Australian Geology and Geophysics*, **9**: 121–31.

Young, G.C., 1991, The first armoured agnathan vertebrates from the Devonian of Australia. In Chang, M.M., Liu, Y.H. and Zhang, G.R. (eds), *Early Vertebrates and Related Problems of Evolutionary Biology*, Science Press, Beijing: 67–85.

Young, G.C. and Zhang G.R., 1992, Structure and function of the pectoral joint and operculum in antiarchs, Devonian placoderm fishes, *Palaeontology*, **35** (2): 443–64.

Zhang, G., 1978, The antiarchs from the Early Devonian of Yunnan, *Vertebrata Palasiatica*, **16**: 147–186.

Zhang, G., 1984, New form of Antiarchi with primitive brachial process from Early Devonian of Yunnan, *Vertebrata Palasiatica*, **22**: 81–91.

Zhang, G. and Liu, S., 1978, Fossil *Bothriolepis* from Yujiang Formation of Kwangsi, *Vertebrata Palasiatica*, **16**: 4–6.

Zhang, G. and Liu, Y., 1991, A new Antiarchi from the Upper Devonian of Jiangxi, China. In Chang, M.M., Liu, Y.H. and Zhang, G.R. (eds), *Early Vertebrates and Related Problems of Evolutionary Biology*, Science Press, Beijing, China: 67–85.

Zhang, G. and Wang, S., 1992, A New Antiarchi from the Silurian of Qujing, Yunnan. (in press).

Zhang, M. (Chang Meeman), 1980, Preliminary note on a Lower Devonian antiarch from Yunnan, China, *Vertebrata Palasiatica*, **18**: 179–90.

Zhang, M. and Yu, X., 1981, A new crossopterygian, *Youngolepis praecursor* gen. et sp. nov. from Lower Devonian of East Yunnan, China, *Scientia Sinica*, **24**: 89–97.

Zhang, M. and Yu, X., 1987, A nomen novum for *Diabolichthys* Chang et Yu 1981, *Vertebrata Palasiatica*, **25**: 79.

Zhao, X., 1986, Paleogeographic map of Late Devonian of China. In Wang Hongzen (ed.), '*Atlas of Paleogeography of China*', Cartographic Publishing House, Beijing: 57–8.

Chapter 11

PALAEOZOIC VERTEBRATE BIOSTRATIGRAPHY OF SOUTH-EAST ASIA AND JAPAN*

John A. Long

INTRODUCTION

Modern South-east Asia comprises four major terranes during Palaeozoic times. The Shan-Thai (or Sibumasu) terrane is separated to the west from the West Burma terrane by the West Burma fault, and from the Indo-China terrane to the east by the Nan-Uttradit suture (Burrett *et al.*, 1990; Figure 11.1). The South China terrane borders these three terranes to the north. The diverse well-preserved fish faunas of the South China terrane are discussed by Wang (Chapter 10), although recent finds from northern Vietnam (Tong and Janvier, 1987, 1990; Janvier *et al.*, 1990; Long *et al.*, 1990), also on the South China terrane, are included in this chapter. The West Burma terrane is not discussed here due to lack of Palaeozoic fish remains, although Chhibber (1936) has reported one 'ichthyoduralite' from the Devonian reef sequences near Maymyo, central Burma (locality 1, Figure 11.1).

The Shan-Thai and Indo-China terranes were devoid of Palaeozoic vertebrates until relatively recently. The discovery of an acanthodian scale and a thelodont scale from northern Thailand by Blieck and Goujet (1978; also Blieck *et al.*, 1984) was the first find of a Devonian vertebrate from the Shan-Thai terrane, although discoveries of microvertebrates from western Yunnan, China, the northern limit of Shan-Thai, followed soon after (Wang *et al.*, 1986). Recent finds of more diverse Late Devonian (Famennian) microvertebrates, which occur with abundant conodonts, show an assemblage containing both cosmopolitan and endemic elements (Long and Burrett, 1989; Long, 1990).

The only Late Palaeozoic vertebrates known from Shan-Thai are some isolated shark's teeth discovered in Carboniferous and Permian limestones, and described by Janvier (1981) and by Ingavat and Janvier (1981). The

* A contribution to IGCP Project 328.

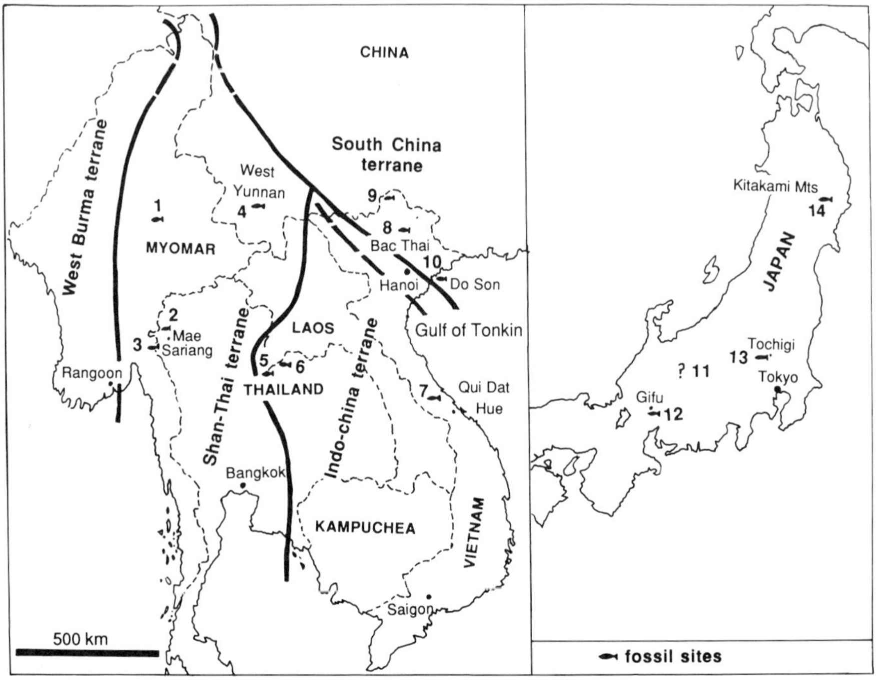

Figure 11.1. Map showing major fossil localities and terrane boundaries referred to in the text.

only Palaeozoic vertebrates known from Vietnam which come from south of the Song Ma suture zone (dividing South China terrane to the north, from Indo-China terrane, south, Figure 11.1) were collected in 1990 by the author and Dr Clive Burrett, from the late Famennian Xom Nha Formation, situated in Qui Dat region of Binh Tri Thien Province. In addition, some vertebrate fragments were also found in conodont residues of the Qui Dat Formation (upper Givetian), of Binh Tri Thien Province, kindly lent to us for study by Dr Pham Kim Ngan. These are, therefore, the only known Palaeozoic vertebrates from the Indo-China terrane. The Famennian assemblage is not well-preserved but includes shark teeth of both stethacanthid and phoebodontid types, as well as actinopterygian scales, whereas the Givetian fauna contains sarcopterygian scales, and many indeterminate fragments of bone. A brief description of the new material from the Famennian of Vietnam is included in this chapter. It is hoped that this may lead to further discoveries of microvertebrates from these poorly-known regions.

In recent years a number of finds of Palaeozoic vertebrates have been described from Japan by Goto (1975, 1984, 1989; Goto and Kuga, 1982;

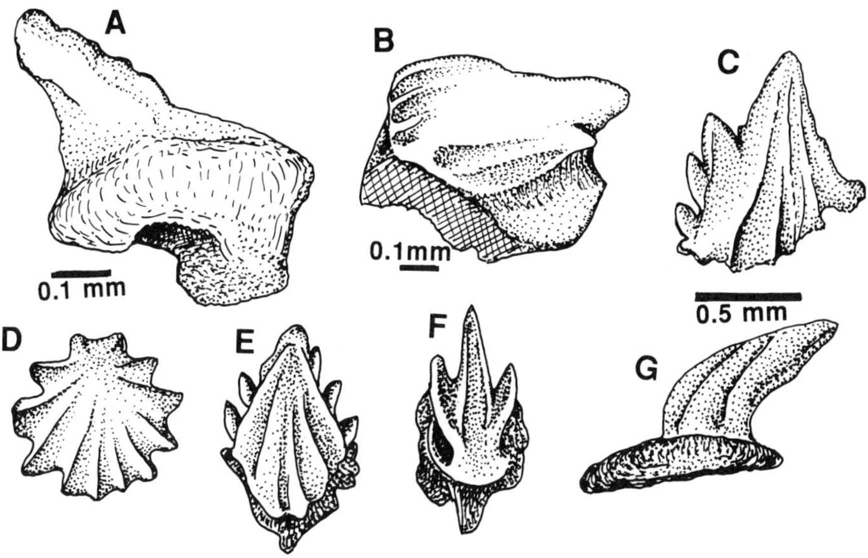

Figure 11.2. A, B, Late Silurian–?Early Devonian microvertebrates from north of Mae Sariang, Thailand. A, thelodont scale; B, *Nostolepis* scale. C–F, thelodont scales from the Middle Devonian of western Yunnan. C–E, *Turinia pagoda*. F, G, *Turinia* sp. (after Wang *et al.*, 1986).

Goto *et al.*, 1988; Okura and Goto, 1988). These are also reviewed in this chapter to complete the overall revision of Asian Palaeozoic vertebrate discoveries.

SHAN-THAI TERRANE

Late Silurian–Early Devonian?

Mae Sariang district, northern Thailand

Two microscopic fish scales were obtained from a limestone sample collected from the road cutting about 20 km north of Mae Sariang, northern Thailand (locality 2, Figure 11.1). These were described by Blieck and Goujet (1978) and Blieck *et al.* (1984). One is an indeterminate thelodont (Figure 11.2A), the other is an acanthodian scale referred to *Nostolepis* sp. (Figure 11.2B). Further collecting by C. Burrett and the author in 1988 tried to sample the same limestones but did not succeed in finding further microvertebrates.

Middle Devonian

Baoshan region, W. Yunnan, China

Thelodonts have been described from the Middle Devonian section of marine sediments at Malutang, Shidian County of Western Yunnan (locality 4, Figure 11.1) by Wang *et al.* (1986). These include the distinctive form of *Turinia pagoda* Wang, Dong and Turner 1986 (Figure 11.2C–E) from the lower part of the Heyuanzhai Formation (upper Givetian), characterised by well-developed marginal lappets projecting from the base of the crown. Long *et al.* (1988) reported *Turinia* cf. *T. pagoda* from the Munyarai-1 well, eastern Officer Basin, Australia in sediments now thought to be early Eifelian age (see also Turner, Chapter 8; Young, Chapter 9). *Turina* spp. (types A and B) were also described from the Eifelian Malutang Formation (Figure 11.2F, G).

A partial skull roof of an Early Devonian macropetalichthyid placoderm was also found in western Yunnan from the marine Guanshang Formation of Yingjiang County (Wang and Cao, 1988). It has been compared with *Wijdeaspis*, a form also known from Russia and Australia.

Late Devonian (upper Famennian)

Mae Sam Lap, northern Thailand

Bulk limestone samples were collected by Burrett and the author from the road cuttings along the track south of Mae Sariang to Mae Sam Lap (locality 3, Figure 11.1) on the Burmese border. A diverse assemblage of microvertebrates was obtained associated with late Famennian conodonts, now regarded as from the *postera-expansa* zone (M. Ginter, pers. comm.). The fauna has been reported by Long and Burrett (1989), Long (1989) and described by Long (1990). The faunal list includes the following chondrichthyans—*Phoebodus australiensis* Long 1990 (Figure 11.3A, B), *Siamodus janvieri* Long 1990 (Figure 11.3C), *Thrinacodus ferox* Turner 1982, *Symmorium* sp. (Figure 11.3D), *Cladodus* cf. *acutus* Agassiz 1843 (Figure 11.3E) and several scale types. Other vertebrates present in the fauna include indeterminate acanthodian and palaeoniscoid scales.

Recent work on phoebodontid shark's teeth by M. Ginter (Poland) suggests that the *P. australiensis* sample from Thailand includes two morphotypes, one of which may constitute a separate species. *P. australiensis* does not occur in Australia, as Ginter and Ivanov (pers. comm.) believe that the specimens referred to by Long (1990) from Australia belong to a new species of *Phoebodus*. *P. australiensis* has since been found in China (Wang and Turner, 1985; Wang, 1989), Poland (Ginter, 1990) and parts of Russia (Ginter, pers. comm.). The unusual multicuspid teeth of *Siamodus* are seen as closely approaching those of some aberrant *Phoebodus* species, and like *Thrinacodus*, should be regarded as a member of the Phoebodontidae (Ginter, pers. comm.).

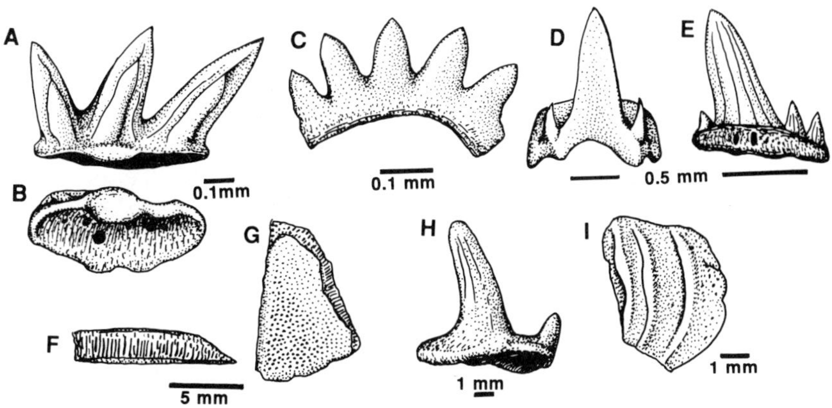

Figure 11.3. Palaeozoic chondrichthyans from Thailand. A–E, Late Devonian (late Famennian) chondrichthyan teeth from near Mae Sam Lap, northern Thailand (Shan-Thai terrane). A, B, *Phoebodus australiensis*, tooth in labial (A) and ventral (B) views; C, *Siamodus janvieri* (labial view); D, *Symmorium* sp. (labial view); E, *Cladodus* cf. *C. acutus* (lingual view). F–H, late Palaeozoic vertebrates from the Indo-China terrane. F, G, Lower Carboniferous psammosteid tooth from Chom Nang quarry, northeastern Thailand, in cross-section (F) and buccal view (G, after Ingavat and Janvier, 1981). H, I Chondrichthyan teeth from Ban Na Chareon, Early Permian, northeastern Thailand; H, *Cladodus* sp. (lingual view, after Janvier, 1981); I, *Deltodus* sp. (buccal view, after Ingavat and Janvier, 1981).

INDO-CHINA TERRANE

Binh Tri Thien Province, central Vietnam

Qui Dat Formation (Givetian)

The outcrop of Qui Dat Formation is best exposed in the hills about 4 kilometres west of Tuyen Hoa, along the road leading out of Tuyen Hoa to the old settlement of Qui Dat (locality 7, Figure 11.1). This same road is also where the samples of the Xom Nha Formation, described below, were also collected, at the bridge crossing the creek near Xom Nha village.

The fauna from here is dated by conodonts (Ngan, in Tong, 1986) as Givetian. None of the vertebrates recovered from this formation are identifiable to generic or specific level. The fauna includes cosmine-covered scales and dermal bones probably attributed to early osteolepiform fishes as they lack the large pores often seen in porolepid or dipnoan cosmine scales (e.g. *Uranolophus*, Campbell and Barwick, 1988). Other fragments include pieces of actinopterygian bone and scales probably belonging to placoderms of indeterminate affinity.

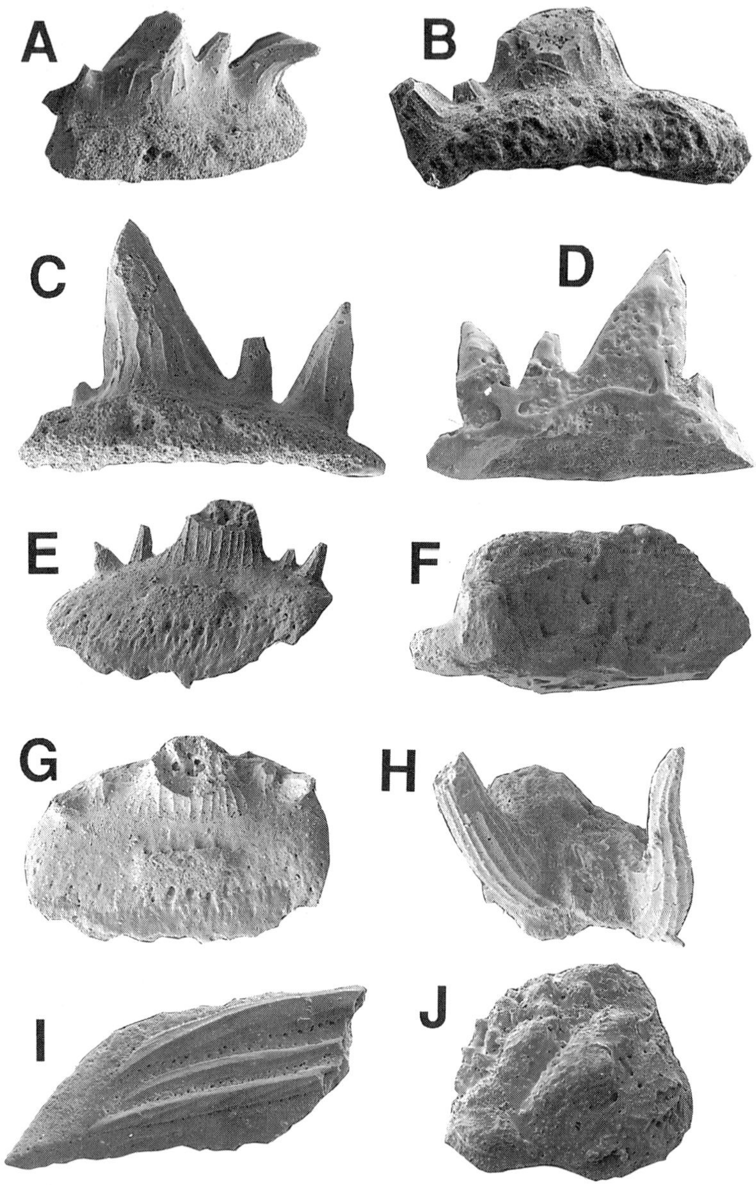
A
B
C
D
E
F
G
H
I
J

Xom Nha Formation (late Famennian)

Material was obtained from a pale blue limestone sample collected under the bridge, from the little creek, near Xom Nha village by the author, Clive Burrett, Pham Kim Ngan and Tran Van Tri in March 1991. The sample yielded abundant conodonts of the *marginifera* zone (M. Ginter, pers. comm.). Rare fish remains include stethacanthid teeth (Figure 11.4A, B, E, F, G) which show the typical broad base, well-developed lingual torus, ribbed central cusp and accessory smaller lateral cusps. They differ from those of *Stethacanthus thomasi* (Turner, 1982) in having fewer lateral cusps (only two each side). One tooth (Figure 11.4H) has been examined by M. Ginter who believes it to be a partial tooth of *Phoebodus* cf. *P. gothicus* (Ginter, 1990). A broad-based cladodont tooth with an asymmetrical central cusp (Figure 11.4C, D) appears similar to *Cladodus* cf. *C. acutus* teeth recorded from Europe and Thailand (Long, 1990). A single scale of a palaeoniscoid (Figure 11.4I) has widely separated rows of linear ganoine ornament closely resembling those of the caudal flank of *Moythomasia* from the Upper Devonian Gogo Formation of Western Australia (Gardiner, 1984). Figure 11.4J shows a scale or dermal denticle from either a placoderm or possibly a chondrichthyan.

Early Carboniferous

Pak Chom Nang quarry, northeastern Thailand

A single tooth from an indeterminate psammosteid holocephalan (Figure 11.3F, G) was found from the Pak Chom Nang limestone (locality 5, Figure 11.1), on the Mekong River (Ingavat and Janvier, 1981).

Early Permian

Ban Na Chareon, northeastern Thailand

A few chondrichthyan teeth have been removed from an unnamed limestone at Ban Na Chareon, Changwat Udon, east of Loei, in northeastern Thailand (locality 6, Figure 11.1), and described by Janvier (1981) and Ingavat and Janvier (1981). Two worn cladodont teeth, referred to as *Cladodus* sp. (Figure 11.3H) are comparable with *Cladodus occidentalis* from the upper Pennsylvanian of North America. A single flat

Figure 11.4. Vertebrate microfossils from the mid Famennian Xom Nha Formation, Binh Tri Thien Province, Vietnam. A, B, E–G, *Stethacanthus* sp. teeth, various views. C, D, *Cladodus* cf. *C. acutus* tooth in lingual (C) and labial (D) views. H, *Phoebodus* cf. *P. gothicus*, partial tooth in buccal view; I, cf. *Moythomasia* scale; J, placoderm or chondrichthyan denticle.

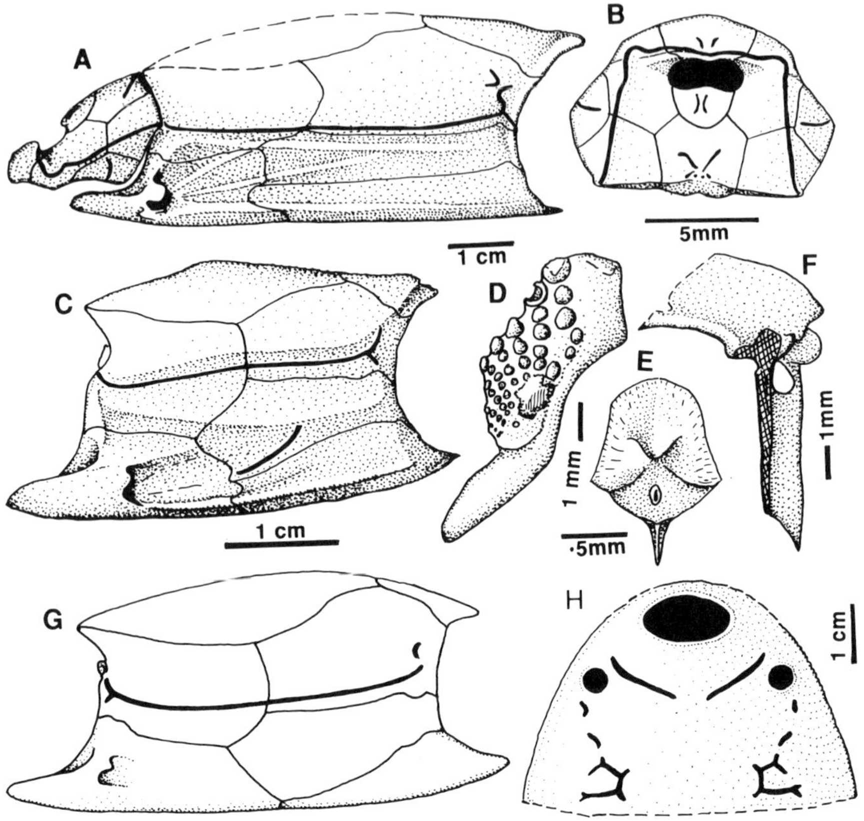

Figure 11.5. Early Devonian vertebrates from northern Vietnam (South China terrane; after Tong and Janvier, 1987, 1990). A, B, *Chuchinolepis dongmoensis*, restored armour in lateral view (A) and restored headshield in dorsal view (B); C, *Yunnanolepis bacboensis*, restored armour in lateral view; D, dipnoan tooth plate, E, unidentified antiarch posterior median dorsal plate in visceral view; F, anterior ventrolateral plate of the sinolepid *Vanchienolepis langsonensis* in dorsal view; G, restored armour of *Yunnanolepis deprati* in lateral view; H, partial armour of *Polybranchiaspis* cf. *P. gracilis* in dorsal view.

crushing tooth was referred to *Deltodus* sp. (Figure 11.3I).

SOUTH CHINA TERRANE

Early Devonian

Bac Bo, near Thai Nguyen, northern Vietnam
A fauna comparable with the diverse Early Devonian assemblages of Southern China has been described from the Sika Formation (Lochkovian)

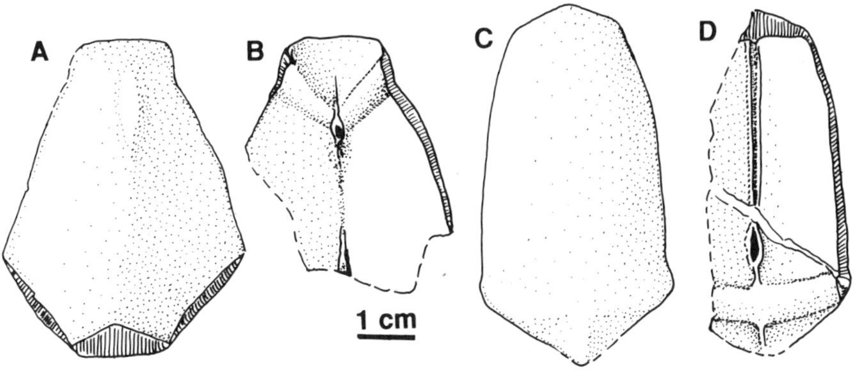

Figure 11.6. The ?Middle–Late Devonian antiarch *Vietnamaspis trii* Long from Do Son Peninsula, northern Vietnam. A, B, anterior median dorsal plate in dorsal (A) and ventral views; C, D, posterior median dorsal plate in same views (after Long *et al.*, 1990).

and Bac Bun Suites (Lochkovian–Pragian), near Trang Xa (Bac Thai Province; locality 8, Figure 11.1) and Dong Mo (Lang Son Province) by Tong and Janvier (1987, 1990). The faunal list is as follows.

The older fauna of the Sika Formation (Figure 11.5) contains the galeaspid *Polybranchiaspis* sp., the antiarchan placoderms *Yunnanolepis* sp. and *Yunnanolepis* cf. *Y. parvus*, and the crossopterygian *Youngolepis* sp. The Bac Bun Suite fauna contains *Polybranchiaspis* sp. *Yunnanolepis bacboensis* Tong and Janvier 1990, *Y. deprati* Tong and Janvier 1990, *Chuchinolepis dongmoensis* Tong and Janvier 1990, a sinolepid *Vanchienolepis langsonensis* Tong and Janvier 1990, the arthrodire ?*Szeaspis* sp., and the sarcopterygians *Youngolepis* cf. *Y. praecursor* Chang and Yu 1981, and toothplates of the early dipnoan. Figure 11.4 shows examples of some of the northern Vietnamese Early Devonian fish fauna.

Khao Loc, Ha Giang Province, northern Vietnam
Polybranchiaspis cf. *P. gracilis* Cao was reported by Tong and Janvier (1987) from Ha Giang Province, near the Chinese border (locality 9, Figure 11.1).

Middle–Late Devonian

Do Son Peninsula, near Haiphong, northern Vietnam
The ?Late Givetian–Frasnian Do Son sandstone exposed along the Do Son Peninsula (locality 10, Figure 11.1) near Haiphong have produced a fauna containing the antiarchan placoderms *Bothriolepis* sp. (Janvier *et al.*, 1987, 1990) and the endemic bothriolepid *Vietnamaspis trii* (Long *et al.*, 1990;

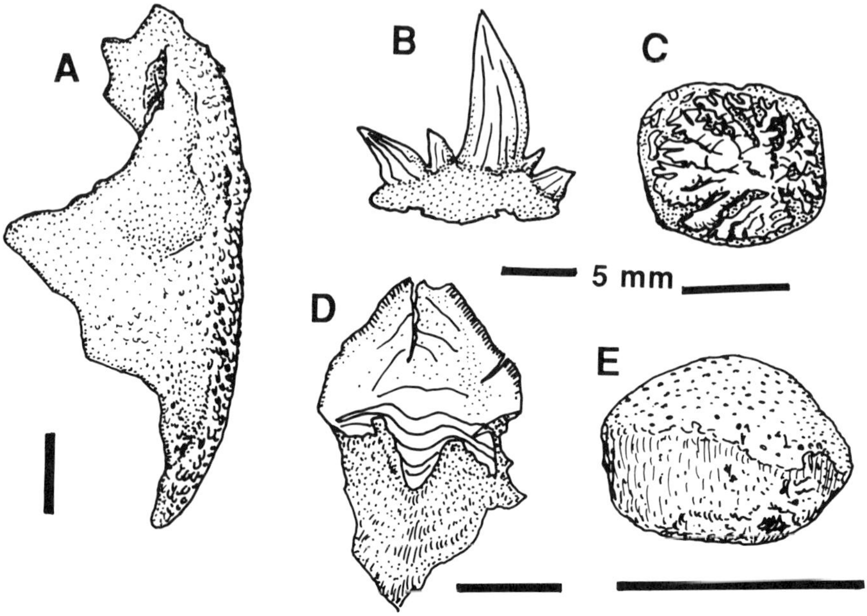

Figure 11.7. Palaeozoic vertebrates from Japan. A, Early Devonian placoderm, cf. *Romundina* sp., partial thoracic armour. B–E, Permian chondrichthyans. B, *Symmorium* sp.; C, *Petrodus* sp. (B, C, from the Asaka Limestone, from Goto *et al.*, 1988); D, *Petalodus* sp. (Nabeyama Formation, after Goto, 1984); E, *Helodus* sp. (from colour atlas of Palaeozoic fish fossils in Japan, 1989, abstract of 43rd Annual meeting of the Association for Geological Collaboration in Japan, pp. 283–4).

Figure 11.6) associated with a groenlandaspid arthrodire (spinal plate), lycophyte plants and eurypterid-like arthropods.

JAPAN

Early Devonian

Central Japan, Fujiki Formation

Okura and Goto (1988) and Goto (1989) figured and described two partial thoracic armours from central Japan (locality 11, Figure 11.1). The specimens, which included spinal plates from acanthothoracid placoderms which have been referred to as *Romundina* sp. (Figure 11.7A), also known from the Gedinnian of Arctic Canada (Ørvig, 1975).

Early–Middle Permian

Central Japan
A single tooth of the petalodont chondrichthyan *Petalodus* sp. (Figure 11.7D) was described from the black limestones of the Middle Permian Nabeyama Formation near Kuzuu by Goto (1984). Isolated chondrichthyan scale and tooth types were described from this unit earlier by Goto (1975).

The Middle Permian Asaka Limestone in Gifu Prefecture (locality 12, Figure 11.1) has yielded teeth referred to *Symmorium* sp. (Figure 11.7B), *Petrodus* sp. (Figure 11.7C) cochiodontid indet., and '*Neopetalodus*' (Goto *et al.*, 1988). The edestid sharks are represented in Japan by a tooth row of *Helicoprion bessonowi* Karpinsky from the Middle Permian Yagihara Limestone, described by Yabe (1903), and a tooth row of *Helicoprion* sp. from the contemporaneous Kanokura Formation (Goto *et al.*, 1988). '*Serratodus*' sp. is known by a tooth from the Lower–Middle Permian Ryozensan Limestone (Goto *et al.*, 1988).

Late Permian

Kitakami Mountains, Central Japan
A tooth row of *Helicampodus* sp. was reported from the Late Permian Toyama Formation of the Kitakami Mountains by Uyeno *et al.* (1979).

ACKNOWLEDGEMENTS

Sincere thanks go to Drs Tran Van Tri and Pham Kim Ngan, Geological Research Institute, Hanoi, for their co-operation during two successful field trips in Vietnam. Funding for the Vietnamese trips was through the National Geographic Society. I thank Michael Ginter, Warsaw University, Poland, for looking at the conodonts associated with Vietnamese and Thai microvertebrates. I also thank Dr Clive Burrett, University of Tasmania, whom I worked with on several trips collecting in Thailand and Vietnam, and who gave much helpful discussion on the subject of south-east asian geology; and to Dr Sangad Bunopas, Thai Geological Survey, for co-operation and helpful discussion during fieldwork in Thailand.

REFERENCES

Blieck, A. and Goujet, D., 1978, A propos de nouveau matériel de Thelodontes (Vertébrés Agnathes) d'Iran et de Thaïlande: aperçu sur la répartition géographique et stratigraphique des Agnathes des «regions gondwaniennes» au Paléozoïque moyen, *Annales de la Societe Géologique du Nord*, **XCVII**: 363–72.

Blieck, A., Goujet, D., Janvier, P. and Lelièvre, H., 1984, Microrestes de vertébrés du Siluro-Dévonien d'Algérie, de Turquie et de Thaïlande, *Géobios*, **17**: 851–6.

Burrett, C., Long, J. and Stait, B., 1990, Early–Middle Palaeozoic biogeography of Asian terranes derived from Gondwana. In W.S. McKerrow and C.R. Scotese (eds), *Palaeozoic Palaeogeography and Biogeography, Geological Society Memoir*, **12**: 163–74.

Campbell, K.S.W. and Barwick, R.E., 1988, *Uranolophus*: a reappraisal of a primitive dipnoan, *Memoirs of the Association of Australasian Palaeontologists*, **7**: 87–144.

Chhibber, H.L., 1936, *The Geology of Burma*, McMillan and Co., London.

Gardiner, B.G., 1984, Relationships of the palaeoniscoid fishes, a review based on new specimens of *Mimia* and *Moythomasia* from the Upper Devonian of Western Australia, *Bulletin of the British Museum of Natural History (Geology)*, **37**: 173–428.

Ginter, M., 1990, Late Famennian shark teeth from the Holy Cross Mts, central Poland, *Acta Geologica Polonica*, **40**: 69–81.

Goto, M., 1975, New find of the Permian and Triassic fish fossils in Japan—on the dermal denticles and teeth of fishes discovered from Karasawa area southeast Asio Mountains, *Earth Sciences (Chikyu Kagaku)*, **29**: 72–4, [in Japanese].

Goto, M., 1984, Discovery of a petalodont tooth from the Nabeyama Formation (Middle Permian) in Kinzuu, Tochigi Prefecture, Central Japan, *Earth Science (Chikyu Kagaku)*, **38**: 140–2, [in Japanese].

Goto, M., 1989, Researches on Palaeozoic fish remains in Japan, *Anima*, **199**: 108–11, [in Japanese].

Goto, M. and Kuga, N., 1982, A review on the fossil Chondrichthyes of Mesozoic and Palaeozoic of Japan, *Fossil Club Bulletin*, **14**: 47–53, [in Japanese].

Goto, M., Okura, M. and Ogawa, H., 1988, On teeth and dermal denticles of Chondrichthyes from the Asaka Limestone (Middle Permian), Central Japan, *Earth Science (Chikyu Kagaku)*, **42**: 290–7, [in Japanese].

Ingavat, R. and Janvier, P., 1981, Bradyodont (Chondrichthyes) teeth from the Permian and Carboniferous of Northern Thailand, *Géobios*, **14**: 651–53.

Janvier, P., 1981, On some fish remains from the Permian of northeastern Thailand, *Journal of the Geological Society of Thailand*, **4**: 23–8.

Janvier, P., Blieck, A., Gerrienne, P. and Thanh, T.-D., 1987, Faune et flore de la Formation de Sika (Dévonien inférieur) dans la presqu'ile de Do Son (Vietnam), *Bulletin de la Musee Nationale d'Histoire Naturelle, Paris*, **9**: 291–301.

Janvier, P., Gerrienne, P. and Thanh, T.-D., 1990, Les placodermes, arthropodes et lycophytes des Grès Dévoniens de Do Son (Haiphong, Vietnam), *Géobios*.

Long, J.A., 1989, Discovery of Late Devonian vertebrate microfauna from near Mae Sariang, *Mineral Resources Gazette, Bangkok*, **34**: 25–8.

Long, J.A., 1990, Late Devonian chondrichthyans and other microvertebrate remains from northern Thailand, *Journal of Vertebrate Paleontology*, **10**: 59–71.

Long, J.A. and Burrett, C.F., 1989, Fish from the Upper Devonian of the Shan-Thai terrane indicate proximity to east Gondwana and south China terranes, *Geology*, **17**: 811–13.

Long, J.A., Turner, S. and Young, G.C., 1988, A Devonian fish fauna from subsurface sediments in the eastern Officer Basin, South Australia, *Alcheringa*, **12**: 61–78.

Long, J.A., Burrett, C.F., Ngan, P.K. and Janvier, P., 1990, A new bothriolepid antiarch (Pisces, Placodermi) from the Devonian of Do Son peninsula, northern Vietnam, *Alcheringa*, **14**: 181–94.

Okura, M. and Goto, M., 1988, On the dermal armours from the Fukiji Formation (lower Devonian), Central Japan, *Abstracts of the 137th Regular Meeting of the Palaeontological Society of Japan*: 41.

Ørvig, T., 1975, A new radotinid arthrodire from Arctic Canada, In *Problèmes actuels de paléontologie (évolution des vertébrés)*, Editions du CNRS, Paris, 1975: 41–72.

Tong, D.-T., 1986, *He Devon o Viet Nam*, Nha Xuat Ba'n Khoa Hoc Va Ky Thuat, (*The Devonian System of Vietnam*, Editions Scientific and Technical, Hanoi), 141pp, [in Vietnamese].

Tong, D.-T. and Janvier, P., 1987, Les vertébrés Dévoniens du Vietnam, *Annales de Paléontologie*, **73**: 165–94.

Tong, D.-T. and Janvier, P., 1990, Les vertébrés du dévonien inférieur du Bac Bo oriental (provinces de bac Thai et Lang Son, Vietnam), *Bulletin de Museum Nationale d'Histoire Naturelle, Paris, séries 4*, **12**: 143–223.

Turner, S., 1982, Middle Palaeozoic elasmobranch remains from Australia, *Journal of Vertebrate Paleontology*, **2**: 117–31.

Uyeno, T., Mori, K. and Tazawa, J., 1979, An upper Permian helicoprionid fish from Toyama Formation, Kitakami Mountains, Japan, *Abstracts of the 1979 Regular Meeting of the Palaeontological Society of Japan*.

Wang, S., 1989, Biostratigraphy of vertebrate microfossils. In Ji Qiang *et al.* (eds), *The Dapoushang Section—an excellent section for the Devonian-Carboniferous Boundary Stratotype in China*, Science Press, Beijing, China: 36–8, 103–7.

Wang, S. and Cao, R., 1988, Discovery of Macropetalichthyidae from Lower Devonian in Western Yunnan, *Vertebrata Palasiatica*, **26**: 73–5.

Wang, S., Dong, Z. and Turner, S., 1986, Middle Devonian Turiniidae (Thelodonti, Agnatha) from western Yunnan, China, *Alcheringa*, **10**: 315–25.

Wang, S. and Turner, S., 1985, Vertebrate microfossils of the Devonian–Carboniferous boundary, Muhua Section, Guizhou Province, *Vertebrata Palasiatica*, **23**: 223–34.

Yabe, H., 1903, On a fusulina-limestone with *Helicoprion* in Japan, *Journal of the Geological Society of Tokyo*, **10**: 1–13.

PART THREE

PALAEOZOIC VERTEBRATE BIOGEOGRAPHY

Chapter 12

VERTEBRATE FAUNAL PROVINCES IN THE MIDDLE PALAEOZOIC

Gavin C. Young

INTRODUCTION

The Devonian 'Age of Fishes' represented the first major radiation of vertebrates into fluviatile environments, and was associated with similar radiations onto land and/or its water bodies in a range of invertebrate and plant groups during the Siluro–Devonian (e.g. Gray, 1988). Freshwater fishes have long been regarded as having a special place in biogeographic studies because of their limited dispersal capabilities (e.g. Darlington, 1957). The distribution of fossil fishes which occurred in non-marine deposits therefore attracted attention from the biogeographic viewpoint, but firm conclusions were difficult to reach because of their limited fossil record (e.g. Patterson, 1975). Nevertheless the diverse vertebrate faunas of the Devonian 'Old Red Sandstone' of Europe and North America indicated some provincial patterns which suggested explanation in terms of the palaeogeography of the time. Early contributions to this topic (e.g. Halstead and Turner, 1973) emphasised the Northern Hemisphere regions (the 'Old Red Continent'), and since discoveries early this century (e.g. Woodward, 1921; Hills, 1929, 1931) it was concluded that some widespread taxa typical of the Old Red Sandstone had dispersed to remote parts of the Southern Hemisphere like Antarctica and Australia. Some studies concerned with biostratigraphic utility (e.g. Gross, 1950) emphasised the wide dispersal capabilities of such forms. Others (Turner, 1970; Turner and Tarling, 1982) attempted to relate fish occurrences to proposed reconstructions of global geography for the Middle Palaeozoic.

In the last 25 years detailed systematic studies in extra-European areas such as central and eastern Asia, and the southern continents, have enlarged the taxonomic database, which can now be viewed from a global perspective (Figure 12.1). These new data have revealed a high proportion of previously overlooked endemic taxa in different regions, which necessitated

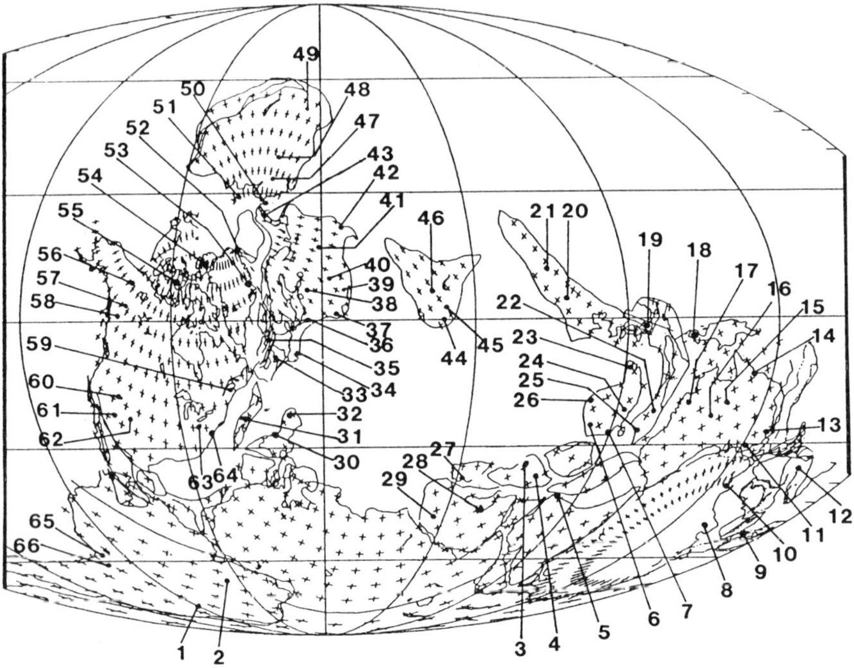

Figure 12.1. Locality map for Devonian vertebrate occurrences (modified from Young, 1990). 1, Tibagi, Parana Basin, Brazil; 2, Parnaiba Basin, Brazil; 3, Khush-Yeilagh, Iran; 4, Hutk-Bidou, Iran; 5, Kashmir and Himachal Pradesh, India (possibly suspect); 6, Qujing, Yunnan, China; 7, northeast Vietnam; 8, Ohio Range, Transantarctic Mountains, Antarctica; 9, Ellsworth Mountains, Antarctica; 10, southern Victoria Land, Antarctica; 11, Grampians, Victoria, Australia; 12, Reefton, New Zealand; 13, southeastern Australia; 14, Broken River, Australia; 15, Georgina Basin, Australia; 16, Amadeus Basin, Australia; 17, Canning Basin, Australia; 18, Kepala Burung, Irian Jaya; 19, Fukuji, central Japan; 20, Ningxia, China; 21, Gansu, China; 22, Mae-Sariang and Mae Sam Lap, Thailand; 23, Nanjing, China; 24, South China foldbelt; 25, Shidian, West Yunnan, China; 26, Szechuan, China; 27, Antalya, Turkey; 28, Qatar; 29, Jauf, Saudi Arabia; 30, Artois-Ardenne, France, Belgium; 31, Knoydart, Nova Scotia; 32, Barrandian Basin, Bohemia; 33, Welsh Borderlands, UK; 34, Rheinisches Schiefergebirge, Germany; 35, Scotland, UK; 36, central Poland; 37, Podolia, Ukraine; 38, Baltic States; 39, Donbas, Donets Basin; 40, Tula, Russian Platform; 41, Sharya, Russian Platform; 42, Ufa, southern Urals; 43, Novaya Zemlya; 44, southern Tien Shan, Kazakhstan; 45, Tashkent, Tien Shan; 46, Teniz, central Kazakhstan; 47, Taymyr, Siberia; 48, Kureyka River, Siberia; 49, Minusa and Tuva Basins, Altai Sayan; 50, Severnaya Zemlya; 51, Kotelny Island; 52, Spitsbergen; 53, east Greenland; 54, Goose Fjord, Ellesmere Island; 55, western Prince of Wales Island; 56, Snake River, Yukon; 57, Glacier Lake, Northwest Territories; 58, Beaver River, southeast Yukon; 59, Escuminac Bay, Quebec, Canada; 60, Utah, Wyoming, USA; 61, Arizona, USA; 62, Colorado, USA; 63, Ohio, USA; 64, Virginia-Pennsylvania, USA; 65, Sicasica, Bolivia; 66, Seripona, Bolivia.

reappraisal of the earlier dispersalist explanation for Devonian vertebrate distributions. Using this new evidence Young (1981) identified five distinctive faunal provinces for vertebrates of Early Devonian age (Figure 12.2A). Previously, Boucot *et al.* (1969) had identified similar patterns of Early Devonian endemism for shallow marine invertebrate faunas, which contrasted with more widespread assemblages in the Late Devonian. A Gondwana origin and Late Devonian–Early Carboniferous dispersal into Europe had been proposed for tetrapods (Panchen, 1977), and some osteolepid rhipidistians (Janvier *et al.*, 1979), and the wide distribution in both northern and southern continents of some Late Devonian placoderms (e.g. *Bothriolepis*, *Remigolepis*, *Phyllolepis*, *Groenlandaspis*) and dipnoans (*Soederberghia*) was noted (Ritchie, 1975; Campbell and Bell, 1977).

There is now a considerable body of evidence supporting an overall pattern of Early Devonian endemism, and Late Devonian cosmopolitanism for vertebrate faunas (Young, 1990), plants (Edwards, 1990), and various shallow marine (benthic) invertebrates (e.g. Johnson and Boucot, 1973; Boucot, 1975; 1988; Blodgett *et al.*, 1990; Boucot and Gray, 1979; Stock, 1990). Here I summarise the latest evidence on Early Devonian vertebrate provinces and their characteristic endemic taxa, and contrast this with broader distribution patterns in the Late Devonian, when faunal communication between Gondwana, Euramerica and China is indicated. The five original Early Devonian provinces correspond closely to five distinctive palaeogeographic regions (Euramerica, Siberia, Tuva, China, and East Gondwana, see numbered headings below). After summarising these vertebrate faunas a brief review is given of less well known vertebrate occurrences from the Devonian of other palaeocontinental blocks and smaller terranes identified in recent palaeogeographic reconstructions (e.g. Scotese, 1986; Scotese and McKerrow, 1990). The concluding section considers some causal explanations which have been proposed to account for this change in pattern, in the context of different palaeogeographic reconstructions.

THE FIVE PROVINCES OF YOUNG (1981)

1. Euramerica

In the Devonian this palaeocontinent had formed by coalescence of Laurentia (north America, northwest Ireland, Scotland, Greenland, etc.) and Baltica (Europe west of the Urals). The Euramerican Province (Figures 12.2A, 3; 'cephalaspid province' of Young, 1981) is characterised by a major group of armoured agnathans, the cephalaspid osteostracans (Figure 12.4B), a group otherwise only known from central Asia (see below). Associated higher taxa include heterostracan agnathans (Early Devonian cyathaspids, pteraspids, etc.; psammosteids in the Givetian–Frasnian), and many species, genera, and some families of placoderms and osteichthyans restricted to the region (Figure 12.4A, C). The northeastern extremity of this fauna is defined by agnathan occurrences in Novaya Zemlya (Mark-

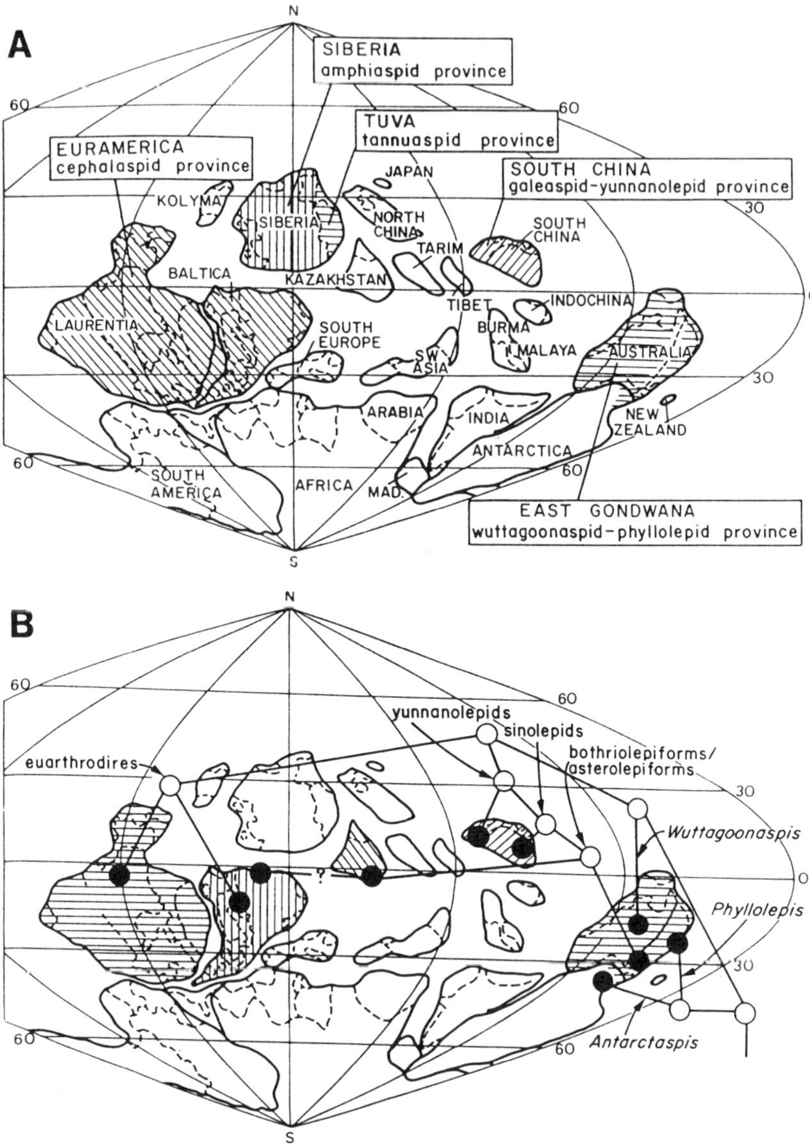

Figure 12.2. A. Five major vertebrate faunal provinces of Young (1981) on a palaeogeographic reconstruction of Heckel and Witzke (1979). B. Cladogram for various placoderms superimposed on the reconstruction of A (from Young, 1981, Figure 12.6). An area cladogram inferred from the taxon cladogram would only be significant if congruent patterns were demonstrated for other groups endemic to the areas concerned.

Kurik and Novitskaya, 1977), and Severnaya Zemlya, which lie inside the margin of the Kara Block, north of the Taymyr Suture (43, 50, Figure 12.3). The southern margin in Europe (34, 37, Figure 12.3) is indicated by osteostracan faunas of Podolia (Janvier, 1985) and the German Rhineland (Friman and Janvier, 1986; Bardenheuer and Janvier, 1990), although the latter is a structurally complex region (e.g. Belliere and Laduron, 1990), and may include elements of the Gondwana fauna (Young, 1987b, 1990; see *Armorica* below). From the Laurentian block, vertebrate evidence from Nova Scotia (Denison, 1955; Dineley, 1967; 31, Figure 12.3) demonstrates that the 'Avalon terrane', considered by Van der Voo (1988) to be attached to Gondwana, was associated with Euramerica by Lochkovian time.

2. *Siberia*

The Siberian block (Blieck and Janvier, Chapter 5) has a diverse endemic vertebrate fauna characterised by the amphiaspid heterostracans, a derived endemic group from Taymyr and the northwestern Siberian platform (48, 51, Figure 12.3; Novitskaya, 1971; Mark-Kurik, 1974), but not distributed across the Taymyr suture, which forms its northern margin. The older (Lochkovian–Pragian) and younger (Emsian) parts of these successions are correlated with those of Euramerica on pteraspid heterostracans and placoderms respectively (Blieck, 1984; Mark-Kurik, 1991). The occurrence of pteraspids suggests that Siberia was close to Euramerica in the Early Devonian. Possible pteraspids from northeastern Siberia mentioned by Cherkesova (1973; see Blieck, 1984; Blieck and Janvier, Chapter 5) come from the Chukotka block, which includes the Kotelny Island fauna, with both Siberian (*Gerronaspis*) and Euramerican affinities (Mark-Kurik, 1974). This block is associated with Laurussia by Ziegler (1988; see also Chapter 5).

The high northern palaeolatitude for Siberia based on palaeomagnetic evidence shown on many reconstructions (e.g. Figure 12.3) has been problematic. Edwards (1990) distinguished the Siluro–Devonian rhyniophytoid plant assemblages of Euramerica from the very endemic Devonian floras from Siberia and Kazakhstan, and she interpreted the Siberian flora as a rare example of high latitude vegetation. However, Heckel and Witzke (1979) discussed palaeoclimatic evidence from Siberia, which has thick carbonate and evaporite sequences, indicating a position in the Devonian tropics (Figure 12.2A). Both Witzke and Heckel (1988) and Witzke (1990) place Siberia close to Euramerica, within the 'carbonate belt' at latitudes of less than 45°.

3. *Tuva*

The Tuva–Minusinsk basins of the Altai Sayan, which lie east of the Irtysh suture, are shown as part of the Siberian block on most Devonian reconstructions (Figure 12.3). Young (1981) proposed a separate 'tannuaspid' province to account for endemic faunas from this region,

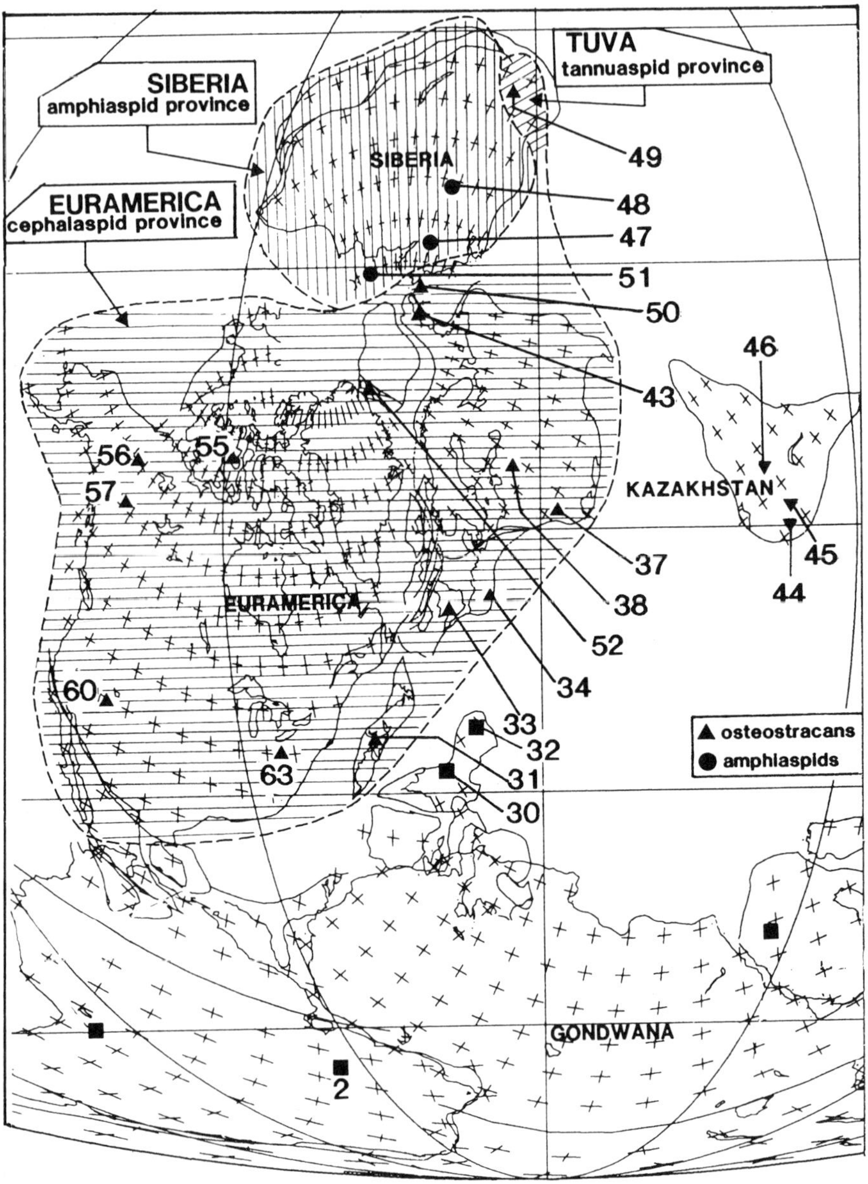
TUVA
tannuaspid province
SIBERIA
amphiaspid province
EURAMERICA
cephalaspid province
SIBERIA
EURAMERICA
KAZAKHSTAN
GONDWANA
49
48
47
51
50
43
46
45
44
37
38
52
34
33
32
31
30
56
55
57
60
63
2
▲ osteostracans
● amphiaspids

which include tannuaspid osteostracans (Figure 12.4D; Afanassieva and Janvier, 1985), early sharks (*Elegestolepis*; Karatajute-Talimaa, 1973), primitive quasipetalichthyid like placoderms (Figure 12.5A), and endemic Late Devonian crossopterygians associated with cosmopolitan placoderms from the Frasnian ('*Bothriolepis cellulosa*', *B. siberica*, *Dipterus*, *Megistolepis*, etc.) and Famennian (*B. siberica*, *Thaumatolepis?*). Noteworthy is the absence of *Phyllolepis* from the latter fauna. This area has the only confirmed occurrences of cephalaspids outside the Euramerican region, and the only ones on the Siberian block (49, Figure 12.3). There is no evidence of vertebrate faunal affinity with the distinctive endemic faunas of Siberia (Blieck and Janvier, Chapter 5).

4. South China

The South China Province of Young (1981) was based on the Early Devonian fish fauna of endemic galeaspid agnathans and yunnanolepid placoderms (Figure 12.2A; see also Wang, Chapter 10). This fauna is currently documented to contain some 50 mainly endemic genera belonging to endemic higher taxa, known from many localities in the Yangtze and Zhujiang regions (Liu, 1991) of the South China Paraplatform (Pan and Dineley, 1988, Figure 7). These include localities in the southeast (South China Fold System) such as southern Guangxi and northern Guangdong Provinces (24, Figure 12.6; Wang, 1984), and possibly Honk Kong (Lee, 1991), a distribution which contradicts the interpretation of Hsu *et al.* (1988; also Sun *et al.*, 1989) that the South Chinese region comprised separate Palaeozoic 'Yangtze' and 'Huanan' terranes. The endemic Early Devonian South China fauna also occurs in northeast Vietnam (7, Figure 12.6), where galeaspids, yunnanolepids, sinolepids and youngolepids have been described (Thanh and Janvier, 1987, 1990). This is north of the Red

---

Figure 12.3. Euramerica, Siberia, and Tuva (cephalaspid, amphiaspid and tannuaspid provinces) on the reconstruction used by Young (1990). Localities discussed in the text (numbered as in Figure 12.1) are: 30, Artois-Ardenne, France, Belgium; 31, Knoydart, Nova Scotia; 32, Barrandian Basin, Bohemia; 33, Welsh Borderlands, UK; 34, Rheinisches Schiefergebirge, Germany; 37, Podolia; 38, Baltic States; 43, Novaya Zemlya; 44, southern Tien Shan; 45, Tashkent, Tien Shan; 46, Teniz, central Kazakhstan; 47, Taymyr, Siberia; 48, Kureyka River, Siberia; 49, Minusa and Tuva Basins, Altai Sayan; 50, Severnaya Zemlya; 51, Kotelny Island; 52, Spitsbergen. The high northern palaeolatitude indicated by palaeomagnetic data for Siberia is contradicted by thick carbonate and evaporite sequences, indicating a position in the Devonian tropics (see text).

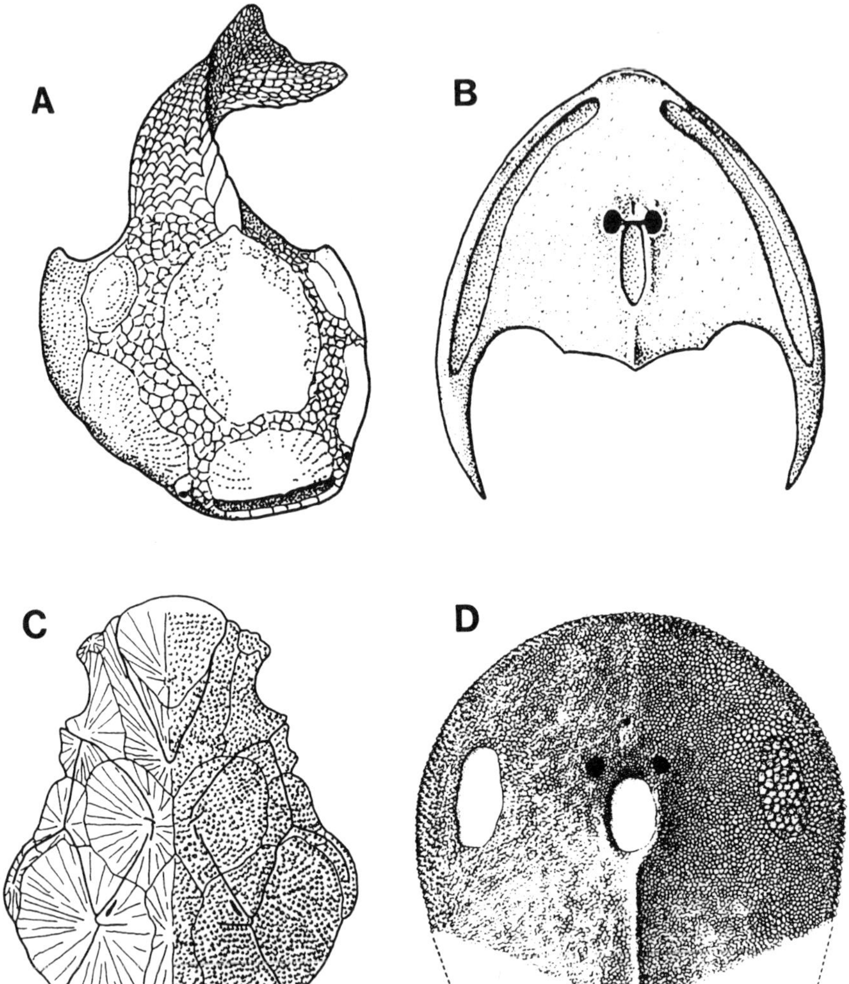

Figure 12.4. Various endemic taxa of Euramerica and Siberia. A, a Late Devonian psammosteid heterostracan; B, an Early Devonian cephalaspid osteostracan; C, a Middle Devonian actinolepid arthrodire; D, the Early Devonian osteostracan *Tannuaspis* [D from Afanassieva and Janvier (1985); A–C from Young (1987b), which gives original sources].

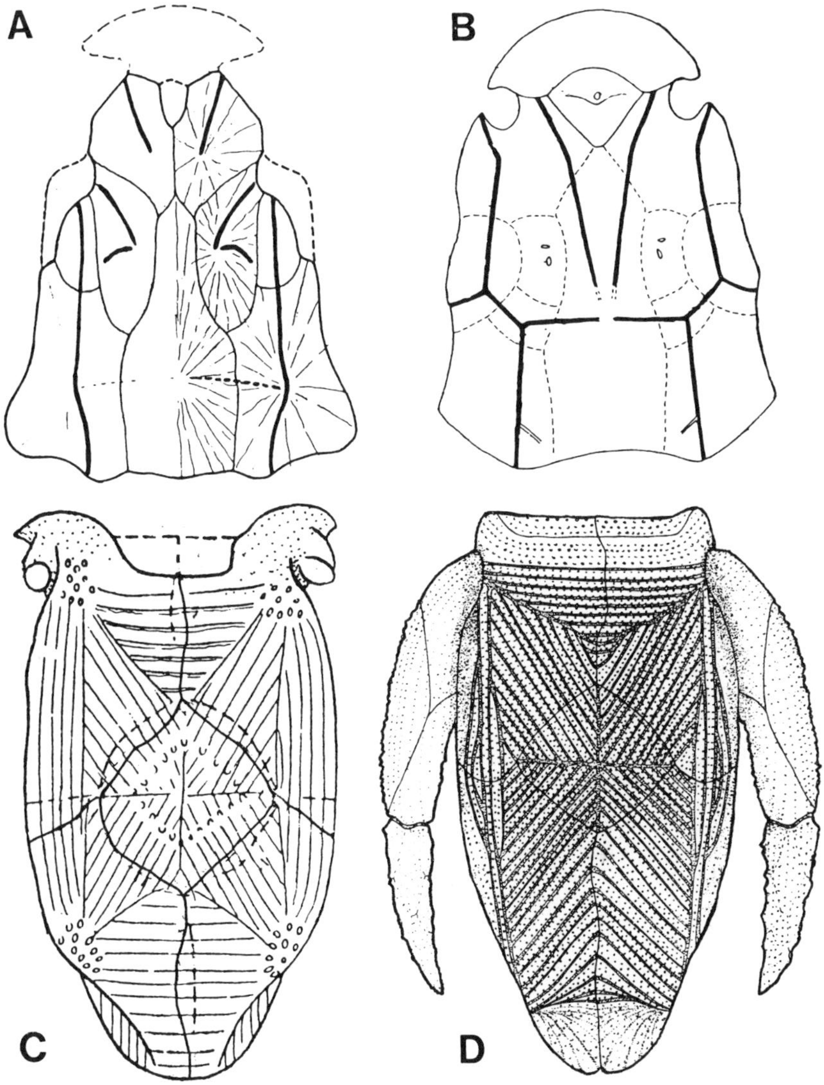

Figure 12.5. Placoderms suggesting affinity between central and east Asia. A, skull pattern of an undescribed genus from the Middle Devonian (?Givetian) of the northern Minusinsk Depression; B, skull of the Chinese genus *Diandongpetalichthys* (from Zhu, 1991, Figure 1B); C, *Stegolepis*, and D, *Hunanolepis*, both ventral views of the trunk-armour, after Malinovskaya (1973) and J.Q. Wang (1991).

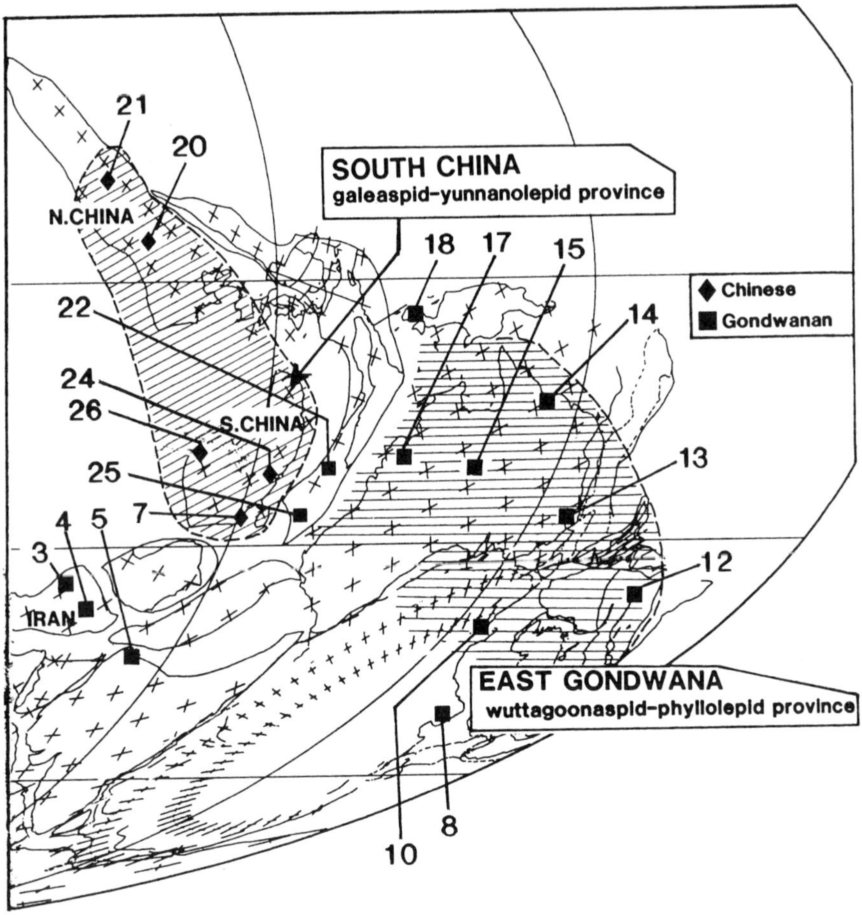

Figure 12.6. East Gondwana, North and South China on the reconstruction used by Young (1990). Localities discussed in the text (numbered as in Figure 12.1) are: 3, 4, Khush-Yeilagh, and Hutk-Bidou, Iran; 5, Kashmir, Himachal Pradesh, India (possibly suspect; see Talent *et al.*, 1988); 7, northeast Vietnam; 8, Ohio Range, Transantarctic Mountains, Antarctica; 10, southern Victoria Land, Antarctica; 12, Reefton, New Zealand; 13, southeastern Australia; 14, Broken River, Australia; 15, central Australia (Georgina and Amadeus Basins); 17, Canning Basin, Australia; 18, Kepala Burung, Irian Jaya; 20, Ningxia, China; 21, Gansu, China; 22, Mae-Sariang and Mae Sam Lap, Thailand; 24, South China foldbelt; 25, Shidian, West Yunnan, China; 26, Szechuan, China;

River–Song Ma suture zone, which forms the southern and southwestern borders of the South China block. A different fauna occurs across this zone within the Sanjiang Fold System in western Yunnan (25, Figure 12.6), which is considered to be part of the Sibumasu (Shan-Thai) terrane (Wang *et al.*, 1986; Burrett *et al.*, 1990). By Late Devonian time the strong endemism is less pronounced, but still evident in the antiarch assemblages of the lower Yangtze valley (Liu and Pan, 1958) and Jiangxi Province (Zhang and Liu, 1991).

The affinities of the non-endemic element of the South China fauna are various. Middle Devonian affinity with Euramerica is indicated by several groups (e.g. the antiarchs *Microbrachius* and *Byssacanthus*), and possible connections with central Asia and Kazakhstan by the quasipetalichthyid features of undescribed placoderms from the Minusinsk area, and similarities in antiarchs (Figure 12.5). Givetian bothriolepid antiarchs (?non-marine) show affinities with eastern Australia and Victoria Land, Antarctica (Young, 1988a), and by the latest Famennian the Gondwana connection is strengthened, with the occurrence of the endemic Chinese antiarch suborder Sinolepidoidei in eastern Australia (Ritchie *et al.*, 1992). Affinities of the major endemic groups are still contentious. Yunnanolepid antiarchs are the most primitive representatives of the group, and may be paraphyletic. Because they appear earlier in South China than elsewhere, an evolutionary origin and dispersal from South China can be hypothesised (discussed by Young, 1984). For galeaspids, Young (1981) showed a scheme of relationships placing them as the sister-group to cephalaspids, from which an area cladogram was deduced which reflected this higher level relationship. Since then Janvier (1981, 1984) and N.Z. Wang (1991) have respectively argued that cephalaspids or galeaspids were the sister-group to gnathostomes, and Young (1991) represented this relationship as a trichotomy, based on evidence of the new agnathan *Pituriaspis* from Australia, which together with the associated *Neeyambaspis* resembles the huananaspid galeaspids in possessing long rostral processes.

A recent Early Devonian reconstruction of South China against the northwest margin of Australia (Scotese, 1986; Young, 1990) places it between about 20° and 30° south latitude (Figure 12.6), compared to the earlier reconstruction of Heckel and Witzke (1979) where it lay just above the equator as an island continent midway between Euramerica/Kazakhstan and Australia (Figure 12.2A). Emsian marine faunas of eastern Australia show some affinity with the placoderms of South China (e.g. the buchanosteid arthrodire *Kweichowlepis*), but in the Siluro–Devonian the continental vertebrate faunas of South China and eastern Australia are very different. Several lines of evidence suggest an equatorial position. As well as isolation, an adequate rainfall to support an extensive system of rivers and lakes is required before a diverse freshwater fish fauna can develop, as is indicated by diversity differences in modern faunas (Young, 1990). The modern freshwater fish faunas of Africa and South America are very diverse (32 and 46 families respectively; Berra, 1981). By comparison, the modern freshwater fish fauna of Australia, with a low rainfall in the subtropical dry belt, contains only 18 families, even though it is an island continent surrounded by ocean barriers. By analogy with the modern Amazon basin, a position straddling the equator for South China, in the equatorial doldrums rain belt, is consistent with other evidence by Hou and Wang (1985), and supported by palaeomagnetic data from Guangxi (Liu *et al.*, 1988) and Yunnan (Fang *et al.*, 1989). A close connection with East Gondwana may have been attained in the Late Devonian/Early Carboniferous (see below).

5. *East Gondwana*

Young (1981) identified a wuttagoonaspid–phyllolepid province in East Gondwana, based on the Emsian–Eifelian endemic placoderm *Wuttagoonaspis* (Ritchie, 1973), and the diversity of the probably related phyllolepids of the Givetian–Famennian (e.g. Long, 1984; Ritchie, 1984; Young, 1988b). The oldest (pre-Emsian) faunas of Australia contain only acanthodians, thelodonts and a new undescribed lungfish (J. Long, pers. comm.), in contrast to the highly diverse faunas of the Lochkovian–Pragian of other continental blocks, such as Euramerica and South China. The position of Australia on current reconstructions in the subtropical dry belt provides an explanation for its impoverished pre-Emsian Devonian fish fauna, as it partly does today for the low diversity of the modern Australian freshwater fish fauna compared to that of southeast Asia or the Amazon basin (see above).

The East Gondwana province (Figure 12.2A) was originally identified only in Australia and possibly Antarctica, but some elements (e.g. antarctilamnid sharks) may have been widespread across the Gondwana landmass (Young, 1987a, 1990). The biostratigraphic distribution of thelodonts and other groups (Young, 1988a, 1989; Turner and Young, 1992) indicates that the endemism of the East Gondwana region persisted into the Givetian–Frasnian. However, Famennian faunas are characterised by close similarity with Euramerica in such faunal elements as Late Devonian placoderms (e.g. *Bothriolepis*, *Remigolepis*, *Groenlandaspis*, phyllolepids) and dipnoans (*Soederberghia*; Ritchie, 1975; Campbell and Bell, 1977), whilst the presence of sinolepid antiarchs is a resemblance to China. Various explanations may be proposed to account for the disappearance of the East Gondwana faunal province (see below).

OTHER PALAEOGEOGRAPHIC UNITS

Kazakhstan

The palaeogeographic setting of this block is uncertain. It may have been separated from Siberia by some 750–1000 km (Talent *et al.*, 1987), but Scotese and McKerrow (1990) suggest it was an extension of the Siberian block similar to the extension of the Malay Peninsula from southeast Asia today. However, there is no evidence of such a relationship in the vertebrate faunas, which tend to be widespread or endemic, but with some indication of affinity with South China in the Middle Devonian as noted above. The acanthothoracid placoderm *Kimaspis* (Mark-Kurik, 1973) from a Lochkovian marine deposit in southern Kazakhstan (44, Figure 12.3), has relatives in Bohemia and Arctic Canada, so may be representative of a more widespread group. Givetian–Frasnian faunas include widespread (*Bothriolepis*) and endemic genera (e.g. antiarchs *Turanolepis*, *Tenizolepis*, and *Stegolepis* (Malinovskaya, 1973, 1989), the latter two having possible close relatives in South China (Zhang and Liu, 1991; J.Q. Wang, 1991). Famennian faunas are poorly known. Pteraspids from Kazakhstan are probably misidentified (Blieck and Janvier, Chapter 5), so apart from widespread taxa there is no indication of a special relationship either to the Siberian block or Euramerica. Palaeoclimatic evidence suggests an equatorial position for Kazakhstan, and Scotese and McKerrow (1990) place it between 20° and 40° north palaeolatitude. Phytogeographically Kazakhstan was distinctive in the Siluro–Devonian (Edwards, 1990), which is consistent with the vertebrate evidence.

North China

The North China block (Sino-Korean Platform) is traditionally separated on geological evidence from South China, as on the palaeocontinental reconstructions used by Young (1981, 1990; Figures 12.2A, 12.6). However, subsequent documentation of the Devonian vertebrate faunas from North China indicates the presence of some South Chinese endemics. From Yumen, Gansu Province, in the Qilian Foldbelt (21, Figure 12.6), comes an antiarch and arthrodire fauna (Wang, 1984; the Qilian biogeographic region of Liu, 1991), and from Ningxia (20, Figure 12.6). Middle Devonian quasipetalichthyid and Late Devonian galeaspid and sinolepid assemblages have been described (Pan *et al.*, 1987). Ritchie *et al.* (1992) have suggested a connection with South China at least by Middle Devonian time on the evidence of Givetian quasipetalichthyids and Late Devonian galeaspids and sinolepids in Ningxia. However, the Ningxia region may have had a separate history to the rest of the North China block as part of the 'Hexizoulang Terrane' (Li *et al.*, 1985).

West Gondwana

Gondwana was the largest landmass of the time, so some form of differentiation in its continental vertebrate faunas might be expected, from north to south (due to palaeoclimatic gradients) and from east to west (if geographic barriers were operative). The broad distribution of vertebrate localities across Gondwana was summarised by Young (1987a), and a detailed review of faunas currently known from West Gondwana (South America plus Africa, and including Armorica–Bohemia and the Middle East) is given by Lelièvre *et al.* (Chapter 7).

Of the major geographic constituents of West Gondwana, faunas from South America are most likely to differ from the better known assemblages from East Gondwana (Australia–Antarctica), because of geographic remoteness. Lelièvre *et al.* (Chapter 7) list 10 localities from South America (including the Malvinas Islands), but only from Bolivia (Janvier and Suarez-Riglos, 1986) is there a consistent vertebrate record ranging in age from latest Silurian to Famennian. The well-known Early Devonian faunas from Bolivia (Goujet *et al.*, 1984, 1985; Gagnier *et al.*, 1988, 1989) suggest some affinity with East Gondwana (e.g. the shark *Antarctilamna*, the thelodont *Turinia*; Young, 1990; Turner and Young, 1992). The abundance of chondrichthyans and acanthodians at the expense of placoderms may be due to high latitude and a colder climate (see Chapter 7). A Siluro–Devonian acanthodian assemblage of Euramerican aspect occurs in the Middle Amazon Basin (Janvier and Melo, 1988), and placoderm–elasmobranch assemblages have recently been discovered in the Late Devonian of both South America and South Africa, but faunal content is not yet established. Young (1987a, b) noted unconfirmed reports in early literature of spines of the acanthodian *Machaeracanthus* from Brazil (2, Figure 12.3), a taxon also found in the Ellsworth Mountains and Ohio Range of Antarctica (8, Figure 12.6), and possibly representing part of a Malvinokaffric element in the vertebrate fauna (Young, 1990). Like the placoderm *Bolivosteus* described by Goujet *et al.* (1985) from Bolivia, which is closely related to *Gemuendina* from the German Rhineland, *Machaeracanthus* is also known from southern Europe. To what extent such taxa can be considered Gondwanan elements, or examples of shallow marine dispersal, is at present unclear (see *Armorica* below). At present there is no clear pattern of north–south differentiation within South America, but this could be obscured by uncertainties about the palaeolatitude and orientation of Gondwana as a whole (see below).

Armorica

Palaeomagnetic evidence from central and southern Europe (Iberian Peninsula, Armorican Massif of France, southern Germany and Bohemia) was considered by Van der Voo (1982, 1983) to indicate a separate continent called ‘Armorica’, attached to Gondwana in the Early Palaeozoic, and drifting independently across the ‘Rheic Ocean’ during the Devonian to collide with Euramerica (Perroud *et al.*, 1984). Young (1986, 1987b) analysed this problem in terms of four competing hypotheses involving three areas (Euramerica, Gondwana, Armorica) and two hypothesised faunal barriers (Figure 12.7). The palaeontological database for vertebrates from these areas is updated by Lelièvre *et al.* (Chapter 7). The large arthrodires from Morocco (Lelièvre, 1984a, b), Bohemia and the German Rhineland (Young, 1987b, Figure 4), and various other taxa (rhenanids *Gemuendina, Bolivosteus*; machaeracanthids, etc.; Young, 1987b, Table 1) were suggested as Gondwana elements which may have been involved in shallow marine dispersal into Euramerica during the Early–Middle Devonian. Similarly, for shallow marine invertebrates there was an increase in faunal affinity between the north Gondwanan benthic faunas (southern Europe and north Africa) and those of northern Europe during the Emsian (e.g. Robardet *et al.*, 1991). As noted above, the osteostracan vertebrate assemblages from the Early Devonian of the German Rhineland indicate Euramerican

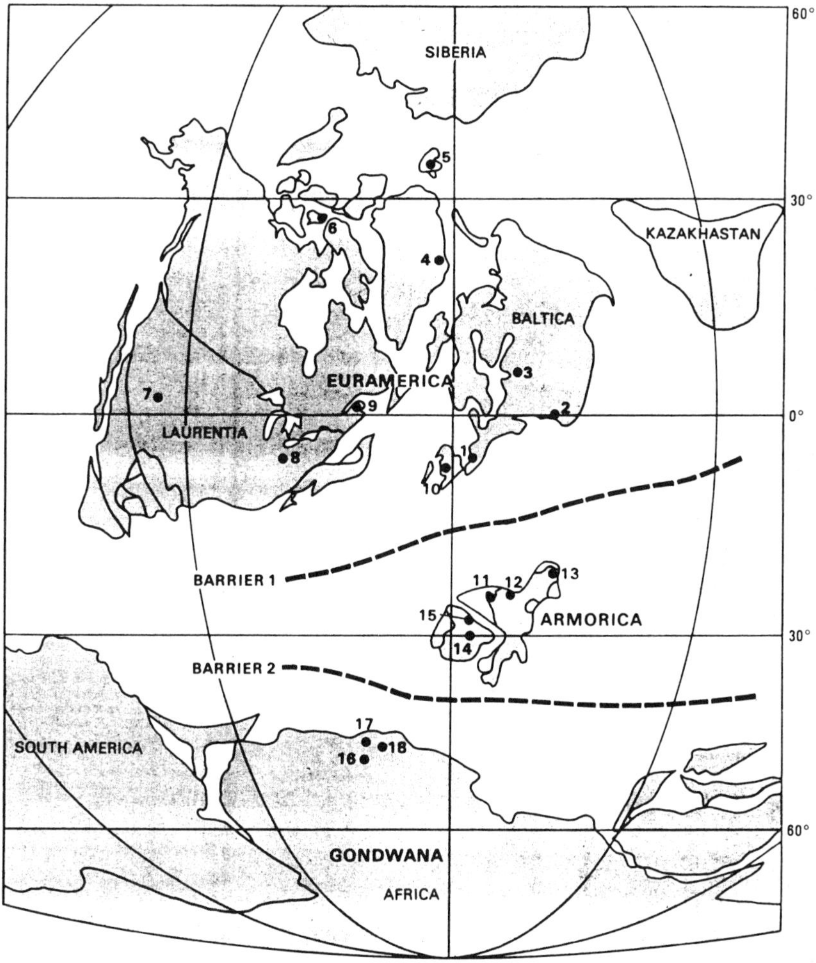

Figure 12.7. Vertebrate localities analysed in relation to the Armorica hypothesis by Young (1987b; Figure 2). Localities are: 1, Artois-Ardenne; 2, Podolia; 3, Baltic States; 4, East Greenland; 5, Spitsbergen; 6, Prince of Wales Island; 7, Utah; 8, Ohio; 9, Gaspe, Quebec; 10, Welsh Borderlands; 11, Armorican Massif; 12, Rheinisches Schiefergebirge; 13, Barrandian Basin, Czechoslovakia; 14, 15, Aragon, Palencia, Spain; 16, Ougarta, Algeria; 17, 18, Akka, Tafilalet, Morocco.

affinities, so the Gondwanan elements may be interpreted as the result of dispersal across shallow marine barriers, or vice versa. Similarly, heterostracan fragments from the Lochkovian of Spain (Blieck, 1982), and Emsian–Eifelian psammosteid heterostracans from Bohemia (Tarlo, 1964) may be examples of shallow marine dispersal indicating proximity of Baltica and Gondwana, or alternatively evidence of composite origin.

Although some palaeomagnetists (e.g. Torsvik *et al.*, 1990, Figure 12) still interpret Armorica as moving independently from Gondwana in the

Siluro–Devonian, other reconstructions (e.g. Scotese and McKerrow, 1990) have accepted the palaeontological evidence indicating that palaeomagnetic data from this area track the movement of the Gondwana landmass as a whole (see discussion of palaeomagnetic data below). Armorican palaeomagnetic data which are reliable show only a transition from high latitude in the Ordovician to equatorial latitudes in the Carboniferous, but the timing of this transition is not indicated, since the previously cited Devonian palaeomagnetic data from the area are all overprinted and, therefore, unreliable (Morzadec *et al.*, 1988).

Kolyma

The endemic acanthothoracid placoderm *Kolymaspis* was described by Buistrov (1956) from the Early Devonian of the Kolyma River—listed as from the Lochkovian Neliudim Formation by Denison (1978), but from the Vechernii Formation according to Obruchev (1964). An early Emsian age (*gronbergi-inversus* conodont Zone; acanthodian assemblage VI) is assigned by Valiukevicius (1988). Young (1980) suggested a possible sister-group relationship between *Kolymaspis* and the slightly younger *Brindabellaspis* from the Emsian of southeastern Australia, thus indicating a possible Gondwana association for Kolyma. The reported occurrence of pteraspids (Cherkesova, 1973; Blieck, 1984), which would indicate Euramerican affinities (Young, 1987a, c), come from the Chukotka block (Blieck and Janvier, this volume), which includes the Kotelny Island fauna (Mark-Kurik, 1974) with Siberian and Euramerican affinities (see above). The locality of *Kolymaspis* is in the vicinity of the Prikolymsk Terrane (Fujita and Newberry, 1983), which, like the Omolon Terrane to the east, may have originated somewhere in the proto-Pacific. The biogeographic affinities of the *Kolymaspis* assemblage, therefore, remains unresolved, and will depend on more reliable assessment of phylogenetic relationships of the endemic taxa.

Shan-Thai (Sibumasu) Terrane

This terrane is regarded as having occupied the northwestern margin of Australia in the Early Palaeozoic, from which it rifted during the Permian (e.g. Metcalfe, 1990). Vertebrate faunas (Long, Chapter 11) include rare acanthodian (*Nostolepis*) and thelodont scales from the Early Devonian of Thailand (22, Figure 12.6) reported by Blieck and Goujet (1978) and Blieck *et al.* (1984), and a Late Devonian chondrichthyan microfauna from nearby Mae Sam Lap on the Burmese border (Long, 1990). From Shidian in West Yunnan (25, Figure 12.6), a Middle Devonian (Givetian) thelodont assemblage associated with undescribed antiarchs and arthrodires was reported by Wang *et al.* (1986). There is also an unconfirmed report of a Devonian 'ichthyodorulite' from near Maymyo, central Burma (Burrett *et al.*, 1990). The species *Turinia pagoda* from Shidian is suggested by Burrett *et al.* (1990) to resemble forms from Australia, and the distribution of the

Late Devonian sharks *Phoebodus australiensis*, *Thrinacodus ferox*, and *Siamodus* in Shan-Thai (Thailand), South China, and East Gondwana (Long and Burrett, 1989; Long, 1990) are consistent with a position close to Australia in the Middle Palaeozoic. However, a Devonian palaeolatitude of 40°S proposed by Fang *et al.* (1989) is difficult to reconcile with available reconstructions, which have northwestern Australia in an equatorial position. Metcalfe (1990) shows two other southeast Asian terranes (Indochina and East Malaya) attached to South China off the northwest margin of Australia in an Early Carboniferous reconstruction (cf. Barr and Macdonald, 1987, who regard closure across the Nan River suture to have been a Permian or Triassic event). As yet there is no Palaeozoic vertebrate evidence from these areas.

CURRENT ISSUES CONCERNING DEVONIAN VERTEBRATE DISTRIBUTIONS

Cladistic representation

Young (1990) noted that the three major subdisciplines (palaeomagnetism, palaeoclimatology, palaeobiogeography) which are the main contributors to Palaeozoic palaeogeographic reconstructions (Van der Voo, 1988) tend to have their data organised and represented in different ways. Apparent polar wander path (APWP) representation facilitates testing of palaeomagnetic data against those palaeoclimatic or palaeobiogeographic data which provide evidence of palaeolatitude. But for biogeographic and other qualitative data sets which provide evidence concerning connections or barriers between regions, a hierarchical rather than APWP representation is appropriate (Young, 1990). Early work on vicariance biogeography (Platnick and Nelson, 1978; Nelson and Platnick, 1980) was applied by Young (1981) to explain distribution patterns using inferred phylogenetic relationships of endemic groups (Figure 12.2B). Later (Young, 1986) it was argued that the various hierarchically organised data sets could be better analysed with reference to area cladograms summarising competing hypotheses in palaeogeography (Figure 12.8). This research program changes the emphasis from process to pattern, from scenarios about possible causes of major events in the earth's history, to determining the degree of congruence between actual data sets organised hierarchically (Young, 1990). The development of this more rigorous approach is an aim of future research, to deal in a more formal structured way with scenarios of causal explanation, by testing competing hypotheses against each other, and with new data as they become available.

Dispersal hypothesis between Gondwana and Euramerica (Frasnian–Famennian)

The Frasnian–Famennian boundary between the two stages of the Late Devonian coincides with a well documented global sea level maximum and

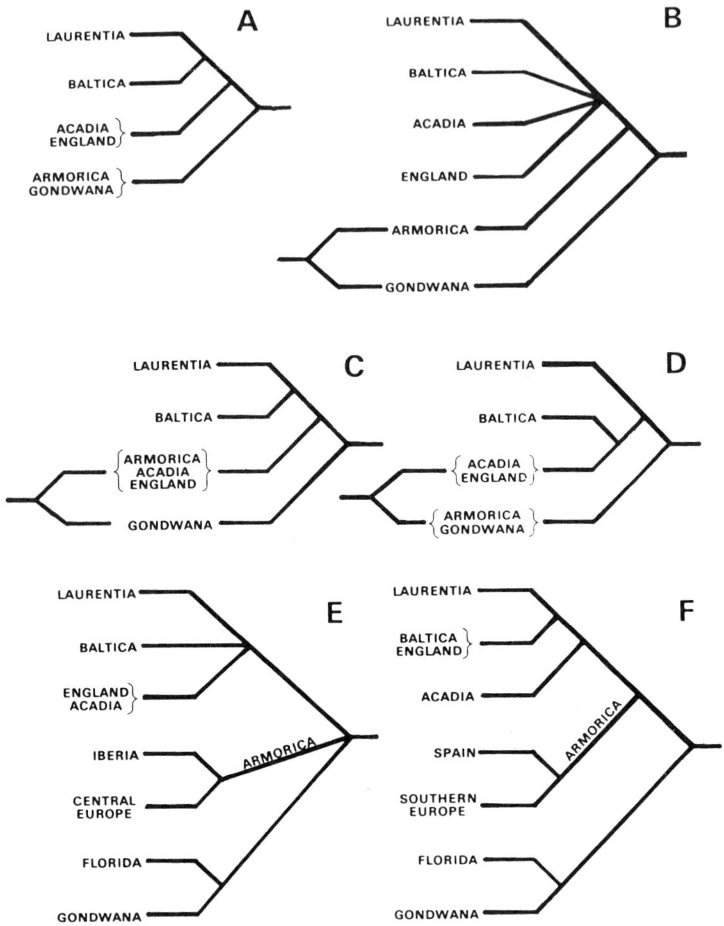

Figure 12.8. Alternative hypotheses of Palaeozoic continental collision sequences represented as cladograms (from Young, 1986).

major extinction event in the marine benthos. However, rather than extinction, continental vertebrate faunas show a marked change in distribution pattern across this boundary, attributed to a major reorganisation of palaeogeography (Young, 1981, 1987b, 1990). The dispersal model is supported by the evidence of much earlier occurrences in East Gondwana for several major vertebrate groups (e.g. freshwater xenacanth sharks, phyllolepid placoderms, large rhizodontid crossopterygian fishes, early tetrapod amphibians), and other evidence from independent data sets of various invertebrates (summarised by Young, 1987b, 1989, 1990).

This marked change in pattern between the Early and Late Devonian may be attributed to intrinsic (evolutionary) or extrinsic causal factors. Dispersal capabilities of aquatic vertebrates may have increased during the initial gnathostome radiation of the Devonian, but a predominantly extrinsic cause (e.g. global change in geography or climate) is suggested by the similar pattern for marine invertebrate faunas of Early Devonian endemism and Late Devonian cosmopolitanism, as discussed above. However, Burrett *et al.* (1990, p. 171) dismissed extrinsic factors because rising sea levels and global cooling were considered to have inhibited continental dispersal, by causing shallow marine seaways (geographic barriers) and temperature barriers leading to development of disjunct populations in different regions. Instead, they argued for intrinsic causal factors: 'the major factor responsible for changing fish distribution patterns . . . in the Late Devonian [was] physiological changes in the fishes themselves'. It was suggested that widespread antiarchs were phyletically more derived members of the respective groups, and therefore capable of physiological adaptation, to disperse along shallow sea pathways. Some widespread antiarchs (e.g. *Bothriolepis*, *Asterolepis*, *Remigolepis*) occasionally occur in marine strata, and therefore can be assumed to have dispersed along coasts. But if the complex biological database built up over many decades of research is to be acknowledged, it is not appropriate to apply a simplistic 'fish can swim' argument to groups with restricted distributions (e.g. phyllolepids) or to early tetrapods (as was argued for example by Thomson, 1980, and Spjeldnaes, 1982). Rather, one must look critically at the evidence of other data sets (palaeomagnetism, palaeoclimate) on which palaeogeographic reconstructions are based (see below).

Palaeogeographic relationship of Gondwana to Euramerica in the Late Devonian

Two competing hypotheses on this question involve differing interpretations of data from palaeomagnetism, palaeoclimate, sea-level change, and extinction and glaciation events in the Middle Palaeozoic. Widely used reconstructions of global palaeogeography which are based mainly on palaeomagnetic data (e.g. Van der Voo, 1988; Kent and Van der Voo, 1990) show close contact between Euramerica and Gondwana in the Early Devonian, and a major equatorial ocean separating the northern and southern landmasses in the Late Devonian (Figure 12.9). This is the opposite of what is indicated by the vertebrate distributions discussed above. The wide equatorial ocean for the Late Devonian places West Gondwana in high southern palaeolatitudes (Figures 12.9B, 12.10B), and under this model coalescence to form the Pangaea supercontinent was much later (Mississippian) than is required by the continental vertebrate dispersal hypothesis just mentioned. Rowley *et al.* (1985) proposed a similar (Carboniferous) timing for major palaeogeographic change, causing changes in oceanic circulation and onset of glaciation in the southern continents.

Alternatively, Copper's (1986) palaeoclimatic and oceanic circulation interpretations suggested a Late Devonian ocean closure, associated with

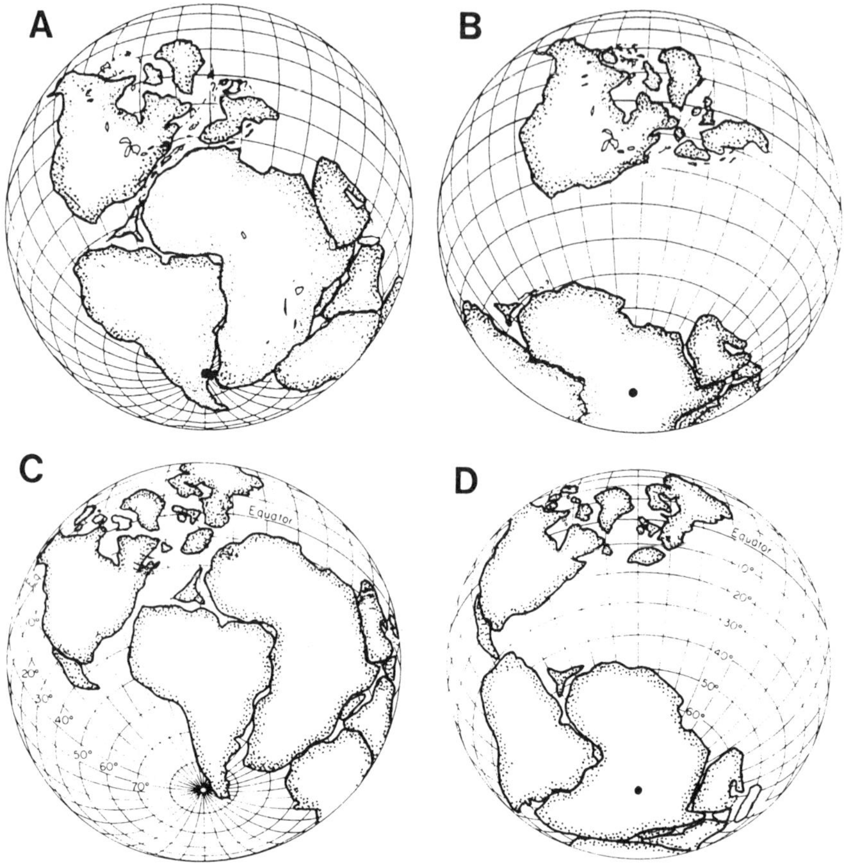

Figure 12.9. Reconstructions based on palaeomagnetic data showing the relationship of Euramerica to West Gondwana. A, B, Early and Late Devonian, after Van der Voo (1988). C, D, Early and Late Devonian, after Kent and Van der Voo (1990). Note that Early Devonian reconstructions (A,C) show close connections between Euramerica and Gondwana, and in the Late Devonian (B,D) they are widely separated, exactly the opposite of the pattern indicated by palaeontological data.

global cooling. This is consistent with evidence from supposed diamictites in Brazil for a Late Devonian glaciation (Caputo and Crowell, 1985), evidence accepted by some (e.g. Veevers and Powell, 1987; Powell and Veevers, 1987), but disputed by Boucot (1988). There is evidence of continental glaciation in the eastern half of the Parnaiba Basin, but the age could be anywhere from mid-Devonian to Early Carboniferous (Boucot, 1988).

Caputo and Crowell (1985) proposed an apparent polar wander path (APWP) based on their palaeoclimate interpretation, which resembles in general form the APWP based on palaeomagnetic data used by Kent and

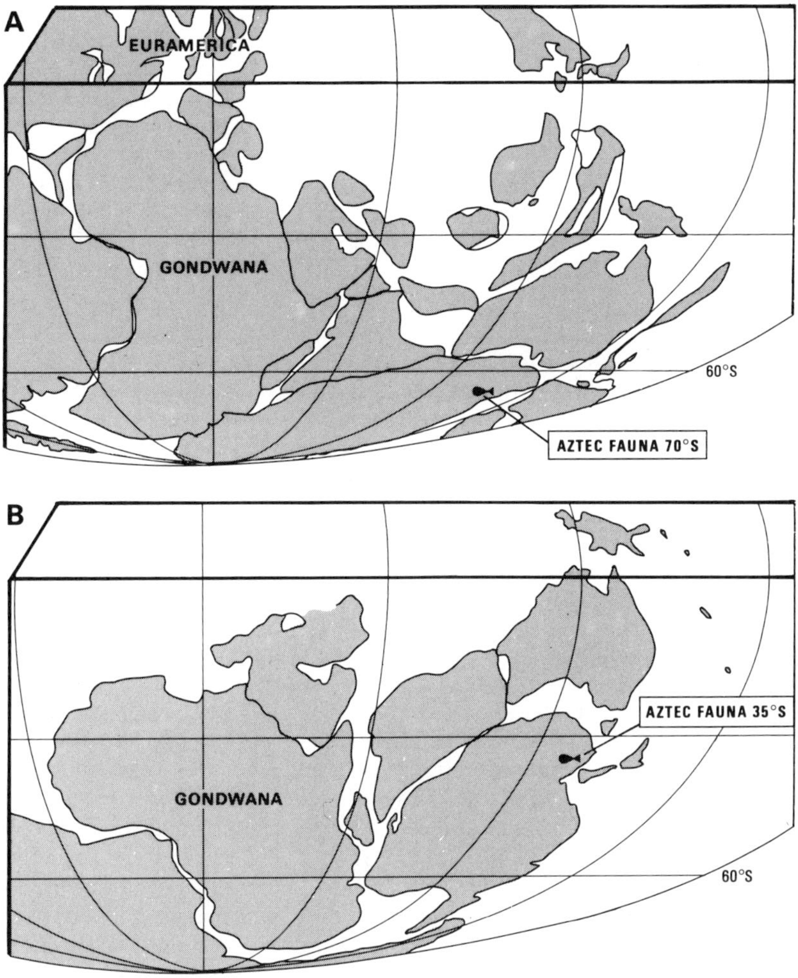

Figure 12.10. Alternative palaeogeographic reconstructions for the Late Devonian, showing very different palaeolatitudes for the Aztec fish fauna of southern Victoria Land, Antarctica (from Young, 1989).

Van der Voo (1990, Figure 2), which has a loop to place the Late Devonian south pole in central Africa. The reconstruction based on this APWP places north Africa at about 50° south latitude in the Late Devonian (Figure 12.9D), and Australia at a similar latitude to the Scotese (1986) reconstruction, which was criticised by Young (1989) because it placed the Aztec fish fauna of Antarctica too far to the south (Figure 12.10). An alternative interpretation of palaeomagnetic data by Klootwijk and Giddings (1988) corresponds well to the APWP based on palaeoclimatic indicators of

Scotese and Barrett (1990, Figure 13), which places the south pole in southern South America in the Late Devonian. Other reconstructions based on palaeoclimatic indicators (Witzke and Heckel, 1988; Scotese and McKerrow, 1990; Witzke, 1990) are in general agreement. New palaeomagnetic data of Bachtadse and Briden (1990) appear consistent with the alternative interpretation.

Regarding overall climatic trends in the Late Devonian, Boucot (1988) proposed that the disappearance of the southern cool temperate Malvinokaffric invertebrate faunas of southern South America, South Africa, and Antarctica resulted from a significant lowering of the global climatic gradient in the late Middle Devonian. A more equable climate in the Late Devonian is supported by the evidence of Streel *et al.* (1990) of uniform flora and palaeoclimate between Euramerica and West Gondwana during the Givetian–Frasnian, and the discovery by Hunicken *et al.* (1988) of Famennian conodonts from northwestern Brazil, where the putative glacials occur. In contrast, Copper (1986) has argued for global cooling to explain extinction events of the Late Devonian.

For vertebrates, Young (1989, 1900) noted the anomalously wide latitudinal distributions on palaeomagnetic global reconstructions of what were probably tropical fishes; in the modern freshwater fish fauna only 8 out of 157 families are distributed more than 50° from the equator in both hemispheres. These wide distributions could be explained by global warming in the Late Devonian, and diversity evidence from the Aztec fauna of Antarctica (over 30 taxa) is consistent with palaeoenvironmental indicators for the Aztec Siltstone suggesting a hot, seasonally wet and dry climate, with extended periods of aridity (McPherson, 1980). A similar problem is presented by the Famennian vertebrate faunas of East Greenland and eastern Australia, which have at least six genera in common, but on the Famennian reconstruction of Scotese (1986) the former is about 20°N, and the latter about 55°S (cf. Heckel and Witzke, 1979, where both lie between 15–30°S). To summarise, these apparently anomalous wide latitudinal distributions can be explained in terms of biotic dispersal, reduced global climatic gradients, or erroneous palaeogeographic base maps. Clearly there are difficulties in synthesising a large body of evidence from disparate data sets to formulate hypotheses about biotic dispersal, global warming and/or major palaeogeographic change, in a way which distinguishes basic from interpreted data, a problem which cladistic analysis should help to overcome.

Dispersal hypothesis between China and East Gondwana (Famennian–Tournaisian)

As noted elsewhere (Young, Chapter 9), the sinolepid antiarch *Grenfellaspis* characterises the youngest continental placoderm assemblage in eastern Australia. This is the only occurrence of this major endemic Chinese group outside of China, and therefore has biogeographic interest (Young, 1981, 1984, 1990; Long, 1983; Burrett *et al.*, 1990). Ritchie *et al.* (1992)

inferred from their earlier history in South China that sinolepids only dispersed to Australia in the latest Devonian. The apparent absence of phyllolepid placoderms from the Devonian of China has been taken to indicate persistent (probably marine) barriers between it and East Gondwana, but since the Grenfell fauna is probably younger than the upper biostratigraphic range of phyllolepids in eastern Australia (Young, Chapter 9), any palaeogeographic changes which facilitated sinolepid dispersal into East Gondwana can be assumed to have postdated the extinction of the Phyllolepida.

This dispersal event could be attributed to chance, but as is the case with the Frasnian–Famennian event discussed above, there is some evidence from other fossil groups (invertebrates) supporting a more general causal explanation. In the latest Devonian–Early Carboniferous close affinity between Australia and China is also indicated by freshwater leaiid ostracods, which are known in the Devonian only from China, but then appear in the Early Carboniferous of northwestern and northeastern Australia (Canning and Drummond Basins). They only become widespread later in the Carboniferous, in Namurian and younger strata (P.J. Jones, pers. comm.). Amongst Early Carboniferous brachiopods the sister taxa *Eochoristes* and *Ectochoristes* (Chu, 1933; Campbell, 1957) show a similar pattern (K.S.W. Campbell, pers. comm.), whilst a conodont species ('*Polylophodonta* sp. A') from the late Famennian Buttons Beds of the Bonaparte Basin of northwestern Australia (Jones, 1985, Figure 2), is probably conspecific with *Rhodalepis polylophodontiformes* of Wang and Yin (1985), only known from the lower–middle *praesulcata* Zone of the Rongxian Formation in Guangxi Province (P.J. Jones, pers. comm.).

This preliminary evidence, combined with the vertebrate distribution of sinolepid antiarchs, suggests that both non-marine and shallow marine organisms may have dispersed in a consistent fashion at or near the Devonian–Carboniferous boundary. However, the palaeogeographic change which may have caused this dispersal event is not yet clear. It is of interest that Metcalfe (1990) has supported Palaeozoic collision between the South China and Indochina terranes by an Early Carboniferous closure across the Song Ma suture (rather than much later Mesozoic closure across the Song Da suture as suggested by other authors). The coincidence in timing might suggest that the vertebrate faunal evidence in the latest Devonian are manifestations of this closure. However, as noted above, Metcalfe (1990, Figure 3a) shows North China as a separate block throughout the Palaeozoic, which is contradicted by the evidence of Middle Palaeozoic vertebrate distributions.

ACKNOWLEDGEMENTS

Visits to China under the Australian Academy of Science–Academia Sinica Scientific Exchange Agreement, to Europe and Russia under an Australian Public Service postgraduate scholarship, and to Britain supported by the Antarctic Society, Amoco, Conoco, Exxon, Shell, Texaco, Mobil, and the

Geological Society of London enabled the comparative studies on which this summary is based. Support from the above is gratefully acknowledged. Dr C. Klootwijk provided advice on palaeomagnetic data and interpretations, Drs J. Long, C. Burrett, and I. Metcalfe commented on southeast Asia, and Dr A. Blieck reviewed an early biogeography manuscript. Dr P.J. Jones and Professor K.S.W. Campbell advised on Devonian–Carboniferous invertebrate faunas, and Dr J.A. Long provided editorial advice and access to unpublished material. Published with the permission of the Director, Australian Geological Survey Organisation, Canberra.

REFERENCES

Afanassieva, O. and Janvier, P., 1985, *Tannuaspis*, *Tuvaspis* and *Ilemoraspis*, endemic osteostracan genera from the Silurian and Devonian of Tuva and Khakassia (USSR), *Géobios*, **18**: 493–506.

Bachtadse, V. and Briden, J.C., 1990, Palaeomagnetic constraints on the position of Gondwana during Ordovician to Devonian times. In W.S. McKerrow and C.R. Scotese (eds), *Palaeozoic biogeography and palaeogeography. Geological Society of London Memoir*, **12**: 43–8.

Bardenheuer, P. and Janvier, P., 1990, Cephalaspiden (Osteostraci) aus dem Unterdevon (Emsium) von Waxweiler (Rheinisches Schiefergebirge, Deutschland), *Neues Jahrbuch für Geologie und Palaontologie Monatscheft*, **1990**: 639–46.

Barr, S.M. and Macdonald, A.S., 1987, Nan River suture zone, northern Thailand, *Geology*, **15**: 907–10.

Belliere, J. and Laduron, D., 1990, Tectonics of the Rhenish Massif (Ardenne and Rheinisches Schiefergebirge), *Annales de la Societe Geologique de Belgique*, **113**: 137–8.

Berra, T.M., 1981, *An Atlas of Distribution of the Freshwater Fish Families of the World*, University of Nebraska Press, 197pp.

Blieck, A., 1982, Les grandes lignes de la biogeographie des heterostraces du Silurian Superieur–Devonien inferieur dans le domaine Nord-Atlantique, *Palaeogeography, Palaeoclimatology, Palaeoecology*, **38**: 283–316.

Blieck, A., 1984, *Les heterostraces pteraspidiformes, agnathes du Silurien–Devonien du continent nord-atlantique et de blocs avoisinants: revision systematique, phylogenie, biostratigraphie*, Cahiers de Paleontologie (section vertebres), Editions du CNRS, Paris, 199pp.

Blieck, A. and Goujet, D., 1978, A propos de nouveau materiel de Thelodontes (Vertebres, Agnathes) d'Iran et de Thailande: apercu sur la repartition geographique et stratigraphique des Agnathes des 'regions gondwaniennes' au Paleozoique moyen, *Annales de la Societe Geologique du Nord*, **97**: 363–72.

Blieck, A., Goujet, D., Janvier, P. and Lelièvre, H., 1984, Microrestes de vertebres du Siluro–Devonien d'Algerie, de Turquie et de Thailande, *Géobios*, **17**: 851–6.

Blodgett, R.B., Rohr, D.M. and Boucot, A.J., 1990, Early and Middle Devonian gastropod biogeography. In W.S. McKerrow and C.R. Scotese (eds), *Palaeozoic Biogeography and Palaeogeography. Geological Society of London Memoir*, **12**: 277–84.

Boucot, A.J., 1975, *Evolution and extinction rate controls*, Elsevier, Amsterdam, 427pp.

Boucot, A.J., 1988, Devonian biogeography: An Update. In N.J. McMillan, A.F. Embry and D.J. Glass (eds), *Devonian of the World*, Canadian Society of Petroleum Geologists, Calgary, **III**: 211–27.

Boucot, A.J. and Gray, J., 1979, *Historical Biogeography, Plate Tectonics, and the Changing Environment*, Oregon State University Press.

Boucot, A.J., Johnson, J.G. and Talent, J.A., 1969, Early Devonian brachiopod zoogeography, *Geological Society of America Special Paper*, **199**: 1–113.

Buistrov, A.P., 1956, *Kolymaspis sibirica* g.n.,s.n., a new representative of lower Devonian Agnatha, *Vestnik Leningrad University, series Geology and Geography*, **18**: 5–13.

Burrett, C.F., Long, J.A. and Stait, B., 1990, Early–Middle Palaeozoic biogeography of Asian terranes derived from Gondwana. In W.S. McKerrow and C.R. Scotese (eds), *Palaeozoic Biogeography and Palaeogeography. Geological Society of London Memoir*, **12**: 163–74.

Campbell, K.S.W., 1957, A Lower Carboniferous brachiopod–coral fauna from New South Wales, *Journal of Paleontology*, **31**: 34–98.

Campbell, K.S.W. and Bell, M.W., 1977, A primitive amphibian from the Late Devonian of New South Wales, *Alcheringa*, **1**: 369–81.

Caputo, M.V. and Crowell, J.C., 1985, Migration of glacial centers across Gondwana during the Paleozoic era, *Geological Society of America Bulletin*, **96**: 1020–36.

Cherkesova, S.V., 1973, Lower Devonian deposits of the Soviet Arctic. In *Third International Symposium of the Silurian–Devonian Boundary and the stratigraphy of the Lower and Middle Devonian, II: Stratigraphy of the Lower and Middle Devonian*, B.V. Nalivkin (ed.), pp. 276–9, Leningrad 1968 (Russian, with English summary).

Chu, S., 1933, Corals and brachiopods of the Kinlin Limestone, *Monographs of the Natural Research Institute, Geology, Nanking (A)*, **2**: 1–58.

Copper, P., 1986, Frasnian–Famennian mass extinction and cold-water oceans, *Geology*, **14**: 835–9.

Darlington, P.J., 1957, *Zoogeography: the geographical distribution of animals*, John Wiley & Sons, New York.

Denison, R.H., 1955, Early Devonian vertebrates from the Knoydart Formation of Nova Scotia, *Fieldiana, Zoology*, **37**: 449–64.

Denison, R.H., 1978, *Placodermi Handbook of Paleoichthyology, volume 2*, H.-P. Schultze (ed.), Gustav Fischer Verlag, Stuttgart, 128pp.

Dineley, D.L., 1967, The Lower Devonian Knoydart faunas, *Journal of the Linnean Society (Zoology)*, **47**: 15–29.

Edwards, D., 1990, Constraints on Silurian and Early Devonian phytogeographic analysis based on megafossils. In W.S. McKerrow and C.R. Scotese (eds), *Palaeozoic biogeography and palaeogeography. Geological Society of London Memoir*, **12**: 233–42.

Fang, Wu, Van der Voo, R. and Liang, Qizhong, 1989, Devonian paleomagnetism of Yunnan Province across the Shan Thai–South China suture, *Tectonics*, **8**: 939–52.

Friman, L. and Janvier, P., 1986, The Osteostraci (vertebrate, Agnatha) from the Lower Devonian of the Rhenish Slate Mountains, with special reference to their anatomy and phylogenetic position, *Neues Jahrbuch für Geologie und Palaontologie, Abhandlungen*, **173**: 99–116.

Fujita, K. and Newberry, J.T., 1983, Accretionary terranes and tectonic evolution of northeast Siberia. In M. Hashimoto, M. and S. Uyeda (eds), *Accretion Tectonics in the Circum-Pacific Regions*, pp. 43–58, Advances in Early and Planetary Sciences Series. Terra Publishing Tokyo, Reidel Dordrecht, Boston, London.

Gagnier, P.-Y., Turner, S., Friman, L., Suarez-Riglos, M. and Janvier, P., 1988, The Devonian vertebrate and mollusc fauna from Seripona (Dept. of Chuquisaca, Bolivia), *Neues Jahrbuch für Geologie und Palaontologie, Abhandlungen*, **176**: 269–97.

Gagnier, P.-Y., Paris, F., Racheboeuf, P.R., Janvier, P. and Suarez-Riglos, M., 1989, Les vertebres de Bolivie: Donnees biostratigraphiques et anatomiques complementaires, *Bulletin de l'Institute Français d'Etudes Andines 1989*, **18**: 75–93.

Goujet, D., Janvier, P. and Suarez-Riglos, M., 1984, Devonian vertebrates from South America, *Nature*, **312**: 311.

Goujet, D., Janvier, P. and Suarez-Riglos, M., 1985, Un nouveau rhenanide (Vertebrata, Placodermi) de la Formation de Belen (Devonien moyen), Bolivie, *Annales de Paleontologie*, **71**: 35–53.

Gray, J., 1988, Evolution of freshwater ecosystems: the fossil record, *Palaeogeography, Palaeoclimatology, Palaeoecology* , **62**: 1–214.

Gross, W., 1950, Die palaeontologische und stratigraphische Bedeutung der Wirbeltierfaunen des Old Reds und der marinen altpalaozoischen Schichten, *Abhandlungen der Deutschen Akademie der Wissenschaften zu Berlin. Mathematisch-naturwissenschaftliche Klasse*, **1949**: 1–130.

Halstead, L.B. and Turner, S., 1973, Silurian and Devonian ostracoderms. In A. Hallam (ed.), *Atlas of Palaeobiogeography*, Elsevier, Amsterdam, pp. 67–79.

Heckel, P.H. and Witzke, B.J., 1979, Devonian world palaeogeography determined from distribution of carbonates and related lithic palaeoclimatic indicators, *Special Paper of the Palaeontological Association*, **23**: 99–123.

Hills, E.S., 1929, The geology and palaeontography of the Cathedral Range and Blue Hills in north western Gippsland, *Proceedings of the Royal Society of Victoria*, **41**: 176–201.

Hills, E.S., 1931, The Upper Devonian fishes of Victoria, Australia, and their bearing on the stratigraphy of the state, *Geological Magazine*, **68**: 206–31.

Hou, H.F. and Wang, S.T., 1985, Devonian palaeogeography of China, *Acta Palaeontologica Sinica*, **24**: 186–97.

Hsu, K.J., Sun, S., Li, J., Chen, H., Pen, H. and Sengor, A.M.C., 1988, Mesozoic overthrust tectonics in south China, *Geology*, **16**: 418–21.

Hunicken, M.A., Gonçalves de Melo, J.H. and Brasil Lemos, V., 1988, Devonian conodonts from the Upper Amazon Basin, northwestern Brazil. In N.J. McMillan, A.F. Embry and D.J. Glass (eds), *Devonian of the World*, Canadian Society of Petroleum Geologists, Calgary, **III**: 479–83.

Janvier, P., 1981, The phylogeny of the Craniata, with particular reference to the significance of fossil 'Agnathans', *Journal of Vertebrate Paleontology*, **1**: 121–59.

Janvier, P., 1984, The relationships of the Osteostraci and Galeaspida, *Journal of Vertebrate Paleontology*, **4**: 344–58.

Janvier, P., 1985, Preliminary description of Lower Devonian Osteostraci from Podolia (Ukranian S.S.R.), *Bulletin of the British Museum of Natural History (Geology)*, **38**: 309–34.

Janvier, P. and Suarez-Riglos, M., 1986, The Silurian and Devonian vertebrate of Bolivia, *Bulletin de l'Institute Français d'Etudes Andines*, Lima, **15**: 73–114.

Janvier, P. and Melo, J.H.G., 1988, Acanthodian fish remains from the Upper Silurian or Lower Devonian of the Amazon Basin, Brazil, *Palaeontology*, **31**: 771–7.

Janvier, P., Termier, G. and Termier, H., 1979, The osteolepiform rhipidistian fish *Megalichthys* in the Lower Carboniferous of Morocco, with remarks on the paleobiogeography of the Upper Devonian and Permo–Carboniferous osteolepidids, *Neues Jahrbuch für Geologie und Palaontologie, Monatscheft*, **1979**: 7–14

Johnson, J.G. and Boucot, A.J., 1973 Devonian brachiopods. In A. Hallam (ed.), *Atlas of Palaeobiogeography*, pp. 89–96, Elsevier, New York.

Jones, P.J., 1985, Treposellidae (Beyrichiacea: Ostracoda) from the latest Devonian (Strunian) of the Bonaparte Basin, Western Australia, *BMR Journal of Australian*

Geology and Geophysics, **9**: 149–62.
Karatajute-Talimaa, V., 1973, *Elegestolepis grossi* gen. et sp. nov., ein neuer typ der placoidschuppe aus dem oberen Silur der Tuwa, *Palaeontographica*, **143A**: 35–50.
Kent, D.V. and Van der Voo, R., 1990, Palaeozoic palaeogeography from palaeomagnetism of the Atlantic-bordering continents. In W.S. McKerrow and C.R. Scotese (eds), *Palaeozoic biogeography and palaeogeography. Geological Society of London Memoir*, **12**: 49–56.
Klootwijk, C. and Giddings, J., 1988, An alternative APWP for the Middle to Late Palaeozoic of Australia—implications for terrane movements in the Tasman Fold Belt, *Ninth Australian Geological Convention, Abstracts*, **21**: 219–220.
Lee, C.M., 1991, The discovery of Devonian Placodermi in Hong Kong and its significance. In M.M. Chang, Y.H. Liu and G.R. Zhang (eds), *Early Vertebrates and Related Problems of Evolutionary Biology*, pp. 131–8, Science Press, Beijing, China, 514pp.
Lelièvre, H., 1984a, *Atlantidosteus hollardi* n.g., n.sp., nouveau Brachythoraci (vertebres, placodermes) du Devonien inferieur du Maroc presaharien, *Bulletin du Museum national d'Histoire naturelle*, **6**: 197–208.
Lelièvre, H., 1984b, *Antineosteus lehmanni* n.g., n.sp., nouveau Brachythoraci du Devonien inferieur du Maroc presaharien. Remarques sur la paleobiogeographie des homosteides de l'Emsien, *Annales de Paléontologie, Vertébrés*, **70**: 115–58.
Li, Y., McWilliams, M., Zhu, H., Tan, C. and He, Z., 1985, Paleozoic paleomagnetic results from the Hexizoulang Terrane, China, *Transactions of the American Geophysical Union EOS*, **66**: 864.
Liu, H., Wang, J., Bai, S. and Bai, Z., 1988, Preliminary study on paleomagnetism of the Lower-Middle Devonian boundary beds from Guangxi, South China. In N.J. McMillan, A.F. Embry and D.J. Glass (eds), *Devonian of the World, Canadian Society of Petroleum Geologists, Calgary*, **III**: 137–9.
Liu, Shi-Fan, 1991, Biogeography of early vertebrates in China. In M.M. Chang, Y.H. Liu and G.R. Zhang (eds), *Early Vertebrates and Related Problems of Evolutionary Biology*, pp. 499–504, Science Press, Beijing, China, 514pp.
Liu, T.S. and Pan, K., 1958, Devonian fishes from the Wutung Series near Nanking, China, *Palaeontologica Sinica (new series C15)*, **141**: 1–41 (Chinese and English).
Long, J.A., 1983, New bothriolepid fish from the Late Devonian of Victoria, Australia, *Palaeontology*, **26**: 295–320.
Long, J.A., 1984, New phyllolepids from Victoria and the relationships of the group, *Proceedings of the Linnean Society of New South Wales*, **107**: 263–308.
Long, J.A., 1990, Late Devonian chondrichthyans and other microvertebrate remains from northern Thailand, *Journal of Vertebrate Paleontology*, **10**: 59–71.
Long, J.A. and Burrett, C., 1989, Fish from the Upper Devonian of the Shan-Thai terrane indicate proximity to east Gondwana and south China terranes, *Geology*, **17**: 811–13.
Malinovskaya, S.P., 1973, *Stegolepis* (antiarchi, Placodermi), a new Middle Devonian genus from central Kazakhstan. *Paleontological Journal*, **7**: 189–99.
Malinovskaya, S.P., 1989, Devonian Bothriolepidae (Placodermi) of Central Kazakhstan, *Ispitatelei otd Geologie 1988*, **63**: 56–70.
Mark-Kurik, E., 1973, *Kimaspis*, a new palaeacanthaspid from the Early Devonian of Central Asia, *Eesti NSV teaduste akadeemia toimetised*, **22**: 322–30.
Mark-Kurik, E., 1974, Discovery of new Devonian fish localities in the Soviet Arctic, *Eesti NSV teaduste akadeemia toimetised*, **23**: 332–5.
Mark-Kurik, E., 1991, Contribution to the correlation of the Emsian (Lower Devonian) on the basis of placoderm fishes, *Newsletters on Stratigraphy*, **25**: 11–23.

Mark-Kurik, E. and Novitskaya, L., 1977, The Early Devonian fish fauna on Novaya Zemlya, *Eesti NSV teaduste akadeemia toimetised*, **26**: 143–9.

McPherson, J.G., 1980, Genesis of variegated redbeds in the fluvial Aztec Siltstone (Late Devonian), southern Victoria Land, Antarctica, *Sedimentary Geology*, **27**: 119–42.

Metcalfe, I., 1990, Allochthonous terrane processes in southeast Asia, *Philosophical Transactions of the Royal Society, Series A*, **331**: 625–40.

Morzadec, P., Paris, F., Plusquellec, Y., Rachebœuf, P.R. and Weyant, M., 1988, Devonian stratigraphy and paleogeography of Armorican Massif (NW France). In N.J. McMillan, A.F. Embry and D.J. Glass (eds), *Devonian of the World, Canadian Society of Petroleum Geologists, Calgary*, **I**: 401–20.

Nelson, G.J. and Platnick, N.I., 1980, A vicariance approach to historical biogeography, *Bioscience*, **40**: 339–43.

Novitskaya, L.I., 1971, *Les amphiaspides (Heterostraci) du Devonien de la Siberie*, Cahiers de Paleontologie, Section Vertebres. Paris: Editions du Centre National de la Recherche Scientifique, 130pp.

Obruchev, D.V., 1964, *Osnovy Paleontologii v. 11, Beschelyostni i Ribi*, Moscow [English translation: *Fundamentals of Paleontology, Agnatha and Pisces*; Israel program for scientific translation, 1967].

Pan, J. and Dineley, D.L., 1988, A review of early (Silurian and Devonian) vertebrate biogeography and biostratigraphy of China, *Proceedings of the Royal Society of London*, **B235**: 29–61.

Pan, J., Hou, F., Cao, J., Gu, Q., Liu, S.Y., Wang, J., Gao, L. and Liu, C., 1987, *Continental Devonian system of Ningxia and its Biotas*, Geological Publishing House, Beijing.

Panchen, A.L., 1977, Geographical and ecological distribution of the earliest tetrapods. In M.K. Hecht (ed.), *Major patterns in vertebrate evolution*, pp. 723–8, Plenum Press, New York.

Patterson, C., 1975, The distribution of Mesozoic freshwater fishes, *Memoires du Museum National d'Histoire Naturelle. Serie A, Zoologie*, 88: 156–74.

Perroud, H., Van der Voo, R. and Bonhommet, N., 1984, Paleozoic evolution of the Armorica plate on the basis of paleomagnetic data, *Geology*, **12**: 579–84.

Platnick, N.I. and Nelson, G.J., 1978, A method of analysis for historical biogeography, *Systematic Zoology*, **27**: 1–16.

Powell, C.McA. and Veevers, J.J., 1987, Namurian uplift in Australia triggered the main Gondwanan glaciation, *Nature*, **326**: 177–9.

Ritchie, A., 1973, *Wuttagoonaspis* gen. nov., an unusual arthrodire from the Devonian of Western New South Wales, Australia, *Palaeontographica*, **143A**: 58–72.

Ritchie, A., 1975, *Groenlandaspis* in Antarctica, Australia and Europe, *Nature*, **254**: 569–73.

Ritchie, A., 1984, A new placoderm, *Placolepis* gen. nov. (Phyllolepidae) from the Late Devonian of New South Wales, Australia, *Proceedings of the Linnean Society of New South Wales*, **107**: 321–53.

Ritchie, A., Wang, S.T., Young, G.C. and Zhang, G.R. (in press), The Sinolepidae, a family of antiarchs (placoderm fishes) from the Devonian of South China and eastern Australia, *Records of the Australian Museum*.

Robardet, M., Groose-Uffenorde, H., Gandl, J. and Rachebœuf, P.R., 1991, Trilobites et ostracodes du Devonien inferieur de la zone d'Ossa-Morena (Espagne), *Géobios*, **24**: 333-48.

Rowley, D.B., Raymond, A., Parrish, J.T., Lottes, A.L., Scotese, C.R. and Ziegler, A.M., 1985, Carboniferous paleogeographic, phytogeographic, and paleoclimatic reconstructions, *International Journal of Coal Geology*, **5**: 7–42.

Scotese, C.R., 1986, Phanerozoic reconstructions: a new look at the assembly of Asia, *University of Texas Institute for Geophysics Technical Report*, **66**, 54pp.
Scotese, C.R. and Barrett, C.F., 1990, Gondwana's movement over the South Pole during the Palaeozoic: evidence from lithological indicators of climate. In W.S. McKerrow and C.R. Scotese (eds), *Palaeozoic biogeography and palaeogeography. Geological Society of London Memoir*, **12**: 75–85.
Scotese, C.R. and McKerrow, W.S., 1990, Revised world maps and introduction. In W.S. McKerrow and C.R. Scotese (eds), *Palaeozoic biogeography and palaeogeography. Geological Society of London Memoir*, **12**: 1–21.
Spjeldnaes, N., 1982, Palaeoecology of *Ichthyostega* and the origin of terrestrial vertebrates. In E. Montanaro Gallitelli (ed.), *Palaeontology, essential of historical geology*, Modena, S.T.E.M., Mucchi, pp. 323–43.
Stock, C.W., 1990, Biogeography of Devonian stromatoporoids. In W.S. McKerrow and C.R. Scotese (eds), *Palaeozoic biogeography and palaeogeography. Geological Society of London Memoir*, **12**: 257–65.
Streel, M., Fairon-Demaret, M. and Loboziak, S., 1990, Givetian–Frasnian phytogeography of Euramerica and western Gondwana based on miospore distribution. In W.S. McKerrow and C.R. Scotese (eds), *Palaeozoic biogeography and palaeogeography. Geological Society of London Memoir*, **12**: 291–6.
Sun, S., Li, J., Chen, H., Peng, H., Hsu, K. and Shelton, J.W., 1989, Mesozoic and Cenozoic sedimentary history of South China, *The American Association of Petroleum Geologists Bulletin*, **73**: 1247–69.
Talent, J.A., Goel, R.K., Jain, A.K. and Pickett, J.W., 1988, Silurian and Devonian of India, Nepal and Bhutan: biostratigraphic and palaeobiogeographic anomalies, *Courier Forschungsinstitut Senckenberg*, **106**: 1–57.
Talent, J.A., Gratsianova, R.T., Shishkina, G.R. and Yolkin, E.A., 1987, Devonian faunas in relation to crustal blocks: Kazakhstan, Mongolia, Northern China, *Courier Forschungsinstitut Senckenberg*, **92**: 225–33.
Tarlo, L.B.H., 1964, Psammosteiformes (Agnatha). A review with descriptions of new material from the Lower Devonian of Poland, *Palaeontologica Polonica*, **13**: 1–135.
Thanh, T.-D. and Janvier, P., 1987, Les vertebres Devoniens du Vietnam, *Annales de Paleontologie*, **73**: 165–94.
Thanh, T.-D. and Janvier, P., 1990, Les vertebres du Devonien inferieur du Bac Bo oriental (provinces de Bac Thai et Lang Son, Viet Nam), *Bulletin du Museum d'Histoire Naturelle, 4 Serie*, **12**: 143–223.
Thomson K.S., 1980, The ecology of Devonian lobe-finned fishes. In A.L. Panchen, (ed.), *The terrestrial environment and the origin of land vertebrate*, Systematics Association Special Volume, **15**: 187–222.
Torsvik, T.H., Smethurst, M.A., Briden, J.C. and Sturt, B.A., 1990, A review of palaeomagnetic data from Europe and their palaeogeographical implications. In W.S. McKerrow and C.R. Scotese (eds), *Palaeozoic biogeography and palaeogeography. Geological Society of London Memoir*, **12**: 25–42.
Turner, S., 1970, Timing of the Appalachian/Caledonian orogen contraction, *Nature*, **227**: 9.
Turner, S. and Tarling, D.H., 1982, Thelodont and other agnathan distributions as tests of Lower Palaeozoic continental reconstructions, *Palaeogeography, Palaeoclimatology, Palaeoecology*, **39**: 295–311.
Turner, S. and Young, G.C., 1992, Thelodont scales from the Middle–Late Devonian Aztec Siltstone, southern Victoria Land, Antarctica, *Antarctic Science*, **4**: 89–105.
Valiukevicius, J., 1988, Correlation of Lower and Middle Devonian deposits of the USSR with Acanthodian assemblages. In N.J. McMillan, A.F. Embry and D.J. Glass (eds), *Devonian of the World, Canadian Society of Petroleum Geologists, Calgary*, **III**: 601–7.

Van der Voo, R., 1982, Pre-Mesozoic paleomagnetism and plate tectonics, *Annals and Reviews of Earth and Planetary Science, 1982*, **10**: 191–220.

Van der Voo, R., 1983, Paleomagnetic constraints on the assembly of the Old Red Continent, *Tectonophysics*, **91**: 271–283.

Van der Voo, R., 1988, Paleozoic paleogeography of North America, Gondwana, and displaced terranes: comparisons of paleomagnetism with paleoclimatology and biogeographical patterns, *Geological Society of America Bulletin*, **100**: 311–24.

Veevers, J.J. and Powell, C.McA., 1987, Late Paleozoic glacial episodes in Gondwanaland reflected in transgressive–regressive depositional sequences in Euramerica, *Geological Society of America Bulletin*, **98**: 475–87.

Wang, C.Y. and Yin, B.A., 1985, An important Devonian–Carboniferous boundary stratotype in neritic facies of Yishan County, Guangxi, *Acta Micropalaeontologica Sinica*, **2**: 28–48.

Wang, J.Q., 1984, Geological and palaeogeographic distribution of Devonian fishes in China, *Vertebrata PalAsiatica*, **22**: 219–29.

Wang, J.Q., 1991, New material of *Hunanolepis* from the Middle Devonian of Hunan. In Chang, M.M., Liu, Y.H. and Zhang, G.R. (eds), *Early Vertebrates and Related Problems of Evolutionary Biology*, pp. 213–47, Science Press, Beijing, China, 514pp.

Wang, N.Z., 1991, Two new Silurian galeaspids (jawless craniates) from Zhejiang Province, China, with a discussion of galeaspid–gnathostome relationships. In M.M. Chang, Y.H. Liu and G.R. Zhang (eds), *Early Vertebrates and Related Problems of Evolutionary Biology*, pp. 41–65, Science Press, Beijing, China, 514pp.

Wang, S.T., Dong, Z.Z. and Turner, S., 1986, Discovery of Middle Devonian Turiniidae (Thelodonti: Agnatha) from western Yunnan, China, *Alcheringa*, **10**: 315–25.

Witzke, B.J., 1990, Palaeoclimatic constraints for Palaeozoic palaeolatitudes of Laurentia and Euramerica. In W.S. McKerrow and C.R. Scotese (eds), *Palaeozoic biogeography and palaeogeography. Geological Society of London Memoir*, **12**: 57–73.

Witzke, B.J. and Heckel, P.H., 1988, Paleoclimatic indicators and inferred Devonian paleolatitudes of Euramerica. In N.J. McMillan, A.F. Embry and D.J. Glass (eds), *Devonian of the World, Canadian Society of Petroleum Geologists, Calgary*, **I**: 49–63.

Woodward, A.S., 1921, Fish-remains from the Upper Old Red Sandstone of Granite Harbour, Antarctica, *British Antarctic ('Terra Nova') Expedition, 1910. Natural History Report (Geology)*, **1**: 51–62.

Young, G.C., 1980, A new Early Devonian placoderm from New South Wales, Australia, with a discussion of placoderm phylogeny, *Palaeontographica*, **167A**: 10–76.

Young, G.C., 1981, Biogeography of Devonian vertebrates, *Alcheringa*, **5**: 225–43.

Young, G.C., 1984, Comments on the phylogeny and biogeography of antiarchs (Devonian placoderm fishes) and the use of fossils in biogeography, *Proceedings of the Linnean Society of New South Wales*, **107**: 443–73.

Young, G.C., 1986, Cladistic methods of Paleozoic continent reconstruction, *Journal of Geology*, **94**: 523–37.

Young, G.C., 1987a, Devonian vertebrates of Gondwana. In G.D. McKenzie (ed.), *Gondwana Six. Stratigraphy, Sedimentology and Paleontology*, pp. 41–50, AGU Geophysical Monograph 41.

Young, G.C., 1987b, Devonian palaeontological data and the Armorica problem, *Palaeogeography, Palaeoecology, Palaeoclimatology*, **60**: 283–304.

Young, G.C., 1987c, Relationships between northern and southern vertebrate faunas during the Middle Palaeozoic. In K.G. McKenzie (ed.), *Shallow Tethys 2*, pp. 79–85, Balkema, Rotterdam.

Young, G.C., 1988a, Antiarchs (placoderm fishes) from the Devonian Aztec Siltstone, southern Victoria Land, Antarctica, *Palaeontographica*, **A202**: 1–125.
Young, G.C., 1988b, New occurrences of phyllolepid placoderms from the Devonian of central Australia, *BMR Journal of Australian Geology and Geophysics*, **10**: 363–76.
Young, G.C., 1989, The Aztec fish fauna of southern Victoria Land—evolutionary and biogeographic significance. In Crame, J.A. (ed.), *Origins and Evolution of the Antarctic Biota*, Geological Society of London Special Publication No. 47: 43–62.
Young, G.C., 1990, Devonian vertebrate distribution patterns, and cladistic analysis of palaeogeographic hypotheses. In W.S. McKerrow and C.R. Scotese (eds), *Palaeozoic biogeography and palaeogeography. Geological Society of London Memoir*, **12**: 243–55.
Young, G.C., 1991, The first armoured agnathan vertebrates from the Devonian of Australia. In Chang, M.M., Liu, Y.H. and Zhang, G.R. (eds), *Early Vertebrates and Related Problems of Evolutionary Biology*, pp. 67–85, Science Press, Beijing, China, 514pp.
Zhang, G.R. and Liu, Y.G., 1991, A new antiarch from the Upper Devonian of Jiangxi, China. In Chang, M.M., Liu, Y.H. and Zhang, G.R. (eds), *Early Vertebrates and Related Problems of Evolutionary Biology*, pp. 195–212, Science Press, Beijing, China, 514pp.
Zhu Min, 1991, New information on *Diandongpetalichthys* (Placodermi: Petalichthyida). In Chang, M.M., Liu, Y.H. and Zhang, G.R. (eds), *Early Vertebrates and Related Problems of Evolutionary Biology*, pp. 179–94, Science Press, Beijing, China, 514pp.
Ziegler, P.A., 1988, Laurussia—the old red continent. In N.J. McMillan, A.F. Embry and D.J. Glass (eds), *Devonian of the World, Canadian Society of Petroleum Geologists, Calgary*, **I**: 15–48.

Chapter 13

BIOGEOGRAPHY OF PALAEOZOIC TETRAPODS

Andrew R. Milner

INTRODUCTION

Palaeozoic tetrapod biogeography—data and problems

The certain Palaeozoic tetrapod record extends for some 132 million years from the Frasnian to the end of the Permian and represents the least well-documented third of the history of tetrapods, not least biogeographically. The known Palaeozoic tetrapod faunas fall into three distinct successive biogeographical patterns, one in the Devonian, one in the Mississippian to Lower Permian, and one in the Upper Permian, each of which presents us with a group of problems requiring explanation. The problems largely stem from the poverty and stochastic nature of the Palaeozoic record. In several cases, single localities have to serve as representation for whole continents, and at such a low level of information, new discoveries can change the database quite radically. As may be noted in the ensuing account, several biogeographically significant new records have been described over the last two decades, which have necessitated revision of the biogeographical history of Palaeozoic tetrapods. This process will undoubtedly continue and no biogeographical narrative should be considered to be more than a set of testable provisional hypotheses. An ever-present complication is the phenomenon of taphonomic overlay, namely that the patterns observed are not purely biogeographic in origin but were also generated by local facies and taphonomic circumstances. Literal interpretation of negative evidence as positive evidence (i.e. interpreting absence of fossils of a tetrapod group from an area as absence of the original group from that area) will always run the risk of later falsification, but in running that risk, it is at least a testable scientific statement. The Palaeozoic tetrapod record is sufficiently poor that trackways and footprints, often preserved in different lithologies from bone, are sometimes a significant component of total distribution data, even if they cannot be placed in a systematic framework within the Tetrapoda.

Cladistic-vicariance analysis is a potentially useful tool for correlating biogeographical and phylogenetic patterns in lower tetrapods (Milner, 1983) and has some application to clarifying Palaeozoic tetrapod biogeography (Long, 1990; this chapter), although there are, at present, very few recognisable examples of Palaeozoic endemism to which the procedures of cladistic-vicariance study could be applied. Long (1990) has used this approach of taking the first stratigraphic appearance of members of various components of a cladogram to argue for centres of endemism for different portions of that cladogram. This is a parsimony argument based on preference for an economical solution, but with the reservation that local taphonomic circumstances may operate for or against local preservation of a particular group and thus bias the apparent first appearance of that group. In this chapter, the suggestion that the stereospondyl clade originated in the European intermontane limnic basins of Euramerica is based on similar reasoning. Another possible application of cladistic analysis to the problems of Palaeozoic tetrapod biogeography is recognition of the sudden appearance of a clade without close relatives in the middle of a relatively dense record. Such a situation might imply immigration or range extension from another faunal province and is suggested in this chapter for the appearance of seymouriamorphs in the Lower Permian of Euramerica.

The following chapter is divided into chronologically based sections, looking at successive periods, reviewing their basic geography and faunas, and discussing the sequential biogeographic patterns and possible faunal provinces of Palaeozoic tetrapods.

DEVONIAN

Background

By the beginning of the Devonian, vascular plants and arthropods had already colonised the land. The Devonian is, at present, believed to be the period during which one group of choanate fishes evolved into digit-bearing tetrapods and began to diversify as semiterrestrial or terrestrial organisms. Most of the restricted evidence pertaining to the earliest tetrapods is of Frasnian–Famennian age, implying a late Devonian appearance, but one earlier Devonian trace-fossil suggests the possibility that tetrapods have a history extending well back through the Devonian. This is discussed first. The published Devonian record is sufficiently poor that trackways are still a major component of the biogeographical evidence (four out of nine localities) although they are not precisely attributable taxonomically. This restricted tetrapod record has to form the basis of any discussion of two biogeographical problems, namely what was the general distribution of Devonian tetrapods and in which geographical region did they originate?

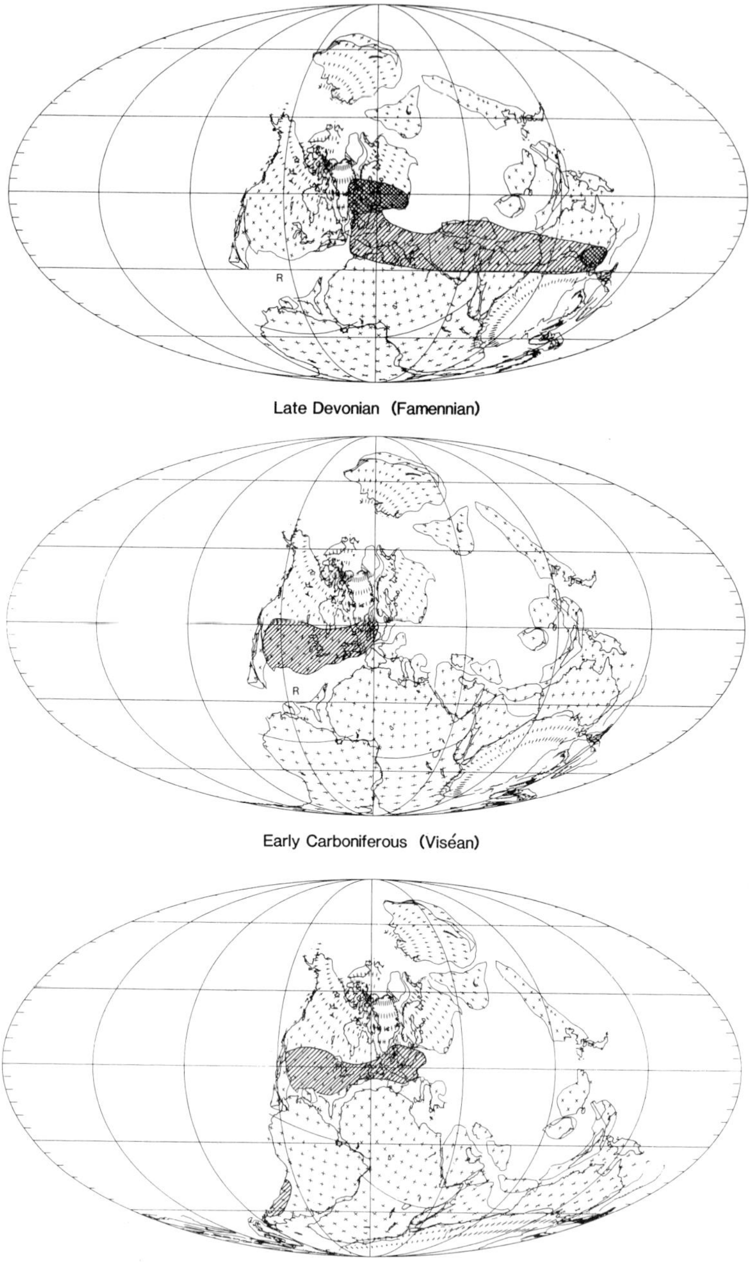

Figure 13.1. Tetrapod distributions from the late Devonian (Famennian) to the Late Carboniferous (Westphalian). Maps after Scotese and McKerrow (1990, Figures 16, 17, 19). On each map, the light shading represents the presumed minimal range of the contemporaneous tetrapod province.

A Middle Devonian tetrapod trackway?

The only indirect evidence for the presence of tetrapods in pre-Frasnian times is the probable trackway from the Grampians Group of Victoria, Australia, (Warren, Jupp and Bolton, 1986). Warren *et al.* suggested this to be of basal Devonian age, but a better age control is provided by the thelodonts which suggest a Gedinnian to basal Frasnian age (Turner, 1986) and hence late Lower to Middle Devonian. If accepted, this implies a Lower Devonian origin of tetrapods despite a total absence of any other evidence for the group prior to the Upper Devonian. An earlier origin for tetrapods is becoming increasingly plausible ecologically with the recent discovery of terrestrial arthropods including a predatory centipede in the late Silurian of England (Jeram *et al.*, 1990) which implies the presence of an extensive soil invertebrate cryptofauna by the Upper Silurian. Although the Australian trackway represents a positive geographical record, it is difficult to accept that the negative data from the rest of the world is of significance. However, this record undoubtedly provides support for conclusions drawn from the Upper Devonian data, discussed below.

Upper Devonian: chronology, geography and climate

The Upper Devonian is here taken to comprise the Frasnian and Famennian stages and to extend for 14 million years from 377 to 363 Ma (Harland *et al.*, 1990, figures rounded). At this time, the principal continents were Gondwana (occupying much of the southern hemisphere but reaching the equator only at its eastern end), Euramerica (bracketing the equator and in contact with the northwest region of Gondwana), Siberia (isolated), Kazakhstan (isolated) and the Chinese continents (either isolated or probably in contact with eastern Gondwana) (Figure 13.1). There is some evidence for global cooling in the late Devonian (Copper, 1986) but equatorial-based fish faunas have wide latitudinal distribution (Young, 1990), indicating that climate zonation was not pronounced.

Characteristic tetrapod groups and their distribution

The Upper Devonian tetrapod fauna is still extremely poorly known and only a few genera have been described. These are *Ichthyostega* (= *Ichthyostegopsis*) and *Acanthostega* from eastern Greenland (e.g. Jarvik, 1980; Clack, 1988), *Tulerpeton* from European Russia (Lebedev, 1984), *Metaxygnathus* from New South Wales, Australia (Campbell and Bell, 1977), and *Obruchevichthys* and similar material from Latvia and Scotland (Ahlberg, 1991). All appear to be stem-tetrapods with no immediate relationship to any group smaller than the remaining Tetrapoda. Footprints have been described from the Upper Devonian of Victoria, Australia (Warren and Wakefield, 1972), Parana Province, Brazil (Leonardi, 1983) and Tarbat Ness, Scotland (Rogers, 1990). The known Upper Devonian distribution, therefore, comprises Euramerica and northern Gondwana from west to east (Figure 13.1). There is as yet no evidence for Devonian tetrapods on the isolated continents of Siberia, Kazakhstan or on the Chinese continents which may have been in direct contact with Gondwana. Most localities fall between 10 degrees north of the Famennian palaeoequator (Greenland, Russia) and 25 degrees south of it (Australia). The Brazilian footprint represents an anomalous exception to this.

Geography of tetrapod origins

From the time of the general acceptance of plate tectonics in the mid-1960s up to the discovery of Australian Devonian footprints to 1972, all tetrapods and footprints from the Devonian and Carboniferous were known from Euramerica and it was generally concluded that the tetrapods originated and initially diversified on that continent (e.g. Panchen, 1973 for a late example of this view). The discoveries in the Devonian of Australia by Warren and Wakefield (1972) and later workers, provided the basis for alternative hypotheses and these soon appeared. Panchen (1977), followed by Janvier (1977) suggested the possibility of a Middle Devonian East Gondwanan origin of tetrapods, followed by an equatorial range extension along the north of Gondwana to Euramerica in the Upper Devonian, succeeded in turn by tetrapod extinction in the original Gondwana area, generated by the late Palaeozoic Gondwana glaciation. The Australian material was slightly older (Frasnian) than the Greenland tetrapods (Famennian) and Panchen and Janvier appear to have adopted Matthew's dispersalist reasoning that the location of the oldest fossil of a group represented the region of origin of that group. However, in the context of such a poor fossil record (just the two records at that time), this could hardly be said to represent positive evidence, and it could equally be argued that either a) tetrapods originated in Euramerica, extending their range briefly to Gondwana but contracting again during the Gondwana glaciation; or b) that tetrapods originated over an area not less than Gondwana and Euramerica combined. A further problem at that time, was that Euramerica and Gondwana appeared to have had a long separate history prior to the Upper Devonian (Cox, 1974), which implied that tetrapods could not have moved from one to the other via land or freshwater during the Devonian, and hence that the earliest tetrapods might have been littoral.

Two recent lines of research have bearing on this problem. Neugebauer (1988), followed by Scotese and McKerrow (1990), have proposed Devonian geographical models in which Euramerica and Gondwana are in close contact, if not collision, throughout the Devonian. The Upper Devonian map in Figure 13.1 is broadly representative of the relationship of the two continents throughout the Devonian. Thus, terrestrial or freshwater-based range extensions of tetrapods could have taken place between the two continents from the Lower Devonian onwards. Long (1990) has recently applied cladistic analysis to the problem of the geography of the origins of the major osteichthyan groups including the tetrapods. Long presented a

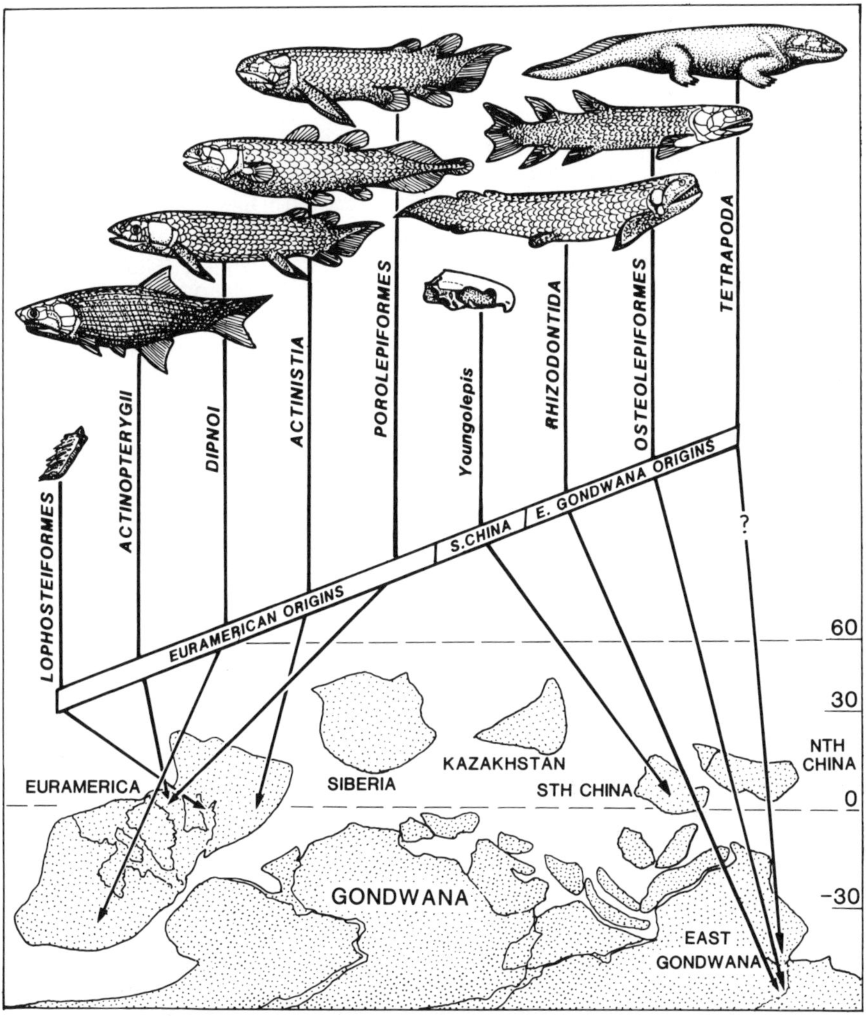

Figure 13.2. Interrelationships of major osteichthyan groups superimposed on a Devonian global reconstruction, showing locations of the most plesiomorphic or oldest members of each osteichthyan group (including Tetrapoda), after Long (1990, Figure 3).

state-of-the-art osteichthyan cladogram and pointed out that the five sequential basal dichotomies of the Osteichthyes gave rise to the Lophosteiformes, Actinopterygii, Dipnoi, Actinistia and Porolepiformes, and that the earliest representatives of all these groups appear in Euramerica which combines to suggest a Euramerican centre for osteichthyan diversification (Figure 13.2). However, one subclade of osteichthyans—the sister-group to the Porolepiformes—comprises successively: *Youngolepis* (first appearance in South China), the Rhizodontida, the Osteolepiformes and the Tetrapoda, all with first appearances in Australia. This combines to suggest an east Gondwanan origin for this clade, with the implication that the stem of this clade originated from a range extension from Euramerica and then diversified in eastern Gondwana. In the absence of a richer record, this is probably as strongly reasoned a case as can be expected, although the recent recognition of *Metaxygnathus*-like material from the Frasnian of Europe (Ahlberg, 1991) weakens the argument slightly.

MISSISSIPPIAN

Background

The Mississippian was the period of tetrapod diversification into many of the major groups characterising the rest of the Palaeozoic. The tetrapod fossil record is very poor and restricted to the Upper Mississippian. It is also restricted to the equatorial belt along the southern region of Euramerica. The principal biogeographical problem presented by the Mississippian Tetrapod Province is whether this restricted distribution is real or an artifact of taphonomic overlay.

Chronology, geography and climate

The Mississippian is here taken to comprise the Tournaisian, Viséan and Namurian A (= Serpukhovian) stages and to extend for 40 million years from 363 to 323 Ma (Harland *et al.*, 1990, figures rounded). The relative positions of the continents had changed only slightly since the Upper Devonian. Euramerica and western Gondwana were drifting northwards, and were in closer contact, the Rheic Ocean between Europe and Gondwana closing at this time. It may be noted that Variscan (central and southern) Europe comprised a series of small terranes which had been part of northern Gondwana earlier in the Palaeozoic, and these appear to have been deformed in the shear zone of the oblique collision between northern Europe and Gondwana. The Variscan–Hercynian Orogeny was the result of this collision and these terranes became an integral part of the Variscan mountain system across late Palaeozoic Europe. They now comprise much of France, Germany and Czechoslovakia. At this time, eastern Gondwana was moving southwards away from the equator. Siberia, Kazakhstan and

North China were probably isolated, but South China may still have been in contact with eastern Gondwana (Figure 13.1).

The consolidation of western Pangaea resulted in gradual emergence of much of Euramerica and northwestern Gondwana, together with alteration to the patterns of oceanic circulation. As a result climatic zonation became more extreme and the southern glaciation developed over much of central Gondwana. In the equatorial region of Euramerica, the climate was non-seasonally warm and wet, and vast swamps developed on lowland plains.

Characteristic tetrapod groups and their distribution

The Mississippian fauna is considered separately here because it is quite distinct from the later Pennsylvanian–Lower Permian fauna (Smithson, 1985). For much of this century, the Mississippian tetrapod fauna was known largely on the basis of a few aquatic or amphibious specialists, such as the stem-tetrapod *Crassigyrinus*, the loxommatids *Loxomma* and *Spathicephalus*, the colosteids, the anthracosaur *Proterogyrinus* and the adelogyrinids. Only recently have new finds started to expand our view of this time. These include the earliest microsaur *Utaherpeton* (at the Mississippian–Pennsylvanian boundary) and, most significantly, the terrestrial assemblage collected at East Kirkton in Scotland (Milner *et al.*, 1986). The latter assemblage includes the earliest terrestrial temnospondyls, aïstopods, anthracosaurs and the probable amniote *Westlothiana* (Smithson and Rolfe, 1990).

Apart from a few undescribed elements from Nova Scotia, tetrapods are not represented in the Tournaisian, and only the distribution of Viséan–early Namurian forms can be discussed. All are found in what was southern Euramerica, ranging from Utah (Carroll, Bybee and Tidwell, 1991) and Iowa (Bolt, 1990) in the west, through Virginia and Nova Scotia, to the Scottish Midland Valley in the east (Figure 13.1). Mississippian tetrapod fossils, including trace fossils, have not been found in any other continent nor in the terranes which later formed central and southern Europe. All known localities range from less than 5 degrees north to 20 degrees south of the Viséan palaeoequator.

The Mississippian Tetrapod Province

The assemblages from the Mississippian localities are too small in number and too ecologically diverse in origin for robust conclusions to be drawn, but they do appear to be consistent with the presence of a single faunal province throughout the known tetrapod range. Godfrey (1988) has summarised the evidence for this. Crassigyrinids, the form genus *Doragnathus*, and the anthracosaur *Proterogyrinus* are known from Scotland and West Virginia; *Spathicephalus* is known from Scotland and Nova Scotia; and the family Colosteidae is known from Scotland, Nova Scotia, West Virginia and, recently, Iowa (Bolt, 1990). The Euramerican tetrapod fauna must be assumed to comprise a single biogeographical province, here called the Mississippian Tetrapod Province. It is premature to attempt to define this province based on the available taxa, no one genus of tetrapod having yet been found throughout this area. However, as no other contemporaneous tetrapod provinces are known, there would seem to be no urgent necessity to choose a defining taxon.

The absence of collected fossil tetrapods from elsewhere is negative evidence of uncertain significance. Siberia and Kazakhstan were isolated continents that had not yet collided with Euramerica and might be expected to either lack tetrapods completely or to have an undiscovered different tetrapod faunal province, perhaps derived from North Gondwana via the Chinese continental fragments. The distribution of the Lower Permian seymouriamorphs can be explained in terms of such a province and this possibility will be discussed below in relation to them. In Gondwana, the late Palaeozoic glaciation had developed by the late Mississippian, and it might be expected that any southern tetrapod province would be species-poor, in comparison with the equatorial tetrapod province. Two decades ago it could be argued that tetrapods originated in Euramerica, that Euramerica and Gondwana had not yet sutured in the Mississippian and, hence, that there was a marine barrier preventing movement between the two continents prior to the Middle Permian (Cox, 1974). It is now clear that neither of these arguments apply. Any explanation for the absence of tetrapod fossils from Mississippian Gondwana, either as relicts or Euramerican immigrants, must be considered in terms of rarity (rather than absence), taphonomic overlay, or lack of appropriate collecting. However, there are no later biogeographical patterns that require us to theorise the existence of a Gondwanan Mississippian tetrapod province.

PENNSYLVANIAN–LOWER PERMIAN

Background

The Pennsylvanian and Lower Permian together present a vast period of time (67 Ma—as long as the Cenozoic) but are considered together here, because a relatively uniform biogeographical pattern prevails throughout, modified by the collision of the Kazakhstan and Siberian continents with the Euramerican region of Pangaea. At least sixty families of Pennsylvanian and Lower Permian tetrapods are known from many rich assemblages, almost entirely from the southern region of Euramerica close to the palaeoequator, but with one family occupying a completely different range in southern Gondwana.

The similarity of the European and North American Pennsylvanian and Lower Permian tetrapod faunas and the conclusion that they represent a single faunal province on a single continent was first drawn by Nopsca (1934) and is now one of the uncontroversial observations of early tetrapod biogeography (Panchen, 1973; Carroll, 1984). Two areas of uncertainty remain: is there any evidence for endemic sub-provinces within the equatorial province, and do seymouriamorphs have a unique biogeographical history centred on Kazakhstan?

Chronology, geography and climate

Pennsylvanian

The Pennsylvanian comprises the Namurian B and C, the Westphalian A to D and the Stephanian A to C, while the Lower Permian comprises the Asselian, Sakmarian, Artinskian and Kungurian. Together they extend for 67 million years from 323 to 256 Ma (Harland *et al.*, 1990, figures rounded). By the Westphalian, the slow rotation of Gondwana and Euramerica had resulted in a collision between the two on a broad front, to give a single continent for the rest of the Palaeozoic. The resultant continent continued to drift northwards across the equator at its west end, while the eastern end drifted south towards the pole. The central European terranes had accreted onto Euramerica, compressed and folded in the sliding contact between Gondwana and Euramerica. The Variscan/Hercynian/Appalachian Orogeny was at its peak, the resultant erosion and sedimentation being responsible for most of the tetrapod-preserving strata of the Pennsylvanian and Permian. By the Westphalian, Kazakhstan and Siberia were in close proximity to Euramerica, while the Chinese continents formed a tenuous arc from eastern Gondwana to Kazakhstan (Figure 13.1). In the Pennsylvanian, the equatorial region of western Pangaea was characterised by coals in the immediate equatorial belt, and evaporites at about 30 degrees north and south of the equator (Parrish *et al.*, 1986). Pennsylvanian glacial tillites are known from South America, Antarctica, South Africa, Madagascar and Australia. Thus the immediate equatorial region was apparently hot and humid, but greater seasonality and some aridity prevailed in the sub-equatorial belts on either side. To the south, much of Gondwana was temperate to polar, but with the assumption that interglacial interludes may have occurred.

Lower Permian

By the Lower Permian, the western region of Pangaea had been assembled and was still drifting northwards. In the early Permian, Siberia and Kazakhstan collided with the Baltic region of Euramerica to complete the northeastern region of Pangaea. The exact position and connection of North and South China remain uncertain but they continued to form an arc over much of the distance between the northeastern and southeastern regions of Pangaea (Figure 13.3). In equatorial Pangaea, the coals were replaced by red beds and evaporites indicating a greater seasonality of climate (Olson and Vaughn, 1970; Parrish *et al.*, 1986). The coal basins vanish and certain tetrapod families vanish with them, although the Euramerican fauna shows substantial continuity through this ecological change. In Gondwana, the glaciations withdrew late in the Lower Permian, to give a temperate and non-seasonally wet climate in which the Ecca Formation of South Africa was laid down.

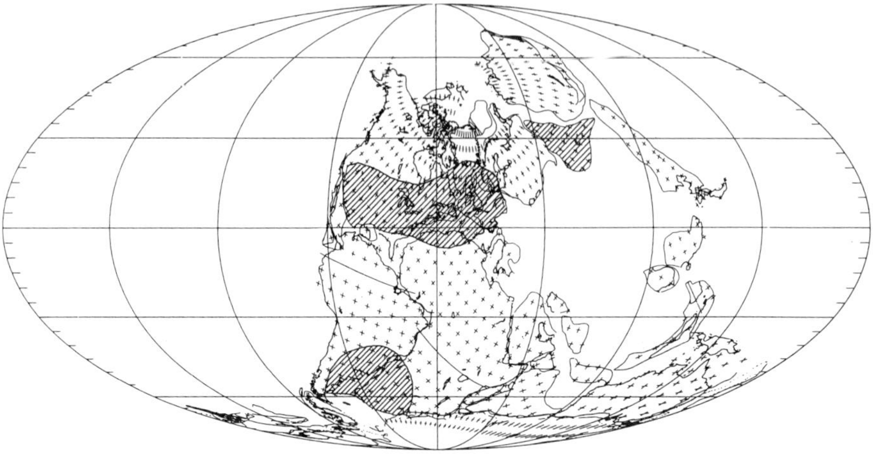

Early Permian (Artinskian)

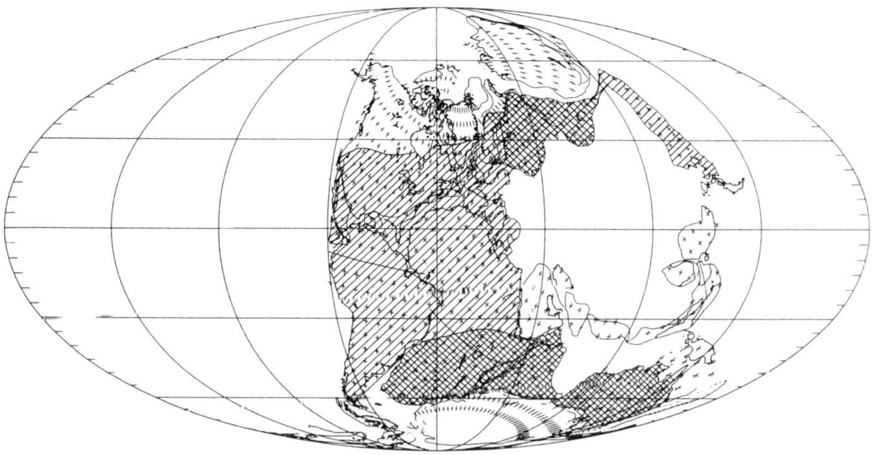

Late Permian (Kazanian)

Figure 13.3. Tetrapod distribution in the Lower Permian (Artinskian) and Upper Permian (Wordian/Capitanian). Maps after Scotese and McKerrow (1990, Figures 20 and 21). On the Lower Permian map, the shaded areas represent the minimum areas of the Edaphosaur-Nectridean Province, the Mesosaur Province, and the area of Kazakhstan where seymouriamorphs have been found. On the Upper Permian map, the light shading represents the minimum area occupied by tetrapods in the late Permian. The darker areas represent the minimum areas of the northern Dvinosaurid-Chroniosuchid Province and the southern Rhinesuchid Province.

Table 13.1 *Pennsylvanian and Lower Permian terrestrial tetrapod families from the Edaphosaurid-Nectridean Province, west and east of the Hercynian-Appalachian 'barrier'. This represents an updating of the data presented by Milner and Panchen (1973). Recent sources for data are identified by number, although much of the information on temnospondyls is taken from the author's database. Where commonality extends to well-characterised genera, these are listed.*

Family	*North America*	*Europe and Nova Scotia*
TEMNOSPONDYLI		
Dendrerpetontidae (1)	–	+
Edopidae (2)	+	–
Eryopidae (3)	+	+ (*Eryops*)
Parioxyidae (3)	+	–
Trematopidae (2, 4)	+	+
Dissorophidae (2)	+	+
Amphibamidae (5)	+	+ (*Amphibamus*)
AÏSTOPODA		
Ophiderpetontidae (6)	+	+ (*Ophiderpeton*)
Phlegethontiidae (6)	+	+ (*Phlegethontia*)
NECTRIDEA		
Scincosauridae (7)	+	+
MICROSAURIA		
Tuditanidae (8)	+	+
Hapsiopareiontidae (8)	+	+
Pantylidae (8)	+	–
Gymnarthridae (8)	+	+
Ostodolepididae (8)	+	–
Rhynchonkidae (8)	+	–
Trihecatontidae (8)	+	–
Hyloplesiontidae (8)	–	+
Brachystelechidae (9)	+	+
ANTHRACOSAURIA		
Gephyrostegidae (10)	+	+
SEYMOURIAMORPHA		
Seymouriidae (2, 4)	+	+
COTYLOSAURIA		
Limnoscelididae (4, 11)	+	+
Tseajaiidae (11)	+	–
Diadectidae (11, 12)	+	+
Solenodonsauridae (10)	–	+
AMNIOTA		
Protorothyrididae (4, 12)	+	+
Captorhinidae (12)	+	– (A)
Bolosauridae (13)	+	– (A)
Acleistorhinidae (13)	+	–
Araeoscelididae (14)	+	+? (B)
Petrolacosauridae (14)	+	+? (B)
Edaphosauridae (15)	+	+ (*Edaphosaurus*)
Caseidae (15)	+	+ (*Casea*)
Sphenacodontidae (15)	+	+ (*Haptodus*)
Varanopseidae (15)	+	–
Eothyrididae (15)	+	–
Ophiacodontidae (15)	+	+

Table 13.2 *Pennsylvanian and Lower Permian aquatic tetrapod families from the Edaphosaur-Nectridean Province, west and east of the Hercynian-Appalachian 'barrier'. This represents an updating of the data presented by Milner and Panchen (1973). Recent sources for data are identified by number, although much of the information on temnospondyls is taken from the author's database. Where resemblance extends to well-characterised genera, these are listed.*

Family	*North America*	*Europe and Nova Scotia*
Loxommatidae (16)	+	+ (*Megalocephalus* x)
Colosteidae (17)	+	+
TEMNOSPONDYLI		
Cochleosauridae (18)	+	+
Trimerorhachidae (2, 19)	+	+
Saurerpetontidae (5)	+	−
Capetus (20)	−	+
Actinodontidae (3)	−	+
Intasuchidae (3)	−	+
Archegosauridae (3)	−	+
Zatrachydidae (21)	+	+ (*Dasyceps*)
Micromelerpetontidae (22)	−	+
Branchiosauridae (22)	?(C)	+
NECTRIDEA		
Urocordylidae (7)	+	+ (*Sauropleura*)
Diplocaulidae (7, 23)	+	+
MICROSAURIA		
Lysorophidae (24)	+	+
Microbrachidae (8)	−	+
ANTHRACOSAURIA		
Eogyrinidae (25)	+	+
Archeriidae (25)	+	−
Anthracosauridae (25)	−	+
SEYMOURIAMORPHA		
Discosauriscidae (22)	−	+

Sources for Tables 13.1 and 13.2
1: Milner, A.R. 1980; 2: Hook 1989; 3: Milner 1990a; 4: Martens 1990; 5: Milner 1982; 6: Baird 1964; 7: Milner, A.C. 1980; 8: Carroll and Gaskill 1978; 9: Carroll 1991; 10: Carroll 1970; 11: Heaton 1980; 12: Clark and Carroll 1973; 13: Ivakhnenko and Tverdokhlebova 1987; 14: Reisz, Berman and Scott 1984; 15: Reisz 1986; 16: Beaumont 1977; 17: Hook 1983; 18: Rieppel 1980; 19: Milner database; 20: Milner and Sequeira in press; 21: Boy 1989; 22: Werneburg 1989; 23: Walter and Werneburg 1988; 24: Boyd 1980; 25: Clack 1987.
Notes: (A). Absent from European Lower Permian but present in Upper Permian of Russia. (B). *Kadaliosaurus* is an araeoscelidian belonging either to the Petrolacosauridae or the Araeoscelididae. (C). The only possible North American branchiosaurids are *Branchiosaurus darrahi* supposedly from Montrose near Pittsburgh and *Branchiosaurus* sp. indet. from Mazon Creek. The former is of less certain provenance than the literature indicates and may be from Autun. The latter is too poorly preserved to stand as a certain biogeographic record.

Characteristic tetrapod groups

The early–mid Pennsylvanian tetrapod fauna is the first rich community represented in the record, with at least forty families recognised. Apart from some Mississippian relicts such as the loxommatids and the colosteids, the tetrapods fall into six major groups: temnospondyls, aïstopods, nectrideans, microsaurs, anthracosaurs and amniotes. The Westphalian fauna has recently been reviewed by Milner (1987) and the families are listed in the composite Pennsylvanian–Lower Permian family lists in Tables 13.1 and 13.2.

The Upper Pennsylvanian–Lower Permian tetrapod fauna was made up largely of the same major groups as the earlier Pennsylvanian fauna, and is similarly diverse although the relative diversity of individual groups changes with time. This fauna corresponds to the 'Edaphosaur Empire' of Anderson and Cruickshank (1978) who named a chronological succession of Permo–Triassic tetrapod fauna after the prevalent herbivores. This usage is modified below. Three of the major tetrapod groups are each represented by a limited range of relict forms; the nectrideans by the boomerang-headed diplocaulids (Figure 13.4) and the tiny scincosaurid *Sauravus*, the aïstopods by the late phlegethontiid *Sillerpeton*, and the anthracosaurs by the late embolomere *Archeria* (Figure 13.4). However, the temnospondyls, microsaurs and amniotes had increased in diversity and were clearly successful, together with the seymouriamorphs which appear in the Lower Permian without obvious Pennsylvanian relatives. The Lower Permian amphibian faunas of Europe have recently been reviewed by Werneburg (1989).

The Edaphosaurid–Nectridean Tetrapod Province

Throughout the Pennsylvanian and Lower Permian, the vast majority of tetrapod-producing localities are still restricted to Euramerica (or the Euramerican region of Pangaea), although now including the accreted terranes of Southern Europe. The latter, including what is now Germany and Czechoslovakia, appear to have become part of Euramerica during the Mississippian. By the early Pennsylvanian, tetrapods range from Arizona in the west (Thayer, 1985) to the Ruhr in the east (Boy and Bandel, 1973). From the Westphalian C onwards they occur in Czechoslovakia (Turek, 1989) (Figure 13.1). Within Euramerica, the west–east Lower Permian range is not significantly different (Figure 13.3) from that of the Pennsylvanian.

Because Euramerica and Gondwana are now known to have been in contact by the Lower Permian, it might be predicted that Euramerican biota would occur in equatorial Gondwana. One probable example of this was reported several years ago but has received little publicity. Diplocaulid nectrideans from the lower Argana Formation of Morocco, were reported by Dutuit (1976) but have not subsequently been described. They include complete skulls of a form almost identical to the genus *Diplocaulus* (Figure 13.4) from the Texas Lower Permian. The age of this part of the Argana

Figure 13.4. Diorama depicting the lowland terrestrial and freshwater elements of the Edaphosaur-Nectridean Province found in the Lower Permian of Texas.
1. *Eryops*, a temnospondyl; 2. *Dimetrodon*, a sphenacodontid pelycosaur; 3. *Captorhinus*, a captorhinid; 4. *Seymouria*, a seymouriamorph; 5. *Archeria*, an anthracosaur; 6. *Diplocaulus*, a nectridean; 7. *Trimerorhachis*, a temnospondyl; 8. *Lysorophus*, a microsaur.

Figure 13.5. Diorama depicting the intermontane terrestrial and freshwater elements of the Edaphosaur-Nectridean Province found in the Lower Permian of Germany.
1. *Edaphosaurus*, an edaphosaurid pelycosaur; 2. a seymouriamorph (unnamed); 3. a captorhinid (unnamed); 4. a trematopid temnospondyl (unnamed); 5. *Micromelerpeton*, a temnospondyl; 6. *Sclerocephalus*, a temnospondyl; 7. *Discosauriscus*, a seymouriamorph; 8. *Apateon*, a branchiosaurid temnospondyl.

Formation is not known but the nectridean itself suggests a Lower Permian age. Now that it is accepted that the Euramerica–Gondwana suture was completely established by the Lower Permian, it is clear that the Moroccan region of North Africa was wedged against southeastern North America on the equator, so there was no apparent barrier to the Euramerican biota being present in equatorial Gondwana. The equatorial fauna is thus not strictly a Euramerican fauna. For this reason, I have not referred to this province as simply the Euramerican Province because this is no longer accurate, nor have I used the term equatorial province as this would not preclude the separate South China continent which is also equatorial. I have modified the term 'Edaphosaur Empire' of Anderson and Cruickshank (1978). Because the edaphosaurs (Figure 13.5) only occur in terrestrial Stephano–Permian assemblages, I have added the nectrideans as a defining group. Nectrideans occur in almost all Pennsylvanian aquatic assemblages and many Lower Permian ones including the Moroccan material. At our present state of knowledge, the term 'Edaphosaur–Nectridean Province' defines this province as adequately as any pair of groups.

Almost all localities fall within 10 degrees of the palaeoequator. This restricted nature of these Pennsylvanian and Lower Permian distributions has been obscured by the tendency of earlier workers to aggregate all the Palaeozoic 'labyrinthodont' (temnospondyl and anthracosaur amphibian) data into one latitudinal diversity curve (Irving and Brown, 1964, 1966; Brown, 1968; Cifelli, 1980). Their aggregate curves suggested a high diversity of labyrinthodonts at low latitudes declining to a low diversity at high latitudes in the Palaeozoic. Had these authors separated their Upper Permian data from the earlier data, then it would have been clear that in the Pennsylvanian/Lower Permian, the entire diversity of amphibians was at low latitudes, whereas in the Upper Permian they were spread thinly over all latitudes.

Cosmopolitan and endemism within Euramerica

In 1973, Milner and Panchen suggested the possibility of developing regional endemism within the Euramerican tetrapod faunas of the Pennsylvanian and Lower Permian. A comparison of faunas from the west and east of the Hercynian–Appalachian mountain range suggested to us that two distribution patterns were visible. First, the terrestrial tetrapod families appeared to be uniformly present across the entire range, the known European fauna representing a small sample of the North American fauna. Second, the aquatic tetrapod families appeared to show a degree of endemism in each of the two areas, and it was possible to characterise a western assemblage based on material from the USA, and an eastern assemblage based on material from Europe and Nova Scotia. Since we put forward this interpretation, new material has been described, old material had been reinterpreted and taxonomic revisions have taken place. Tables 13.1 and 13.2 represent a summary of the present situation.

Our first hypothesis, that there was a single cosmopolitan terrestrial fauna in the Edaphosaurid–Nectridean Province, has been supported by subsequent discoveries—notably a caseid pelycosaur from France (Reisz, 1986), trematopids, seymouriids, limnoscelidids and protorothyridids from Thuringia (Martens, 1990) (see also Figure 13.5), a scincosaurid from Ohio (A.C. Milner, 1980) and a brachystelechid from Texas (Carroll, 1991). As may be seen in Table 13.1, a few 'North American' terrestrial families remain to be discovered in Europe but some of these are known from only a single specimen (Trihecatontidae, Tseajaiidae, Acleistorhinidae) or two specimens (Eothyrididae) in North America. Their absence in the smaller European assemblage can hardly be considered significant.

Our second hypothesis has fared less well in the light of subsequent data. Several families of aquatic forms, which we suggested might be North American endemics, have been described or reported from Europe (Table 13.2), notably lysorophid microsaurs (Boyd, 1980), a long-horned diplocaulid based on a body imprint (Walter and Werneberg, 1988), and a possible trimerorhachid (*Dawsonerpeton*, being restudied by M. Manuorova, Prague). Cochleosaurids which seemed to be restricted to the eastern part of the continent, certainly occur at Linton, Ohio (personal observation). Only two North American families remain to be found in Europe, the Saurerpetontidae and the Archeriidae, and this can be explained in terms of the smaller representation of lowland flood-plain assemblages in Europe. There remains a series of amphibian families found in and around the interconnected valleys and intermontane lakes of the Variscan fold-mountains across what is now France, Germany and Czechoslovakia (Werneburg, 1988). These families include the Actinodontidae (*Sclerocephalus*) (Figure 13.5), Intasuchidae (*Chelider-peton*—see Milner, 1990a), Archegosauridae, Micromelerpetontidae (Figure 13.5), Branchiosauridae (Figure 13.5), Microbrachidae and possibly Disco-sauriscidae (Figure 13.5), although the peculiar distribution of the last family is discussed further below. Of the above families only the Branchiosauridae may be present in North America but I have doubts about the provenance of the only well-preserved specimen purported to have been collected in North America. There are no comparable intermontane lake assemblages from North America, so that these European forms might be genuine geographical endemics (based on ecological restriction) or their absence from North America might simply represent absence of suitable facies. As I concluded in my discussion of the Westphalian part of this fauna (Milner, 1987), there seem to have been lowland/coastal plain assemblages and intermontane limnic assemblages through the Pennsylvanian and Permian. If one takes the view that these are genuine endemics to the European intermontane limnic system, it suggests that the stereospondyl clade of the Temnospondyli originated in this region. In recent provisional cladistic accounts of temnospondyl relationships (Milner, 1990a, 1990b), I argued that the Actinodontidae, Intasuchidae and Archegosauridae were the basal branches on the stem of the Stereospondyli and all these groups make their first and major appearances in the Lower Permian of Europe. As the basal three families of the stereospondyl clade all make their first appearance in this one region, it is possible that this is the region of endemism for this clade.

The Mesosaurid Province of South Gondwana

A second tetrapod faunal province existed across parts of interglacial southern Gondwana during the Pennsylvanian and Lower Permian. It is poorly known and probably limited in diversity. The record may be summarised as follows.

The family Mesosauridae has been described from the white band at the base of the Ecca Formation of South Africa and Namibia, the contemporary Irati Formation in Brazil, and the Mangrullo Formation in Uruguay. The mesosaurid genera *Mesosaurus*, *Stereosternum* and *Brazilosaurus* represent a family of specialised long-snouted planktonivorous amniotes, associated with a wide range of fishes in both regions. No other tetrapods occur with them, and they have not been found in the equatorial assemblages of Euramerica and elsewhere. The name 'Mesosaurid Province' is the only logical one for this tetrapod province. The Karroo mesosaurs are found in the 'White Band' which immediately overlies Dwyka tillites and represents a Lower Permian immediate postglacial situation. The localities are between 40 and 60 degrees south of the Lower Permian palaeoequator (Figure 13.3).

Two tetrapod trace fossils have been described from apparently Pennsylvanian strata in Gondwana. The first is a trackway of a long-toed tetrapod from the probable Upper Carboniferous of Pampa del Chanar in San Juan Province, Argentina, described by Frenguelli (1950) and named *Orchesteropus atavus*. The second is a tetrapod trackway from the Carboniferous Cahinches Formation of northern Chile, described by Bell and Boyd (1986). This Formation is unconformably overlain by the Las Represas Formation which bears Lower Permian molluscs. The precise age is uncertain within the Carboniferous, fish scales suggesting a Mississippian age while K-Ar dating of granitoid intrusives suggest a later Carboniferous age. The trackway appears to have been made in lake margin and presumably represents an interglacial interlude, occurring at about 50 degrees south of the palaeoequator.

These two trackways and the mesosaurs comprise the only evidence for the Mesosaurid Tetrapod Province. The trackways differ from one another and it seems unlikely that more than one of them was produced by a mesosaur, so it may be presumed that other tetrapods were present, but have yet to be found.

Two erroneous records

Two other records of temnospondyl amphibians have been attributed to the Lower Permian of Gondwana but both are probably younger and do not belong to this faunal province but to a later one. One such record comprises the temnospondyls *Archegosaurus* and *Lysipterygium* from the *Gangamopteris* Beds of Kashmir. Several temnospondyl and palaeoniscoid specimens have been described from these beds. The general resemblance of the assemblage is to the Autunian limnic assemblages of France and Germany, and the *Gangamopteris* Beds have generally been taken to be Lower Permian. However, Werneburg and Schneider (pers. comm.) are preparing a full reassessment of this material and have concluded that the temnospondyls are more derived than those from Europe and show tendencies towards the Upper

Permian archegosaurs and rhinesuchids. They conclude that the *Gangamopteris* Beds may be latest Lower Permian or even basal Upper Permian. These amphibians may represent an early phase of the post-glaciation range extension of tetrapods into Gondwana.

The second such record is that of the archegosaurid temnospondyl *Prionosuchus plummeri* from the Pedra de Fogo Formation of Brazil. This was reported by Price (1948) who used the combination of archegosaur and ctenacanth sharks from this formation to identify it as Lower Permian. Cox and Hutchinson (1991) have reconsidered this material and concluded it to be Upper Permian. *Prionosuchus* is most similar to the Upper Permian *Platyoposaurus* from Russian and ctenacanth sharks are now also known from the Upper Permian.

The seymouriamorphs and Kazakhstan

It is generally accepted that Kazakhstan collided with the Euramerican region of Pangaea early in the Permian. Within the Edaphosaur–Nectridean Tetrapod Province, faunal evidence for such a collision could be expected to take the form of the appearance of radically new elements in the Lower Permian assemblages, without obvious Pennsylvanian relatives. The seymouriamorph clade is not just the only group to conform to this pattern, but also occurs in ambiguously dated deposits in Kazakhstan.

Seymouriamorph distribution

The Seymouriamorpha and related batrachosaurs show a unique distribution pattern in the Permian which suggests an equally unique biogeographical history. There are two Permian families, the larger terrestrial Seymouriidae (Figure 13.4) and the smaller aquatic Discosauriscidae (Figure 13.5). The Seymouriidae are known from the Lower Permian of New Mexico, Texas and Germany, and the Upper Permian of Russia. The discosauriscids have been known for many years from the Lower Permian of Czechoslovakia, Germany and France, but have never been found in North America. However, three genera of discosauriscid of uncertain age have also been described from a restricted area of central Asia, in beds which have produced no other tetrapods. These are *Utegenia* from Kazakhstan, *Ariekanerpeton* from Tadzhikistan (both reviewed by Ivakhnenko, 1987) and *Urumqia* from northern Xinjiang in western China (Zhang *et al.*, 1984). These localities fall within the Kazakhstan continent which sutured with Euramerica in the Lower Permian. The European discosauriscids are all of Lower Permian age, and by approximate correlation with them, the central Asian localities are assumed to be late Carboniferous or Lower Permian although there appears to be no independent corroboration for this. They could precede or post-date the suture of Kazakhstan with Euramerican Pangaea. At noted above, both seymouriids and discosauriscids appear suddenly partway through the Lower Permian of Euramerica without any obvious earlier relatives. Werneberg (1989) has used the appearance of discosauriscids in the Autunian of Europe as a stratigraphical marker defining Zone 5 of six stratigraphical zones based on small tetrapods.

Biogeographical hypotheses

There are two plausible hypotheses to explain these patterns, a Carboniferous seymouriamorph province hypothesis and a taphonomic overlay hypothesis.

The seymouriamorph province hypothesis states that the stem-seymouriamorphs were isolated from other tetrapods in north Gondwana in the latest Devonian, and reached Kazakhstan via the Chinese continents in the early Mississippian, where they diversified in isolation, after the late Palaeozoic glaciation made much of Gondwana uninhabitable. This would explain the presence of seymouriamorphs in isolation in Kazakhstan, if they prove to be older than the suture with Euramerica. Phylogenetically it would be chronologically plausible—the presence of an amniote in the late Mississippian of Scotland (Smithson and Rolfe, 1990) demands the existence of seymouriamorphs (an earlier dichotomy) by the late Mississippian at the latest. In the early Permian, Kazakhstan collided with Euramerica, and permitted some seymouriamorph groups to extend their range over Euramerica. Discosauriscids appear suddenly in Europe in the Asselian, as do seymouriids in North America. Such a hypothesis explains why there are no Carboniferous seymouriamorphs, or plausible relatives, in the rich Euramerican coal-swamp faunas. It is based on a relatively literal reading of the presences and absences in the fossil record and it demands the possibility of tetrapod range extension from Devonian Gondwana, across South China and North China to Kazakhstan.

The taphonomic overlay hypothesis states that seymouriamorphs were present in Euramerica throughout the Carboniferous as part of the Edaphosaur–Nectridean Tetrapod Province, but have not been collected. They then diversified and became successful in the Lower Permian, and they extended their ranges into Kazakhstan as soon as it sutured with Euramerica, where they happen to be the only group represented at first. This hypothesis would effectively attribute the unusual chronological and geographical distribution to taphonomic and ecological overlay producing a deceptive fossil record, but would not demand any series of terrestrial connections from Gondwana across South and North China to Kazakhstan, during the Carboniferous.

At present, there seems to be no further data that bears directly on this problem. The discovery of unequivocally Carboniferous seymouriamorphs in either Kazakhstan or Euramerica is the test that will falsify one hypothesis or the other.

UPPER PERMIAN

Background

Upper Permian tetrapod faunas show a quantum change from those of the Lower Permian, although there are some conservative elements in Russia (Ivakhnenko, 1990). It is increasingly clear that the best known Lower Permian faunas are mostly Asselian–Sakmarian and that the Artinskian and

Kungurian are poorly represented outside Texas. The dramatic faunal changes represent a late Lower Permian–early Upper Permian 'gap' in the record. The successive suturing of Kazakhstan, Siberia, North China and South China provided the possibility for tetrapod range extensions across the northern hemisphere, while the gradual amelioration of the Gondwanan climate following the withdrawal of the glaciation permitted a massive range extension to the south. The tetrapod record is locally rich in South Africa and Russia but relatively scrappy and stochastic elsewhere. Thus, although there was a global tetrapod distribution by the end of the Permian, evidence for post-range extension, vicariance-generated provinces is still tenuous.

Chronology, geography and climate

The Upper Permian is here taken to comprise the Ufimian, Guadelupian and Lopingian and extends for 11 million years from 256–245 Ma (Harland *et al.*, 1990). By the Upper Permian, western and southern Pangaea were functionally a single continent. North China had collided with Siberia, South China was possibly tenuously connected to North China, and series of small terranes (Turkey, Iran, Tibet and Shan Thai-Malaya) had rifted from the northern coast of southeastern Pangaea and were drifting northwards across Tethys, possibly as an elongate small continent (Figure 13.3). In equatorial Pangaea, the Euramerican region continued to become even more seasonal and arid, while in the higher palaeolatitudes of Russia and South Africa, the climate was temperate to warm and wet. Shelf seas with characteristic coastal faunas include the European Zechstein Sea and the Madagascar–Kenya seaway.

Characteristic tetrapod groups

By the Upper Permian, the aïstopods and microsaurs have ceased to appear in the fossil record, the nectrideans are represented by one relict diplocaulid and the anthracosaurs are represented by one relict family—the Chroniosuchidae. The temnospondyls continued to be a diverse group, some confamilial with Lower Permian relatives (Eryopidae, Dissorophidae, Actinodontidae, Archegosauridae) and others sufficiently morphologically distinct to be placed in different families (Dvinosauridae, Brachyopidae, Peltobatrachidae, Rhinesuchidae). The batrachosaurs were represented by a range of uniquely Upper Permian families (Leptorophidae, Enosuchidae, Lanthanosuchidae, Nycteroleteridae—see Ivakhnenko, 1987). Primitive amniotes included relict captorhinids, procolophonians, and the uniquely Upper Permian mesosaurs and pareiasaurs, but most amniotes were either diapsids or synapsids. The diapsids had differentiated into archosauromorphs (protorosaurs and proterosuchids), lepidosauromorphs (younginids and tangasaurs) and enigmatic forms such as the gliding coelurosauravids and the aquatic claudiosaurs. The synapsids were represented by some relict caseid pelycosaurs and the vast therapsid radiation of titanosuchians, tapinocephalids and gorgonopians (all restricted to the Upper Permian), together with early parts of the dicynodont, therocephalian and cynodontian radiations.

General distribution of Upper Permian tetrapods

For the first time in the fossil record, tetrapods appear to have achieved a tangible global distribution in the Upper Permian (Figure 13.3). They have been found not only in Europe and North America, but also in South America, Africa, Madagascar, India, China and Australia (see Anderson and Cruickshank, 1978, for documentation). Most of the tetrapods found in the Upper Permian of Gondwanaland have immediate relatives in the Lower Permian of Euramerica and the obvious conclusion is that they represent a range extension of the Edaphosaur–Nectridean Province following the withdrawal of the Gondwana glaciation during the Lower Permian. The temnospondyl assemblage from Kashmir probably represents an early instance of this range extension. None of the known Upper Permian Gondwana tetrapods are relatives of the mesosaurs so there is no direct evidence of surviving elements of the mesosaurid province. A literal interpretation of the fossil record suggests that the Upper Permian range extension was a gradual process (Anderson and Cruickshank, 1978; Parrish *et al.*, 1986). In the early Upper Permian (Guadaloupian), tetrapods had apparently extended their range to the Urals and to southern Africa and India. Only in the later Upper Permian (Lopingian) do tetrapods appear in the fossil record of South America, Australia and China. The late South American record may well be an artifact of taphonomy or collection as there was substantial physical continuity with Africa at this time. The late appearance of tetrapods in Australia and China may be genuine. North China and South China appeared to have finally sutured with Siberia and Kazakhstan in the late Permian (Scotese and McKerrow, 1990, p. 17), and it is at this time that therapsids appear in the Chinese fossil record, suggesting a range extension of the north Pangaean fauna into China. The late appearance of tetrapods in Australia may be a taphonomic artifact but may be due to the peripheral position of that area. Even in the Lower Triassic, the Australian fauna appears deficient in amniotes and it is possible that range extensions across southern Gondwana were subject to ecological filtering, even in the absence of an ice-cap. The only Upper Permian records from Australia are two brachyopid temnospondyls (*Bothriceps*) and some probable tetrapod trails (Warren, 1972).

Cosmopolitanism and endemism

The Upper Permian tetrapod faunas of North America, India, Australia and South America comprise only a few specimens, while those from East Africa, Madagascar, western Europe and China have only 10–20 taxa. The only really rich assemblages which can form the basis for biogeographical comparison are those from Russia and South Africa. Among the amniotes, there is a high degree of commonality at a family level and little evidence for endemism or provinciality, suggesting continuity between the two and thus implying that the equatorial belt was not entirely seasonally arid (Olson, 1984; Parrish *et al.*, 1986). Among the widely distributed amniote families are the following (source: Anderson and Cruickshank, 1978, unless otherwise stated): Pareiasauridae (Russia, Europe, South Africa and China), Nyctiphruretidae (Russia, Madagascar and South Africa—Ivakhnenko, 1987), Coelurosauravidae (Europe and Madagascar—Evans and Haubold, 1987), Rubidgeidae (Russia and South Africa—Tatarinov, 1977), Procynosuchidae (Russia, Europe and South Africa—Sues and Boy, 1988), Anteosauridae (Russia and South Africa), Robertiidae (South Africa and China—King, 1990) and Endothiodontidae (South Africa, India and South America—King, 1990). Clearly the greater part of Pangaea did not include mutually exclusive terrestrial tetrapod provinces. Anderson and Cruickshank (1978) defined the early Upper Permian fauna as the 'Tapinocephalid Empire' and the later fauna as the 'Dicynodontid Empire'. They labelled some aquatic communities as provinces but it is not at all clear what the extent of these locally collected communities is, and whether they are primarily geographical or ecological entities. Their 'Protorosaurid Aquatic Province' comprises only *Protorosaurus* itself from the Zechstein of Germany and England. Their 'Tangasaurid Aquatic Province' is based on the Tangasauridae from Madagascar and Kenya.

The non-amniote tetrapods appear to show two biogeographic patterns suggesting, perhaps, they were more susceptible to isolation at this time. The Upper Permian amphibians of Russia and those of Gondwana are almost entirely mutually exclusive with the exception of one shared family—the Archegosauridae represented by several genera in Russia, by *Archegosaurus* in the basal Upper Permian of Kashmir and by *Prionosuchus* and *Platyoposaurus* in Brazil. The Russian amphibian fauna includes many forms which are confamilial with their Euramerican Lower Permian antecedents, such as the Eryopidae, Dissorophidae, Actinodontidae and Archegosauridae (Milner, 1990b). The Dvinosauridae and Chroniosuchidae are sufficiently morphologically characteristic to be defined as families but are probably late relatives of the Trimerorhachidae and Eogyrinidae respectively. A second category of amphibians are not relatable to earlier Euramerican families but may be part of the seymouriamorph radiation which appears to be centred in this region. These include not only late seymouriids (*Kotlassia*) but also the Leptorophidae, Enosuchidae, Lanthanosuchidae, Nycteroleteridae and Tokosauridae. This group probably represents a clade which may have originated and diversified in Kazakhstan in the Pennsylvanian or within Euramerica in the mid-Permian. None of these forms have yet been found outside Russia. The northern amphibian province of the Upper Permian appears very much to have a dual input from the Edaphosaur–Nectridean Province (see also Ivakhnenko, 1990) and from the seymouriamorph clade which may or may not have been a province.

The tetrapod range extensions into post-glaciation Gondwana appear to have included few amphibians and these seem to have immediately developed vicariantly. Evidence for the Euramerican origin of some of the Gondwanan temnospondyls comes not only from their demonstrable cladistic relationship (Milner, 1990b) but also from the actinodontid and archegosaurid found in the basal Upper Permian of Kashmir. However, there is no evidence for any chroniosuchid, seymouriamorph or related batrachosaur in Gondwana, nor most of the above-listed temnospondyl families. The Upper Permian Gondwanan families are the Rhinesuchidae (Africa and India), the Peltobatrachidae (Africa), the Brachyopidae (Australia) and (from the Lower Triassic but needing an Upper Permian ancestor) the Micropholidae (South Africa and Australia) (Milner, 1990b). All can ultimately be related to earlier temnospondyls but show no immediate relationship to the Upper Permian forms in Russia (Milner, 1990b). There seem to be two distinct amphibian faunal provinces, even if amniote provinciality is less certain. Because so many of the amphibian families are only represented in a single area, I am reluctant to suggest definitive families for defining geographical provinces. Using those families which might define faunal provinces most specifically in time and space, I suggest that the northern province be referred to as the Dvinosaurid–Chroniosuchid Province, and the southern province be referred to as the Rhinesuchid Province.

CONCLUSIONS

The following conclusions represent the simplest interpretations of the available distributional data for Palaeozoic tetrapods.

1. Tetrapods are most likely to have arisen in the equatorial region of eastern Gondwana in the Devonian. They then extended their range across all of equatorial Gondwana and Euramerica by the Upper Devonian.
2. From the base of the Mississippian to the end of the Lower Permian, (121 Ma—almost a third of tetrapod history), our direct knowledge of tetrapods is almost entirely restricted to Euramerica or the Euramerican component of western Pangaea.
3. In the Mississippian, the available material suggests the presence of a single homogenous tetrapod province.
4. In the Pennsylvanian and Lower Permian, most known tetrapods still fall within a single province here termed the Edaphosaur–Nectridean Province and encompassing equatorial Euramerica and Gondwana. The Mesosaurid Province is present in the Lower Permian of southern Gondwana, represented by the Mesosauridae and some trackways.
5. The distribution pattern of Permian seymouriamorphs suggests the possibility of a unique Carboniferous biogeographical history for this group. It is possible that the seymouriamorph clade developed within the Kazakhstan continent and extended its range into Euramerican Pangaea in the Lower Permian.

6. In the Upper Permian, collision and suturing of the northern continents and withdrawal of the Gondwanan glaciation permitted tetrapod range extension over most, if not all continental areas. There is no good evidence for more than one tetrapod province among the amniotes but the amphibian-grade tetrapods appear to have differentiated by vicariance into a northern Dvinosaurid–Chroniosuchid Province and a southern Rhinesuchid Province.

REFERENCES

Ahlberg, P.E., 1991, Tetrapod or near-tetrapod fossils from the Upper Devonian of Scotland, *Nature*, **354**: 298–301.

Anderson, J.M. and Cruickshank, A.R.I., 1978, The biostratigraphy of the Permian and the Triassic. Part 5. A review of the classification and distribution of Permo–Triassic tetrapods, *Palaeontologia Africana*, **21**: 15–44.

Baird, D., 1964, The aïstopod amphibians surveyed, *Breviora*, No. **206**: 1–17.

Beaumont, E.H., 1977, Cranial morphology of the Loxommatidae (Amphibia: Labyrinthodontia), *Philosophical Transactions of the Royal Society of London Series B*, **280**: 29–101.

Bell, C.M. and Boyd, M.J., 1986, A tetrapod trackway from the Carboniferous of northern Chile, *Palaeontology*, **29**: 519–26.

Bolt, J.R., 1990, Mississippian vertebrates from Iowa, *National Geographic Research*, **6**: 339–54.

Boy, J.A., 1989, Über einige vertreter der Eryopoidea (Amphibia: Temnospondyli) aus dem europäischen Rotliegend (?höchstes Karbon-Perm) 2. *Acanthostomatops*, *Paläontologische Zeitschrift*, **63**: 133–51.

Boy, J.A. and Bandel, K., 1973, *Bruktererpeton fiebigi* n.gen n.sp. (Amphibia: Gephyrostegida). The first tetrapod from the Carboniferous of the Rhein–Ruhr District (Namurian B; West Germany), *Paläontographica* **145A**: 39–77.

Boyd, M.J., 1980, A lysorophid amphibian from the Coal Measures of northern England, *Palaeontology*, **23**: 925–9.

Brown, D.A., 1968, Some problems of distribution of late Paleozoic and Triassic terrestrial vertebrates, *Australian Journal of Science*, **30**: 434–45.

Campbell, K.W.S. and Bell, M.W., 1977, A primitive amphibian from the Late Devonian of New South Wales, *Alcheringa*, **1**: 369–81.

Carroll, R.L., 1970, The ancestry of reptiles, *Philosophical Transactions of the Royal Society of London, Series B*, **257**: 267–308.

Carroll, R.L., 1984, Problems in the use of terrestrial vertebrates for zoning in the Carboniferous. In P.K. Sutherland and W.L. Manger, *Neuvième Congrès International de Stratigraphie et de Géologie du Carbonifère, Compte Rendu*, Vol. 2. Southern Illinois University Press, Carbondale and Edwardsville: 135–47.

Carroll, R.L., 1991, *Batropetes* from the Lower Permian of Europe—a microsaur, not a reptile, *Journal of Vertebrate Paleontology*, **11**: 229–42.

Carroll, R.L. and Gaskill, P., 1978, The Order Microsauria, *Memoir of the American Philosophical Society*, **126**: 1–211.

Carroll, R.L., Bybee, P. and Tidwell, W.D., 1991, The oldest microsaur (Amphibia), *Journal of Paleontology*, **65**: 314–22.

Cifelli, R.L., 1980, A reassessment of labyrinthodont paleolatitudinal distributions, *Palaeogeography, Palaeoclimatology, Palaeoecology*, **30**: 121–31.

Clack, J.A., 1987, *Pholiderpeton scutigerum* Huxley, an amphibian from the

Yorkshire Coal Measures, *Philosophical Transactions of the Royal Society of London, Series B*, **318**: 1–107.

Clack, J.A., 1988, New material of the early tetrapod *Acanthostega* from the Upper Devonian of East Greenland, *Palaeontology*, **31**: 699–724.

Clark, J. and Carroll, R.L., 1973, Romeriid reptiles from the Lower Permian, *Bulletin of the Museum of Comparative Zoology, Harvard University*, **144**: 353–407.

Copper, P., 1986, Frasnian–Famennian mass extinction and cold-water oceans, *Geology*, **14**: 835–9.

Cox, C.B., 1974, Vertebrate palaeodistributional patterns and continental drift, *Journal of Biogeography*, **1**: 75–94.

Cox, C.B. and Hutchinson, P., 1991, Fishes and amphibians from the Late Permian Pedra de Fogo Formation of northern Brazil, *Palaeontology*, **34**: 561–73.

Dutuit, J.-M., 1976, Découverte d'Amphibiens Lépospondyles dans la série inférieure de la formation rouge d'Argana (atlas occidental marocain), *Comptes Rendues de l'Academie des Sciences, Ser.D*, **283**: 1733–4.

Evans, S.E. and Haubold, H., 1987, A review of the Upper Permian genera *Coelurosauravus*, *Weigeltisaurus* and *Gracilisuchus* (Reptilia, Diapsida), *Zoological Journal of the Linnean Society*, **90**: 275–303.

Frenguelli, J., 1950, Ichnites en el Paleozoico superior del Oeste argentino, *Revista de la Asociacion Geologica Argentina*, **5**: 136–48.

Godfrey, S.J., 1988, Isolated tetrapod remains from the Carboniferous of West Virginia, *Kirtlandia*, **43**: 27–36.

Harland, W.B., Armstrong, R.L., Cox, A.V., Craig, L.E., Smith, A.G. and Smith, D.G., 1990, *A geologic time scale 1989*, Cambridge University Press, Cambridge.

Heaton, M.J., 1980, The Cotylosauria: a reconsideration of a group of archaic tetrapods. In A.L. Panchen (ed.), *The terrestrial environment and the origin of land vertebrates*, Academic Press, London: 497–552.

Hook, R.W., 1983, *Colosteus scutellatus* (Newberry), a primitive temnospondyl amphibian from the Middle Pennsylvanian of Linton, Ohio, *American Museum Novitates*, **No. 2270**: 1–41.

Hook, R.W., 1989, Stratigraphic distribution of tetrapods in the Bowie and Wichita Groups, Permo–Carboniferous of north-central Texas. In R.W. Hook (ed.), *Permo–Carboniferous Paleontology, Lithostratigraphy, and Depositional Environments of North-Central Texas*, **47–53**. Field Trip Guide Book No. 2, 49th Annual Meeting of the Society of Vertebrate Paleontology, Austin, Texas.

Irving, E. and Brown, D.A., 1964, Abundance and diversity of the labyrinthodonts as a function of paleolatitude, *American Journal of Science*, **262**: 689–708.

Irving, E. and Brown, D.A., 1966, Reply to Stehli's discussion of labyrinthodont abundance and diversity, *American Journal of Science*, **264**: 488–96.

Ivakhnenko, M.F., 1987, (Permian parareptiles of the USSR), *Trudy Paleontologicheskogo Instituta, Akademiya Nauk SSSR*, **223**: 1–59 (in Russian).

Ivakhnenko, M.F., 1990, Elements of the early Permian tetrapod faunal assemblages of eastern Europe, *Paleontological Journal*, **24**: 104–12.

Ivakhnenko, M.F. and Tverdokhlebova, G.I., 1987, A revision of the Permian bolosauromorphs of eastern Europe, *Paleontological Journal*, **21**: 93–100 (in English).

Janvier, P., 1977, Vertébrés dévoniens de deux nouveux gisements du Moyen-Orient. Le problème des relations intercontinentales au Paléozoique moyen vu à la lumière de la paléobiogéographie des Rhipidistiens ostéolepiformes et des premiers Tétrapodes, *Annales de la Societe Geologique du Nord*, **47**: 373–82.

Jarvik, E.A., 1980, *Basic structure and evolution of the vertebrates*, Volume 1, Academic Press, London and New York.

Jeram, A.J., Selden, P.A. and Edwards, D., 1990, Land animals in the Silurian: arachnids and myriapods from Shropshire, England, *Science*, **250**: 658–61.

King, G., 1990, *The dicynodonts. A study in palaeobiology*, Chapman and Hall, London and New York.

Lebedev, O.A., 1984, The first find of a Devonian tetrapod in the USSR, *Doklady Akademiy Nauk SSSR*, **278**: 1470–3 (in Russian).

Leonardi, G., 1983, *Notopus petri* nov.gen., nov.sp.: une empreinte d'amphibien du Devonien au Parana (Bresil), *Géobios*, **16**: 233–9.

Long, J.A., 1990, Heterochrony and the origin of tetrapods, *Lethaia*, **23**: 157–66.

Martens, T., 1990, First evidence of terrestrial tetrapods with North American faunal elements in the red beds of Upper Rotliegendes (Lower Permian, Tambach Beds) of the Thuringian Forest (GDR)—first results, *Acta Musei Reginae-Hradecensis Series A*, **22** (1989): 99–104.

Milner, A.C., 1980, A review of the Nectridea (Amphibia). In A.L. Panchen (ed.), *The terrestrial environment and the origin of land vertebrates*, Academic Press, London: 377–405.

Milner, A.R., 1980, The temnospondyl amphibian *Dendrerpeton* from the Upper Carboniferous of Ireland, *Palaeontology*, **23**: 125–41.

Milner, A.R., 1982, Small temnospondyl amphibians from the Middle Pennsylvanian of Illinois, *Palaeontology*, **25**: 635–64.

Milner, A.R., 1983, The biogeography of salamanders in the Mesozoic and early Caenozoic, a cladistic–vicariance model. In R. Sims, J. Price and P. Whalley (eds), *Evolution, time and space, the emergence of the biosphere*. Systematics Association Special Volume 23. Academic Press, London and New York: 431–68.

Milner, A.R., 1987, The Westphalian tetrapod fauna: some aspects of its geography and ecology, *Journal of the Geological Society of London*, **144**: 495–506.

Milner, A.R., 1990a, The relationships of the eryopoid-grade temnospondyl amphibians from the Permian of Europe, *Acta Musei Reginae–Hradecensis, Series A*, **22**: 131–7.

Milner, A.R., 1990b, The radiations of temnospondyl amphibians. In P.D. Taylor and G.P. Larwood (eds), *Major Evolutionary Radiations*. Systematics Association Special Volume 42. Clarendon Press, Oxford: 321–49.

Milner, A.R. and Panchen, A.L., 1973, Geographical variation in the tetrapod faunas of the Upper Carboniferous and Lower Permian. In D. H. Tarling and S.K. Runcorn (eds), *Implications of Continental Drift to the Earth Sciences*, Vol. 1. Academic Press, London: 353–68.

Milner, A.R. and Sequeira, S.E.K. (in press), *Capetus* and the problems of primitive temnospondyl relationships, *Pollichia*.

Milner, A.R., Smithson, T.R., Milner, A.C., Coates, M.I. and Rolfe, W.D.I., 1986, The search for early tetrapods, *Modern Geology*, **10**: 1–28.

Neugebauer, J., 1988, The Variscan plate tectonic evolution: an improved 'Iapetus model', *Schweizerisch Mineralogische und Petrographische Mitteilungen*, **68**: 313–33.

Nopsca, F., 1934, The influence of geological and chronological factors on the distribution of non-marine reptiles and Stegocephalia, *Quarterly Journal of the Geological Society of London*, **90**: 76–140.

Olson, E.C., 1984, Nonmarine vertebrates and Late Paleozoic climates. In P.K. Sutherland and W.L. Manger, *Neuvième Congrès International de Stratigraphie et de Géologie du Carbonifère, Compte Rendu*, Vol. 5. Southern Illinois University Press, Carbondale and Edwardsville: 403–14.

Olson, E.C. and Vaughn, P.P., 1970, The changes of terrestrial vertebrates and climates during the Permian of North America, *Forma et Functio*, **3**: 113–38.

Panchen, A.L., 1973, Carboniferous tetrapods. In A. Hallam (ed.), *Atlas of*

Palaeobiogeography, Elsevier, Amsterdam, London, New York: 117–25.

Panchen, A.L., 1977, Geographical and ecological distribution of the earliest tetrapods. In M.K. Hecht, P.C. Goody and B.M. Hecht (eds), *Major patterns in vertebrate evolution*, Plenum, New York: 723–38.

Parrish, J.M., Parrish, J.T. and Ziegler, A.M., 1986, Permian–Triassic Paleogeography and Paleoclimatology and implication for therapsid distribution. In N. Hotton III, P.D. MacLean, J.J. Roth and E.C. Roth (eds), *The Ecology and Biology of Mammal-like Reptiles*, Smithsonian, Washington and London: 109–31.

Price, L.I., 1948, A labyrinthodont amphibian from the Pedra de Fogo Formation, State of Maranhão, *Boletim Divisao de Geologia e Mineralogia, Brasil*, **124**: 1–34.

Reisz, R.R., 1986, *Teil 17A, Pelycosauria. Handbuch der Paläoherpetologie*, P. Wellnhofer (ed.), Gustav Fischer, Stuttgart.

Reisz, R.R., Berman, D.S. and Scott, D., 1984, The anatomy and relationships of the Lower Permian reptile *Araeoscelis*, *Journal of Vertebrate Paleontology*, **4**: 57–67.

Rieppel, O., 1980, The edopoid amphibian *Cochleosaurus* from the Middle Pennsylvanian of Nova Scotia, *Palaeontology*, **23**: 143–50.

Rogers, D.A., 1990, Probable tetrapod tracks rediscovered in the Devonian of North Scotland, *Journal of the Geological Society of London*, **147**: 746–8.

Scotese, C.R. and McKerrow, W.S., 1990, Revised World maps and introduction. In W.S. McKerrow and C.R. Scotese (eds), *Palaeozoic Palaeogeography and Biogeography. Geological Society Memoir* No. **12**: 1–21.

Smithson, T.R., 1985, Scottish Carboniferous amphibian localities, *Scottish Journal of Geology*, **21**: 123–42.

Smithson, T.R. and Rolfe, W.D.I., 1990, *Westlothiana* gen. nov.: naming the earliest known reptile, *Scottish Journal of Geology*, **26**: 137–8.

Sues, H.-D. and Boy, J.A., 1988, A procynosuchid cynodont from central Europe, *Nature*, **331**: 523–4.

Tatarinov, L.P., 1977, A new *Gorgonops* from Upper Cretaceous (sic) deposits of Vologda Province, *Paleontological Journal*, **11**: 227–32.

Thayer, D.W., 1985, New Pennsylvanian lepospondyl amphibians from the Swisshelm mountains, Arizona, *Journal of Paleontology*, **59**: 684–700.

Turek, V., 1989, Fish and amphibian trace fossils from Westphalian sediments of Bohemia, *Palaeontology*, **32**: 623–43.

Turner, S., 1986, Vertebrate fauna of the Silverband Formation, Grampians, western Victoria, *Proceedings of the Royal Society of Victoria*, **98(2)**: 53–62.

Walter, H. and Werneburg, R., 1988 Über liegespuren (Cubichnia) aquatischer Tetrapoden (?Diplocauliden, Nectridea) aus den Rotteroder Schichten (Rotliegendes, Thuringer Wald/DDR), *Freiberger Forschungshefte*, Ser.C, 419: 96–106.

Warren, A., 1972, Queensland trace fossils pose a 230 million year old problem, *Australia Natural History*, March 1972: 160–2.

Warren, A., Jupp, R. and Bolton, B., 1986, Earliest tetrapod trackway, *Alcheringa*, **10**: 183–6.

Warren, J.W. and Wakefield, N.A., 1972, Trackways of tetrapod vertebrates from the Upper Devonian of Victoria, Australia, *Nature*, **238**: 469–70.

Werneburg, R., 1988, Paläobiogeographie der labyrinthodonten Amphibien im Oberkarbon und Rotliegenden Mitteleuropas, *Zeitschrift für geologischen Wissenschaft*, **16**: 929–32.

Werneburg, R., 1989, Labyrinthodontier (Amphibia) aus dem Oberkarbon und Unterperm Mitteleuropas—systematik, phylogenie und biostratigraphie, *Freiberger*

Forschungshefte Ser.C, **436**: 7–57.

Young, G.C., 1990, Devonian vertebrate distribution patterns and cladistic analysis of palaeogeographic hypotheses. In W.S. McKerrow and C.R. Scotese (eds), *Palaeozoic Palaeogeography and Biogeography. Geological Society Memoir* No. **12**: 243–55.

Zhang, F., Li, Y. and Wan, X., 1984, A new occurrence of Permian Seymouriamorphs in Xinjiang, China, *Vertebrata Palasiatica*, **22**: 294–304.

Systematic index

Numbers in italics indicates figure

A. australis 189, 195, 197
A. guizhouensis 270
A. -type scales *12*, 189, 194
Aanthaspis 123
Acanthodes *12*, 153, 159, 162, 269
Acanthodopsis 60
Acanthoides dublinensis 97, 191
Acanthostega 20, 117, 327
Acanthothoraci 57
Aceraspis *70*, 74
Acleistorhinidae 341
Acrospirifer 254
Acrotomaspis 80
Actinodontidae 341, 354–7
Actinolepis 80
Adelogyrinids 331
Aesturaichthys 153
Aethaspis 126
Ageleodus 187, 195, 197
A. pectinatus 187
Agnatha 5, 107
Aïstopods 337
Alaspis 75, 79, 124
Aldingeria 121
Allocryptaspis 155
Amatheolepis 80
?Amblypteridae 159
Amphiaspids 89, 92
Amphiaspis *90*
Amphoraspis *90*
Anarcestes lateseptatus 148
Anarthraspis 126
Anaspids 8, 57, 68, 71, 75, 82
Anchipteraspis 70
Andrejevichthys 158, 167
Andreolepis heddei 14, 15
Androichthys 153
Angarichthys *91*, 96, 119
A. hyperboreus *91*
Anglaspis *12, 70,* 80
Anomalichthys 122
Antarctasipidae 60
Antarctaspis 217, 295
A. mcmurdoensis 248
Antarctilamna 143, 167, 220, 229, 306
A. prisca *11*, 186, 225, 248, 185
A. seriponensis *144*, 146
A. sp. *160*, 164
Antarctolepis gunni 248
Antarctonchus glacialis 248
Anteosauridae 347
Anthracosaurs 337
Antiarchi, indet. *284*
Antineosteus lehmani 148, *149*, 151, 166
Apalolepis 80
Apateon *339*
Aphataspis 80
Arandaspididae 57, 72
Arandaspis 177
A. prionotolepis *178*, 177
Arauzia 155
A. federicoi *154*
Archegosauridae 341, 245, 347
Archegosaurus 342, 347
Archeria 337
Arctaspis 119, 80
Arctolepis 119, 80
Argyriaspis 89, *90*
Ariekanerpeton 343
Artenolepis (= *Holonema*) 156, 193
Asiacanthus 255, 268
Asiaspis *9*, 264, 268
Aspidichthys 120, 150, 162–3, 165
Aspidin 7
Asterolepididae 60, 126, 188

Asterolepis 99, 112, 119, 126, 128–30, 141, 163, 267, 254
Asterosteidae 60
Asterosteus 146
Astraspids 57, 72, 95
Ateleaspis *70*, 74
Athenaegis 7
Atlantidosteus hollardi 148, *149*, 166–7, 224, 247
Atopacanthus 98, 121
Auchenaspis 74
Australichthys 153
Australolepis
A. seddoni *183*, 191, 193–4, 250
A. sp., Poland 194
Austrophyllolepis 217, 229–31
A. edwini 217, 229–31, 248
A. ritchiei 217, 249, 221
A. youngi 217, 249, 221
A. sp. 248
Axinaspis 80

Barameda decipiens 127
Baringaspis 125
Barrydalaspis theroni *152*
Barwickia downunda *227*, 249
Beelarongia patrichae 249
Beloceras kayseri 150
Belonaspis *70*, 80
Benneviaspidida 74
Benneviaspis *70*, 74, 80
Birkenia *70*
Bogdanovia 93
Bolivacanthus sagitalis *144*, 146
Bolivosteus 145, 166, 306
B. chacomensis 143, *144*
Boothiaspis 89
Boreaspididae 74
Boreaspis *6*, 80
Bothriceps 346
Bothriolepididae 60
Bothriolepis 14, 78–9, 92, 96, 98–9, 109, 113, 124–6, 129, 153, 159, 166, 183, 210, 216, 226–31, 233–6, 247–251, 254, 265–6, 268, 270–1, 296, 304, 305, 311
B. alexi 217, 228, 248
B. alvesensis 116
B. antarctica 248
B. askinae 217, 248, *218*
B. barretti 228, 248, *218*
B. bindareei 230
B. canadensis 157, *158*
B. cellulosa 299
B. cristata 116
B. cullodenensis 217, 230, 249, *218*
B. fergusoni 217, 230, 249, *218*
B. gigantea 116
B. gippslandiensis 217, 230, 249, *218*
B. groenlandica 117–8
B. hydrophila 116
B. karawaka 217, 228, 248, *218*
B. kohni 217, 248
B. kwangtungensis 260, 265
B. lochangensis 265
B. lohesti 120
B. macphersoni 217, 228, 248, *218*
B. mawsoni 217, 228, 248, *218*
B. minor 123
B. nitidus 123
B. niushoushanensis 254, 260, *265*
B. ornata 118
B. portalensis 217, 226, 229, 248, *218*
B. shaokwanensis 229, 265
B. siberica 299
B. sinensis 260, 265–6
B. tatongensis 217, 229, 248, *218*
B. tungseni 260, 265–6
B. turanica 93
B. vuwae 217, 248, *218*
B. yunnanensis 260, 265–6
B. sp., *Do son* 285
Brachyopidae 345, 348
Branchiosauridae 341
Branchiosaurus 336
B. darrahi 336
Brazilosaurus 342
Breizosteus armoricensis *154*, 120
Briagalepis 217
B. warreni 231, 249, *218*
Brindabellaspis 14, 98, 192, 308
Bruntonichthys multidens 250
Bryantolepis 126
Buchanosteidae 58, 269
Buchanosteus 259
B. confertituberculatus 162, 188, 192, *221*, 222
Bullerichthys fascidens 250
Byssacanthoides 225
Byssacanthus 303
'B.' (Onchus) aff. arcuatus 153

Campbellodus 14
C. decipiens 250
?Campodus 198
Camptaspis utahensis 75
Campylodus? delgadoi 153, *154*
Camuropiscidae 60
Camuropiscis concinnus 249
C. laidlawi 249
Canadapteraspis 80
Canobius sp. 159, 161
Canowindra grossi 251, *227*
Captorhinus *338*
Cartieraspis 123
Caseodus 187
Cephalaspidida 74

Cephalaspis 74, 141, 80
'C.' gabrielsi 75
C. lyelli 67
'C.' wyomingensis 75
'C.' sp. 80
Changolepis 259, 262
C. tricuspidatus 261
Changxingaspis gui 258
Cheiracanthoides 99, 155, 191–2, 222, 262, 269
C. comptus 97, 162, 189, 190, 193, 248
Cheiracanthus 124, 157
Cheirolepids 17, 59
Cheirolepis 15, 108, 124, 156
Cheliderpeton 341
Chiastodus 187
Chirodipterids 18, 58
Chirodipterus 17
C. australis 223, 250
C. sp. 159, 161, 163
Chroniosuchidae 345, 347
Chuanheiolepis jiangyouensis 257, 259
Chuchinolepis dongmoenesis 284, 285
Cladodus c. *C. acutus 281–2*, 283
C. occidentalis 283
C. sp. 281, 212
Cladolepis 155, *185*, 194
Cladoselache 11, 110
Climatiformes 107
Climatius 124, 143, 146
C. enodicosta 144, 146
Coccosteidae 60, 148
Coccosteus 113, 121
Cochliodont indet. 159
Coelurosauravidae 347
Colosteidae 331
Conchodus 116
Coronodontids 59
Corvaspis 80
Cosmacanthus malcolmsoni 116
Cosmine (tissue) 5, 17, 92, 281
Crassidonta 198
Crassigyrinus 331
Ctenacanthus 113, 147, 150, 163, 197
C. clarkei 150
C. devonicus 150
C. venustus 150
C. cf. C. crenulatus 159
C. sp. 267
Ctenaspis 80
Ctenodus sp. 153, 251
Ctenurella 130
C. gardineri 250
C. gladbachensis 121
Culmacanthidae 54
Culmacanthus 231, 248
C. antarcticus 225–6, *227*
C. pambulensis 225, *227*
C. stewarti 180, 189, 225, 249, *227*
Cyathaspidiformes 72
Cyclodiscaspis ctenus 258
Cyrtospirifer verneuili 152

Davelaspis 80
Dawsonerpeton 341
Dayaoshania 252, 262–3, 268
Dayongaspis 252, 255, 268
Deirosteus 96
Deltodus sp. *281*, 284
Denisonosteus 192, 223
D. weejasperensis 221
Devonchus 147
Diabolepis speratus 262, 268
Diademaspis 70, 74, 80
Dialipina 92, 98
D. markae 91
D. orvigi 16
Diandongpetalichthys 264, *301*
Dianolepis 164, 167, 254, *265*, 268
D. liui 265–6
Dickosteus 127
Dicksonosteus 80
Dicranaspis 80
Dictyosteus 119
'Dicynodont Empire' 347
Didymaspis 74
Dimetrodon 338
Dinaspidella 80
Dinichthyidae 60
Dinichthys 122, 123
D. belgicus 120
Diplacanthidae 60, 108
Diplacanthus 124, 157, 190
Diplocaulus 337, *338*
Diplocercides 160, 163
Diplocercids 59
Diplodus 260, 270
D. cf. *priscus* 267
Dipnorhynchidae 57
Dipteridae indet. *284*, 18
Dipterus 17, 93, 96, 126, 162, 299
D. angustus 123
D. controversus 123
D. fleischeri 123
D. fourmarieri 120
D. nelsoni 120
D. sekwoodi 123
D. ch. *valencienessi* 121
?D. sp. 163
Discosauriscidae 341, 343
Discosauriscus 339
Dissorophidae 345–7
Doliodus 124
Dongfangaspis 268
D. qujingensis 264
Doragnathus 331

Doryaspis 70, 80
Dotaspis 90
Drepanaspis 70
Dunkleosteus marsaisi 150
  D. sp. 150, 265
Duyunolepis 268
  D. paoyangensis 254
Dvinosaurid-Chroniosuchid Province 334, 348–9
Dvinosauridae 345, 347
Dwykia 153

Eastmanosteus 268, 270
  E. calliaspis 250, 266
  E. yunnanensis 260, 266
  E. sp. 148, 162–3
Ectinaspis 80
Ectochoristes 315
Eczematolepis 155
Edaphaspis 90
Edaphosaur-Nectridean Province 334, 335–41, 343–4, 346–8
Edaphosaurus 339
Eglonaspis 70, 89
Elegantaspis 80
Elegestolepis 11, 92, 96, 184, 299
  E. conica 91
Elgonaspis 90
Ellopetalichthys 125
Empedaspis 90
Endeiolepis 70, 71, 75, 81, 124
Endothiodontidae 347
Enosuchidae 345, 347
Eochoristes 315
Eoctenodus 248–9
Eogyrinidae 347
Eothyrididae 341
Errivaspis 70
Eryopidae 345, 347
Eryops 338
Eryptychiidae 57, 72
Escuminaspis 75, 124
Eskimaspis 125
Eugaleaspida 252
Eugaleaspis 259
  E. changi 264
  E. xujiachongensis 264
Euleptaspids 80
Euphanerops 70, 71, 75, 124
Eupleurogmus cresswelli 189
Eurycaraspis 266
Euthenodon 123, 251
Eusthenopteron 20, 124, 126, 130
  E. dalgleishensis 116
Euthacanthidae 60
Euthacanthus 155

Fallacosteus turnerae 220, 250, *221*
Favosites 148
Fleurantia 124
Fleurantiids 58
Fungulodus rotundus 267

Gabreyaspis 90
Galeaspids 57, 82
Gampsolepis 247
Ganoine (tissue) 16, 17
Ganorhynchus 120–1
  G. beecheri 123
  G. caucasius 161
  G. sp. 94
Gaspeaspis 123
Gemuendenaspids 58
Gemuendina 306
Gerdalepis 156
Gerronaspis 90, 98, 297
Gigantaspis 80
Glyptolepis 79, 124–5, 130
  G. benedini 120
  G. radians 120
Glyptopomus 116, 123
  G. kinnairdi 116, 120
  G. minor 116
Gogodipterus paddyensis 250
Gogonasus 233
  G. andrewsae 250
Golshanichthys asiastica 160, 163–4
Gomphonchus 99, 155, 179, 258
  G. pluriformis 146
  G. sp. 80
Goodradigbeeon 148, 166, 188, 192, *188*
Grazosteus hoernesi 157, *154*
Grenfellaspis 217, 314
  G. branagani 236, 251
Griphognathus 17, 130
  G. whitei 250
Groenlandaspis 117–8, 129, 159, 161, 165, 210, 228, 231, 234–6, 296
  G. antarctica 228, 248
  G. seni 154, 157
  G. sp. 152, *158*, 159, 188, 231, 248, 286
Grossius aragonensis 154, 155
Gualepis 11, 184, 185, 252, 262, 269
Guerichosteus 122
Gunaspis 89
Gustavaspis 74, 80
Gyracanthides 152, 167, 197, 228
  G. murrayi 189
  G. warreni 248
  G. sp. *152*, 164
Gyracanthus 124, 155
  G. seriponensis 144, *146*
  G. sherwoodi 123
  G. sp. 146
Gyroptychius 118, 121, 128, 167
  G.? australis 223, 247

G. groenlandicus 117, 121
G. milleri 117
?G. sp. 164

Hanyangaspidae 255
Hangyaspis goudingshanensis 255, *258*
Hanilepis 252
H. wangi 261, 258
Haplacanthus 148, 151, 190
Hariostius 122
Harrytoombsia elegans 249
Heimenia 80, 119
Heintzosteus 80, 119
Helenolepis 89, 96
Helenolepis multicostata 90
Helicampodus sp. 287
Helicoprion 198
H. bessonowi 287
H. sp. 287
Helodus sp. *287*
Hemicyclaspis 6, *70*, 74
Herasmius 80
Hercynolepis 150, 155
Heterogaspis 80
Heterosteus 117, 118
Heterostraci 7, 71, 72, 89
Hibernaspis 89, *90*
Hildenaspis 80
'*Hillsaspis*' 230
Hirella 70, 74
Hoelaspis 74, 80
Hohsienolepis hsientuensis 268
Hollardosteus marocanus 148
Holmaspidella 80
Holmesella 195
Holodipterus 58, 130
H. gogoensis 250
H. longi 250
Holonema 126, 150, 156, 159, 162, 164, 165, 166, 193, 266, 270
H. radiatum 154, 156, 164
H. rugosum 123
H. westolli 249
H. cf. *radiatum* 120, *154*, 156–7, *158*, 163
Holoptychioids 19, 236
Holoptychius 19, 236
H. americanus 123
H. flemingi 116, 120
H. giganteus 123
H. halli 123
H. murchisoni 120
H. nobilissimus 116
H. radiatus 123
H. sp. 162, 164
'*Homacanthus*' 124
Homalacanthus 124
Homostius 80, 117–8
Howella 254
Howidipterus donnae 249
Howittacanthus kentoni 189, 225, 249
Howqualepis 167, 225
H. rostridens 249, *180*
H. sp. *158*
Hunanaspis 268
Hunanolepis tieni 254, *265*, 266, 268, 271, *301*
Hyrcnaspis bliecki 160, 167
Hyrenia 123

Iberolepis 155
I. aragonensis 154
Ichthyostega (=*Ichthyostegopsis*) 117, 327
Ilemoraspis 89, 96
I. kirkinskayae 90
Incisoscutum ritchiei 250
Intasuchidae 341
Iranorhynchus seyedemamii 160, 163–4
Irregulareaspis 80
Ischnacanthida (Ischnacanthiformes) 60, 143, 197
Ischnacanthus gracilis 127
?I. scheii 127
I. sp. *91*

Jaymoytius 8, *70*, 71, 75
Jiangxilepidae 254
Jiangxilepis 254, 268
J. longibrachius 267
Jiuchengia longioccipita 260, 268

Kadaliosaurus 336
Kawalepis 269
K. comptus 258, *261*
Kendrickichthys cavernosus 250
Kiaeraspididae 74
Kiaeraspis 70, 80
Kimaspis 92, 98, 304
K. tienshanica 91, 93
Kimberleyichthys
K. bispicatus 249
K. whybrowi 249
Koharalepis jarviki 227, 248
Kolymaspis 308
K. siberica 91
Kolypaspis 123
Kosoraspis peckai 156
Kotlassia 347
Kujanowiaspis 113, 120
Kunmingolepis lucaowanensis 266
Kureykaspis 73, *90*
Kweichowlepis 254, 264, 270, 303
K. sinensis 259

Laccognathus 162
Lambdodus 187
Larnovaspis 80

Lasanius 70
Lataspis 80
Lathanosuchidae 345, 347
Latimeria 17, 18
Latirostraspis 253, 255
  L. chaohuensis 255
Latocamurus coulthardi 250
Latvius 130
  L. decherti 121
Laurentaspis 123
Laxaspis qujingensis 261, 264
  L. rostrata 261
Lecaniaspis 89, *90*
Legendrelepis 75
Lehmanosteus 80, 119
Leonodus 10, 155
  L. carlsi *154*
Lepidaspis 72, 80
Leptophoridae 345
Lianhuashanolepis 254
Ligulalepis toombsi 14, *16*, 190–2, 222
  L. yunnanensis 258, 261
Litoptychus bryanti 126
Liujiangolepis 254, 262, 268
  L. suni 263
Loganellia *70*, 75, 89
  L. tuvaensis *90*
'*Logania*' 93, 96
  L. incompta 93
  L. mokalenkoae 95
  L. sibirica 95
  cf. *cruciformis* 95
Lophosteiformes 15, 330
Lophosteus 14, 15
  L. superbus 15
Loxomma 331
Luetkeichthys 96
Lunalepis 155
  L. leonensis *154*
Lunaspis 121, 191, 259, 264, 269
Lysipterygium 342
Lysorophus 338

Machaeracanthus 122, 126, 141, 147, 151, 153, 156, 175 181, 189, 192, 262, 306
  M. cf. *M. peracutus* 148
  M. stonehousenesis 179
  M. sp. *154*, 156
Machairaspis 74, 80
Macropetalichthys 120, 125
Mahalalepis 226, *227*
  M. resima 248
Mansfieldiscus 197
Manticoceras 150
Maplemillia 194
Marsdenichthys 231
  M. longioccipitus *180*, 249
McMurdodus 184, 223, 225
  M. featherensis 184, 225
  M. whitei 184, 225, 247
Mediaspis 119
Megalichthyids 20, 157
Megalichthys 197
Megistolepis 92, 96, 299
  M. klementzi *91*
Melanognathus canadensis 127
Melanosteus occitanus *154*, 156
Mentzichthys 153
Mesacanthidae 60
Mesacanthus 124
Mesmiodmus 197
Mesosaurus 342
Metaxygnathus 327, 330
Microbrachius 115–6, 128–9, 268, 270, 303
  M. sinensis 260, 268
Micromelerpeton *339*
Micropholidae 348
Miltaspis 80
Mimetaspis 74, 80
Mimia toombsi *16*, 225, 250
Modoliopsis 153
Monarolepis 192, 217
  M. verrucosa 247
Monaspis 119
Mongolepis 92, 98, 184
  M. rozmanae *91*
Moythomasia 109, 130, 145, 159, 163, 165, 190, 195, 225
  M. durgaringa 16, 250
  M. nitida 121
  ?M. sp. 143, 283
Murchisonia 153
Murrindalaspis *13*, 188, 192
  M. bairdi 247
  M. wallacei 247
Mylostomod indet. 150

Nanpanaspis microculus *258*, 264
Nawagiapis 217, 225
  N. wadae *218*, 248
Naxilepis 252
  N. gracilis 258, *261*
Nectaspis 74, 80
Neeyambaspis 223, 303
Neosiacanthus 255, 268
Neopetalichthys 264
Neopetalodus 287
Neosinacanthus 268
Nesides? heiligenstockensis 121
Nikolivia 80
Nodonchus rectus 146
Norilaspis 94
Norselaspis 74, 80
Nostolepis 258–9, *261*, **269, 308**
  N. costata 162, 175, 189–90

N. striata 179, 181
N. sp. *279*
'*N.*' sp. 80
Notopetalichthys hillsi 247
Notopus petri 147
Notorhizodon mackelveyi 288, 248
Novogonatodus 197
Nowakia elegans - zone 151
Nycteroleteridae 345, 347
Nyctiphruretidae 347

Obolus 153
Obruchevichthys 327
Oervigia 117, 157, *158*
Ohioaspis 97, 157, 158
Ohiolepis 14, 155, *185*, 191–2, 222, 269
?O. xitunensis 261
Olbiaspis 90
Onchus 156
O. punctuatus 146
O. rectus 123
O. sicaeformis 146
O. striatus 153
O. sp. 146, 164
O. sp. 80
Onychodontids 18, 59, *149*
Onychodus 18, 96, 119, 157, 159, 165
O. firouzi 163
O. obliquimarginatus 121
O. sp. 164–5, 250
Orchesteropus atavus 342
Orientolepis 254
O. neokwangsiensis 259
Orodus varicostatus 150
O. sp. 150, 197
?Orthacanthus 195
Orvikuina 80, 108, 119, 165
Osteolepiformes 20, 330
Osteolepis 125, 128
Osteostraci 6, 68, *70*, 71–2, 80, 89

Pageauaspis 123
Pambulaspis 228
P. antarctica 228, 248
P. cobahandrensis 218, 232, 249
Panderichthys 116
Panxiosteus 60, 268
P. ocullus 260
Parameteoraspis 70, 74, 80
Paraplesiobatidae 58
Pareiasauridae 347
Pattenaspis 74, 80
Peilepis 262, 269
P. solida 261
Pelaspis 90
Peltobatrachidae 345, 348
Pelurgaspis 90
Perichondral bone 3
Persacanthus kermanensis 158, *160*, 163
Petalodus 286, 287
P.? daihuanensis 260, 267
Petrodentine (tissue) 17
Petrodus 286, 287
Phaneropleuron andersoni 116
Pharangolepis 70
Phlebolepis 9, 70, 75
Phlyctaenius 124
Phoebodus 11, 122, 186, 195, 260, 267
P. australiensis 186, 280, 281, 308
P. politus 150, 194, 267
P. cf. *gothicus 282*
P. sp. 150, 186, 195, 260, 267
Phyllolepids 54, 315
Phyllolepis 118, 129, 210, 295–6, 299
P. orvini 118
P. undulata 120
P. woodwardi 116
Phymolepis cuifshangensis 259
Pillararhynchus 262, 270
P. longi 250
Pinguosteus thulborni 250
Pituriaspida 10
Pituriaspis 223, 270, 303
P. doylei 247
Placodermi (introduction) 13
Placolepis 217
P. budawangensis 221, 233, 250
Platyethmoidea 226
P. antarctica 226, 248
Platyoposaurus 343, 347
Platysomidae 159, 165
Plectrodus mirabilis 153, 154
Plourdosteus 128, 164, 167
Polybranchiaspis 9, 261, 268
P. liaojiashanensis 261, 264
P. cf. *gracilis 284*, 285
Polymerolepis 184
Poraspis 70, 80
Porolepidae 57
Porolepis 19
P. kureikensis 92
P. taimyrica 92
P. sp. 91
Porophoraspis 177, *178*
Powichthys 17, 19, 125, 270
Prescottaspis 113
Prionosuchus 343, 347
P. plummeri 347
Pristicladodus springeri 194
Procephalaspis 74
Procondylolepis 256, 268
P. qujingensis 259
Procynosuchidae 347
Prosarctaspis 90
Protacrodus 11, 165, 194, 267, 269, 270
P. sp. 150, 163

P. wellsi 194
Proterogyrinus 331
'*Protodus*' 80
Protopteraspis 70, 80
Protorosaurus 347
Pruemolepis 155
Psammolepis 116
Psammosteids 58, 72, 79
Psammosteus 78, 116, 125
Psephaspis 126
Pteraspids 71, 147
Pteraspis *7*, 147
Pterichthyodes 92, 116
Pterygolepis 70
Ptychodus 120
Ptyctodontids 14, *91*
Ptyctodus 121, 130, 151, 163
Pucapampella rodrigae 143, *144*, 145
Putoranaspis 90
Pycnosteus 70, *72*

Qataraspis deprofundis *160*, 162
Qingmenaspis microculus 264
Quasipetalichthys 268, 271
Q. haikouensis 266
Qujinolepis – see also Chuchinolepis 254, 268
Q. gracilis 259

Radotina 92, 191
R. kosoriensis *154*
Remigolepis 116, 129, 210, 216, 234–6, 250–1, 254, 267–71, 296, 303, 311
R. major 260, 267
R. microcephala 260, 267
R. xiangshanensis 260, 267
R. xixiaensis 260, 267
R. zhongingensis 260, 265, 267
R. zhongweiensis 260, 267
Retzia 153
Rhachiosteus pterygiatus 121
Rhamphodopsis 128
Rhinesuchidae 345, 348
Rhinodipterus 58, 163–5, 167
Rhinopteraspis *70*
R. crouchi 93, 96
Rhizodus 19
Rhynchodipteridae 163
Rhynchodipterus elginensis 116
Rhynchodus 163
Robertiidae 347
Rockycampacanthus 189, 225, 163
Rolfosteus canningensis 250
Romundina 98, *286*

Saccabambaspis 7, 57, 177, 178
Sagenodids *18*
Sanchaspis magalarostrata *258*, 264
Sanqiaspis 268
S. zhaotongensis *258*
Sauravus 337
Sauripterus 113
Scaumenacia 124
Schizosteus perneri *154*, 156
Sclerocephalus *339*, 341
Scolenaspis 74, 80
Securiaspis 74, 80
Selenosteidae 60, 150
Serratodus 287
Seymouria *338*
Seymouriidae 343
Sharpeys fibres (tissue) 17
Sherbonaspis 217, 223
S. hillsi *218*
Shimenolepis graniferans *258*
Shirolepis 92
Siamodus *11*, 260, 266, 269, 309
S. janvieri *11*, *281*
Siberiaspis *90*
Sigaspis 80
Sigerdia 80
Sillerpeton 337
Silurolepis 252, 262, 268
S. platydorsalis 255
Sinacanthus 166, 255, 268
?S. boliviensis *144*, 146
?S. sp. 146
Sinogaleaspis 253, 255, 268
Sinolepidae 255, 308
Sinolepis 254, 262, 265, 268, 271
S. macrocephala 263, 267
S. szei 263
S. wutungensis 267
Sinopetalichthys 254, 264, 268
Skamolepis 165, *185*, 186, 192, 222
S. fragilis 186
Soederberghia 296, 303
Soetendalichthys 153
Sorbitorhynchus 252, 262, 268, 270
Spathicephalus 331
Spatulaspis 80
Spirifer verneuili 151
Stegolepis 92, 93, 99, *301*
S. jugata *91*
Stegotrachelus 16, 108, 116, 121
Steinaspis 94
Stemmatodus 187
Stensioella 121
?S. heintzi 162
Stensioellidae 58
Stereosternum 342
Sterropterygion 126
Stethacanthus thomasii 150, 195, *199*, 283
S. sp. 267, *281*
Stomiahykus thalodus 127
Strepsodus 159, 161
Striacanthus 248, 249

Strunius 18
S. rolandi 121, *160*, 163–4
S. walteri 121
Sundeyichthys 153
Surcaudalus rostratus 198
"Svalbardaspis" 80
Symmorium 281
Szeaspis (= Szelepis) 268

Taemasacanthus 189, 225
Taemasosteus 148, 166
Tafilalichthys lavocati *149*, 150
Tangasauridae 347
Tannauspididae 89, 99
Tannuaspis 89, 96, 300
T. levenkoi *90*
Tareyaspis *91*
Tartuosteus 72, 80
Tegaspis 74
Tegeolepis 109
Tenizolepis 93, 305
Tesakoviaspis 95
Thaumatolepis 92, 96, 299
Thelodonti 8, 75, 87, 89, 95
Thelodus 9, 75, 183, *279*
T. sculptilis 77
T. sinensis 252, 258, *261*
Thoralodus cabrieri 156
Thrinacodus 186, 252, 269–70
T. ferox 122, 150, 186, 195, 199, 260, 267, 270, 308
T. nanus 186
Thursius 121, 124, 128
Thyestes 7
Titanichthys termieri *149*, 150
Tokosauridae 247
Torosteus *13*
T. pulchellus 250
T. tuberculatus 250
Traquairaspis *70*, 80
Tremataspis *70*
Trematosteidae 60
Triazeugacanthus 124
Trihecatontidae 341
Tsuifengshanolepis diantungensis 259
Tubonasus lennardensis 259
Tulerpeton 327
Turanolepis 93, 305
Turinia 75, 76
T. antarctica 226, 248
T. asiatica 252, 259, 261
T. australiensis 175, 181, *182*, 183, 190–2
T. fuscina 181, *182*, 183
T. gondwana *144*, 146, 166
T. hutkensis *160*
T. pagei 155
T. pagoda *9*, 260, 266, *279*
T. polita 181, 183, 192, 308
T. cf. *T. hutkensis* 165
T. cf. *T. pagoda* 266, 280
T. sp. 260, *279*
Turiniidae 54, 58
Tuvaspis 89, 96

Uranolophus 126, 281
Urumqia 343
Utaherpeton 331
Utegenia 343

Vanchienolepis langsonensis *284*, 285
Venustodus 187
Vietnamaspis trii 285
Vorobjevaia 228

Waengsjoeaspis 74, 80
Wataaspis 74
Watsonacanthus 94
W. oervigi 97
Watsonosteus 77, 128
Weigeltaspis 80
Westlothiana 331
Westralichthys uwagadensis 220, *221*
Wetteldorfia *12*
Wijdeaspis 96, 139–40, 262, 268
W. warooensis 247
Williamsaspis bedfordi 246
Williamson, canals of 17
Willomorichthys 153
Wudinolepis 254, 268
W. weni 260
Wumenshanaspis 268
Wurungulepis 217, *218*, 224
W. denisoni 247
Wuttagoonaspis 192, 217, 222, 223, 226, 295
W. fletcheri *188*, *221*, 247

Xenacanthus *11*
Xichonolepis 254. 262. 266
X. qujingensis 263, 266
Xinanpetalichthys 264, 268
Xiushuiaspis 253, 255, 268
Xylacanthus 80

Yangaspis jinningensis 266
Yinosteus major 266
Youngacanthus *12*, 259
Youngolepis 17, 19, 92, 256, 259, 262, 268, 270, 285, 330
Y. praecursor 264, 285
Yunlongolepis 257
Y. liui 259
Yunnanogaleaspis 259
Y. major 264
Yunnanolepids 259
Yunnanolepis 254, 256–7, 259, 268, 285
Y. bacboensis *284*

Y. chii 256–7, 259, 264
Y. deprati *284*
Y. liui 259
Y. parvus *265*, 285
Yvonaspis 75, 124

Zamponiopteron 143, 145
Z. falciformis 143
Z. spinifera 143
Z. triangularis 143, *144*
Zascinaspis 80
Zenaspidida 74
Zenaspis 74, 80
Zhangilepis aspritilis 259

Stratigraphic and locality index

Abava 128
Acadia 68
Achanarras Limestone 117, 128
Afghanistan *26*, 39, 139, 141, 165, 270
Africa *26*, 27, *28*, 29, *142*, 147–53, 303, 305, 313–4, 333, 342, 345–8
Air ring complex 29
Alaska *69*, 88
Algeria *142*, 151, 165, 305, 307
Alves Beds 128
Amadeus Basin 176, 183, 208, 222, 226, 302
Amata 128
Amazon Basin 140–1, *142*, 147
Amphitheatre Fm 190
Angara 87, 139–40
Angers Limestone 153
Anglo-Welsh Basin *111*, 113, 122
Anjou 156
Antalya 157
Antarctica *26*, 27, *28*, 29, 32, 54, 146, 183, 208, 210, 213, 223, 226–9, 313–4, 333
Aouniet-Ouenine Fm 151
Appalachian Basin 122, 127, 130
Arabia *26*, 40, 140, 160, 162
Arctic-North Atlantic Caledonian Fold belt *69*
Argana Fm 337
Argentina 141, 342
Arizona 126
Armiagan Tepe 159
Armorica 31, 35–6, 41, 45, 139, 141, 153, 166, 259, 305–7
Armorican Quartzite 36
Armutgizlek Tepe 157–8
Artois 82, 120
Arukula 128
Asaka Limestone 286–7
Ashkidah Fm 151
Askiz Fm 96
Atholville Beds 124
Australia 7–8, 10, *26*–7, *28*, 40–3, 68, 98, 146, 157, 167, *209*, 247, 259, 270–1, 280, 293, 303, 308, 327–8, 333, 346, 348; *see also* Chapters 8 and 9
Austria 156–7
Avalon-Meguma Terranes 69
Avalonia *26*, 35, 68
Avon River Group 230
Azerbaidjan 161, 164–5
Aztec fauna 226–8, 248
Aztec Siltstone 214, 226, 228–9, 313–4

Bac Bo 284
Bac Bun Suite 262, 285
Bailiuping Fm 262
Baldwin Fm 194
Baltic region *26*, 34, 35, 76, 78, 93, 114, 128–30, 140, 333
Baltica *26*, 31, *69*, 334, 45, 68, *106*, 139, 156
Ban Na Chareon 283
Baramba Limestone 197–8
Barentia 93, 106
Bathurst Island 125–6
Baton River Beds 189
Battery Point Fm 123–4
Bauska beds 78
Baywulla Fm 197
Béchar Basin 148, 151
Bega Batholith 233
'Belek' facies 159, 161
Belen Fm 141, 143
Belgium 129, 270
Bell Point Limestone 222
Bellbird Creek Fm 214–5, 233
Bely Kamen beds 93
Bentong Group 39
Bergisch-Gladbach 18, 121, 130
Beverley Inlet Fm 125
Bindaree Rd-Freestone Creek fauna 230, 249
Bird Fjord Fm 125
Blue Fjord Fm 126
Blue Range Rm 230

Bogan Gate Sandstone Group 181
Boghole Beds 118, 128
Bohemia 139–40, 156, 167, 305–7
Bohemian Massif 156
Bokkeveld Group 152–3
Bolivia 141, *142*, 146–7, 165, 177, 183, 186
Bonaparte Basin 177, 197, 208, 216
Boogledie Fm 192
Booral Fm 195
Boothia Uplift region 112, 124
Borneo *26*, 39
Boyd Volcanic Complex 214, 232–3
Bratsk 95
Brazil 141, *142*, 147, 237
Britain 35, 67, 74, 76, 121–2, 186; *see also* Scotland, Ireland, Avalonia, Old Red Continent
British Colombia 127
Brittany *111*, 112, 120, 141, 156–7
Broken River Fm 183, 193
Buchan Group 184, 192, 210, 222
Bundock Creek Fm 216
Bunga Beds 186, 229, 232
Burdekin Basin 176
Burges Fm 224
Burma 277, 308
Burtnieki 128

Cahinches Fm 342
Caithness Flagstones Group 128
Caledonian Basins 106
California 126
Canada 75, 82, 89, 123–7, 130, 184, 286, 305–7; *see also* Gaspé, Québec, Euramerica
Cann River Beds 195
Canning Basin 175, 177, 179, 181, 183, 190, 193, 208, 215, 219, 222, 235, 302
Cape Philips Fm 125
Cape Storm Fm 124
Carmichael Sandstone 177
Carnarvon Basin 183, 193, 223
Carozo Fm 155
Carribuddy Fm 181
Catavi Fm 143
Cathyasian Flora Province 44
Catombal Group 216
Catskill Delta facies 122–3
Chacoma 145
Chakass 68, 89, 96, 99
Changlai Fm 267
Chile 342
China 10, 14, 26, 54, 146, 180–1, 183–6, 188, 236, 252–71, 280, 296, 304, 314–5, 327–8, 343; *see also* Chapter 10, South China, North China, Kazakhstan
Chinese Realm 79
Chotek Limestone 156
Chuandong Fm 265–6
Chukotka block 68, *69*, 88, 99, 297, 308
Cimmeria 39, 140
Cleveland Shale 110
Cockatoo Fm 216
Colorado 126
Comerong Volcanics 29, 32, 215, 234
Cookeys Plains Fm 181
Coopers Creek Fm 189, 191
Cordilleran Basins 112, 126–7
Crainswick Fm 127
Cravens Peak Beds 183–4, 192–3, 222
Crouchi zone 93, 125
Cruziana facies 177
Cuifengshan Fm 264
Czechoslovakia 120, 156, 307, 330, 337, 341, 343

Dahepo Fm 267
Dale Fm. China 262
Dangarfield Fm 197
Danlin Fm 254
Dapoushang Section 255, 267, 270
Darling Basin 176
Dasaiba Fm 267
Dayaoshan Group 254
Dayaoshan Oldland *253*
Deering Siltstone 226
Delorme Fm 127
Denizli Fm 159
Derriwong Beds 190
Deua Fm 214
Diahua Fm 267
Do Son, Vietnam 285
Dolomite Member 266
Donovan Basic Complex 233
Dra Valley 148, 150
Drake Bay Fm 124–5
Dulcie Sandstone 234
Dulladerry Rhyolite 191
Dura Den Fm 113, 116, 120
Dwyka Tillites 342

East Avalonia 30, 35
East Gondwana 76, 98, 105, 295–6, 302–4, 309–10, 314–5; *see also* Chapters 8, 9
East Greenland 26, 33, 68, *69*, 109, 112, 117–8, 129, 235, 314, 327–8
Ecca Fm 333, 342
Ellesmere Island 125, 127
Emeishan Basalt 38
Emi-Magri Fm 151
Escuminac Fm 58, 124, 130
Estonia 67, 79, 82
Ettrema Limestone 215
Eugaleaspid-Polybranchiaspid-Yunnanolepid realm 270
Euramerica 6, 61, 87, 104–5, *106*, 107–31,

111, 167, 215, 295–8, 303–6, 309, 311–4, 328, 331–3, 337, 340, 343–8; *see* Chapters 4, 5 and 6

Fairfield Group 186, 197
Falkland Islands 141, 147
Fentou Fm 255
Fram Fm 125
France 74, 112, 139–41, 157, 270, 306, 330; *see also* Armorica
Franklinian Basins 124
Fujiki Fm 286

Gabo Island Granite 233
Gangamopteris Beds 247
Ganxi Fm 262
Ganzi Sea Trough *253*
Garra Limestone Fm 190, 192
Gaspé Peninsula 79, 123–4
Gauja 128
Gebre Fm 159
Gebze Fm 159
Gedongguan Beds 270
Genoa River Beds 235
Georgina Basin 175–6, 183, 208, 222
Germany 18, 111–2, 121, 259, 270, 306, 330, 341, 343
Ghadames Basin ch. 7
Gilif Hills ring complex 27, 28
Gogo Fm 16, 159, 193, 211, 232, 249, 270, 283
Gogo Fm faunal list 249–50
Gondwana 25, 27, 30, 42–5, 298, 309, 311–33, 337, 344, 346–9; *see also* Chapters 7, 8, 9
Gotland 15, 33, 78
Gneudna Fm 183, 193, 234, 250
Grampians Gp 175, 237
Graveyard Creek Fm 179
Grenfell fauna 236, 251
Grey Hoek Fm 118
Groenlandaspis Series 118
Guandi Fm 255, 259
Guanshang Fm 280
Guizhou Province 252, 254–5
Gumhole Fm 195

Haikou Fm 253, 265–6
Harajicca Sandstone 226
Hargreaves Member 216
Hatchery Creek Conglomerate 183, 192–3, 223–4
Hatchery Creek fauna 223, 247
Hell Gate Fm 125
Hervey Group 216, 234–5
Hexian Fm 254
Hexizoulang Terrane 305
Heyuanzhai Fm 266, 280
Hollands Creeek Rhyodacite 229
Holmeswood Shale 198
Holy Cross Mts 36, 112, 121, 130, 184, 194
Horn Valley Siltstone 177
Hornelen Field 117
Huamampampa Fm 143
Hunan Terrane 37
Hunter Siltstone 236, 270

Iapetus *34*, 35
Iberic Peninsula 140
Idaho 126
India 346–8
Indo-China Terrane 37, 38, 277–83
Innuitian Fold Belt *69*
Iowa 186
Iquiri Fm 145
Iran 8, *26*, 39, 120, 141, 146, 160, 162, 345
Iraq 162
Irati Fm 342
Ireland 186
Irian Jaya 175, 302
Italy 157

Jack Limestone 179, 190
Jandiatuba Basin 147
Japan 278, 286–7
Jauf Fm 162
Jerula Fm 189, 191
Jesse Limestone 186, 192
Jiamusi-Bureinsk Block 87, *88*
John O'Groats Group 116
Junggar-Hinggan Sea *253*
Junggar Oldland *253*

Kameruka Granodiorite 233
Kangdian Oldland *253*
Kanokura Fm 287
Kara-Tajmyr *88*, 99, 106, 297
Kashmir 348
Kaskaskia Sequence 127
Kazakhstan *26*, 36, 37, 43–5, 87, 88, 93, 99, 297–8, 303–4, 327, 330, 333, 343–5
Kemeri 128
Kerman 162–5
Kernavé 128
Ketleri Beds 79
Khan Yunis Fm 234
Khush-Yeilagh Fm 164, 186, 302
Kinderhook Fm 186
Kirawin Shale 222
Knobby Sandstone 234
Kolyma Block 68, *88*, 91, 99, 308
Köprülu Fm 159
Kotelny Island 92, 98–9, 297, 308
Kotuy River 94, 96
Kununnurra Meber 216
Kureyka Basin 94, 96

Kurtdojnurus Fm (Turkey) 159
Kuznetsk Basin 96

Labrador *69*
LaGard Fm 124
Lambie Group 216
Lankey Limestone 175, 184
Latvia 121, 186
Laurasia 87
Laurentia 6, *26*, 30, 33, 40, 45, 68, *69*, *106*, 183
Laureuropa 87
Laurussia 41, 44, 99, 297
Leigutai Fm 267
Leivu 128
Lhasa Block 39, 140
Lianhua shan Fm 259
Libya 151
Lomonosov Fold Belt *69*
Long Swamp Creek Fm 214
Ludlow Bone Bed 113
Luesma Fm 155

Ma'anshan Section 262
Madagascar *26*, 333, 346
Mae Sam Lap 280, 302, 308
Mae Sariang 279, 302
Maecuru Fm 147
Maider 148
Malaysia 38, 39, 304
Malutang Fm 280
Manacapuru Fm 147
Manchuria 87
Mangrullo Fm 342
Mansfield Group 189, 197
Martins Well Limestone 188, 189, 191
Maturin Basin 147
Melby fish bed 117
Meloola Volcanics 181
Melville Islands 125
Merrimbula Group 194, 214–5, 233, 234
Mianaliacun Fm 259
Mimer valley Fm 118, 120
Mineral Hill Volcanics 181
Minusinsk Basin 92
Mirrabooka Fm 179
Mongolia 87, 98, 99, 184
Mongolo-Okhotsk Block 87, 88
Montana 126
Morocco 141, 148, 166, 305–307, 337
Mostyn Vale Fm 186, 194, 216
Mountain Dam Limestone 191
Moyuela Fm 155
Mt. Frome Limestone 186, 192, 193
Mt. Howitt 189, 249
Mt. Ida Fm 220
Mt. Knowles Limestone 192
Muhua Section 270
Mulga Downs Group 183, 192, 210, 222–3
Munabia Sandstone 183, 223, 250
Munyarai-1 190, 192, 280
Murrumbidgee Group 186
Murzuk Basin 151
Myrtevale Fm 186, 194

Nabeyama Fm 286–7
Nagaoling Fm 261
Nairn Fm 118
Nakhitchevan 161
Namibia 342
Narrien Range 195
Narva Beds 164
Netteltons Creek fish bed 215, 233
Nevada 126, 184
New Guinea 175
New Mexico 126
New Siberian Islands *69*
New York 127
New Zealand 174–5, 208, 302
Noguerras Fm 155
Norilsk 94, 96
North America *26*, 33, 122–3, 126–7, 187–8, 195, 198; *see also* Euramerica, Laurentia, Old Red Continent
North Atlantis 68
North China Oldland *253*
North China Terrane 10, 31, 37, 40, 43–4, 98, 305, 315, 344–5
Norway 75, 78, 82, 117
Nova Scotia 340
Novaya Zemlya 68, *69*, 106, 112, 122, 296
Nyurol Basin 93

Officer Basin 208, 222
Ohio 127
Ohio Range 302
Old Red Continent 68, 82, 98–9, 110, 211
Old Red Sandstone 67, 76, 104, 107, 121, 123, 126, 219, 293
Omolon Block *88*, 308
Orcadian Basin 113, 115–7
Orkney 113
Ouan Kasa Fm 151
Ougata Chain 151

Pacoota Sandstone 177
Pak Chom Nang Limestone 283
Pambula River fauna 232–3, 249
Pangaea 44–5, 332–3, 345, 347–8
Panthalassa *45*
Papilio Fm 186, 193, 225
Parke Siltstone 183, 226
Parnaiba Basin 147, 312
Parnu 128
Pedra de Fogo Fm 343
Peel Sound Fm 124–5

Piker Hills Fm 234
Pimenteira Fm 147
Pingyipou Group 259
Pitinga Fm 147
Plattenkalk 121, 130
Plavinas Beds 121
Podolia 68, 74, 78, 106, 112, 120–1, 125, 297
Point Hibbs Limestone 184, 191
Poland 184, 280
Ponta Grossa Fm 147
Port Philomel Beds 141, 147
Portugal 141, 153
Prikolymsk Terrane 308
Prince of Wales Island 124–5
Psammites de Montfort 120

Qiantang Block 39, 140
Qiantang Oldland *253*
Québec 79, 81, 122, 123–4
Qui Dat Fm 278, 281
Qujing Basin 259, 270
Qujing Fm 266

Raub-Bentong Suture 39
Raymond Fm 187, 197
Receptaculites Limestone 186
Red Bay Group 118
Reefton 189
Remigolepis Series 118
Represas Fm 342
Rezekne 128
Rheic Ocean 166, 306, 330
Rhenohercynian Basins 112, 120
Rhineland 112, 121
Ringerike-Rudstangen 78
Rongxi Fm 255
Rongxian Fm 315
Rousey bed 118
Russia 33, 194, 327–8, 336, 344–8
Ryozensan Limestone 287

Saaremaa 78–9
Salair 96–7
San Pedro Redbeds 36
Sandwick fish bed 117
Sanmentan Fm 254
Santa Rosa Fm 146, 186
Sarclet Group 128
Scat Craig 128
Scotland 68, *69*, 74–5, 112–13, 128–9, 153, 159, 327
Sea-levels, Devonian 215–16
Seripona 146
Severnaya Zemlya 68, *69*, 87–8, 93, 99
Shan-Thai Terrane (= Sibumasu) *26*, 32, 37, 38, 43, 277–81, 302, 308, 345
Shetland 113
Shixiagou Fm 265–6
Shizishan Fm 262
Siberia *26*, 31, 36, 42–3, 87, *88*, 89, 95, 198, 295–9, 327, 332, 345
Sibumasu *see* Shan-Thai Terrane
Sicasica Fm 141, 143
Silverband Fm 175, 181
Snetnaya Gora Beds 130
Snobs Creek Rhyolite 214
Snowblind Bay Fm 124
Snowy River Volcanics 29, 32
South Africa 26, 142, 152, 347–8
South America 7, *26*–7, *28*, 140–1, 166, 305, 314, 346–7
South China Terrane 32, 37–8, 40–4, 57, 98, 105, 139, 295, 299–300, 302–5, 308, 315, 340, 344–6
Spain 35–6, 141, 155, 181, 184, 270, 307
Spitsbergen 68, 74, 78–9, 93, 99, *111*, 112, 118–19, 122, 141, 186, 306
Stairway Sandstone 177
Star Basin 176
Stokes Siltstone 177
Svalbard *69*

Tadzhikistan (Tadjikistan) 193, 343
Taemas 189, 192, 211, 222
Taemas-Buchan fauna 210, 213, 222, 246–7
Taemas Limestone 192, 224
Tafilalt 148–51
Tafilalt-Maider Basin 148
Taggerty-Howitt fauna 230, 249
Tajmyr 68, 87, 89, 93–4, 297
Talagrouna Fm 152
Tarabuco Fm 143
Tarim *26*, 31, 37, 44, *88*
Tasman Foldbelt 208
Tasmania 40, 174, 184, 191
Tatong-South Blue Range fauna 229, 248
Taylors Hill Fm 191
Teddy Mount Fm 186, 194–5
Terra Cotta Sandstone 228
Thailand 193, 270, 277–81, 308–9; *see also* Shan-Thai Terrane
Thurso Flagstone Group 118
Tibet *26*, 37, 39, 88, 345
Tien Shan 92–3
Timan *26* 82, 130
Tokya-Dewal anticline 165
Toyama Fm 287
Trarut Fm 151
Trebatov Limestone, Bohemia 156
Trombeitas Group 147
Trundle Beds 184, 191
Tumblong Oolite 175, 183
Turkey *26*, 39, 139, 140, 157–8, 161, 167, 345
Tuva (= Tuvin) 8, *26*, 68, 89, 91–2, 96, 99, 105, 295, 297–8
Twofold Bay fauna 234, 250

Twofold Bay Fm 214, 234–5

United States 126, 341; *see also* Euramerica
Urals 82, 87, 98, 122
Uruguay 342
Uryum beds 93
Ustrednaya Beds 130
Utah 126
Utting Calcarenite 197
Uzbekistan 93, 99

Vadja 128
Vas'uganje 93
Vechernii Fm 308
Venezuela *142*, 147
Vietnam 259, 262, 277–8, 299, 302
Virgin Hills Fm 194
Vogti horizon 125

Wangyou Fm 267, 270
Wellington Rhyolite 230
Welsh Basin 35
Welsh Borderland 78–9, 82
West Bondwana 42, 98, 139–167, 305, 312
West Yunnan 280, 302, 308
Westwood Creek Member 216
Wijde Bay Fm 118
Windellama Limestone 190
Windmill Limestone 184
Winduck Group 223
Witteberg Group 153
Wood Bay Fm 79, 118–20
Worange Point fauna 235, 251
Worange Point Fm 29, 43, 214, 235–6
Wudang Fm 254, 264
Wujiahe Fm 255
Wuttagoonaspis fauna 222, 225, 247
Wutung Fm 254, 267
Wyoming 126

Xiashan Fm 264
Xiejiawan Fm 262
Xikuangshan Fm 267
Xishancun Fm 259, 261
Xitun Fm 259, 262
Xom Nha Fm 278, 281–4
Xuishan Fm 258
Xujiachong Fm 259

Yagihara Limestone 287
Yangtze Block 37
Yangtze Oldlands *253*, 255
Yangtze Valley 254–5, 303
Yea Fm 219
Yellow Drum Sandstone 234
Yujiang Fm 259
Yunnan 38, 183–4, 193, 254–262, 264–6
Yunnanolepid-Polybranchiaspid province 270

Zhongning Fm 254, 263, 267
Ziyang County 255

Library of Congress Cataloging-in-Publication Data
Palaeozoic vertebrate biostratigraphy and biogeography / edited by John A. Long.—Johns Hopkins paperbacks ed.
p. cm.
"First published in 1993 by Belhaven Press . . . London"—T.p. verso.
Includes bibliographical references and index.
ISBN 0-8018-4779-6 (pbk. : acid-free paper)
1. Vertebrates, Fossil. 2. Paleontology—Paleozoic. 3. Paleontology, Stratigraphic. I. Long, John A., 1957– .
QE841.P34 1994 93-11551
566—dc20 CIP